NONLINEAR CONTROL of ELECTRIC MACHINERY

CONTROL ENGINEERING

A Series of Reference Books and Textbooks

Editor

NEIL MUNRO, PH.D., D.SC.

Professor
Applied Control Engineering
University of Manchester Institute of Science and Technology
Manchester, United Kingdom

1. Nonlinear Control of Electric Machinery, *by Darren M. Dawson, Jun Hu, and Timothy C. Burg*

Additional Volumes in Preparation

Series Introduction

Many textbooks have been written on control engineering, describing new techniques for controlling systems, or new and better ways of mathematically formulating existing methods to solve the ever-increasing complex problems faced by practicing engineers. However, few of these books fully address the applications aspects of control engineering. It is the intention of this new series to redress this situation.

The series will stress applications issues, and not just the mathematics of control engineering. It will provide texts that not only contain an exposé of both new and well-established techniques, but also will present detailed examples of the application of these methods to the solution of real-world problems. The authors will be drawn from both the academic world and the relevant applications sectors.

There are already many exciting examples of the application of control techniques in the established fields of electrical, mechanical (including aerospace), and chemical engineering. We have only to look around in today's highly automated society to see the use of advanced robotics techniques in the manufacturing industries; the use of automated control and navigation systems in the many artifacts available to the domestic consumer market; and the reliable supply of water, gas, and electrical power to the domestic consumer and to industry. However, there are currently many challenging problems that could benefit from wider exposure to the applicability of control methodologies, and the systematic systems-oriented basis inherent in the application of control techniques.

This new series will present books that draw on expertise from both the academic world and the applications domains, and will be useful not only as academically recommended course texts but also as handbooks for practitioners in many applications domains.

Neil Munro

NONLINEAR CONTROL of ELECTRIC MACHINERY

Darren M. Dawson
Clemson University
Clemson, South Carolina

Jun Hu
Seltrol, Inc.
Greenville, South Carolina

Timothy C. Burg
S. E. Huffman Corporation
Clover, South Carolina

CRC Press
Taylor & Francis Group
Boca Raton London New York

CRC Press is an imprint of the
Taylor & Francis Group, an **informa** business

CRC Press
Taylor & Francis Group
6000 Broken Sound Parkway NW, Suite 300
Boca Raton, FL 33487-2742

First issued in paperback 2019

ISBN-13: 978-0-8247-0180-2 (hbk)
ISBN-13: 978-0-367-40050-7 (pbk)

Visit the Taylor & Francis Web site at
http://www.taylorandfrancis.com

and the CRC Press Web site at
http://www.crcpress.com

To My Parents

Jack and Carol Dawson

D. M. D.

To My Wife Jane

J. H.

To My Wife Karen

T. C. B.

Preface

In this book, we present several nonlinear control algorithms for a benchmark mechanical system actuated by several different types of electric machines. The motivation for this book came from what we like to call the *robot control man's dilemma.* Specifically, the typical robotics application requires an electric machine to turn a robotic link along a desired position trajectory. The procedure for designing a control algorithm to achieve a desired position tracking performance specification typically involves the following steps: (1) develop the nonlinear model for the mechanical subsystem dynamics of the robot manipulator and (2) develop a nonlinear torque input control algorithm (*i.e.*, assume that the link actuators are torque sources) which provides a means of theoretically quantifying the link position control objective. After the nonlinear algorithm has been developed, the control engineer usually performs simulations and discovers that the algorithm works like a dream. He then decides that experimental validation is required; however, he soon learns that most robots are not capable of implementing his control algorithm because of some or all of the following reasons: i) an archaic computer system - the control hardware is not capable of implementing his nonlinear control algorithm, ii) torque or current amplifiers not included - the control algorithm is designed for the mechanical subsystem dynamics; hence, torque or current amplifiers are required for implementation, iii) a geared robot - the encoder is mounted on the motor side of the gear; hence, the link position measurements are unknown if significant gear transmission dynamics exist, iv) what no velocity measurements? - most direct drive robots are only equipped with link position measurements to save cost; furthermore, since tachometer-based velocity measurements are usually very noisy, they are usually not worth including. Fortunately, there are some direct drive robots equipped with DSP computer systems and torque amplifiers which allow for control implementation[1].

After a year or so of implementing control algorithms on the above robotic hardware, the typical control engineer may find himself wondering what a torque amplifier actually is. For the simple permanent brushed dc

[1]Typically, the user still has to manufacture the velocity signal with a nonlinear observer, a backwards difference algorithm, or some other sort of filter.

motor, torque and current are linearly related; hence, a torque (or current) amplifier is often just a high-gain current feedback loop. However, for multi-phase electric machines with more complicated dynamics (*e.g.*, the switched reluctance motor), the use of the phrase torque amplifier is unclear. That is, does the phrase torque amplifier mean that several high-current feedback loops are used in conjunction with one or more of the following: i) a look-up table which provides torque based on the rotor position and the winding configuration, ii) a commutation strategy which is developed from a model of the static torque transmission equation (*i.e.*, the relationship with mechanical torque as the output and per-phase winding currents and rotor position as the inputs), or iii) special purpose, *secret* electrical circuitry[2].

To summarize the above problem, we can simply state that motor drive manufacturers are attempting to neglect the motor's electrical subsystem dynamics. However, many researchers now believe that these neglected actuator dynamics have restricted the development of high-performance motion tracking drive systems for many types of applications. The association of neglected electrical actuator dynamics and poor tracking performance can be heuristically explained by viewing the motor's inductive and resistive effects as a low pass filter with voltage the input and current the output. From a systems engineering viewpoint, the motor drive manufacturers are approximating[3] this low pass filter as an all pass filter. Since nonlinear controllers proposed for high-performance tracking often contain high frequency content, the neglected electrical dynamics essentially filter out the high frequency portion of the required control input. Not surprisingly, poor position tracking performance may be observed. Therefore, we believe that improved position tracking performance, and hence improved mechanical system positioning, can be achieved by explicitly including the electrical actuator dynamics during the control synthesis.

With the above comments in mind, this book presents several nonlinear algorithms for load (or rotor) position trajectory following for several different types of electric machinery. The control algorithms are based on the nonlinear, *full-order* electromechanical dynamics. That is, the nonlinearities are not linearized away or neglected; furthermore, the electrical subsystem dynamics are not reduced into a static set of equations as is commonly done in more traditional motor control approaches. To substantiate the performance and the design of the approach, each control algorithm is accompanied with a proof of stability. In addition, the position tracking performance for twenty of the twenty-three algorithms presented are validated via experimental results. To facilitate the experimental set-up, a simple robotic type load was utilized as a benchmark mechanical load; how-

[2]The secret electronics may be standard high-gain current circuitry found in any power electronic textbook.

[3]The use of high-gain current feedback usually makes this approximation better.

ever, extensions[4] are possible for a variety of mechanical loads. It should also be noted that since almost any realistic velocity control objective can be formulated as a position control problem, the proposed control methodology can be utilized for load velocity tracking applications.

The first part of the book is devoted to the construction of full state feedback (FSFB), exact model knowledge and adaptive controllers. For all of the motors, except for the induction motor, FSFB means that measurements of load position, load velocity, and electrical winding current are required for control implementation (For the induction motor, rotor flux measurements are also required). In Chapter 1, all of the necessary mathematical background needed for following the proofs in the book is given. We attempted to prepare this chapter to ensure that no further system theory reading would be required for individuals whose specialty lies in other areas such as power electronics or robotics. In Chapter 2, we develop nonlinear FSFB controllers for the permanent magnet brushed dc (PMBDC) motor. In Chapters 3, 4, 5, and 6, the controllers are systematically modified to account for the unique dynamics associated with the permanent magnet stepper (PMS) motor, the brushless dc (BLDC) motor, the switched reluctance (SR) motor, and the induction motor. A small portion of the material presented in these chapters is repeated; however, this repetition allows the reader to read Chapters 1 and 5 independently without having to hunt down notation or concepts in other chapters.

The second part of the book is devoted to the construction of exact model knowledge controllers which do not require full state feedback. Since accurate position measurements are required for commutation of motors such as the SR, BLDC, and PMS motors, we believe it is prudent to examine model-based observers which use load position measurements. Hence, the primary focus of the second part of the book is the incorporation of model-based observed information into the control architecture. The elimination of load velocity measurements is desirable since a load velocity signal obtained from "filtering" the load position may provide bad velocity estimates with sample rates of one millisecond and low resolution encoders (*e.g.*, 500 lines). In addition, the elimination of electrical current measurements for PMDC and PMS motors reduces the complexity of the control electronics while measurement of rotor f l ux in the induction machine is impractical. Based on this rational, we present output feedback (OFB) controllers for the PMBDC, the PMS, and the BLDC motors in Chapters 7, 8, and 9, respectively (*i.e.*, these controllers only require measurement of load position). In Chapter 10, we present partial state feedback (PSFB) controllers for the SR and BLDC motors, (*i.e.*, these controllers require measurement of load position and electrical winding current). Finally, in Chapter 11, we present a PSFB controller for the induction motor (*i.e.*, this

[4]Extensions to other mechanical loads is especially true for the full state feedback controllers developed in the first section of the book.

controller requires measurement of load position, load velocity, and stator winding current). It should be emphasized that we explicitly show that the use of the observed information in the controller structure does not compromise the position tracking performance. That is, the corresponding stability proofs take into account the effects of the observed information on the performance of the closed-loop system.

The third part of the book is devoted to a variety of more advanced topics. In Chapter 12, an adaptive PSFB controller is presented for the PMS, the BLDC, and the SR motor. This controller compensates for parametric uncertainty throughout the entire electromechanical system while only measuring load position and electrical current. In Chapter 13, we develop a *sensorless*[5] velocity tracking control for the separately excited dc motor. This control algorithm achieves velocity tracking with measurement of only electrical current despite saturation effects in the field's magnetic circuit. Finally, in Chapter 14, we present an additional PSFB controller for the induction motor which eliminates load velocity measurements and rotor flux measurements. The appendices contain some auxiliary information and some more recent, induction motor control algorithms.

[5]Recent motor control papers often refer to a controller which uses only electrical current measurements as sensorless.

Contents

List of Figures

Chapter 1

Mathematical Background

1.1 Introduction

In this book, we will present several types of algorithms for position-velocity tracking control of nonlinear electromechanical systems. Roughly speaking, these controllers can be divided into two classes: exact model knowledge control and adaptive control. The exact model knowledge controllers are defined as the name implies. That is, this type of approach utilizes the dynamic model of the electromechanical system as part of the feedforward control algorithm; hence, the model must be known exactly. The exact model knowledge controller also provides insight into how modeling uncertainty in the electromechanical system can be dealt with. That is, at least for the full state feedback case, the exact model knowledge controller provides the skeleton from which an adaptive controller can be designed to cope with parametric uncertainty.

In this chapter, we will develop some mathematical tools which will be used to analyze the stability of the closed-loop systems under the proposed controllers. First, we develop a simple Lyapunov-like [1] lemma which can be used to study the performance of the exact model knowledge controllers presented in the subsequent chapters. To illustrate the control technique, we design a tracking controller for a first-order nonlinear system. The Lyapunov-like lemma is then used to show that the tracking error is driven to zero exponentially fast. Second, we present several definitions and lemmas which can be used to study the performance of adaptive control algorithms. The aforementioned exact model knowledge controller which was formulated for the first-order scalar system is then redesigned as an adaptive controller via the use of two dynamic parameter update laws. We then illustrate how the stability tools can be used to show that the tracking error goes to zero asymptotically fast. Lastly, since mechanical systems are often second-order, we present several lemmas which can be used to analyze the closed-loop stability of a second-order system as though it is a first-order system. These lemmas substantially reduce the complexity of the stability analysis for the controllers which are designed for the third-order electromechanical systems presented in the subsequent chapters.

1.2 Exact Model Knowledge Control

As we discussed earlier, we will be developing several exact model knowledge tracking controllers for electromechanical systems. Since this problem involves the design of tracking controllers for nonlinear systems, we will utilize several Lyapunov-like stability lemmas [1] to prove that the tracking error has a Global Exponential Stability (GES) property [1]. To illustrate the meaning of GES tracking, we will utilize the following lemma for the control of a simple dynamic system.

Lemma 1.1 [1]

Let $V(t)$ be a non-negative scalar function of time on $[0,\infty)$ which satisfies the differential inequality

$$\dot{V}(t) \leq -\gamma V(t) \tag{1.1}$$

where γ is a positive constant and $\dot{V}(t) \triangleq \frac{d}{dt}V(t)$. Given (1.1), then

$$V(t) \leq \underline{e}^{-\gamma t}V(0) \qquad \forall t \in [0,\infty) \tag{1.2}$$

where $\underline{e}$ is the base of the natural logarithm.

Proof. It is easy to see that (1.1) can be rewritten in the following form

$$\dot{V}(t) + \gamma V(t) \leq 0. \tag{1.3}$$

Given the structure of the differential inequality in (1.3), we define the following differential equation

$$\dot{V}(t) + \gamma V(t) = -s(t) \tag{1.4}$$

where $s(t)$ must be a non-negative scalar function as a result of (1.3). After rewriting (1.4) in the form

$$\dot{V}(t) = -\gamma V(t) - s(t), \tag{1.5}$$

we can use standard linear control results [2] to solve the differential equation given by (1.5) in the following form

$$V(t) = \underline{e}^{-\gamma t}V(0) - \underline{e}^{-\gamma t}\int_0^t \underline{e}^{\gamma\tau}s(\tau)d\tau. \tag{1.6}$$

Note that since $s(t)$ is non-negative for all time, we can use (1.6) to obtain the following upper bound on the solution of $V(t)$

$$V(t) \leq e^{-\gamma t} V(0) \tag{1.7}$$

which is the result stated in (1.2). □

1.2.1 Design Example

To illustrate how Lemma 1.1 can be used to analyze the stability of a closed-loop system, we now consider the design of an exact model knowledge tracking controller for the first-order, scalar system given by

$$\dot{x} = -ax^3 - b\sin(t) + u \tag{1.8}$$

where $x(t)$ is the state of the system, $u(t)$ is the control input, and a, b are known constant parameters.

The control objective for the system given by (1.8) is to design a controller $u(t)$ to force the state variable $x(t)$ to track a desired trajectory denoted by $x_d(t)$. We will assume that $x_d(t)$ and its first time derivative are continuous and bounded for all time. Since we are solving the tracking problem, we will denote the associated tracking error $e(t)$ as

$$e = x_d - x. \tag{1.9}$$

The controller will be designed to drive the tracking error denoted by (1.9) to zero.

Before we develop the controller and examine the stability of the resulting closed-loop tracking error system, we will rewrite the system dynamics in terms of the tracking error. That is, by utilizing (1.9), we can rewrite (1.8) as

$$\dot{e} = w - u \tag{1.10}$$

where the auxiliary variable $w(t)$ is defined by

$$w = \dot{x}_d + ax^3 + b\sin(t). \tag{1.11}$$

To ensure that the tracking error is GES, we will utilize the controller

$$u = ke + w \tag{1.12}$$

where k is a positive, constant control gain. After substituting (1.12) into (1.10), we have the following closed-loop system

$$\dot{e} = -ke. \tag{1.13}$$

1.2.2 Tracking Error Stability

By utilizing a Lyapunov-like approach [1] and Lemma 1.1, we can develop an exponentially decaying bound for the transient behavior of the tracking error. The following theorem illustrates this concept.

Theorem 1.1

The tracking error is GES in the sense that

$$|e(t)| \leq |e(0)| \, \underline{e}^{-kt} \qquad \forall t \in [0, \infty) \tag{1.14}$$

where k was introduced in (1.12), and $|\cdot|$ denotes the standard absolute value operator.

Proof. First, we define the following non-negative function

$$V(t) = \frac{1}{2} e^2(t). \tag{1.15}$$

Differentiating (1.15) with respect to time yields

$$\dot{V}(t) = e(t)\dot{e}(t). \tag{1.16}$$

Substituting (1.13) into (1.16) yields

$$\dot{V}(t) = -ke^2(t). \tag{1.17}$$

It is easy to see that $\dot{V}(t)$ in (1.17) can be upper bounded as

$$\dot{V}(t) \leq -2kV(t) \tag{1.18}$$

where (1.15) has been utilized. Applying Lemma 1.1 with $\gamma = 2k$ to (1.18) yields

$$V(t) \leq V(0)\underline{e}^{-\gamma t}. \tag{1.19}$$

From (1.15), we have that

$$V(t) = \frac{1}{2} e^2(t) \qquad \text{and} \qquad V(0) = \frac{1}{2} e^2(0). \tag{1.20}$$

Substituting (1.20) appropriately into the left-hand and right-hand sides of (1.19) allows us to form the following inequality

$$\frac{1}{2} e^2(t) \leq \frac{1}{2} e^2(0)\underline{e}^{-\gamma t}. \tag{1.21}$$

After noting that $\gamma = 2k$ and $e^2(t) = |e(t)|^2$, we can solve for $|e(t)|$ in (1.21) to obtain the bound on $|e(t)|$ given in (1.14). □

Remark 1.1

From (1.14), we know that $e(t)$ is bounded for all time; therefore, since we have assumed that $x_d(t)$ is bounded for all time, we can use (1.9) to state that $x(t)$ is bounded for all time. From the definition of the control given by (1.12) and (1.11), we can state that the controller $u(t)$ can be bounded for all time in the form

$$|u| \leq k\,|e| + |\dot{x}_d| + |a|\,\left|x^3\right| + |b|\,|\sin(t)|\,. \tag{1.22}$$

Note, that since $x_d(t)$, $\dot{x}_d(t)$, and $x(t)$ are all bounded, we can use (1.22) to state that the controller $u(t)$ is bounded for all time.

Remark 1.2

It is interesting to note the controller gain k introduced in (1.12) can be increased to obtain a "faster" transient response performance as delineated by the right-hand side of (1.14).

Remark 1.3

If one examines the form of the closed-loop tracking error system given by (1.13), one might question the appropriateness of using Lemma 1.1 for examining the stability of a linear system. However, we will show later (and throughout the book for that matter) that a similar approach to that given above can also be used to examine the stability of nonlinear, closed-loop, tracking error systems.

1.3 Adaptive Control

As illustrated by the previous design example, the exact model knowledge controller requires that the system parameters (*i.e.*, the parameters a and b in (1.8)) are known. Often, there is some uncertainty with regard to the exact value of any given parameter in an electromechanical system; therefore, it would be desirable to redesign the controller of (1.12) to compensate for parametric uncertainty online. That is, one is tempted to rewrite the controller of (1.11) and (1.12) as

$$u = ke + \dot{x}_d + \hat{a}x^3 + \hat{b}\sin(t) \tag{1.23}$$

where $\hat{a}$ and $\hat{b}$ are used to denote estimates of the actual parameters a and b, respectively. The primary question is how should the parametric estimates (*i.e.*, $\hat{a}$ and $\hat{b}$) be defined. That is, should they be constant, best guess estimates of the corresponding parameters, or should they be adjusted online by some update law that monitors the tracking error term defined in (1.9)? Of course, since we are concerned with the study of dynamical systems, we must be wary of any approach which does not examine the stability of the corresponding closed-loop system under a proposed control.

One method for adjusting the parameters online is called adaptive control. As far as this book is concerned, adaptive control will be defined as the compensation for parametric uncertainty via the use of a nonlinear controller which contains dynamic, parametric update laws. These update laws are generated such that the stability of the closed-loop system is preserved. In fact, as we will show later, the update laws are designed to facilitate the desired stability result. Before we can design our first adaptive controller, we must present the necessary background mathematics which encompasses uniform continuity, Barbalat's Lemma [1], and some minor extensions.

Definition 1.1 [3]

Let $\Re_+$ denote the set of non-negative real numbers. A scalar function $f(t) : \Re_+ \rightarrow \Re$ is uniformly continuous if for each positive number ϵ_o, there exists a positive number δ_o such that

$$|f(t) - f(t_1)| < \epsilon_o \qquad \text{for} \qquad \max\{0, t_1 - \delta_o\} < t < t_1 + \delta_o \tag{1.24}$$

where t_1 is a specific instant of time.

Definition 1.2 [4]

Consider a scalar function $f(t) : \Re_+ \rightarrow \Re$. Let the 2-norm (denoted by $\|\cdot\|_2$) of $f(t)$ be defined as

$$\|f(t)\|_2 = \sqrt{\int_0^\infty f^2(\tau)\, d\tau}. \tag{1.25}$$

If $\|f(t)\|_2 < \infty$ then we say that the function $f(t)$ belongs to the subspace L_2 of the space of all possible functions (*i.e.*, $f(t) \in L_2$). Let the ∞-norm (denoted by $\|\cdot\|_\infty$) of $f(t)$ be defined as

$$\|f(t)\|_\infty = \sup_t |f(t)|. \tag{1.26}$$

If $\|f(t)\|_\infty < \infty$ then we say that the function $f(t)$ belongs to the subspace L_∞ of the space of all possible functions (*i.e.*, $f(t) \in L_\infty$).

Definition 1.3 [5]

If the scalar function $f(t) \in L_2$ and the scalar function $g(t) \in L_2$ then we can state the following inequalities

$$\|f(t) + g(t)\|_2 \leq \|f(t)\|_2 + \|g(t)\|_2 \tag{1.27}$$

and

$$\int_0^\infty \sqrt{|f(\tau)|}\sqrt{|g(\tau)|}\, d\tau \leq \sqrt{\int_0^\infty |f(\tau)|\, d\tau}\sqrt{\int_0^\infty |g(\tau)|\, d\tau}. \tag{1.28}$$

The inequality given by (1.27) is often referred to as Minkowski's inequality while the inequality given by (1.28) is a special case of Holder's inequality. The above definitions are utilized during the proof of the subsequent theorems and lemmas presented in the rest of this chapter.

Lemma 1.2 [6]

If a function $f(t)$ is a uniformly continuous function on $[0, \infty)$ and if the integral

$$\lim_{t\to\infty} \int_0^t |f(\tau)|\, d\tau \tag{1.29}$$

exists and is finite then

$$\lim_{t\to\infty} |f(t)| = 0. \tag{1.30}$$

Proof. [6] If

$$\lim_{t\to\infty} |f(t)| \neq 0, \tag{1.31}$$

then we can find some positive instant of time denoted by t_1 such that

$$|f(t_1)| \geq 2\epsilon \tag{1.32}$$

for some positive number ϵ. Now since $f(t)$ is uniformly continuous as delineated by the theorem statement then we can use Definition 1.1 to state that there exists some positive number δ such that

$$|f(t+\tau) - f(t)| < \epsilon \quad \text{for } t \geq 0 \text{ and } 0 \leq \tau \leq \delta \tag{1.33}$$

where ϵ is the same number defined in (1.32), and τ is some positive constant.

Now as an aside, it is easy to see that

$$|f(t)| - |f(t_1)| \geq -|f(t) - f(t_1)|; \tag{1.34}$$

therefore, we can use (1.34) to form the following inequality

$$|f(t)| \geq |f(t_1)| - |f(t) - f(t_1)|. \tag{1.35}$$

We can utilize (1.32), (1.33), and (1.35) to obtain the following inequality

$$|f(t)| > \epsilon \quad \text{for } t_1 \leq t \leq t_1 + \delta. \tag{1.36}$$

To complete the proof, we now use (1.36) to note that

$$\left| \int_{t_1}^{t_1+\delta} |f(t)| \, dt \right| = \int_{t_1}^{t_1+\delta} |f(t)| \, dt > \int_{t_1}^{t_1+\delta} \epsilon dt = \epsilon\delta. \tag{1.37}$$

Now without loss of generality, let $t_1 = 0$; hence, we can use (1.37) to form the following inequality

$$\int_0^\delta |f(t)| \, dt > \epsilon\delta \tag{1.38}$$

which means that $\lim_{\delta \to \infty} \int_0^\delta |f(t)| \, dt$ does not exist. Hence, we have contradicted the theorem hypothesis given by (1.29); therefore, we know our original statement that $\lim_{t\to\infty} |f(t)| \neq 0$ given by (1.31) is not true. Hence, (1.30) is indeed true. □

Since it is often awkward to check the uniform continuity of $f(t)$ as required by the statement of Lemma 1.2, we reformulate this continuity condition in terms of a check on the time derivative of $f(t)$.

Lemma 1.3 [1]

If $\frac{d}{dt} f(t) \triangleq \dot{f}(t)$ is bounded for $t \in [0, \infty)$ then $f(t)$ is uniformly continuous for $t \in [0, \infty)$.

Proof. First, if $\dot{f}(t)$ is bounded for $t \in [0, \infty)$ then $f(t)$ is continuous [3] for $t \in [0, \infty)$. Second, since $\dot{f}(t)$ is bounded and $f(t)$ is continuous for $t \in [0, \infty)$, we can use the Mean Value Theorem [3] to state that there exists two positive numbers t_0 and t_1 where t_0 is between t and t_1 such that

$$f(t) - f(t_1) = \dot{f}(t_0)(t - t_1). \tag{1.39}$$

Since $\dot{f}(t)$ is bounded for all time, we can state $\left|\dot{f}(t)\right| < R$ where R is a positive constant; therefore, we can use (1.39) to establish the following inequality

$$|f(t) - f(t_1)| < R\,|t - t_1|. \tag{1.40}$$

From (1.40), we can see for each positive number ϵ_o, there exists a positive number $\delta_o = \epsilon_o/R$ such that

$$|f(t) - f(t_1)| < \epsilon_o \qquad \text{for} \qquad \max\{0, t_1 - \delta_o\} < t < t_1 + \delta_o; \tag{1.41}$$

hence, $f(t)$ is uniformly continuous. □

Based on Lemma 1.2 and Lemma 1.3, we can state the following corollary which will prove useful during the development of the subsequent adaptive controllers.

Corollary 1.1 [4]

Consider a scalar function $g(t) : \Re_+ \rightarrow \Re$. If $g(t) \in L_\infty$, $\dot{g}(t) \in L_\infty$, and $g(t) \in L_2$ then

$$\lim_{t \rightarrow \infty} g(t) = 0. \tag{1.42}$$

Proof. If we let $f(t) = g^2(t)$ in (1.29), we have

$$\lim_{t \rightarrow \infty} \int_0^t g^2(\tau)\, d\tau$$

which exists and is finite since $g(t) \in L_2$ (See Definition 1.2). Furthermore, since $\dot{g}(t) \in L_\infty$, we can use Lemma 1.3 to state that $g(t)$ (and hence $g^2(t)$) is uniformly continuous. Hence, we can use Lemma 1.2 to state the result given by (1.42). □

1.3.1 Design Example

To illustrate how the above mathematical lemmas can be used to analyze the stability of an adaptive controller, we now reconsider the design of an adaptive tracking controller for the first-order system given by (1.8). The control objective is the same as before; however, now the parameters a and b are considered to be unknown constants. If we use the controller given by (1.23), we can form the closed-loop tracking error system by substituting (1.23) into (1.10) to yield

$$\dot{e} = -ke + \tilde{a}x^3 + \tilde{b}\sin(t) \tag{1.43}$$

where

$$\tilde{a} = a - \hat{a} \quad \text{and} \quad \tilde{b} = b - \hat{b} \tag{1.44}$$

are used to denote the parameter estimation errors. To ensure the stability of the closed-loop system given by (1.43), the estimates (*i.e.*, $\hat{a}$ and $\hat{b}$) in (1.23) are adjusted according to the adaptive update laws given by

$$\hat{a} = \int_0^t e(\tau)x^3(\tau)\, d\tau \quad \text{and} \quad \hat{b} = \int_0^t e(\tau)\sin(\tau)\, d\tau \tag{1.45}$$

which can be rewritten in terms of the parameter error of (1.44) as follows

$$\dot{\tilde{a}} = -ex^3 \quad \text{and} \quad \dot{\tilde{b}} = -e\sin(t). \tag{1.46}$$

The actual form of the update laws given by (1.45) is motivated by the subsequent stability proof which ensures asymptotic tracking.

1.3.2 Tracking Error Stability

By utilizing a Lyapunov-like approach and Corollary 1.1, we can illustrate how the adaptive update laws of (1.45) have been designed to ensure that all signals in the controller and the system remain bounded. In addition, we will show that the adaptive controller ensures asymptotic tracking for any finite set of initial conditions. The following theorem illustrates this concept.

Theorem 1.2

The controller of (1.23) and (1.45) provides asymptotic tracking in the sense that

$$\lim_{t\to\infty} e(t) = 0. \tag{1.47}$$

Proof. To begin the analysis, we select the following non-negative function

$$V(t) = \frac{1}{2}e^2 + \frac{1}{2}\tilde{a}^2 + \frac{1}{2}\tilde{b}^2 \tag{1.48}$$

where e, $\tilde{a}$, and $\tilde{b}$ are defined in (1.9) and (1.44). Differentiating (1.48) with respect to time yields

$$\dot{V}(t) = e\dot{e} + \tilde{a}\,\dot{\tilde{a}} + \tilde{b}\,\dot{\tilde{b}}\,. \tag{1.49}$$

Substituting (1.43) into (1.49) yields

$$\dot{V}(t) = e\left(-ke + \tilde{a}x^3 + \tilde{b}\sin(t)\right) + \tilde{a}\,\dot{\tilde{a}} + \tilde{b}\,\dot{\tilde{b}} \tag{1.50}$$

which can be written as

$$\dot{V}(t) = -ke^2 + \tilde{a}\left(ex^3 + \dot{\tilde{a}}\right) + \tilde{b}\left(e\sin(t) + \dot{\tilde{b}}\right) \tag{1.51}$$

after combining common terms. From (1.51), we can see that the adaptive update laws of (1.46) have been selected such that $\dot{V}(t)$ can be written as

$$\dot{V}(t) = -ke^2. \tag{1.52}$$

We now illustrate how to use the above analytical technique to illustrate that all signals in the adaptive controller and the system remain bounded during closed-loop operation. From the form of (1.52), we can see that $\dot{V}(t)$ is negative or zero; hence, we know from calculus that $V(t)$ given in (1.48) is either decreasing or constant. Since $V(t)$ is non-negative, it is lower bounded by zero; hence, from the form of $V(t)$, we know that $e(t) \in L_\infty$, $\tilde{a}(t) \in L_\infty$, and $\tilde{b}(t) \in L_\infty$ (See Definition 1.2). By assumption, we know that $x_d(t)$ is bounded ($i.e.$, $x_d(t) \in L_\infty$) and a, b are constant; hence, from the explicit definitions of $e(t)$, $\tilde{a}(t)$, and $\tilde{b}(t)$ given in (1.9) and (1.44), we know that $x(t) \in L_\infty$, $\hat{a}(t) \in L_\infty$, and $\hat{b}(t) \in L_\infty$. Finally, we can use (1.43) to show that $\dot{e}(t) \in L_\infty$; therefore, $\dot{x}(t) \in L_\infty$ since $\dot{x}_d(t) \in L_\infty$.

We now illustrate how Corollary 1.1 can be used to show that the tracking error goes to zero. First, we integrate both sides of (1.52) with respect to time to yield

$$\int_0^\infty \frac{dV(\tau)}{d\tau}\, d\tau = -\int_0^\infty ke^2(\tau)\, d\tau. \tag{1.53}$$

If we evaluate the integral on the left-hand side of (1.53), we obtain

$$\sqrt{V(0) - V(\infty)} = \sqrt{k}\sqrt{\int_0^\infty e^2(\tau)\, d\tau} \tag{1.54}$$

after some minor algebraic operations. Since $\dot{V}(t) \leq 0$ as illustrated by (1.52), $V(t)$ of (1.48) is decreasing or constant; hence, $V(0) \geq V(\infty) \geq 0$. We now use this information and (1.54) to obtain the following inequality

$$\sqrt{\int_0^\infty e^2(\tau)\, d\tau} \leq \sqrt{\frac{V(0)}{k}} < \infty \tag{1.55}$$

which indicates according to Definition 1.2 that $e(t) \in L_2$. Since $e(t) \in L_\infty$, $\dot{e}(t) \in L_\infty$, and $e(t) \in L_2$, we can invoke Corollary 1.1 to obtain the result given by (1.47). □

Remark 1.4

While Theorem 1.2 indicates that the tracking error (*i.e.*, $e(t)$) goes to zero, it does not indicate that the parameter errors (*i.e.*, $\tilde{a}(t)$ and $\tilde{b}(t)$) go to zero. Rather, the above proof illustrates that the adaptive controller only ensures that the parameter errors remain bounded.

1.4 Additional Control Design Tools

Since this book is concerned with the development of controllers for electromechanical systems, the nature of the physical dynamics will involve a second-order mechanical dynamical system cascaded with an electrical dynamical subsystem (in most cases the electrical system is first-order). To simplify the development of the subsequent controllers, we will introduce a variable transformation which allows the mechanical subsystem to be analyzed as though it is a first-order system. That is, throughout the rest of this book, we will denote the position tracking error as

$$e = q_d - q \tag{1.56}$$

where $q_d(t)$ is used to represent the desired load position trajectory and $q(t)$ is used to represent the actual load position. We will also denote the velocity tracking error as

$$\dot{e} = \dot{q}_d - \dot{q} \tag{1.57}$$

where $\dot{q}_d(t)$ is used to represent the desired load velocity trajectory and $\dot{q}(t)$ is used to represent the actual load velocity. Based on the above definition, we introduce the filtered tracking error variable [1] as

$$r = \dot{e} + \alpha e \tag{1.58}$$

where α is a positive scalar constant which is used for weighting the position tracking error.

Remark 1.5

Throughout the book, we will assume that $q_d(t)$ and its first three derivatives with respect to time are all bounded functions of time. This assumption on the smoothness of the desired trajectory is required because $q_d(t)$, $\dot{q}_d(t)$, $\ddot{q}_d(t)$, and $\dddot{q}_d(t)$ are all utilized as part of the model-based control strategies.

During the development of the controllers given in the subsequent chapters, we will develop several stability results for the filtered tracking error variable $r(t)$; hence, these results must be translated into meaningful stability results for the position tracking error $e(t)$ and the velocity tracking error $\dot{e}(t)$. The following lemmas provide for this translation of information.

Lemma 1.4

Given the differential equation of (1.58), if $r(t) \in L_\infty$ then $e(t) \in L_\infty$ and $\dot{e}(t) \in L_\infty$.

Proof. The solution to the differential equation of (1.58) is given by

$$e(t) = \mathrm{e}^{-\alpha t} e(0) + \mathrm{e}^{-\alpha t} \int_0^t \mathrm{e}^{\alpha\tau} r(\tau)\, d\tau \tag{1.59}$$

which can be upper bounded by

$$|e(t)| \leq |e(0)| + \mathrm{e}^{-\alpha t} \int_0^t \mathrm{e}^{\alpha\tau} |r(\tau)|\, d\tau$$

or

$$|e(t)| \leq |e(0)| + \sup_t (|r(t)|)\, \mathrm{e}^{-\alpha t} \int_0^t \mathrm{e}^{\alpha\tau}\, d\tau. \tag{1.60}$$

After evaluating the integral on the right-hand side of (1.60), we have

$$|e(t)| \leq |e(0)| + \sup_t (|r(t)|)\, \frac{1 - \mathrm{e}^{-\alpha t}}{\alpha} \tag{1.61}$$

which illustrates that $e(t) \in L_\infty$ if $r(t) \in L_\infty$. Finally, we can use (1.58) to obtain the following upper bound on $\dot{e}$

$$|\dot{e}(t)| \leq \alpha |e(t)| + |r(t)|$$

which can be written as

$$|\dot{e}(t)| \leq \alpha |e(0)| + \sup_t (|r(t)|) (1 - \mathrm{e}^{-\alpha t}) + |r(t)| . \tag{1.62}$$

The inequality given by (1.62) illustrates that $\dot{e}(t) \in L_\infty$ if $r(t) \in L_\infty$. □

Lemma 1.5

Given the differential equation of (1.58), if $r(t)$ is exponentially stable in the sense that

$$|r(t)| \leq \beta_0 \mathrm{e}^{-\beta_1 t} \tag{1.63}$$

where β_0 and β_1 are positive constants then $e(t)$ and $\dot{e}(t)$ are exponentially stable in the sense that

$$|e(t)| \leq \mathrm{e}^{-\alpha t} |e(0)| + \frac{\beta_0}{\alpha - \beta_1} \left(\mathrm{e}^{-\beta_1 t} - \mathrm{e}^{-\alpha t} \right) \tag{1.64}$$

and

$$|\dot{e}(t)| \leq \alpha \mathrm{e}^{-\alpha t} |e(0)| + \frac{\alpha \beta_0}{\alpha - \beta_1} \left(\mathrm{e}^{-\beta_1 t} - \mathrm{e}^{-\alpha t} \right) + \beta_0 \mathrm{e}^{-\beta_1 t}. \tag{1.65}$$

Proof. The solution to the differential equation of (1.58) is given by

$$e(t) = \mathrm{e}^{-\alpha t} e(0) + \mathrm{e}^{-\alpha t} \int_0^t \mathrm{e}^{\alpha \tau} r(\tau)\, d\tau \tag{1.66}$$

which can be upper bounded by use of (1.63) as

$$|e(t)| \leq \mathrm{e}^{-\alpha t} |e(0)| + \mathrm{e}^{-\alpha t} \int_0^t \mathrm{e}^{\alpha \tau} \beta_0 \mathrm{e}^{-\beta_1 \tau}\, d\tau. \tag{1.67}$$

After evaluating the integral on the right-hand side of (1.67), we obtain (1.64) which illustrates that $e(t)$ is exponentially stable if $r(t)$ is exponentially stable. Finally, we can use (1.58) to obtain the following upper bound on $\dot{e}$

$$|\dot{e}(t)| \leq \alpha |e(t)| + |r(t)| \tag{1.68}$$

which can be upper bounded by use of (1.64) and (1.63) as given in (1.65). The inequality given by (1.65) illustrates that $\dot{e}(t)$ is exponentially stable if $r(t)$ is exponentially stable. □

Lemma 1.6

Given the differential equation of (1.58), if $r(t) \in L_\infty$, $r(t) \in L_2$, and $r(t)$ converges asymptotically in the sense that

$$\lim_{t\to\infty} r(t) = 0 \tag{1.69}$$

then $e(t)$ and $\dot{e}(t)$ converge asymptotically in the sense that

$$\lim_{t\to\infty} e(t),\ \dot{e}(t) = 0 \tag{1.70}$$

Proof. The first part of the proof involves showing that $e(t) \in L_2$. We begin by noting the solution to the differential equation of (1.58) is given by

$$e(t) = z_i(t) + z_s(t) \tag{1.71}$$

where

$$z_i(t) = \underline{e}^{-\alpha t} e(0) \qquad \text{and} \qquad z_s(t) = \int_0^t \underline{e}^{-\alpha(t-\tau)} r(\tau)\, d\tau. \tag{1.72}$$

From Definition 1.2 and (1.72), we have that

$$\|z_i(t)\|_2 = \sqrt{\int_0^\infty \underline{e}^{-2\alpha\tau} e^2(0)\, d\tau} = \frac{|e(0)|}{\sqrt{2\alpha}} < \infty; \tag{1.73}$$

hence, $z_i(t) \in L_2$. From (1.72), we have the following upper bound on $z_s(t)$

$$|\, z_s(t)| \le \int_0^t \left|\underline{e}^{-\alpha(t-\tau)}\right| |r(\tau)|\ d\tau \tag{1.74}$$

which can be rewritten as

$$|\, z_s(t)| \le \int_0^t \sqrt{\left|\underline{e}^{-\alpha(t-\tau)}\right| r^2(\tau)}\ \sqrt{\left|\underline{e}^{-\alpha(t-\tau)}\right|} d\tau. \tag{1.75}$$

Applying Holder's inequality given in (1.28) to the right-hand side of (1.75) yields the upper bound for $z_s(t)$

$$z_s(t) \le \sqrt{\int_0^t \left|\underline{e}^{-\alpha(t-\tau)}\right| r^2(\tau)\ d\tau}\ \sqrt{\int_0^t \left|\underline{e}^{-\alpha(t-\tau)}\right|\ d\tau}. \tag{1.76}$$

Letting $\sigma = t - \tau$ in the second integral of (1.76) allows the following computation to be performed

$$\sqrt{\int_0^t \left|\underline{e}^{-\alpha(t-\tau)}\right|\ d\tau} = \sqrt{\int_0^t \underline{e}^{-\alpha\sigma}\ d\sigma} \le \sqrt{\frac{1}{\alpha}}; \tag{1.77}$$

hence, substituting the upper bound given by (1.77) into (1.76) and squaring both sides of the resulting expression yields the upper bound for $z_s^2(t)$

$$z_s^2(t) \leq \frac{1}{\alpha} \int_0^t \left| \underline{e}^{-\alpha(t-\tau)} \right| r^2(\tau)\, d\tau. \tag{1.78}$$

Integrating both sides of the inequality given in (1.78) with respect time yields

$$\int_0^\infty z_s^2(t)\, dt \leq \frac{1}{\alpha} \int_0^\infty \int_0^t \left| \underline{e}^{-\alpha(t-\tau)} \right| r^2(\tau)\, d\tau dt \tag{1.79}$$

which can be manipulated, by changing the order of integration on the right-hand side, into the following form

$$\int_0^\infty z_s^2(t)\, dt \leq \frac{1}{\alpha} \int_0^\infty r^2(\tau) \int_\tau^\infty \left| \underline{e}^{-\alpha(t-\tau)} \right| dt d\tau. \tag{1.80}$$

Letting $\sigma = t - \tau$ in the inner most integral on the right-hand side of (1.80) yields

$$\int_0^\infty z_s^2(t)\, dt \leq \frac{1}{\alpha} \int_0^\infty r^2(\tau) \int_0^\infty \left| \underline{e}^{-\alpha\sigma} \right| d\sigma d\tau \tag{1.81}$$

which can be manipulated, after utilizing the upper bound given by (1.77) and square rooting both sides, into the following form

$$\sqrt{\int_0^\infty z_s^2(t)\, dt} \leq \frac{1}{\alpha} \sqrt{\int_0^\infty r^2(\tau)\, d\tau}. \tag{1.82}$$

By the hypothesis of Lemma 1.6, $r(t) \in L_2$; therefore, it is obvious from Definition 1.2 and (1.82) that $z_s(t) \in L_2$.

We now utilize (1.71) and Minkowski's inequality given in (1.27) to obtain the following inequality

$$\|e(t)\|_2 = \|z_i(t) + z_s(t)\|_2 \leq \|z_i(t)\|_2 + \|z_s(t)\|_2 . \tag{1.83}$$

Since $z_i(t) \in L_2$ and $z_s(t) \in L_2$, $e(t) \in L_2$ as a direct result of (1.83) and Definition 1.2. Continuing, we note that $r(t) \in L_\infty$ by the hypothesis of Lemma 1.6; therefore, $e(t) \in L_\infty$ and $\dot{e}(t) \in L_\infty$ by use of Lemma 1.4. Since $e(t) \in L_2$, $e(t) \in L_\infty$, and $\dot{e}(t) \in L_\infty$, we can use Corollary 1.1 to state that $\lim_{t\to\infty} e(t) = 0$. Since $\lim_{t\to\infty} e(t) = 0$ and $\lim_{t\to\infty} r(t) = 0$ (*i.e.*, by the hypothesis of Lemma 1.6), we can use (1.68) to show that $\lim_{t\to\infty} \dot{e}(t) = 0$. □

Remark 1.6

It is interesting to note that the proof of Lemma 1.6 can be used to show that if $r(t) \in L_\infty$ and $r(t) \in L_2$ then $\lim_{t\to\infty} e(t) = 0$. That is, we do not require $\lim_{t\to\infty} r(t) = 0$ to show that $\lim_{t\to\infty} e(t) = 0$.

In addition to the above filtered tracking error lemmas, the following, assorted, mathematical lemmas will play an important role during the stability analysis of the proposed controllers.

Lemma 1.7 [7]

Let $A \in \Re^{n \times n}$ be a real, symmetric, positive definite matrix; therefore, all of the eigenvalues of A are real and positive. Let $\lambda_{\min}\{A\}$ and $\lambda_{\max}\{A\}$ denote the minimum and maximum eigenvalues of A, respectively, then for any $x \in \Re^n$

$$\lambda_{\min}\{A\} \|x\|^2 \leq x^T A x \leq \lambda_{\max}\{A\} \|x\|^2$$

where $\|\cdot\|$ denotes the standard Euclidean norm. The proof of this lemma (often referred to as the Rayleigh-Ritz Theorem) can be found in any standard linear algebra textbook.

Lemma 1.8

If a function $N_d(x, y) \in \Re^1$ satisfies the following relationship

$$N_d = \Omega(x)xy - k_n \Omega^2(x) x^2 \tag{1.84}$$

where $x \in \Re^1$, $y \in \Re^1$, $\Omega(x) \in \Re^1$ is a function dependent only on x, and $k_n \in \Re^1$ is a positive constant then $N_d(x, y)$ can be upper bounded as follows

$$N_d \leq \frac{y^2}{k_n}. \tag{1.85}$$

The bounding of $N_d(x, y)$ in the above manner is often referred to as nonlinear damping [8] since a nonlinear control function (*e.g.*, $k_n \Omega^2(x) x$) can be used to "damp-out" an unmeasurable quantity (*e.g.*, y) multiplied by a known, measurable nonlinear function (*e.g.*, $\Omega(x)$).

Proof. First, we utilize the form of (1.84) to obtain the following upper bound for $N_d(x, y)$

$$N_d \leq |\Omega(x)| |x| |y| - k_n |\Omega(x)|^2 |x|^2 \tag{1.86}$$

which can be rewritten as

$$N_d \leq |\Omega(x)| |x| \left(|y| - k_n |\Omega(x)| |x|\right). \tag{1.87}$$

We now examine two cases. First, if $|y| < k_n |\Omega(x)| |x|$ then (1.87) can be used to show that $N_d \leq 0$. Second, if $|y| \geq k_n |\Omega(x)| |x|$ then (1.87) can be used to show that

$$N_d \leq |\Omega(x)| |x| |y|. \tag{1.88}$$

Since for this case $|y| \geq k_n |\Omega(x)| |x|$ or

$$|\Omega(x)| |x| \leq \frac{|y|}{k_n}, \tag{1.89}$$

N_d of (1.88) can be further upper bounded by use of (1.89) as that given by (1.85). □

Lemma 1.9

If the time derivative of a function $x(t) : \Re^1_+ \to \Re^1$ satisfies the following relationship

$$|\dot{x}(t)| \leq \beta_0 \underline{e}^{-\beta_1 t} \tag{1.90}$$

where β_0 and β_1 are positive constants then $x(t) \in L_\infty$ (Note it is assumed that $\dot{x}(t) \in L_\infty$ since $x(t)$ is assumed to differentiable).

Proof. First, we note the following equality

$$x(t) = \int_0^t \frac{dx(\tau)}{d\tau} d\tau + C_o \tag{1.91}$$

where C_o is a scalar constant. From (1.91), we can form the following upper bound on $x(t)$

$$|x(t)| \leq \int_0^t \left| \frac{dx(\tau)}{d\tau} \right| d\tau + |C_o| \tag{1.92}$$

which can be further upper bounded by use of (1.90) as

$$|x(t)| \leq \int_0^t \beta_0 \underline{e}^{-\beta_1 \tau} d\tau + |C_o| . \tag{1.93}$$

After evaluating the integral on the right-hand side of (1.93), we obtain

$$|x(t)| \leq \frac{\beta_0 \left(1 - \underline{e}^{-\beta_1 t}\right)}{\beta_1} + |C_o| \tag{1.94}$$

which illustrates that $x(t) \in L_\infty$ (See Definition 1.2). □

Lemma 1.10

Let $V(t) \in \Re^1$ be a non-negative function of time on $[0, \infty)$ which satisfies the differential inequality

$$\dot{V} \leq -\gamma V + \epsilon \tag{1.95}$$

where $\gamma \in \Re^1$ and $\epsilon \in \Re^1$ are positive constants. Given (1.95), then

$$V(t) \leq \underline{e}^{-\gamma t} V(0) + \frac{\epsilon}{\gamma} \left(1 - \underline{e}^{-\gamma t}\right) \qquad \forall t \in [0, \infty) \tag{1.96}$$

where $\underline{e}$ is the base of the natural logarithm.

Proof.

It is easy to see that (1.95) can be rewritten in the following form

$$\dot{V} + \gamma V - \epsilon \leq 0. \tag{1.97}$$

Given the structure of the differential inequality given by (1.97), we define the following differential equation

$$\dot{V} + \gamma V - \epsilon = -s \tag{1.98}$$

where $s(t) \in \Re^1$ must be a non-negative function as result of (1.97). After rewriting (1.98) in the form

$$\dot{V} = -\gamma V + \epsilon - s, \tag{1.99}$$

we can use standard linear control results [2] to solve the differential equation given by (1.99) in the following form

$$V(t) = \underline{e}^{-\gamma t} V(0) + \epsilon \underline{e}^{-\gamma t} \int_0^t \underline{e}^{\gamma \tau} d\tau - \underline{e}^{-\gamma t} \int_0^t \underline{e}^{\gamma \tau} s(\tau) d\tau. \tag{1.100}$$

Note that since $s(t)$ is non-negative for all time, we can use (1.100) to obtain the following upper bound on the solution of $V(t)$

$$V(t) \leq \underline{e}^{-\gamma t} V(0) + \epsilon \underline{e}^{-\gamma t} \int_0^t \underline{e}^{\gamma \tau} d\tau. \tag{1.101}$$

After evaluating the integral on the right-hand side of (1.101), we can obtain the result stated in (1.96). □

Lemma 1.11

If the filtered tracking error defined in (1.58) can be bounded as follows

$$|r(t)| \leq \sqrt{A + B\underline{e}^{-kt}} \tag{1.102}$$

where k is a positive scalar, A and B are scalar constants such that $A+B \geq 0$, then the position tracking error defined in (1.56) can be bounded as follows

$$|e(t)| \leq \underline{e}^{-\alpha t} |e(0)| + \frac{a}{\alpha}\left(1 - \underline{e}^{-\alpha t}\right) + \frac{2b}{2\alpha - k}\left(\underline{e}^{-kt/2} - \underline{e}^{-\alpha t}\right) \tag{1.103}$$

where

$$a = \sqrt{|A|} \quad \text{and} \quad b = \sqrt{|B|}. \tag{1.104}$$

Proof.

From (1.102), we have the following upper bound for the filtered tracking error

$$|r(t)| \leq \sqrt{\left(a + b\underline{e}^{-kt/2}\right)^2 - 2ab\underline{e}^{-kt/2}} \tag{1.105}$$

where a and b are defined in (1.104). From (1.105), we can obtain a larger upper bound for the filtered tracking error as

$$|r(t)| \leq a + b\underline{e}^{-kt/2}. \tag{1.106}$$

Now it is easy to show by standard linear control arguments [2] and the definition of the filtered tracking error given in (1.58) that the position tracking error can be bounded as follows

$$|e(t)| \leq \underline{e}^{-\alpha t} |e(0)| + \int_0^t \underline{e}^{-\alpha(t-\tau)} |r(\tau)| \, d\tau \tag{1.107}$$

which can be further upper bounded as

$$|e(t)| \leq \underline{e}^{-\alpha t} |e(0)| + \int_0^t \underline{e}^{-\alpha(t-\tau)} \left(a + b\underline{e}^{-k\tau/2}\right) d\tau \tag{1.108}$$

after substituting (1.106) into the right-hand side of (1.107). After performing the integration on the right-hand-side of (1.108), $|e(t)|$ can be upper bounded by the expression given in (1.103). □

1.5 Summary

In this chapter, we have attempted to forge a small set of tools that can be used in the design of controllers for electromechanical systems. For one familiar with traditional methods of motor control, one might inquire how these mathematical definitions and lemmas can be used to synthesize controllers for electric machines which exhibit many structural differences (*e.g.*, different number of phases, different winding configurations, *etc.*). Although these structural differences are manifested in various ways with regard to the differential equations governing the motion of the electromechanical system, a control design procedure based on the lemmas in this chapter can be developed for a simple electromechanical system (*i.e.*, a permanent brushed dc motor turning a mechanical load). This control design procedure can then be systematically modified to develop controllers for the permanent stepper motor, the brushless dc motor, the switched reluctance motor, and the induction motor.

Bibliography

[1] J. Slotine and W. Li, *Applied Nonlinear Control*, Englewood Cliffs, NJ, Prentice Hall, 1991.

[2] T. Kailaith, *Linear Systems*, Englewood Cliffs, NJ, Prentice Hall, 1980.

[3] G. Thomas and R. Finney, *Calculus and Analytic Geometry*, Reading, MA, Addison and Wesley, 1982.

[4] S. Sastry and M. Bodson, *Adaptive Control*, Englewood Cliffs, NJ, Prentice Hall, 1989.

[5] M. Vidyasagar, *Nonlinear Systems Analysis*, Englewood Cliffs, NJ, Prentice Hall, Inc., 1978.

[6] H. Khalil, *Nonlinear Systems*, Upper Saddle River, NJ, Prentice Hall, 1996.

[7] R. Horn and C. Johnson, *Matrix Analysis*, Cambridge, Cambridge University Press, 1985.

[8] M. Krstic, I. Kanellakopoulos, and P. Kokotovic, *Nonlinear and Adaptive Control Design,* John Wiley & Sons, 1995.

Chapter 2

BDC Motor (FSFB)

2.1 Introduction

From a controls perspective, all motor drive systems can be decomposed into three distinct parts: (i) an electrical subsystem, (ii) a mechanical subsystem, and (iii) an algebraic torque coupling, see Figure 2.1. Since the electrical subsystem dynamics are inherently faster than the associated mechanical subsystem dynamics, the electrical dynamics are often neglected during the control design procedure. An example of this control simplification is the common assumption that torque (current) is a control input for the brushed dc (BDC) motor under high gain current feedback. This modeling assumption may yield acceptable system performance for many low-performance drive applications such as blower motors, conveyors, and fans; however, high-performance drive applications such as robotics, machine tools, rolling mills, and flying shears require high-bandwidth drives which are insensitive to variations in the operating conditions. Under these conditions, "simplified" controllers may fail to perform satisfactorily. The association of neglected actuator dynamics and poor tracking performance can also be explained heuristically. The motor's electrical dynamics act as a low pass filter having voltage as an input and current as an output. From a systems engineering perspective, the drive manufacturers are attempting to approximate this low pass filter as an all pass filter. The problem with this approach is that advanced controllers designed for high-performance applications often contain substantial high frequency content; therefore, the neglected actuator dynamics act to filter out the high frequency component of the required control signal. Not surprisingly, the result is degraded position tracking performance. Given the above argument, it is reasonable to believe that improved position tracking performance can be achieved by explicitly including the electrical dynamics in the overall control synthesis.

Fortunately, the design of controllers for electromechanical systems is facilitated by the recent development of nonlinear controllers for several

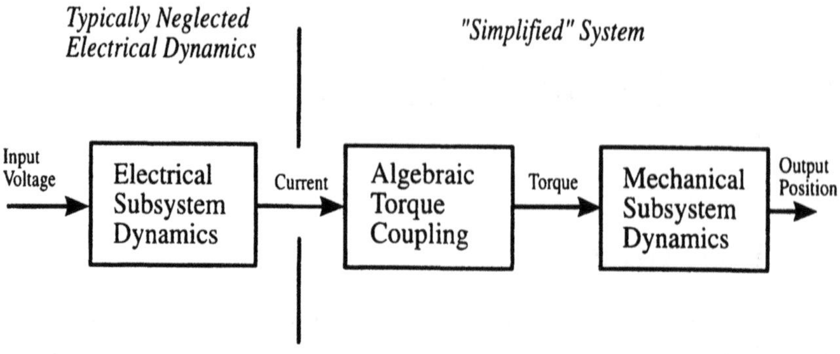

Figure 2.1: Block Diagram for Motor Drive Systems

classes of linear and nonlinear systems. The interested reader is referred to [1], [2], [3], and [4] for some of the recent work in nonlinear control design. One of the most fundamental achievements of the above work along with other system theory research is the extension of nonlinear control design techniques for a broader class of systems. For example, it is now possible, through the use of ingenious ideas such as integrator backstepping [4], [5], [6], [7], and nonlinear damping [4], to design realistic controllers for nonlinear electromechanical systems such as an electric motor actuating a mechanical load. The word realistic is used to emphasize the fact that this new brand of nonlinear controllers is capable of meeting the required control objective: i) for the third order electromechanical systems (*i.e.*, the second order mechanical subsystem cascaded with a first order electrical subsystem), ii) in the presence of parametric uncertainty throughout the entire system, and iii) with only true system state measurements (*i.e.*, position, velocity, and electrical current).

In this chapter, we use the integrator backstepping technique to develop full state feedback (FSFB), position tracking controllers for a brush dc (BDC) motor driving a mechanical load. In this approach, an intermediate tracking objective can be formed between the electrical and mechanical partitions of the control problem, allowing the propagation of the controller, designed for the mechanical subsystem by traditional nonlinear control techniques, back through the electrical subsystem in order to formulate a voltage control input. Specifically, we first view the motor as a torque source and thus design a torque control input to ensure that the mechanical load follows a desired position trajectory. Since the developed motor torque is a function of the electrical winding current, the torque input controller can then be restated as a desired current trajectory. The voltage control input is then formulated to force the electrical winding current to follow the desired current trajectory. That is, the electrical dynamics are taken into account through the current tracking objective, and hence the position control objective is embedded inside of the current tracking objective. Therefore, if the voltage control input can be designed to guarantee that the actual current tracks the desired current, then the position control objective will also be guaranteed.

To illustrate the use of the backstepping technique, an exact model knowledge controller is first designed to yield global exponential position tracking. The exact model knowledge is then redesigned under the guise of an adaptive controller to compensate for parametric uncertainty while yielding global asymptotic position tracking. We then illustrate how the adaptive controller can be modified to reduce the overparameterization [8] phenomenon often associated with the adaptive control technique. Finally, experimental results are presented to illustrate the performance and feasibility of implementing the nonlinear algorithms.

2.2 System Model

The dynamics of most electromechanical systems can be separated into three distinct parts: (i) a dynamic mechanical subsystem, which for the purposes of this discussion includes a position dependent load and the motor rotor, (ii) a dynamic electrical subsystem which includes all of the motor's relevant electrical effects, and (iii) a static relationship which represents the conversion of electrical energy into mechanical energy. The mechanical subsystem dynamics for a position-dependent load (See Figure 2.2) actuated by a brush dc motor are assumed to be of the form [9]

$$M\ddot{q} + B\dot{q} + N\sin(q) = I \tag{2.1}$$

where M denotes the constant lumped inertia, B denotes the constant friction coefficient, N denotes the constant lumped load term, $q(t)$ is the angular load position (and hence the position of the motor rotor), $\dot{q}(t)$ is the angular load velocity, $\ddot{q}(t)$ is the angular load acceleration, and $I(t)$ is the rotor current (Note that the parameters M, B, and N described in (2.1) are defined to include the effects of the torque coefficient constant which characterizes the electromechanical conversion of rotor current to torque).

The electrical subsystem dynamics for the brushed dc motor are assumed to be

$$L\dot{I} = v - RI - K_B\dot{q} \tag{2.2}$$

where L is the constant rotor inductance, R is the constant rotor resistance, K_B is the constant back-emf coefficient, and $v(t)$ is the input control voltage. For the controllers developed in this chapter, we will assume that the true states (*i.e.*, q, $\dot{q}$, and I) are all measurable.

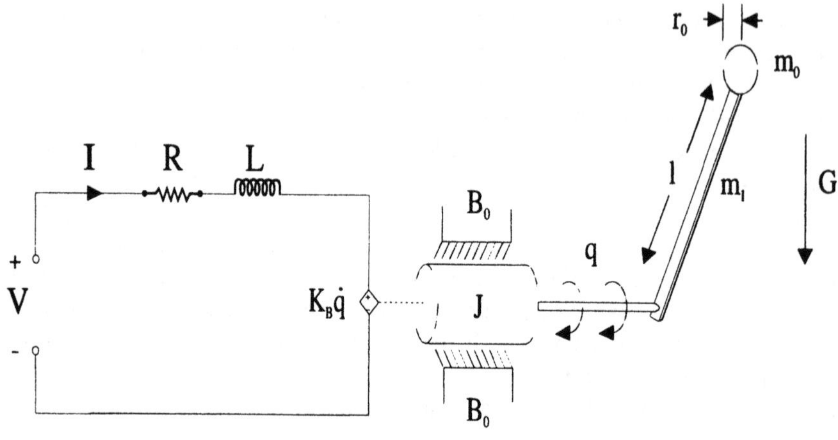

Figure 2.2: Schematic Diagram of a PMBDC Motor/Load System

Remark 2.1

Over the years, adaptive control researchers have attacked a general class of nonlinear systems which encompasses the above electromechanical dynamics. During the study of these types of systems, the dynamic equations are often written in ways that describe the differences between various types of systems. For example, the dynamics given by (2.1) and (2.2) can also be written in the following state-space form

$$
\begin{aligned}
\dot{x}_1 &= x_2 \\
\dot{x}_2 &= b_1 x_3 + \theta_1 \sin(x_1) + \theta_2 x_2 \\
\dot{x}_3 &= b_o u + \theta_3 x_2 + \theta_4 x_3
\end{aligned}
$$

where $x_1 = q$, $x_2 = \dot{q}$, $x_3 = I$, $u = v$, $b_o = 1/L$, $b_1 = 1/M$, $\theta_1 = -N/M$, $\theta_2 = -B/M$, $\theta_3 = -K_B/L$, and $\theta_4 = -R/L$. In the above form, we can state that the electromechanical dynamics represent a relative degree three system (*i.e.*, with q as the output and u as the input) with an uncertainty level (see [11] page 322) of one.

Remark 2.2

Closely related to the BDC motor is the wound stator DC motor configured in a separately excited fashion. The wound field motor features a stator coil used to establish a magnetic field within the motor (recall that the BDC motor uses a permanent magnet in the stator to create a similar field). The connection of the stator and rotor coils relative to the electrical supply(s) will yield different speed and torque characteristics. Three common connections are commercially available: series connected, shunt connected, and separately excited. It is the separately excited case, in which there are independent electrical supplies exciting each of the coils, that behaves much like the BDC motor. The typical dynamic model of the separately excited wound field motor is given by

$$M\ddot{q} + B\dot{q} + N\sin(q) = (K_{\tau}L_{s}I_{s})\, I_{r}\ , \tag{2.3}$$

$$L_{r}\dot{I}_{r} = v_{r}\ - R_{r}I_{r} - (K_{B}L_{s}I_{s})\, \dot{q}, \tag{2.4}$$

and

$$L_{s}\dot{I}_{s} = v_{s}\ - R_{s}I_{s}, \tag{2.5}$$

where the subscript "s" refers to the electrical quantities in the stator coil and the subscript "r" denotes the electrical quantities associated with the rotor coil. Since the stator circuit of (2.5) is completely decoupled from $\dot{q}(t)$ and $I_r(t)$, we can see from (2.5) that a constant stator voltage, v_s, results in a constant stator current, $I_s(t)$, for all possible values of $\dot{q}(t)$ and $I_r(t)$. Equations (2.3) and (2.4) then appear identical to (2.1) and (2.2) if we consider $K_{\tau}L_{s}I_{s}(t)$ to be the torque constant and $K_{B}L_{s}I_{s}(t)$ to be the back-emf constant. Hence, a separately excited dc motor control strategy would involve the servoing of a constant stator current and the application of the rotor voltage input controllers described in this chapter.

2.3 Control Objective

Given full state measurement (*i.e.*, q, $\dot{q}$, and I), the control objective is to develop load position tracking controllers for the electromechanical dynamics of (2.1) and (2.2). To begin the development, we define the load position tracking error $e(t)$ as

$$e = q_d - q \tag{2.6}$$

where $q_d(t)$ represents the desired load position trajectory, and $q(t)$ was defined in (2.1). We will assume that $q_d(t)$ and its first, second, and third derivatives are all bounded functions of time. To simplify the control formulation and the stability analysis, we define the filtered link tracking error $r(t)$ [10] as

$$r = \dot{e} + \alpha e \tag{2.7}$$

where α is a positive, constant controller gain. As explained in Chapter 1, use of the filtered tracking error allows us to analyze the second-order dynamics of (2.1) as though it is a first-order system.

To form the open-loop filtered tracking error system, we differentiate (2.7) with respect to time and rearrange terms to yield

$$\dot{r} = (\ddot{q}_d + \alpha\dot{e}) - \ddot{q}. \tag{2.8}$$

Multiplying (2.8) by M and substituting the mechanical subsystem dynamics of (2.1) yields the filtered tracking error dynamics as shown

$$M\dot{r} = M(\ddot{q}_d + \alpha\dot{e}) + B\dot{q} + N\sin(q) - I. \tag{2.9}$$

To reduce the algebraic overhead, the right-hand side of (2.9) can be rewritten as

$$M\dot{r} = W_\tau\theta_\tau \ - \ I \tag{2.10}$$

where the known regression matrix [12] $W_\tau(q, \dot{q}, t) \in \Re^{1\times 3}$ is given by

$$W_\tau = \left[\begin{array}{ccc} \ddot{q}_d + \alpha\dot{e} & \dot{q} & \sin(q) \end{array}\right] \tag{2.11}$$

and the parameter vector $\theta_\tau \in \Re^3$ is given by

$$\theta_\tau = \left[\begin{array}{ccc} M & B & N \end{array}\right]^T. \tag{2.12}$$

Considering the structure of the electromechanical systems given by (2.1) and (2.2), we are only free to specify the rotor voltage $v(t)$. In other words, the mechanical subsystem error dynamics lack a true current (torque) level control input. For this reason, we shall add and subtract the desired current trajectory $I_d(t)$ to the right-hand side of (2.10), as shown

$$M\dot{r} = W_\tau\theta_\tau - I_d + \eta_I \tag{2.13}$$

where $\eta_I(t)$ represents the current tracking error perturbation to the mechanical subsystem dynamics of the form

$$\eta_I = I_d \ - \ I. \tag{2.14}$$

Remark 2.3

Later, the desired current trajectory $I_d(t)$ will be designed such that it would provide good load position tracking for the mechanical dynamics alone (*i.e.*, assuming it could be applied directly to the load). We will show later in the development that $I_d(t)$ is actually embedded inside of an overall control strategy which is designed at $v(t)$, the motor terminal voltage input. We also note that the above procedure of adding and subtracting the control input $I_d(t)$ is often referred to as integrator backstepping [4].

If the current tracking error term $\eta_I(t)$ in (2.13) was equal to zero, then $I_d(t)$ could be designed to achieve good load position tracking utilizing standard control techniques (*e.g.*, [10]). Since the current tracking error is not equal to zero in general, we must design a voltage control input at $v(t)$ which compensates for the effects of $\eta_I(t)$ in (2.13). To accomplish this control objective, the dynamics of the current tracking error are needed. Taking the time derivative of the current tracking error in (2.14) and then multiplying by L yields

$$L\dot{\eta}_I = L\dot{I}_d - L\dot{I}. \tag{2.15}$$

Substituting the right-hand side of (2.2) for $L\dot{I}$ in (2.15) results in the current tracking error dynamics as shown

$$L\dot{\eta}_I = L\dot{I}_d + RI + K_B\dot{q} - v. \tag{2.16}$$

Remark 2.4

The open-loop dynamics of (2.13) and (2.16) represent the system for which we will design an exact model knowledge controller and an adaptive controller. That is, in the subsequent sections, we will formulate a desired current trajectory for $I_d(t)$ and the voltage control input for $v(t)$ to ensure good load position tracking.

2.4 Exact Model Knowledge Controller

Given exact model knowledge, we now design a position tracking controller for the open-loop dynamics of (2.13) and (2.16); moreover, we formulate the closed-loop electromechanical system which will be used for the stability analysis. The first step in the procedure is to design the desired current trajectory $I_d(t)$ in a certainty equivalence fashion [12] for the mechanical dynamics of (2.13). That is, we select $I_d(t)$ as

$$I_d = W_T\theta_T + k_s r \tag{2.17}$$

where $W_T(q, \dot{q}, t)$ and θ_T were previously defined in (2.11) and (2.12), respectively, and k_s is a positive, constant controller gain. Substituting (2.17) into the open-loop dynamics of (2.13) yields the closed-loop filtered tracking error dynamics as shown

$$M\dot{r} = -k_s r + \eta_I. \tag{2.18}$$

Now that we have designed the desired current trajectory $I_d(t)$, we can complete the open-loop system description for the current tracking error dynamics. That is, we can calculate the term $\dot{I}_d(t)$ in (2.16) by taking the time derivative of (2.17) to yield

$$\dot{I}_d = \dot{W}_T\theta_T + k_s\dot{r}. \tag{2.19}$$

Substituting (2.8) and the time derivative of (2.11) into the right-hand side of (2.19) yields

$$\dot{I}_d = M\left(\dddot{q}_d + \alpha\left(\ddot{q}_d - \ddot{q}\right)\right) + B\ddot{q} + N\dot{q}\cos(q) + k_s\left(\ddot{q}_d - \ddot{q} + \alpha\dot{e}\right) \qquad (2.20)$$

which is in terms of measurable states (*i.e.*, $q(t)$ and $\dot{q}(t)$), known functions, known constant parameters, and the unmeasurable load acceleration $\ddot{q}(t)$. We can solve for $\ddot{q}(t)$ from (2.1) in the following form

$$\ddot{q} = -\frac{B}{M}\dot{q} - \frac{N}{M}\sin(q) + \frac{1}{M}I. \qquad (2.21)$$

Substituting for $\ddot{q}(t)$ from the right-hand side of (2.21) into (2.20), we can write $\dot{I}_d(t)$ in terms of measurable states (*i.e.*, $q(t)$, $\dot{q}(t)$, and $I(t)$), known functions, and known constant parameters in the following manner

$$\begin{aligned} \dot{I}_d = \; & M\left(\dddot{q}_d + \alpha\ddot{q}_d\right) + N\dot{q}\cos(q) + k_s\left(\ddot{q}_d + \alpha\dot{e}\right) \\ & + \left(B - M\alpha - k_s\right)\left(-\frac{B}{M}\dot{q} - \frac{N}{M}\sin(q) + \frac{1}{M}I\right). \end{aligned} \qquad (2.22)$$

Substituting the expression for $\dot{I}_d(t)$ of (2.22) into (2.16) yields the final open-loop model for the current tracking error in the form

$$L\dot{\eta}_I = w_e - v \qquad (2.23)$$

where the auxiliary scalar variable $w_e(q, \dot{q}, I, t)$ is given by

$$\begin{aligned} w_e = \; & L\left[M\left(\dddot{q}_d + \alpha\ddot{q}_d\right) + N\dot{q}\cos(q) + k_s\left(\ddot{q}_d + \alpha\dot{e}\right)\right] + RI + K_B\dot{q} \\ & + L\left(B - M\alpha - k_s\right)\left(-\frac{B}{M}\dot{q} - \frac{N}{M}\sin(q) + \frac{1}{M}I\right). \end{aligned} \qquad (2.24)$$

The second step in the procedure involves the design of the voltage control input $v(t)$ for the open-loop system of (2.23). Given the structure of (2.23) and (2.18), we define the voltage control input as

$$v = w_e + k_e\eta_I + r \qquad (2.25)$$

where k_e is a positive, constant control gain. Substituting (2.25) into the open-loop dynamics of (2.23) yields the closed-loop current tracking error dynamics in the form

$$L\dot{\eta}_I = -k_e\eta_I - r. \qquad (2.26)$$

Remark 2.5

To facilitate the subsequent stability analysis, the right-hand side of (2.25), and therefore the right-hand side of (2.26), contains a filtered tracking error term (*i.e.*, $r(t)$). This additional control term has been added to compensate for the interconnection terms associated with the torque transmission term (*i.e.*, $\eta_I(t)$) in (2.18).

The dynamics given by (2.18) and (2.26) represent the electromechanical closed-loop system for which the stability analysis is performed while the exact model knowledge controller given by (2.17) and (2.25) represents the control input which is implemented at the voltage terminals of the motor. Note that the desired current trajectory $I_d(t)$ is embedded (in the guise of the variable $\eta_I(t)$) inside of the voltage control input $v(t)$. The theorem given below delineates the performance of the closed-loop system under the proposed control.

Theorem 2.1

The proposed exact knowledge controller ensures that the filtered tracking error goes to zero exponentially fast for the electromechanical dynamics of (2.1) and (2.2) as shown

$$\|x(t)\| \leq \sqrt{\frac{\lambda_2}{\lambda_1}}\, \|x(0)\|\, \mathrm{e}^{-\gamma t} \qquad \forall t \in [0, \infty) \tag{2.27}$$

where

$$x = \begin{bmatrix} r & \eta_I \end{bmatrix}^T \in \Re^2, \tag{2.28}$$

$$\lambda_1 = \min\{M, L\}, \qquad \lambda_2 = \max\{M, L\}, \tag{2.29}$$

and

$$\gamma = \frac{\min\{k_s, k_e\}}{\max\{M, L\}}. \tag{2.30}$$

Proof. First, we define the following non-negative function

$$V(t) = \frac{1}{2}Mr^2 + \frac{1}{2}L\eta_I^2 = \frac{1}{2}x^T \begin{bmatrix} M & 0 \\ 0 & L \end{bmatrix} x. \tag{2.31}$$

where $x(t)$ was defined in (2.28). By applying Lemma 1.7 in Chapter 1 to the matrix term in (2.31), we can form the following upper and lower bound on $V(t)$ as follows

$$\frac{1}{2}\lambda_1 \|x(t)\|^2 \leq V(t) \leq \frac{1}{2}\lambda_2 \|x(t)\|^2 \tag{2.32}$$

where λ_1 and λ_2 were defined in (2.29). Differentiating (2.31) with respect to time yields

$$\dot{V}(t) = rM\dot{r} + \eta_I L\dot{\eta}_I. \tag{2.33}$$

Substituting (2.18) and (2.26) into (2.33) yields

$$\dot{V}(t) = -k_s r^2 + r\eta_I - k_e\eta_I^2 - r\eta_I = -x^T \begin{bmatrix} k_s & 0 \\ 0 & k_e \end{bmatrix} x. \tag{2.34}$$

After applying Lemma 1.7 in Chapter 1 to the matrix term in (2.34), $\dot{V}(t)$ in (2.34) can be upper bounded as

$$\dot{V}(t) \leq -\min\{k_s,\, k_e\} \left\|x(t)\right\|^2. \tag{2.35}$$

From (2.32), it easy to see that $\left\|x(t)\right\|^2 \geq 2V(t)/\lambda_2$; hence, $\dot{V}(t)$ in (2.35) can be further upper bounded as

$$\dot{V}(t) \leq -2\gamma V(t) \tag{2.36}$$

where γ was defined (2.30). Applying Lemma 1.1 in Chapter 1 to (2.36) yields

$$V(t) \leq V(0)\underline{e}^{-2\gamma t}. \tag{2.37}$$

From (2.32), we have that

$$\frac{1}{2}\lambda_1 \left\|x(t)\right\|^2 \leq V(t) \qquad \text{and} \qquad V(0) \leq \frac{1}{2}\lambda_2 \left\|x(0)\right\|^2. \tag{2.38}$$

Substituting (2.38) appropriately into the left-hand and right-hand sides of (2.37) allows us to form the following inequality

$$\frac{1}{2}\lambda_1 \left\|x(t)\right\|^2 \leq \frac{1}{2}\lambda_2 \left\|x(0)\right\|^2 \underline{e}^{-2\gamma t}. \tag{2.39}$$

Using (2.39), we can solve for $\left\|x(t)\right\|$ to obtain the result given in (2.27). □

Remark 2.6

Using the result of Theorem 2.1, the control structure, and the electromechanical model, it is straightforward to illustrate that all signals remain bounded during closed-loop operation. Specifically, from (2.27) and (2.28), we know that $r(t) \in L_\infty$ and $\eta_I(t) \in L_\infty$. From Lemma 1.4 in Chapter 1, we know that if $r(t) \in L_\infty$ then $e(t) \in L_\infty$ and $\dot{e}(t) \in L_\infty$. Since $q_d(t) \in L_\infty$ and $\dot{q}_d(t) \in L_\infty$ by assumption then (2.6) can be utilized to show that $q(t) \in L_\infty$ and $\dot{q}(t) \in L_\infty$. From the definition of $I_d(t)$ given in (2.17) and the fact that $\ddot{q}_d(t) \in L_\infty$ by assumption, it is now easy to show that $I_d(t) \in L_\infty$; hence, since $\eta_I(t) \in L_\infty$, we can use (2.14) to state that

$I(t) \in L_\infty$. Using the structure of the voltage control input of (2.25), the above information, and the assumption that $\dddot{q}_d(t) \in L_\infty$, we now know that $v(t) \in L_\infty$. Finally, the electromechanical dynamics of (2.1) and (2.2) can be used to show that $\ddot{q}(t) \in L_\infty$ and $\dot{I}(t) \in L_\infty$.

Remark 2.7

From the result given by (2.27), we can form the following upper bound on the filtered tracking error

$$|r(t)| \leq \sqrt{\frac{\lambda_2}{\lambda_1}} \|x(0)\| \, \underline{e}^{-\gamma t} \qquad \forall t \in [0, \infty). \tag{2.40}$$

From (2.40), it is now straightforward to see that the premise of Lemma 1.5 in Chapter 1 is satisfied; hence, we know that position tracking error (*i.e.*, $e(t)$) and the velocity tracking error (*i.e.*, $\dot{e}(t)$) both go to zero exponentially fast. As an added bonus, we can also use the result given by (2.27) to form the following upper bound on the current tracking error

$$|\eta_I(t)| \leq \sqrt{\frac{\lambda_2}{\lambda_1}} \|x(0)\| \, \underline{e}^{-\gamma t} \qquad \forall t \in [0, \infty); \tag{2.41}$$

hence, the actual current (*i.e.*, $I(t)$) tracks the desired current (*i.e.*, $I_d(t)$) exponentially fast.

Remark 2.8

It is interesting to note the constant γ defined in (2.30) can be increased (*i.e.*, by increasing the controller gains k_s and k_e) to obtain "faster" transient response performance for the filtered tracking error and the current tracking error as delineated by the right-hand side of (2.40) and (2.41), respectively.

2.5 Adaptive Controller

Under the constraint of parametric uncertainty, we now design an adaptive position tracking controller for the open-loop dynamics of (2.13) and (2.16); moreover, we formulate the closed-loop electromechanical error system which will be used for the stability analysis (for the general theory the reader is referred to [1]). The first step in the procedure is to design an adaptive desired current trajectory $I_d(t)$ for the mechanical dynamics of (2.13). That is, we select $I_d(t)$ [10] as

$$I_d = W_\tau \hat{\theta}_\tau + k_s r \tag{2.42}$$

where $W_\tau(q, \dot{q}, t)$ was defined in (2.11), $\hat{\theta}_\tau(t) \in \Re^3$ represents a dynamic estimate for the unknown parameter vector θ_τ defined in (2.12), and k_s is a positive, constant controller gain. The parameter estimate $\hat{\theta}_\tau(t)$ defined in (2.42) is updated online according to the following adaptation law

$$\hat{\theta}_\tau = \int_0^t \Gamma_\tau W_\tau^T(\sigma) r(\sigma)\, d\sigma \tag{2.43}$$

where $\Gamma_\tau \in \Re^{3\times 3}$ is a constant, positive definite, diagonal adaptive gain matrix. Defining the mismatch between $\hat{\theta}_\tau(t)$ and θ_τ as

$$\tilde{\theta}_\tau = \theta_\tau - \hat{\theta}_\tau, \tag{2.44}$$

allows the time derivative of the parameter observation error to be written in terms of the adaptation law of (2.43) as

$$\dot{\tilde{\theta}}_\tau = -\Gamma_\tau W_\tau^T r. \tag{2.45}$$

Substituting (2.42) into the open-loop dynamics of (2.13) yields the closed-loop filtered tracking error dynamics, as shown

$$M\dot{r} = W_\tau \tilde{\theta}_\tau - k_s r + \eta_I. \tag{2.46}$$

Now that we have designed the adaptive desired current trajectory $I_d(t)$, we can complete the open-loop system description for the current tracking error dynamics. That is, we can calculate the term $\dot{I}_d(t)$ in (2.16) by taking the time derivative of (2.42) to yield

$$\dot{I}_d = \dot{W}_\tau \hat{\theta}_\tau + W_\tau \dot{\hat{\theta}}_\tau + k_s \dot{r}. \tag{2.47}$$

Substituting (2.8), and the time derivatives of (2.11) and (2.43) into the right-hand side of (2.47) yields

$$\begin{aligned} \dot{I}_d = \; & \hat{M}\left(\dddot{q}_d + \alpha\left(\ddot{q}_d - \ddot{q}\right)\right) + \hat{B}\ddot{q} + \hat{N}\dot{q}\cos(q) \\ & + W_\tau \Gamma_\tau W_\tau^T r + k_s\left(\ddot{q}_d - \ddot{q} + \alpha\dot{e}\right) \end{aligned} \tag{2.48}$$

where $\hat{M}(t)$, $\hat{B}(t)$, and $\hat{N}(t)$ denote the scalar components of the vector $\hat{\theta}_\tau(t)$ (*i.e.*, $\hat{\theta}_\tau = \begin{bmatrix} \hat{M} & \hat{B} & \hat{N} \end{bmatrix}^T$). Note that $\dot{I}_d(t)$ of (2.48) is in terms of measurable states (*i.e.*, $q(t)$ and $\dot{q}(t)$), known functions, and the unmeasurable quantity $\ddot{q}(t)$. After substituting for $\ddot{q}(t)$ from the right-hand side of (2.21) into (2.48), we can write $\dot{I}_d(t)$ in terms of measurable states (*i.e.*, $q(t)$, $\dot{q}(t)$, and $I(t)$), known functions, and unknown constant parameters. Substituting this expression for $\dot{I}_d(t)$ into (2.16) and then performing the necessary algebra yields a linear parameterized open-loop model of the form

$$L\dot{\eta}_I = W_1\theta_1 - v \tag{2.49}$$

where the known regression matrix $W_1(q, \dot{q}, I, \hat{\theta}_\tau, t) \in \Re^{1\times 6}$ and the unknown constant parameter vector $\theta_1 \in \Re^6$ are explicitly defined as follows

$$\theta_1 = \begin{bmatrix} \frac{L}{M} & \frac{LB}{M} & R & K_B & \frac{LN}{M} & L \end{bmatrix}^T, \tag{2.50}$$

$$W_1 = \begin{bmatrix} W_{11} & W_{12} & W_{13} & W_{14} & W_{15} & W_{16} \end{bmatrix}, \tag{2.51}$$

$$W_{11} = \hat{B}I - k_s I - \alpha \hat{M} I, \quad W_{12} = k_s \dot{q} - \hat{B}\dot{q} + \alpha \hat{M}\dot{q},$$

$$W_{13} = I, \quad W_{14} = \dot{q}, \quad W_{15} = k_s \sin(q) - \hat{B}\sin(q) + \alpha\hat{M}\sin(q),$$

and

$$W_{16} = \hat{M}\dddot{q}_d + \alpha\hat{M}\ddot{q}_d + W_\tau \Gamma_\tau W_\tau^T r + k_s \ddot{q}_d + k_s \alpha \dot{e} + \hat{N}\dot{q}\cos(q).$$

Remark 2.9

Due to the above backstepping procedure, the parameter vector θ_1 of (2.50) consists of parameters from both the electrical and mechanical subsystems. Therefore, the dimension of θ_1 is larger than the parameter vector obtained for a linear parameterization of the electrical subsystem dynamics. This phenomenon is often referred to as overparameterization [4] in the controls literature. In [8], a method which resolves this problem was presented. In the next section of this chapter, we redesign the adaptive controller to reduce the overparameterization problem.

The second step in the design procedure involves the design of the voltage control input $v(t)$ for the open-loop system of (2.49). Given the structure of (2.49) and (2.46), we define the input voltage controller as

$$v = W_1 \hat{\theta}_1 + k_e \eta_I + r \tag{2.52}$$

where k_e is a positive, constant control gain, and $\hat{\theta}_1(t) \in \Re^6$ is a dynamic estimate of the unknown parameter vector θ_1. The parameter estimates are updated on-line by the following adaptation law

$$\hat{\theta}_1 = \int_0^t \Gamma_e W_1^T(\sigma)\eta_I(\sigma)\, d\sigma \tag{2.53}$$

where $\Gamma_e \in \Re^{6\times 6}$ is a constant positive definite, diagonal adaptive gain matrix. If we define the mismatch between θ_1 and $\hat{\theta}_1(t)$ as

$$\tilde{\theta}_1 = \theta_1 - \hat{\theta}_1, \tag{2.54}$$

then the adaptation law of (2.53) can be written in terms of the parameter error as

$$\dot{\tilde{\theta}}_1 = -\Gamma_e W_1^T \eta_I. \tag{2.55}$$

Substituting (2.52) into the open-loop dynamics of (2.49) yields the closed-loop current tracking error dynamics in the form

$$L\dot{\eta}_I = W_1\tilde{\theta}_1 - k_e\eta_I - r. \tag{2.56}$$

Remark 2.10

As done for the exact model knowledge controller, the voltage control input of (2.52) contains a filtered tracking error term to facilitate the stability proof.

The dynamics given by (2.45), (2.46), (2.55), and (2.56) represent the electromechanical closed-loop system for which the stability analysis is performed while the adaptive controller given by (2.42), (2.43), (2.52), and (2.53) represents the controller which is implemented at the voltage terminals of the motor. Similar to the exact model knowledge controller, the adaptive desired current trajectory $I_d(t)$ is embedded (in the guise of the variable $\eta_I(t)$) inside of the voltage control input $v(t)$. The theorem given below delineates the performance of the closed-loop system under the proposed control.

Theorem 2.2

The proposed adaptive controller ensures that the filtered tracking error goes to zero asymptotically for the electromechanical dynamics of (2.1) and (2.2) as shown

$$\lim_{t\to\infty} r(t) = 0. \tag{2.57}$$

Proof. First, define the following non-negative function

$$V = \frac{1}{2}Mr^2 + \frac{1}{2}L\eta_I^2 + \frac{1}{2}\tilde{\theta}_\tau^T\Gamma_\tau^{-1}\tilde{\theta}_\tau + \frac{1}{2}\tilde{\theta}_1^T\Gamma_e^{-1}\tilde{\theta}_1. \tag{2.58}$$

Taking the time derivative of (2.58) with respect to time yields

$$\dot{V} = rM\dot{r} + \eta_I L\dot{\eta}_I + \tilde{\theta}_\tau^T\Gamma_\tau^{-1}\dot{\tilde{\theta}}_\tau + \tilde{\theta}_1^T\Gamma_e^{-1}\dot{\tilde{\theta}}_1 \tag{2.59}$$

where the facts that: i) scalars can be transposed and ii) Γ_τ, Γ_e are diagonal matrices, have been used. Substituting the error dynamics of (2.45), (2.55), (2.56), and (2.46) into (2.59) yields

$$\dot{V} = -k_s r^2 - k_e\eta_I^2 + rW_\tau\tilde{\theta}_\tau - \tilde{\theta}_\tau^T W_\tau^T r + \eta_I W_1\tilde{\theta}_1 - \tilde{\theta}_1^T W_1^T\eta_I. \tag{2.60}$$

Since any scalar quantity can be transposed, (2.60) can be simplified to yield

$$\dot{V} = -k_s r^2 - k_e\eta_I^2. \tag{2.61}$$

From the form of (2.61), we can see that $\dot{V}(t)$ is negative or zero; hence, we know from calculus that $V(t)$ given in (2.58) is either decreasing or constant. Since $V(t)$ is non-negative, it is lower bounded by zero; hence, from the form of $V(t)$, we know that $r(t) \in L_\infty$, $\tilde{\theta}_\tau(t) \in L^3_\infty$, $\eta_I(t) \in L_\infty$, and $\tilde{\theta}_1(t) \in L^6_\infty$. Since $r(t) \in L_\infty$, Lemma 1.4 in Chapter 1 can be used to show that $e(t) \in L_\infty$ and $\dot{e}(t) \in L_\infty$; therefore, since $q_d(t) \in L_\infty$ and $\dot{q}_d(t) \in L_\infty$, (2.6) can be utilized to show that $q(t) \in L_\infty$ and $\dot{q}(t) \in L_\infty$. Since $\tilde{\theta}_\tau(t) \in L^3_\infty$ and $\tilde{\theta}_1(t) \in L^6_\infty$, we can use (2.44) and (2.54) to show that $\hat{\theta}_\tau(t) \in L^3_\infty$ and $\hat{\theta}_1(t) \in L^6_\infty$. From the definition of $I_d(t)$ given in (2.42) and the fact that $\ddot{q}_d(t) \in L_\infty$ by assumption, it is know easy to show that $I_d(t) \in L_\infty$; hence, since $\eta_I(t) \in L_\infty$, we can use (2.14) to state that $I(t) \in L_\infty$. Using the structure of the voltage control input of (2.52), the above information, and the assumption that $\dddot{q}_d(t) \in L_\infty$, we can now show that $v(t) \in L_\infty$. The electromechanical dynamics of (2.1) and (2.2) can be used to show that $\ddot{q}(t) \in L_\infty$ and $\dot{I}(t) \in L_\infty$. Finally, we can use the above information, (2.46) and (2.56) to illustrate that $\dot{r}(t) \in L_\infty$ and $\dot{\eta}_I(t) \in L_\infty$.

We now illustrate how Corollary 1.1 in Chapter 1 can be used to show that filtered tracking error goes to zero. First, we integrate both sides of (2.61) with respect to time to yield

$$\int_0^\infty \frac{dV(\sigma)}{d\sigma}\, d\sigma = -\int_0^\infty \left(k_s r^2(\sigma) + k_e \eta_I^2(\sigma)\right)\, d\sigma. \tag{2.62}$$

If we simplify the left-hand side of (2.62), we obtain

$$\sqrt{V(0) - V(\infty)} = \sqrt{\int_0^\infty \left(k_s r^2(\sigma) + k_e \eta_I^2(\sigma)\right)\, d\sigma} \tag{2.63}$$

after some minor algebraic operations. Since $\dot{V}(t) \leq 0$ as illustrated by (2.61), $V(t)$ of (2.58) is decreasing or constant; hence, $V(0) \geq V(\infty) \geq 0$. We now use this information and (2.63) to obtain the following inequality

$$\sqrt{k_s \int_0^\infty r^2(\sigma)\, d\sigma} \leq \sqrt{\int_0^\infty \left(k_s r^2(\sigma) + k_e \eta_I^2(\sigma)\right)\, d\sigma} \leq \sqrt{V(0)} < \infty \tag{2.64}$$

which indicates according to Definition 1.2 in Chapter 1 that $r(t) \in L_2$. Since $r(t) \in L_\infty$, $\dot{r}(t) \in L_\infty$, and $r(t) \in L_2$, we can invoke Corollary 1.1 in Chapter 1 to obtain the result given by (2.57). □

Remark 2.11

From the result given by (2.57) and the proof of Theorem 2.2, we know that $r(t) \in L_\infty$, $r(t) \in L_2$, and $\lim_{t\to\infty} r(t) = 0$; hence, we can use Lemma

1.6 in Chapter 1 to state that $\lim_{t\to\infty} e(t) = 0$ and $\lim_{t\to\infty} \dot{e}(t) = 0$. As an added bonus, we form a result similar to (2.64) for $\eta_I(t)$ as follows

$$\sqrt{k_e \int_0^\infty \eta_I^2(\sigma)\, d\sigma} \leq \sqrt{\int_0^\infty \left(k_s r^2(\sigma) + k_e \eta_I^2(\sigma)\right) d\sigma} \leq \sqrt{V(0)} < \infty \tag{2.65}$$

which indicates that $\eta_I(t) \in L_2$. Since $\eta_I(t) \in L_\infty$, $\dot{\eta}_I(t) \in L_\infty$, and $\eta_I(t) \in L_2$, we can invoke Corollary 1.1 in Chapter 1 to state $\lim_{t\to\infty} \eta_I(t) = 0$; hence, the actual current (*i.e.*, $I(t)$) tracks the adaptive desired current trajectory (*i.e.*, $I_d(t)$) asymptotically fast.

Remark 2.12

The nonlinear system given by (2.1) and (2.2) is a "mild" nonlinear system in that the only nonlinearity $\sin(q)$ is globally Lipschitz and bounded; therefore, linear high-gain control is applicable. In general, however, the adaptive backstepping designs of [1] do not impose growth restrictions on the nonlinearities in the system.

Remark 2.13

Due to the simplicity of the dynamic equations given by (2.1) and (2.2), measurements of $\dot{q}(t)$ and $I(t)$ can be eliminated if we employ the adaptive output feedback control structures of [13], [2], and [14]. However, the output feedback controllers will require far more online computation than the controllers developed in this chapter.

2.6 Reduction of Overparameterization

From the dynamics given by (2.1) and (2.2), we can see that there are only six parameters; however, in the adaptive controller presented above, there are nine adaptive update laws. That is, $\hat{\theta}_\tau(t) \in \Re^3$ of (2.43) contains three dynamic estimates while $\hat{\theta}_1(t) \in \Re^6$ of (2.53) contains six dynamic estimates. We now present a modification [8] of the adaptive controller which reduces this overparameterization problem. Note, during the development given below, all variables are as defined originally unless otherwise stated.

First, the controller $I_d(t)$ defined in (2.42) is exactly the same and hence the closed-loop filtered tracking error system for the mechanical system is the same as that given by (2.46). The adaptation law for θ_τ is now changed to

$$\dot{\hat{\theta}}_\tau = \Gamma_\tau \left(W_\tau^T r + Y_2^T \eta_I\right) = -\dot{\tilde{\theta}}_\tau \tag{2.66}$$

where $\eta_I(t)$ is defined in (2.14), and $Y_2(t)$ is defined as

$$Y_2 = \begin{bmatrix} 0 & W_{12} & W_{15} \end{bmatrix} \in \Re^{1\times 3}, \tag{2.67}$$

and $W_{12}(t)$, $W_{15}(t)$ were defined in (2.51). Now proceeding as before, we parameterize the dynamics for $\eta_I(t)$ in the following way

$$L\dot{\eta}_I = W_{11}b + Y_1\theta_e + bY_2\theta_\tau - v \tag{2.68}$$

where $W_{11}(t)$ was defined in (2.51),

$$b = \frac{L}{M}, \;\; \theta_e = \begin{bmatrix} L & R & K_B \end{bmatrix}^T \in \Re^3, \;\; Y_1 = \begin{bmatrix} W_{16} & W_{13} & W_{14} \end{bmatrix} \in \Re^{1\times 3},$$

and $W_{13}(t)$, $W_{14}(t)$, $W_{16}(t)$ were defined in (2.51). Note that the $W_\tau\Gamma_\tau W_\tau^T r$ term in $W_{16}(t)$ will now change to $W_\tau\Gamma_\tau\left(W_\tau^T r + Y_2^T\eta_I\right)$ as a result of the adaptive update law modification of (2.66). Based on the parameterized error systems given by (2.46) and (2.68), the new voltage control input is given by

$$v = \hat{b}\left(Y_2\hat{\theta}_\tau + W_{11}\right) + Y_1\hat{\theta}_e + \hat{b}r + k_e\eta_I \tag{2.69}$$

with the adaptive update laws given by

$$\dot{\hat{b}} = \Gamma_b\left(r\eta_I + W_{11}\eta_I + Y_2\hat{\theta}_\tau\eta_I\right) = -\dot{\tilde{b}} \tag{2.70}$$

and

$$\dot{\hat{\theta}}_e = \Gamma_e Y_1^T\eta_I = -\dot{\tilde{\theta}}_e \tag{2.71}$$

where $\tilde{b} = b - \hat{b}$, $\tilde{\theta}_e = \theta_e - \hat{\theta}_e$, Γ_b is a positive adaptive gain constant, and $\Gamma_e \in \Re^{3\times 3}$ is a diagonal, positive definite, adaptive gain matrix. Note from (2.66), (2.70), and (2.71), we can see that there are only seven, as opposed to nine, parameter update laws, and hence the overparameterization problem has been reduced.

After substituting (2.69) into (2.68) and multiplying the resulting equation by $\frac{M}{L}$, we can form the following closed-loop equation for $\eta_I(t)$ as

$$M\dot{\eta}_I = \frac{1}{b}\left(-\hat{b}r - k_e\eta_I + Y_1\tilde{\theta}_e + W_{11}\tilde{b} + Y_2\hat{\theta}_\tau\tilde{b}\right) + Y_2\tilde{\theta}_\tau. \tag{2.72}$$

Now, to prove the stability result, we use the non-negative function

$$V(t) = \frac{1}{2}Mr^2 + \frac{1}{2}M\eta_I^2 + \frac{1}{2}\tilde{\theta}_\tau^T\Gamma_\tau^{-1}\tilde{\theta}_\tau + \frac{1}{2b}\tilde{\theta}_e^T\Gamma_e^{-1}\tilde{\theta}_e + \frac{1}{2b}\Gamma_b^{-1}\tilde{b}^2. \tag{2.73}$$

After differentiating (2.73) with respect to time and substituting (2.46), (2.72), (2.66), (2.70), and (2.71), we can simplify the resulting expression to obtain

$$\dot{V}(t) = -k_s r^2 - \frac{1}{b} k_e \eta_I^2. \tag{2.74}$$

From (2.73) and (2.74), we can use the same arguments used in the proof of Theorem 2.2 to state the same results as those claimed by Theorem 2.2.

2.7 Experimental Results

The hardware used to implement the nonlinear controllers consists of the following: (1) a TMS320-C30 DSP System Board from Spectrum Signal Processing, (2) a 486 ISA based host PC, (3) a DS-2 I/O board from Integrated Motions Inc., (4) a Techron Model 7571 linear power amplifier, (5) a Baldor Model 3300 dc motor with a 1000 count encoder, (6) a 7 $V/krpm$ tachometer, (7) a Microswitch Hall effect current sensor, and assorted interfacing electronics and hardware. The TMS320C30 DSP System Board consists of a Texas Instruments TMS320C30 floating point DSP running at 33MHz (60ns instruction cycle), two channels of 16-bit I/O (ADCs and DACs), 128 Kwords of static RAM, and a 16-bit high-speed parallel expansion bus (DSP-Link) which supports transfer rates up to 10 MBytes/sec. The DSP System Board serves as the computational engine for the entire data acquisition and control system. It provides the necessary interfacing for the tachometer, hall effect sensor, and linear power amplifier. The DS-2 I/O board provides the necessary encoder interface (quadrature decoding), and communicates with the DSP System Board through the DSP-Link. Since the DSP-Link is completely independent of the 486's ISA bus, the host PC is free to perform other tasks. Power is supplied to the motor by a single channel linear power amplifier which is capable of outputting up to 1000W at 100V with a power bandwidth of 0 to 40KHz. See Figure 2.3 for a block diagram of the experimental set-up.

The software associated with the data acquisition and control system consists of the following: (1) Matlab, a Windows based analysis program from The MathWorks, and (2) a user-developed interface between the Simulink program and the DSP System Board. This seamless interface makes it possible to download and execute "*C*" code algorithms on the DSP from the Matlab command prompt (*i.e.*, within the Matlab environment). Control simulation and implementation are performed using the same user-defined "*C*" code blocks. These algorithms are executed like normal Matlab functions or script files. The interface essentially provides a two way data pipeline between the Matlab environment and the DSP. This interface provides the user with Matlab access to all of the data available from the data acquisition and control system and hence allows the system

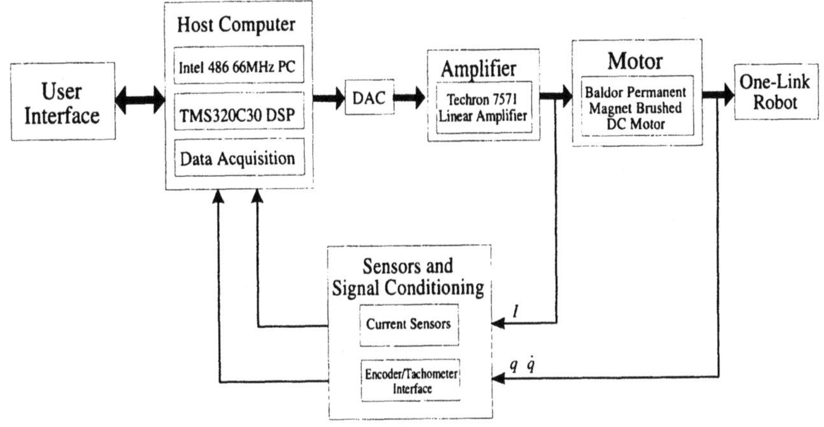

Figure 2.3: Block Diagram of PMBDC Motor Experimental Setup

data to be manipulated via Matlab's analytical capabilities. All control algorithms are run with a sampling rate of 1ms; however, we have run our system at higher sampling rates (*e.g.*, 200μs).

The brushed dc motor was directly connected to a one-link robot arm to mimic the mechanical system given by (2.1) (See Figure 2.2). For this type of mechanical system, the lumped mechanical parameters of (2.1) were calculated to be

$$\begin{aligned} M &= \frac{J}{K_\tau} + \frac{m_1 l^2}{3K_\tau} + \frac{m_0 l^2}{K_\tau} + \frac{2m_0 r_o^2}{5K_\tau}, \\ N &= \frac{m_1 l G}{2K_\tau} + \frac{m_0 l G}{K_\tau}, \quad \text{and} \quad B = \frac{B_o}{K_\tau}, \end{aligned} \tag{2.75}$$

where J is the rotor inertia, m_1 is the link mass, m_0 is the load mass, l is the link length, r_0 is the radius of the load, G is the gravity coefficient, B_o is the coefficient of viscous friction at the joint, and K_τ is the coefficient which characterizes the electromechanical conversion of armature current to torque.

After several standard test procedures and measurements, we determined that the electromechanical system parameters of (2.75) and (2.2) were approximately given by

$$J = 1.625 \times 10^{-3} \text{ kg-m}^2/rad, \quad m_1 = 0.506 \text{ kg},$$

$$r_0 = 0.023 \text{ m}, \quad m_0 = 0.434 \text{ kg},$$

$$l = 0.305 \text{ m}, \quad G = 9.81 \text{kg-m}/\sec^2,$$

$$R = 5.0\ \Omega, \quad B_o = 16.25 \times 10^{-3} \text{ N-m-s/rad},$$

$$L = 25.0 \times 10^{-3} \text{ H}, \quad K_\tau = K_B = 0.90 \text{ N-m/A}.$$

The above electromechanical parameter values were used as the initial conditions for the adaptive controller update laws and as the constant best guess parameter values for the exact model knowledge controller.

The desired position trajectory was selected as

$$q_d(t) = \frac{\pi}{2}\left(1 - \underline{e}^{-0.1t^3}\right)\sin\left(\frac{8\pi}{5}t\right) \text{ rad} \tag{2.76}$$

which has the advantageous property that

$$q_d(0) = \dot{q}_d(0) = \ddot{q}_d(0) = \dddot{q}_d(0) = 0.$$

The desired position trajectory and the actual position trajectory are shown in Figure 2.4.

2.7.1 Exact Model Knowledge Control Experiment

The exact model knowledge controller of (2.17) and (2.25) was found to yield the best position tracking error performance with the following controller gain values

$$k_s = 8, \quad k_e = 8, \quad \text{and} \quad \alpha = 80.$$

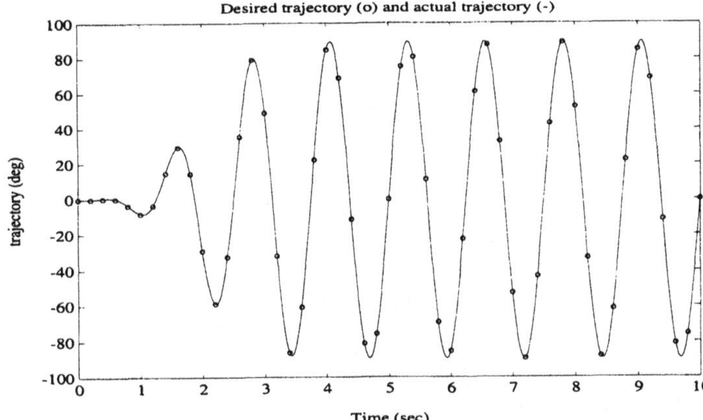

Figure 2.4: Desired and Actual Position for Exact Knowledge Controller

The resulting position tracking error is given in Figure 2.5.

Note that the maximum position tracking error is approximately within ± 0.35 degrees. The corresponding rotor current and voltage are shown in Figures 2.6 and 2.7, respectively.

2.7.2 Adaptive Control Experiment

The adaptive controller of (2.42), (2.43), (2.52), and (2.53) was found to yield the best position tracking error performance with the following controller gain values

$$k_s = 10, \qquad k_e = 1, \qquad \alpha = 35,$$

$$\Gamma_\tau = diag\{0.01, 5, 5\}, \quad \text{and } \Gamma_e = 0.01 I_{6\times 6}$$

where $I_{6\times 6}$ is the 6×6 identity matrix, and $diag\{\cdot\}$ is used to denote the elements of a diagonal matrix. The adaptation update laws were implemented with a standard trapezoid-type integration rule while the desired load trajectory was selected as in (2.76). The desired position trajectory and the actual position trajectory are shown in Figure 2.8. The resulting position tracking error is given in Figure 2.9. Note that the maximum position tracking error is approximately within ± 0.5 degrees. The corresponding rotor current and voltage are shown in Figures 2.10 and 2.11, respectively.

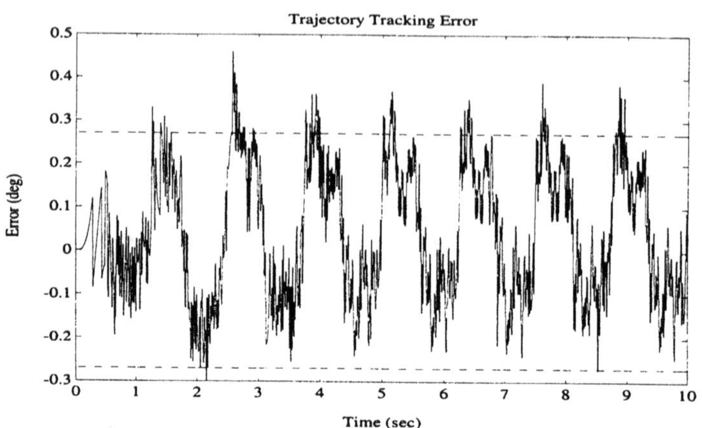

Figure 2.5: Position Tracking Error for Exact Knowledge Controller

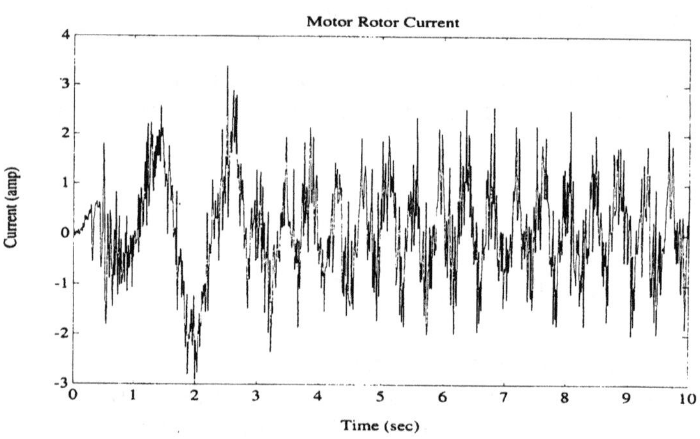

Figure 2.6: Rotor Current for Exact Knowledge Controller

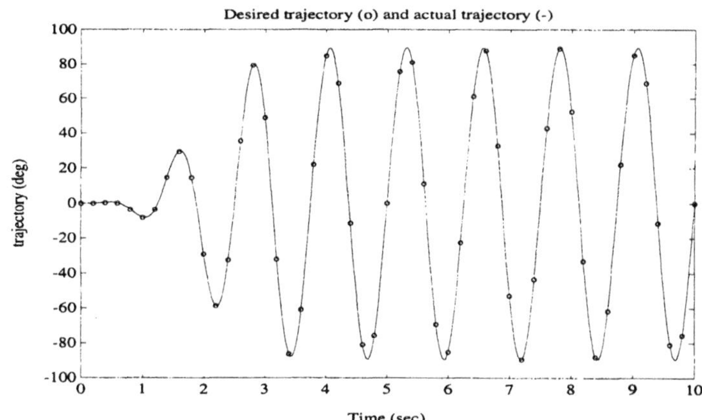

Figure 2.7: Desired and Actual Position for Adaptive Controller

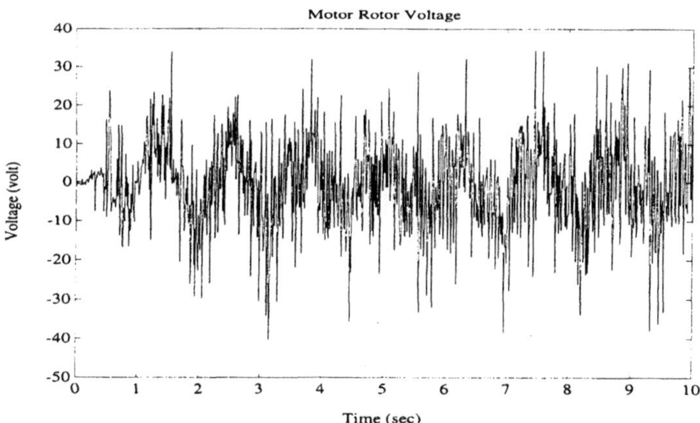

Figure 2.8: Rotor Voltage for Exact Knowledge Controller

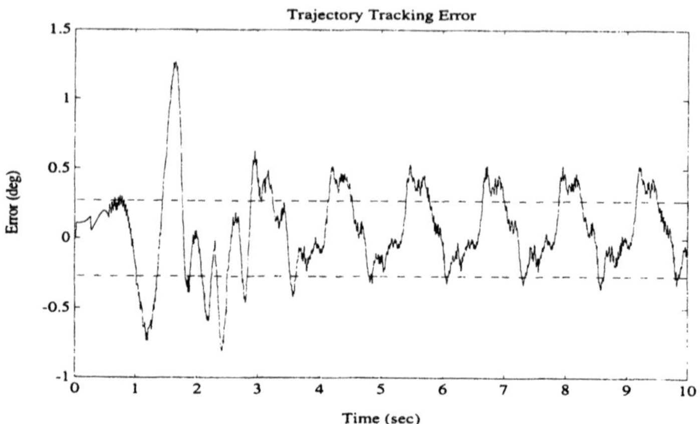

Figure 2.9: Position Tracking Error for Adaptive Controller

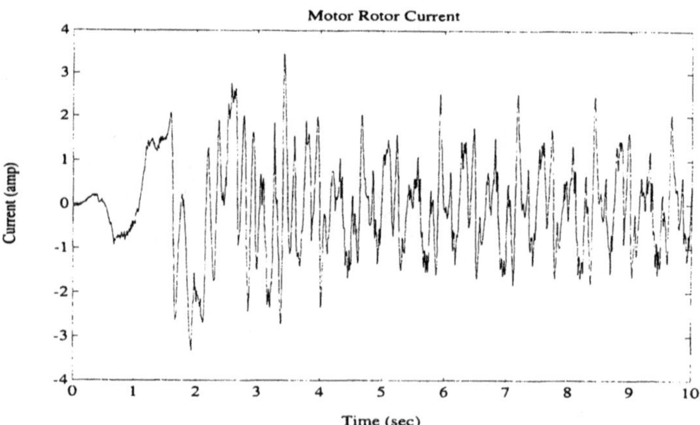

Figure 2.10: Rotor Current for Adaptive Controller

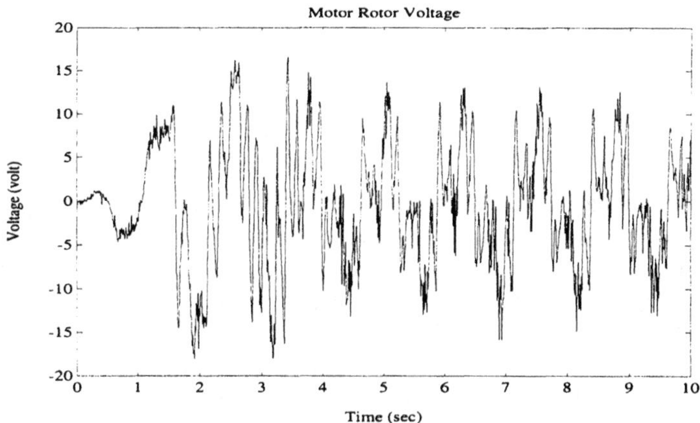

Figure 2.11: Rotor Voltage for Adaptive Controller

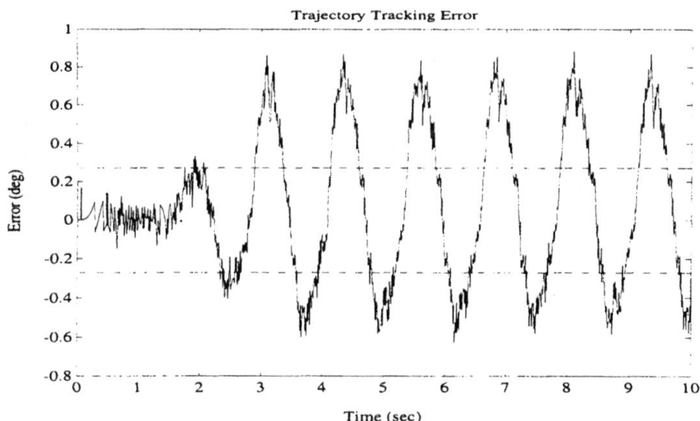

Figure 2.12: Position Tracking Error for Linear Controller

2.7.3 Linear Control Experiment

For comparison purposes, we also implemented a high gain, proportional-derivative outer loop controller in conjunction with a high gain current feedback inner loop controller. That is, the input control voltage controller was taken to be

$$v = k_e \left(k_v \dot{e} + k_p e - I \right)$$

where the controller gains were tuned to $k_e = 7$, $k_v = 8$, and $k_p = 80$ (these controller gain values achieved the best position tracking error performance for our experimental setup). The resulting position tracking error is given in Figure 2.12. Note that the maximum position tracking error is approximately ± 0.9 degrees. For brevity, the actual control voltage and current were omitted; however, in comparison with the input voltage of Figure 2.7, the control voltage and armature current were approximately the same amplitude with an even higher frequency content.

Remark 2.14

From the above experimental results, we can make the following observations. First, with regard to the steady-state tracking error, the adaptive controller outperformed the linear controller by approximately a factor of two; moreover, the linear controller expended a control effort which was substantially more than the adaptive controller. Second, while the exact model knowledge controller outperformed the adaptive controller, the exact model knowledge controller required substantially more control effort to do so. Third, from a practical point of view, the adaptive controller does not exhibit the noisy, chatter-like behavior present with both the exact model knowledge and the linear controller; therefore, from this perspective, the behavior of the adaptive controller is more desirable since chattering voltages and currents can ruin the brushes and commutators and hence shorten the longevity of the motor. For the sake of brevity, plots of the parameter estimates were not included for the adaptive control experiments; however, during the experiemental trials, we noted that the standard gradient algorithm tends to be fairly robust for tracking control experiments. That is, we did not observe excessive parameter drift during the experimental trials.

Bibliography

[1] I. Kanellakopoulos, P. Kokotovic, and A. Morse, "Systematic Design of Adaptive Controllers for Feedback Linearizable Systems", *IEEE Transactions on Automatic Control*, Vol. 36, No. 11, pp. 1241-1253, 1991.

[2] R. Marino and P. Tomei, "Global Adaptive Output-Feedback Control of Nonlinear Systems, Part I: Linear Parameterization", *IEEE Transactions on Automatic Control*, Vol. 38, No.1, pp. 17-31, Jan., 1993.

[3] L. Praly, G. Bastin, J. Pomet, and Z. Jiang, "Adaptive Stabilization of Nonlinear Systems", *Lecture Notes in Control and Information Sciences*, Vol. 160, Springer-Verlag, Berlin, Germany, pp. 347-433, 1991.

[4] M. Krstic, I. Kanellakopoulos, and P. Kokotovic, *Nonlinear and Adaptive Control Design,* John Wiley & Sons, 1995.

[5] P. Kokotovic and H. Sussmann, "A Positive Real Condition for Global Stabilization of Nonlinear Systems", *Systems & Controls Letters*, Vol. 13, pp. 125-133, 1989.

[6] J. Tsinias, "Sufficient Lyapunov-Like Conditions for Stabilization", *Mathematics of Control, Signals, and Systems,* Vol. 2, pp. 343-357, 1989.

[7] C. Byrnes and A. Isidori, "New Results and Examples in Nonlinear Feedback Stabilization", *Systems & Controls Letters*, Vol. 12, pp. 437-442, 1989.

[8] M. Krstic, I. Kanellakopoulos, and P. Kokotovic, "Adaptive Nonlinear Control Without Overparameterization", *Systems & Control Letters*, Vol. 19, pp. 177-185, 1992.

[9] M. Spong and M. Vidyasagar, *Robot Dynamics and Control*, New York: John Wiley and Sons, Inc., 1989.

[10] J. Slotine and W. Li, *Applied Nonlinear Control*, Englewoods Cliff, NJ: Prentice Hall Co., 1991.

[11] P. Kokotovic, Editor, "Foundations of Adaptive Control", *Lecture Notes in Control and Information Sciences* , Vol. 160, Springer-Verlag, Berlin, Germany, 1991.

[12] S. Sastry and M. Bodson, *Adaptive Control: Stability, Convergence, and Robustness*, Englewoods Cliff, NJ: Prentice Hall Co., 1989.

[13] I. Kanellakopoulos, P.V. Kokotovic, and A.S. Morse, "Adaptive Output-Feedback Control of Systems with Output Nonlinearities," *IEEE Transactions on Automatic Control*, vol. 37, pp. 1266-1282, 1992.

[14] M. Krstic, I. Kanellakopoulos, and P. Kokotovic, "Nonlinear Design of Adaptive Controllers for Linear Systems", *IEEE Transactions on Automatic Control,* Vol. 39, No. 4, pp. 738-752, Apr., 1994.

Chapter 3

PMS Motor (FSFB)

3.1 Introduction

Traditionally, brushed DC (BDC) motors have been used in industrial positioning systems. One reason being is that BDC motors are essentially linear systems and thus easy to control. But in recent years, the development of precision nonlinear controllers has encouraged the use of brushless machines in place of BDC motors. The permanent magnet stepper (PMS) motor has become a popular alternative for many high performance motion control applications for several reasons. These reasons include: better reliability due to the elimination of mechanical brushes, better heat dissipation as there are no windings located on the rotor, thus resulting in a 100% duty cycle, higher torque-to-inertia ratio due to a lighter rotor, again as a result of a rotor without windings, and lower prices [1], [2], [3]. While the PMS motor complicates the control problem by coupling multi-input nonlinear dynamics into the overall electromechanical system dynamics, as compared with the single-input dynamics of the BDC motor, it is possible to develop an accurate dynamic model. Unfortunately, as far as control design is concerned, the PMS motor/load electromechanical dynamics exhibit: i) nonlinear terms, which are products of the electrical current and trigonometric terms, in the static torque transmission equation, and ii) nonlinear terms, which are products of rotor velocity and trigonometric terms, in the electrical subsystem dynamics.

In this chapter, we apply the adaptive integrator backstepping tools [4] [5] to the design of a position tracking controller for the PMS motor driving a mechanical load. The original model of the PMS motor is used for the nonlinear controller development, in contrast to the standard DQ transformed model used in [3], [2], [6], [7]. Therefore, a closed form, differentiable commutation [1] strategy must be fused into the backstepping procedure. First,

[1] The term commutation strategy is often used to describe the method for developing the desired current signals which are based on the rotor position, the desired torque

we view the motor as a torque source and thus design a desired torque signal to ensure that the load follows the desired position trajectory. Since the developed motor torque is a function of rotor position and the electrical winding currents, we utilize a simple sinusoidal commutation strategy to restate the desired torque signal as a set of desired current trajectories. The voltage control inputs are then formulated to force the electrical winding currents to follow the desired current trajectories. That is, the electrical dynamics are taken into account through the current tracking objective, and hence the position tracking control objective is embedded inside the current tracking objective. Therefore, if the voltage control input can be designed to guarantee that the actual currents track the desired currents, then the load position will follow the desired position trajectory.

To motivate the structure of the adaptive controller and illustrate the use of the backstepping technique, an exact model knowledge controller is first designed to yield global exponential load position tracking. The exact model knowledge controller is then redesigned via an adaptive approach in order to compensate for parametric uncertainty. The result is an adaptive controller that produces global asymptotic load position tracking. We then illustrate how the adaptive controller can be modified to reduce the over-parameterization [8] [9] phenomenon often associated with many adaptive controllers. Finally, experimental results are presented to illustrate the performance and feasibility of implementing the nonlinear control algorithms.

3.2 System Model

The dynamics of the load position for a two-phase permanent magnet stepper motor can be described by a set of differential equations as in [1] or [10]. Such a representation allows for a distinct segregation of the mechanical and electrical components of the system dynamics. The dynamics are decomposed into one mechanical subsystem and two electrical subsystems that are coupled by the torque transmission and back-emf terms. The coupling between the subsystems is an integral part of motor operation. The dynamics of the mechanical subsystem for a position-dependent load (see Figure 3.1) actuated by a permanent magnet stepper motor are assumed to be of the form [11] [3]

$$M\ddot{q} + B\dot{q} + N\sin(q) + K_D\sin(4N_r q) = \sum_{j=1}^{2} -\sin(x_j)I_j \tag{3.1}$$

where $q(t)$, $\dot{q}(t)$, and $\ddot{q}(t)$ represent the load position, velocity, and acceleration, respectively. The constant parameter M denotes the mechanical inertia of the motor rotor and the connected load, B represents the viscous friction coefficient, and N denotes the constant lumped load term. The

signal, and the static torque transmission model.

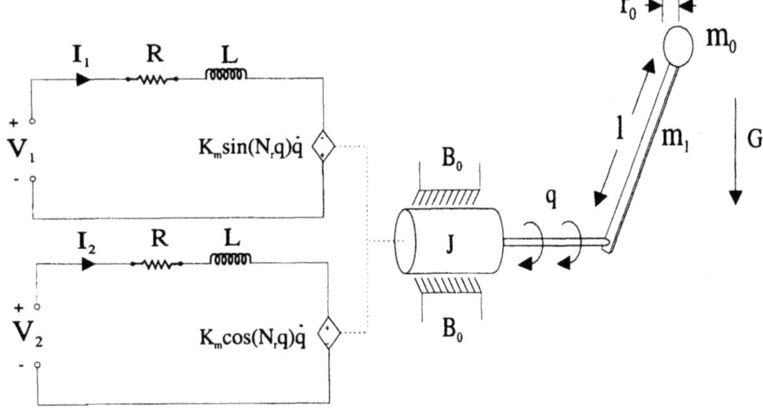

Figure 3.1: Schematic Diagram of a PMS Motor/Load System

term $K_D \sin(4N_r q)$ is used to model the detente torque, and K_D is usually referred to as the detente torque constant. The terms $\sum_{j=1}^{2} - \sin(x_j)I_j$ can be considered as torque inputs originating in the electrical subsystems. Note that the parameters M, B, N, and K_D in (3.1) are defined to include the effects of the torque coefficient constant K_m which characterizes the electromechanical conversion of electrical winding currents to torque. That is, the original mechanical parameters in (3.1) have been divided by the torque constant K_m. In (3.1), I_j denotes a particular phase current, and x_j is given by

$$x_j = N_r q - (j-1)\frac{\pi}{2} \tag{3.2}$$

in which N_r accounts for the number of teeth on the rotor.

The current dynamics for the two electrical subsystems are described by [12]

$$L\dot{I}_j = v_j - RI_j + K_m \dot{q} \sin(x_j), \qquad j = 1, 2 \tag{3.3}$$

where v_j is the voltage input to a particular phase. The constant electrical parameters R and L describe the winding resistance and inductance, respectively. The back-emf term $K_m \dot{q} \sin(x_j)$ can be considered as inherent feedback from the mechanical subsystem. The electrical subsystems described by the parameters R, L, and K_m are assumed to be the same for each of the two phases. The interconnection of the subsystems is illustrated in Figure 3.1 where the system inputs are the voltages v_j, and the output is the load position q.

Remark 3.1

The reader should note that the following assumptions regarding the modeling of the electromechanical system were utilized. First, we made the common assumption that the magnetic material is linear. Second, the above model neglects the variation of stator inductance with position (Note this assumption is also a common assumption for the PMS motor). Since we are interested in using the PMS motor for relatively low speed operation, we have included the detente torque term in the model. This additional sinusoidal term in the mechanical subsystem is caused by the permanent magnet on the rotor aligning itself along the directions of minimum reluctance.

3.3 Control Objective

Given full state measurements (*i.e.*, q, $\dot{q}$, I_1, and I_2), the control objective is to develop load position tracking controllers for the electromechanical dynamics of (3.1) through (3.3). To begin the development, we define the load position tracking error $e(t)$ as

$$e = q_d - q \tag{3.4}$$

where $q_d(t)$ represents the desired load position trajectory and $q(t)$ was defined in (3.1). We will assume that $q_d(t)$ and its first, second, and third derivatives are all bounded functions of time. To simplify the control formulation and the stability analysis, we define the filtered load tracking error $r(t)$ [13] as

$$r = \dot{e} + \alpha e \tag{3.5}$$

where α is a positive, constant control gain. As explained in Chapter 1, use of the filtered tracking error allows us to analyze the second-order dynamics of (3.1) as though it is a first-order system.

To form the open-loop filtered tracking error dynamics, we differentiate (3.5) with respect to time and rearrange terms to yield

$$\dot{r} = (\ddot{q}_d + \alpha\dot{e}) - \ddot{q}. \tag{3.6}$$

Multiplying (3.6) by M and substituting the mechanical subsystem dynamics of (3.1) yields the filtered tracking error dynamics as shown

$$M\dot{r} = M(\ddot{q}_d + \alpha\dot{e}) + B\dot{q} + N\sin(q) + K_D\sin(4N_r q) + \sum_{j=1}^{2}\sin(x_j)I_j. \tag{3.7}$$

To prepare for the controller developments, the right-hand side of (3.7) can be rewritten as

$$M\dot{r} = W_\tau \theta_\tau + \sum_{j=1}^{2} \sin(x_j) I_j \tag{3.8}$$

where the known regression matrix [14] $W_\tau(q, \dot{q}, t) \in \Re^{1\times 4}$ is given by

$$W_\tau = \begin{bmatrix} \ddot{q}_d + \alpha\dot{e} & \dot{q} & \sin(q) & \sin(4N_r q) \end{bmatrix} \tag{3.9}$$

and the parameter vector $\theta_\tau \in \Re^4$ is given by

$$\theta_\tau = \begin{bmatrix} M & B & N & K_D \end{bmatrix}^T. \tag{3.10}$$

Considering the structure of the electromechanical systems given by (3.1) through (3.3), we are only free to specify the two phase voltages, v_1 and v_2. In other words, the mechanical subsystem error dynamics lack true current (torque) level control inputs. For this reason, we add and subtract the desired current trajectories, I_{dj}, to the right-hand side of (3.8), to yield

$$M\dot{r} = W_\tau \theta_\tau + \sum_{j=1}^{2} \sin(x_j) I_{dj} - \sum_{j=1}^{2} \sin(x_j)\eta_j \tag{3.11}$$

where η_j represents the current tracking error perturbations to the mechanical subsystem dynamics of the form

$$\eta_j = I_{dj} - I_j. \tag{3.12}$$

If the current tracking error terms η_j in (3.11) were equal to zero, then I_{dj} could be designed to achieve good load position tracking by utilizing standard control techniques (*e.g.*, [13]) with a proper commutation strategy. Since the current tracking errors are not equal to zero in general, we must design voltage control inputs at v_j which compensate for the effects of η_j in (3.11). To accomplish this control objective, the dynamics of the current tracking errors are needed. Taking the time derivative of the current tracking error in (3.12) and then multiplying the result by L yields

$$L\dot{\eta}_j = L\dot{I}_{dj} - L\dot{I}_j. \tag{3.13}$$

Substituting the right-hand side of (3.3) for $L\dot{I}_j$ into (3.13) results in the open-loop current tracking error dynamics as shown

$$L\dot{\eta}_j = L\dot{I}_{dj} + RI_j - K_m\dot{q}\sin(x_j) - v_j. \tag{3.14}$$

3.4 Commutation Strategy

Unlike the brushed DC motor where the commutation of the electrical windings is done by a mechanical commutator (*i.e.*, the brushes), commutation of the PMS motor must be incorporated into the controller design. In order to generate the appropriate current in each electrical phase, we propose the following continuous, differentiable commutation strategy for the PMS motor

$$I_{dj} = -\tau_d \sin(x_j), \quad \text{for} \quad j = 1,\ 2 \tag{3.15}$$

where τ_d is the desired torque designed to force the load to track the desired position trajectory. The commutation strategy in (3.15) can be thought of as the means by which the task of moving the load is shared by the two electrical phases. To illustrate the motivation for the form of (3.15), suppose we set $I_j = I_{dj}$ so that we can rewrite (3.1) as

$$M\ddot{q} + B\dot{q} + N\sin(q) + K_D \sin(4N_r q) = \sum_{j=1}^{2} -\sin(x_j) I_{dj}. \tag{3.16}$$

After substituting the proposed commutation strategy of (3.15) into (3.16), we have

$$M\ddot{q} + B\dot{q} + N\sin(q) + K_D \sin(4N_r q) = \tau_d \tag{3.17}$$

where the fact that $\sum_{j=1}^{2} \sin^2(x_j) = 1$ has been used to obtain (3.17). From (3.17), we can see that the proposed commutation strategy has been developed such that the desired torque trajectory inherently becomes the control input to the mechanical subsystem and hence can be designed to ensure that the load follows the desired position trajectory. Of course due to the electrical dynamics, we can not guarantee that $I_j = I_{dj}$; therefore, we are motivated to design the voltage control inputs to force the actual currents to track the desired per phase winding current, that is, to force the currents tracking error η_j of (3.12) to zero. To this end, we substitute (3.15) into (3.11) to obtain the open-loop dynamics for the filtered tracking error r as

$$M\dot{r} = W_\tau \theta_\tau - \tau_d - \sum_{j=1}^{2} \sin(x_j)\eta_j \tag{3.18}$$

where the fact $\sum_{j=1}^{2} \sin^2(x_j) = 1$ has been utilized.

3.5 Exact Model Knowledge Controller

Based on exact model knowledge of the electromechanical system and full state feedback, we now design a position tracking controller for the open-loop dynamics of (3.18) and (3.14). As stated in the introduction, we first view the motor as a torque source and design the desired torque τ_d to ensure that the load tracks the desired position trajectory. Next, through the commutation strategy of (3.15), the desired torque is transformed into two desired current trajectories. Finally, we specify the voltage control inputs to make the stator winding currents track the desired current trajectories. The closed-loop error systems will then be formulated from the proposed control for use in the ensuing stability analysis.

As mentioned above, we first design the desired torque signal to force the load along the desired position trajectory. That is, we specify the desired torque $\tau_d(t)$ to drive $r(t)$ to zero. To this end, we select $\tau_d(t)$ as

$$\tau_d = W_\tau \theta_\tau + k_s r \tag{3.19}$$

where W_τ and θ_τ were previously defined in (3.9) and (3.10), respectively, and k_s is a positive constant control gain. Substituting (3.19) into the open-loop dynamics of (3.18) yields the final filtered tracking error dynamics as shown

$$M\dot{r} = -k_s r - \sum_{j=1}^{2} \sin(x_j)\eta_j. \tag{3.20}$$

Given τ_d of (3.19), the desired phase currents, I_{dj}, can be directly calculated from the commutation strategy in (3.15). To complete the open-loop system description for the current tracking error dynamics, we now calculate the term $\dot{I}_{dj}$ required in (3.14) by taking the time derivative of (3.15) to yield

$$\dot{I}_{dj} = -\dot{\tau}_d \sin(x_j) - \tau_d \cos(x_j) N_r \dot{q} \tag{3.21}$$

where $\dot{\tau}_d$ is found from (3.19) as

$$\dot{\tau}_d = \dot{W}_\tau \theta_\tau + k_s \dot{r}. \tag{3.22}$$

Substituting the time derivative of (3.9) and (3.6) into the right-hand side of (3.22) yields

$$\begin{aligned} \dot{\tau}_d = \; & M\left(\dddot{q}_d + \alpha\left(\ddot{q}_d - \ddot{q}\right)\right) + B\ddot{q} + N\dot{q}\cos(q) \\ & + 4N_r \dot{q} K_D \cos(4N_r q) + k_s\left(\ddot{q}_d - \ddot{q} + \alpha\dot{e}\right) \end{aligned} \tag{3.23}$$

which expresses $\dot{\tau}_d$ in terms of measurable states (*i.e.*, q and $\dot{q}$), known functions, known constant parameters, and the unmeasurable load acceleration $\ddot{q}$. From (3.1), we can solve for $\ddot{q}$ in the following form

$$\ddot{q} = -\frac{B}{M}\dot{q} - \frac{N}{M}\sin(q) - \frac{K_D}{M}\sin(4N_r q) + \frac{1}{M}\sum_{j=1}^{2} - \sin(x_j) I_j. \tag{3.24}$$

Substituting for $\ddot{q}$ from the right-hand side of (3.24) into (3.23), we can write $\dot{I}_{dj}$ in terms of measurable states (*i.e.*, q, $\dot{q}$, I_1, and I_2), known functions, and known constant parameters in the following form

$$\begin{aligned}
\dot{I}_{dj} = \; & -\left(M\left(\dddot{q}_d + \alpha\ddot{q}_d\right) + N\dot{q}\cos(q) + \right. \\
& \left. 4N_r\dot{q}K_D\cos(4N_rq)\right)\sin(x_j) \\
& -k_s\left(\ddot{q}_d + \alpha\dot{e}\right)\sin(x_j) - \tau_d\cos(x_j)N_r\dot{q} \\
& +\sin(x_j)\left(B - M\alpha - k_s\right)\left(\frac{B}{M}\dot{q}\right. \\
& \left. +\frac{N}{M}\sin(q) + \frac{K_D}{M}\sin(4N_rq)\right) \\
& -\sin(x_j)M^{-1}\left(B - M\alpha - k_s\right)\sum_{j=1}^{2}\left(-\sin(x_j)I_j\right).
\end{aligned} \tag{3.25}$$

Substituting the expression for $\dot{I}_{dj}$ of (3.25) into (3.14) yields the final open-loop model for the current tracking error in the form

$$L\dot{\eta}_j = w_j - v_j \tag{3.26}$$

where the auxiliary scalar variable $w_j(q, \dot{q}, I_1, I_2, t)$ is given by

$$\begin{aligned}
w_j = \; & -L\left(M\left(\dddot{q}_d + \alpha\ddot{q}_d\right) + N\dot{q}\cos(q) + \right. \\
& \left. 4N_r\dot{q}K_D\cos(4N_rq)\right)\sin(x_j) \\
& -Lk_s\left(\ddot{q}_d + \alpha\dot{e}\right)\sin(x_j) - L\tau_d\cos(x_j)N_r\dot{q} + RI_j \\
& -K_m\dot{q}\sin(x_j) + L\sin(x_j)\left(B - M\alpha\right. \\
& \left.-k_s\right)\left(\frac{B}{M}\dot{q} + \frac{N}{M}\sin(q) + \frac{K_D}{M}\sin(4N_rq)\right) \\
& +L\sin(x_j)\left(B - M\alpha - k_s\right)M^{-1}\sum_{j=1}^{2}\left(\sin(x_j)I_j\right).
\end{aligned} \tag{3.27}$$

We now design the voltage control inputs, v_j, for the open-loop system of (3.26). Given the structure of (3.26) and (3.20), we define the voltage control inputs as

$$v_j = w_j + k_j\eta_j - \sin(x_j)r \tag{3.28}$$

where k_j is a positive, constant control gain. Substituting (3.28) into the open-loop dynamics of (3.26) yields the closed-loop current tracking error dynamics in the form

$$L\dot{\eta}_j = -k_j\eta_j + \sin(x_j)r. \tag{3.29}$$

The dynamics given by (3.20) and (3.29) represent the electromechanical closed-loop system for which the stability analysis is performed while the exact model knowledge controller given by (3.19), (3.15), and (3.28) represents the control input which is implemented at the voltage terminals of the motor. Note that the desired current trajectories, I_{dj}, are embedded (in the guise of the variable η_j) inside of the voltage control inputs, v_j. The theorem given below delineates the performance of the closed-loop system under the proposed control.

Theorem 3.1

The proposed exact knowledge controller for the electromechanical dynamics of (3.1) and (3.3) ensures that the filtered tracking error goes to zero exponentially fast as shown

$$\|x(t)\| \leq \sqrt{\frac{\lambda_2}{\lambda_1}} \|x(0)\| \, e^{-\gamma t} \qquad \forall t \in [0, \infty) \tag{3.30}$$

where

$$x = \begin{bmatrix} r & \eta_1 & \eta_2 \end{bmatrix}^T \in \Re^3, \tag{3.31}$$

$$\lambda_1 = \min\{M, L\}, \qquad \lambda_2 = \max\{M, L\}, \tag{3.32}$$

and

$$\gamma = \frac{\min\{k_s, k_1, k_2\}}{\max\{M, L\}}. \tag{3.33}$$

Proof. First, we define the following non-negative function

$$V(t) = \frac{1}{2}Mr^2 + \frac{1}{2}\sum_{j=1}^{2} L\eta_j^2 = \frac{1}{2}x^T diag\{M, L, L\}\, x \tag{3.34}$$

where $x(t)$ was defined in (3.31), and $diag\{\cdot\}$ represents a diagonal matrix in which the braced terms are the diagonal elements. Application of Lemma 1.7 in Chapter 1 to the matrix term in (3.34), provides the following upper and lower bounds on $V(t)$

$$\frac{1}{2}\lambda_1 \|x(t)\|^2 \leq V(t) \leq \frac{1}{2}\lambda_2 \|x(t)\|^2 \tag{3.35}$$

where λ_1 and λ_2 were defined in (3.32). Differentiating (3.34) with respect to time yields

$$\dot{V}(t) = rM\dot{r} + \sum_{j=1}^{2} L\eta_j\dot{\eta}_j. \tag{3.36}$$

Substituting the closed-loop dynamics from (3.20) and (3.29) into (3.36) yields

$$\dot{V}(t) = -k_s r^2 - \sum_{j=1}^{2} k_j\eta_j^2 = -x^T diag\left\{k_s, k_1, k_2\right\} x. \tag{3.37}$$

The time derivative of $V(t)$ in (3.37) can be upper bounded by applying Lemma 1.7 in Chapter 1 to the matrix term in (3.37), as

$$\dot{V}(t) \leq -\min\{k_s, k_1, k_2\} \left\|x(t)\right\|^2. \tag{3.38}$$

From (3.35), it easy to see that $\left\|x(t)\right\|^2 \geq 2V(t)/\lambda_2$; hence, $\dot{V}(t)$ in (3.38) can be further upper bounded as

$$\dot{V}(t) \leq -2\gamma V(t) \tag{3.39}$$

where γ was defined (3.33). Applying Lemma 1.1 in Chapter 1 to (3.39) yields

$$V(t) \leq V(0)\mathrm{e}^{-2\gamma t}. \tag{3.40}$$

From (3.35), we have that

$$\frac{1}{2}\lambda_1 \left\|x(t)\right\|^2 \leq V(t) \quad \text{and} \quad V(0) \leq \frac{1}{2}\lambda_2 \left\|x(0)\right\|^2. \tag{3.41}$$

Substituting (3.41) appropriately into the left-hand and right-hand sides of (3.40) allows us to form the following inequality

$$\frac{1}{2}\lambda_1 \left\|x(t)\right\|^2 \leq \frac{1}{2}\lambda_2 \left\|x(0)\right\|^2 \mathrm{e}^{-2\gamma t}. \tag{3.42}$$

Using (3.42), we can solve for $\left\|x(t)\right\|$ to obtain the result given in (3.30). □

Remark 3.2

Using the result of Theorem 3.1, the control structure, and the electromechanical model, it is straightforward to illustrate that all signals remain bounded during closed-loop operation. Specifically, from (3.30) and (3.31), we know that $r(t) \in L_\infty$ and $\eta_j(t) \in L_\infty$. From Lemma 1.4 in Chapter 1, we know that if $r(t) \in L_\infty$ then $e(t) \in L_\infty$ and $\dot{e}(t) \in L_\infty$. Since $q_d(t) \in L_\infty$ and $\dot{q}_d(t) \in L_\infty$ by assumption then (3.4) can be utilized to show that $q(t) \in L_\infty$ and $\dot{q}(t) \in L_\infty$. From the definition of $\tau_d(t)$ given in (3.19) and the fact that $\ddot{q}_d(t) \in L_\infty$ by assumption, it is now easy to show that $\tau_d(t) \in L_\infty$ and therefore $I_{dj}(t) \in L_\infty$ from (3.15); hence, since $\eta_j(t) \in L_\infty$, we can use (3.12) to state that $I_j(t) \in L_\infty$. Using the structure of the voltage control inputs of (3.28) and the above information, we now know that $v_j(t) \in L_\infty$. Finally, the electromechanical dynamics of (3.1) and (3.3) can be used to show that $\ddot{q}(t) \in L_\infty$ and $\dot{I}_j(t) \in L_\infty$.

Remark 3.3

From the result given by (3.30), it is now straightforward to see that the premise of Lemma 1.5 in Chapter 1 is satisfied; hence, we know that position tracking error (*i.e.*, $e(t)$) and the velocity tracking error (*i.e.*, $\dot{e}(t)$) both go to zero exponentially fast.

3.6 Adaptive Controller

Given the dynamics of (3.1) through (3.3), we now design an adaptive position tracking controller under the constraint of parametric uncertainty. That is, without any apriori knowledge of the electromechanical system parameters, we achieve our stated control objectives based on measurements of load position, load velocity, and the electrical winding current. The step for designing the adaptive controller are the same as those used in the development of the exact model knowledge with introduction of adaptation mechanisms for the uncertain system parameters (for a general overview of the adaptive control design procedure, the reader is referred to [4]).

The first step in the procedure is to design a desired adaptive torque signal τ_d for the mechanical tracking error dynamics of (3.18) as [13]

$$\tau_d = W_\tau \hat{\theta}_\tau + k_s r \tag{3.43}$$

where $W_\tau \in \Re^{1\times 4}$ was defined in (3.9), $\hat{\theta}_\tau(t) \in \Re^4$ represents a dynamic estimate of the unknown parameter vector θ_τ defined in (3.10), and k_s is a positive constant controller gain. The parameter estimate $\hat{\theta}_\tau$ defined in (3.43) is updated online according to the following adaptation law

$$\hat{\theta}_\tau = \int_0^t \Gamma_\tau W_\tau^T(\sigma) r(\sigma)\, d\sigma \tag{3.44}$$

where $\Gamma_\tau \in \Re^{4\times 4}$ is a constant, positive definite, diagonal adaptive gain matrix. If we define the mismatch between $\hat{\theta}_\tau$ and θ_τ as

$$\tilde{\theta}_\tau = \theta_\tau - \hat{\theta}_\tau, \tag{3.45}$$

then the adaptation law of (3.44) can be written in terms of the parameter error as

$$\dot{\tilde{\theta}}_\tau = -\Gamma_\tau W_\tau^T r. \tag{3.46}$$

Substituting (3.43) into the open-loop dynamics of (3.18) yields the closed-loop filtered tracking error dynamics, as shown

$$M\dot{r} = W_{\tau}\tilde{\theta}_{\tau} - k_s r - \sum_{j=1}^{2} \sin(x_j)\,\eta_j. \tag{3.47}$$

The desired current trajectories can be found by direct application of the commutation strategy of (3.15) and the proposed desired torque signal given in (3.43). We can now complete the open-loop system description for the current tracking error dynamics. We first calculate the term $\dot{I}_{dj}$ in (3.14) by taking the time derivative of (3.15) to yield

$$\dot{I}_{dj} = -\dot{\tau}_d \sin(x_j) - \tau_d \cos(x_j) N_r \dot{q}. \tag{3.48}$$

Given (3.43), $\dot{\tau}_d$ can be calculated as

$$\dot{\tau}_d = \dot{W}_{\tau}\hat{\theta}_{\tau} + W_{\tau}\,\dot{\hat{\theta}}_{\tau} + k_s\dot{r}. \tag{3.49}$$

Substituting the time derivatives of (3.9) and (3.44) along with (3.6) into the right-hand side of (3.49) yields

$$\begin{aligned}\dot{\tau}_d = \; & \hat{M}\left(\dddot{q}_d + \alpha\left(\ddot{q}_d - \ddot{q}\right)\right) + \hat{B}\ddot{q} \\ & + \hat{N}\dot{q}\cos(q) + 4N_r\dot{q}\hat{K}_D\cos(4N_r q) \\ & + k_s\left(\ddot{q}_d - \ddot{q} + \alpha\dot{e}\right) + W_{\tau}\Gamma_{\tau}W_{\tau}^T r\end{aligned} \tag{3.50}$$

where $\hat{M}$, $\hat{B}$, $\hat{N}$, and $\hat{K}_D$ denote the scalar components of the vector $\hat{\theta}_{\tau}$ (*i.e.*, $\hat{\theta}_{\tau} = \begin{bmatrix} \hat{M} & \hat{B} & \hat{N} & \hat{K}_D \end{bmatrix}^T$). Note that $\dot{\tau}_d$ of (3.50) is in terms of measurable states (*i.e.*, q and $\dot{q}$), known functions, and the unmeasurable quantity $\ddot{q}$. After substituting for $\ddot{q}$ from the right-hand side of (3.24) into (3.50), we can write $\dot{\tau}_d$ in terms of measurable states (*i.e.*, q, $\dot{q}$, I_1, and I_2), known functions, and unknown constant parameters. Substituting this expression for $\dot{\tau}_d$ into (3.48) and then substituting the resulting expression into (3.14) yields the following linear parameterized open-loop model for the current error dynamics as

$$L\dot{\eta}_j = W_j\theta - v_j \tag{3.51}$$

where the known regression matrix $W_j(q, \dot{q}, I_1, I_2, \hat{\theta}_{\tau}, t) \in \Re^{1\times 7}$ and the unknown constant parameter vector $\theta \in \Re^7$ are explicitly defined as follows

$$\theta = \begin{bmatrix} \dfrac{L}{M} & \dfrac{LB}{M} & R & K_m & \dfrac{LN}{M} & \dfrac{LK_D}{M} & L \end{bmatrix}^T, \tag{3.52}$$

$$W_j = \begin{bmatrix} W_{j1} & W_{j2} & W_{j3} & W_{j4} & W_{j5} & W_{j6} & W_{j7} \end{bmatrix}, \tag{3.53}$$

$$W_{j1} = -\sin(x_j)\left(\hat{B} - \alpha\hat{M} - k_s\right)\sum_{j=1}^{2} -\sin(x_j)\, I_j,$$

$$W_{j2} = \sin(x_j)\left(\hat{B} - \alpha\hat{M} - k_s\right)\dot{q}, \quad W_{j3} = I_j,$$

$$W_{j4} = -\dot{q}\sin(x_j), \quad W_{j5} = \sin(x_j)\left(\hat{B} - \alpha\hat{M} - k_s\right)\sin(q),$$

$$W_{j6} = \sin(x_j)\left(\hat{B} - \alpha\hat{M} - k_s\right)\sin(4N_r q)$$

and

$$\begin{aligned} W_{j7} = \; & -\sin(x_j)\left(\hat{M}\left(\dddot{q}_d + \alpha\ddot{q}_d\right) + \hat{N}\dot{q}\cos(q) + 4N_r\dot{q}\hat{K}_D\cos(4N_r q)\right) \\ & -\sin(x_j)\left(k_s\left(\ddot{q}_d + \alpha\dot{e}\right) + W_\tau\Gamma_\tau W_\tau^T r\right) \\ & -\left(W_\tau\hat{\theta}_\tau + k_s r\right)\cos(x_j)\, N_r\dot{q}. \end{aligned}$$

The second step in the design procedure involves the design of the voltage control inputs v_j for the open-loop system of (3.51). Given the structure of (3.47) and (3.51), we define the voltage input controllers as

$$v_j = W_j\hat{\theta} + k_j\eta_j - \sin(x_j)r \tag{3.54}$$

where k_j is a positive control gain, and $\hat{\theta}(t) \in \Re^7$ is a dynamic estimate of the unknown parameter vector θ. The parameter estimates are updated online by the following adaptation law

$$\hat{\theta} = \int_0^t \Gamma\sum_{j=1}^{2}\left(W_j^T(\sigma)\eta_j(\sigma)\right) d\sigma \tag{3.55}$$

where $\Gamma \in \Re^{7\times 7}$ is a constant positive definite, diagonal adaptive gain matrix. If we define the mismatch between θ and $\hat{\theta}$ as

$$\tilde{\theta} = \theta - \hat{\theta}, \tag{3.56}$$

then the adaptation law of (3.55) can be written in terms of the parameter error as

$$\dot{\tilde{\theta}} = -\Gamma\sum_{j=1}^{2} W_j^T\eta_j. \tag{3.57}$$

Substituting (3.54) into the open-loop dynamics of (3.51) yields the closed-loop current tracking error dynamics in the form

$$L\dot{\eta}_j = W_j\tilde{\theta} - k_j\eta_j + \sin(x_j)r. \tag{3.58}$$

The dynamics given by (3.46), (3.47), (3.57), and (3.58) represent the electromechanical closed-loop system for which the stability analysis is performed while the adaptive controller given by (3.15), (3.43), (3.44), (3.54), and (3.55) represents the controller which is implemented at the voltage terminals of motor. Similar to the exact model knowledge controller, the adaptive desired current trajectories I_{dj} are embedded (in the guise of the variable η_j) inside of the voltage control inputs v_j. The theorem given below delineates the performance of the closed-loop system under the proposed control.

Theorem 3.2

The proposed adaptive controller for the electromechanical dynamics of (3.1) through (3.3) ensures that the filtered tracking error goes to zero asymptotically as shown

$$\lim_{t\to\infty} r(t) = 0. \tag{3.59}$$

Proof. First, define the following non-negative function

$$V = \frac{1}{2}Mr^2 + \frac{1}{2}\sum_{j=1}^{2} L\eta_j^2 + \frac{1}{2}\tilde{\theta}_\tau^T\Gamma_\tau^{-1}\tilde{\theta}_\tau + \frac{1}{2}\tilde{\theta}^T\Gamma^{-1}\tilde{\theta}. \tag{3.60}$$

The derivative of (3.60) with respect to time is

$$\dot{V} = rM\dot{r} + \sum_{j=1}^{2}\eta_j L\dot{\eta}_j + \tilde{\theta}_\tau^T\Gamma_\tau^{-1}\,\dot{\tilde{\theta}}_\tau + \tilde{\theta}^T\Gamma^{-1}\,\dot{\tilde{\theta}} \tag{3.61}$$

where the facts that: i) scalars can be transposed and ii) Γ_τ, Γ are diagonal matrices, have been used. Substituting the error dynamics of (3.46), (3.57), (3.47), and (3.58) into (3.61) yields

$$\dot{V} = -k_s r^2 - \sum_{j=1}^{2} k_j\eta_j^2 + rW_\tau\tilde{\theta}_\tau - \tilde{\theta}_\tau^T W_\tau^T r + \sum_{j=1}^{2}\left(\eta_j W_j\tilde{\theta} - \tilde{\theta}^T W_j^T\eta_j\right). \tag{3.62}$$

Since any scalar quantity can be transposed, (3.62) can be simplified to yield

$$\dot{V} = -k_s r^2 - \sum_{j=1}^{2} k_j\eta_j^2. \tag{3.63}$$

From the form of (3.63), we can see that $\dot{V}(t)$ is negative or zero; hence, we know from calculus that $V(t)$ given in (3.60) is either decreasing or constant. Since $V(t)$ is non-negative, it is lower bounded by zero; hence, from the form of $V(t)$, we know that $r(t) \in L_\infty$, $\tilde{\theta}_\tau(t) \in L_\infty^4$, $\eta_j(t) \in L_\infty$, and $\tilde{\theta}(t) \in L_\infty^7$. Since $r(t) \in L_\infty$, Lemma 1.4 in Chapter 1 can be used to show that $e(t) \in L_\infty$ and $\dot{e}(t) \in L_\infty$; therefore, since $q_d(t) \in L_\infty$ and $\dot{q}_d(t) \in L_\infty$, (3.4) can be utilized to show that $q(t) \in L_\infty$ and $\dot{q}(t) \in L_\infty$. Since $\tilde{\theta}_\tau(t) \in L_\infty^4$ and $\tilde{\theta}(t) \in L_\infty^7$, we can use (3.45) and (3.56) to show that $\hat{\theta}_\tau(t) \in L_\infty^4$ and $\hat{\theta}(t) \in L_\infty^7$. From the definition of $I_{dj}(t)$ given in (3.15) and the fact that $\ddot{q}_d(t) \in L_\infty$ by assumption, it is know easy to show that $I_{dj}(t) \in L_\infty$; hence, since $\eta_j(t) \in L_\infty$, we can use (3.12) to state that $I_j(t) \in L_\infty$. Using the structure of the voltage control input of (3.54) and the above information, we can now show that $v_j(t) \in L_\infty$. The electromechanical dynamics of (3.1) and (3.3) can be used to show that $\ddot{q}(t) \in L_\infty$ and $\dot{I}_j(t) \in L_\infty$. Finally, we can use the above information, (3.47), and (3.58) to illustrate that $\dot{r}(t) \in L_\infty$ and $\dot{\eta}_j(t) \in L_\infty$.

We now illustrate how Corollary 1.1 in Chapter 1 can be used to show that the filtered tracking error goes to zero. First, we integrate both sides of (3.63) with respect to time to yield

$$\int_0^\infty \frac{dV(\sigma)}{d\sigma}\,d\sigma = -\int_0^\infty \left(k_s r^2(\sigma) + \sum_{j=1}^{2} k_j \eta_j^2(\sigma)\right) d\sigma. \tag{3.64}$$

If we perform the integration on the left-hand side of (3.64), we obtain

$$\sqrt{V(0) - V(\infty)} = \sqrt{\int_0^\infty \left(k_s r^2(\sigma) + \sum_{j=1}^{2} k_j \eta_j^2(\sigma)\right) d\sigma} \tag{3.65}$$

after some minor algebraic manipulation. Since $\dot{V}(t) \leq 0$ as illustrated by (3.63), $V(t)$ of (3.60) is decreasing or constant; hence, $V(0) \geq V(\infty) \geq 0$. We now use this information and (3.65) to obtain the following inequality

$$\sqrt{k_s \int_0^\infty r^2(\sigma)\,d\sigma} \leq \sqrt{\int_0^\infty \left(k_s r^2(\sigma) + \sum_{j=1}^{2} k_j \eta_j^2(\sigma)\right) d\sigma} \leq \sqrt{V(0)} < \infty \tag{3.66}$$

which indicates according to Definition 1.2 in Chapter 1 that $r(t) \in L_2$. Since $r(t) \in L_\infty$, $\dot{r}(t) \in L_\infty$, and $r(t) \in L_2$, we can invoke Corollary 1.1 in Chapter 1 to obtain the result given by (3.59). □

Remark 3.4

From the result given by (3.59) and the proof of Theorem 3.2, we know that $r(t) \in L_\infty$, $r(t) \in L_2$, and $\lim_{t\to\infty} r(t) = 0$; hence, we can use Lemma 1.6 in Chapter 1 to state that $\lim_{t\to\infty} e(t) = 0$ and $\lim_{t\to\infty} \dot{e}(t) = 0$.

3.7 Reduction of Overparameterization

From the dynamics given by (3.1) through (3.3), we can see that there are only seven unknown parameters (N_r is assumed known); however, in the adaptive controller presented above, there are eleven adaptive update laws. That is, $\hat{\theta}_\tau(t) \in \Re^4$ of (3.44) contains four dynamic estimates while $\hat{\theta}(t) \in \Re^7$ of (3.55) contains seven dynamic estimates. We now present a modification [8] of the adaptive controller which reduces this overparameterization problem. Note, during the development given below, all variables are as defined originally unless otherwise stated.

First, the adaptive desired current trajectory I_{dj} defined in (3.15) and τ_d in (3.43) are exactly the same and hence the filtered tracking error system for the mechanical dynamics is given by (3.47). The adaptation law for θ_τ is now changed to

$$\dot{\hat{\theta}}_\tau = \Gamma_\tau \left(W_\tau^T r + \sum_{j=1}^{2} Y_j^T \eta_j \right) = -\dot{\tilde{\theta}}_\tau \tag{3.67}$$

where η_j is defined in (3.12), Y_j is defined as

$$Y_j = \begin{bmatrix} 0 & W_{j2} & W_{j5} & W_{j6} \end{bmatrix} \in \Re^{1\times 4}, \tag{3.68}$$

and W_{j2}, W_{j5}, W_{j6} were defined in (3.53). Proceeding as before, we parameterize the dynamics for η_j in the following manner

$$L\dot{\eta}_j = W_{j1}b + \Omega_j\theta_e + bY_j\theta_\tau - v_j \tag{3.69}$$

where W_{j1} was defined in (3.53),

$$b = \frac{L}{M}, \quad \theta_e = \begin{bmatrix} L & R & K_m \end{bmatrix}^T \in \Re^3, \quad \Omega_j = \begin{bmatrix} W_{j7} & W_{j3} & W_{j4} \end{bmatrix} \in \Re^{1\times 3},$$

and W_{j3}, W_{j4} , W_{j7} were defined in (3.53). Note that the $W_\tau \Gamma_\tau W_\tau^T r$ term in W_{j7} will now change to $W_\tau \Gamma_\tau \left(W_\tau^T r + \sum_{j=1}^{2} Y_j^T \eta_j \right)$ as a result of the adaptive update law modification of (3.67). Based on the parameterized error systems given by (3.47) and (3.69), the new voltage control inputs are given by

$$v_j = \hat{b}\left(Y_j\hat{\theta}_\tau + W_{j1}\right) + \Omega_j\hat{\theta}_e + k_j\eta_j - \hat{b}r\sin(x_j) \tag{3.70}$$

with the adaptive update laws given by

$$\dot{\hat{b}} = \Gamma_b \sum_{j=1}^{2} \left(W_{j1}\eta_j + Y_j\hat{\theta}_\tau\eta_j - r\sin(x_j)\eta_j \right) = -\dot{\tilde{b}} \tag{3.71}$$

and

$$\dot{\hat{\theta}}_e = \Gamma_e \sum_{j=1}^{2} \Omega_j^T \eta_j = -\dot{\tilde{\theta}}_e \tag{3.72}$$

where $\tilde{b} = b - \hat{b}$, $\tilde{\theta}_e = \theta_e - \hat{\theta}_e$, Γ_b is a scalar, positive adaptive gain constant, and $\Gamma_e \in \Re^{3\times 3}$ is a diagonal, positive definite, adaptive gain matrix. Note from (3.67), (3.71), and (3.72), we can see that there are only eight, as opposed to eleven, parameter update laws, and hence the overparameterization problem has been reduced.

After substituting (3.70) into (3.69) and multiplying the resulting equation by $\frac{M}{L}$, we can form the following closed-loop equation for η_j

$$M\dot{\eta}_j = \frac{1}{b}\left(\hat{b}r\sin(x_j) - k_j\eta_j + \Omega_j\tilde{\theta}_e + W_{j1}\tilde{b} + Y_j\hat{\theta}_\tau\tilde{b}\right) + Y_j\tilde{\theta}_\tau. \tag{3.73}$$

Now, to prove the stability result, we use the non-negative function

$$V = \frac{1}{2}Mr^2 + \frac{1}{2}\sum_{j=1}^{2} M\eta_j^2 + \frac{1}{2}\tilde{\theta}_\tau^T\Gamma_\tau^{-1}\tilde{\theta}_\tau + \frac{1}{2b}\tilde{\theta}_e^T\Gamma_e^{-1}\tilde{\theta}_e + \frac{1}{2b}\Gamma_b^{-1}\tilde{b}^2. \tag{3.74}$$

After differentiating (3.74) with respect to time and substituting from (3.47), (3.73), (3.67), (3.71), and (3.72), we can simplify the resulting expression to obtain

$$\dot{V} = -k_s r^2 - \frac{1}{b}\sum_{j=1}^{2} k_j\eta_j^2. \tag{3.75}$$

From (3.74) and (3.75), we can use the same arguments used in the proof of Theorem 3.2 to state the same results as those claimed by Theorem 3.2.

3.8 Experimental Results

The basic hardware setup was described in the experimental section of Chapter 2. The only hardware modifications to the basic setup are an additional linear amplifier to power the second electrical phase and an additional current sensor to monitor the current in this second phase (See Figure 3.2). The software was modified to accommodate the new control algorithms and the sampling time was set to 500μ sec.

The Aerotech 310SMB3 (manufactured by Eastern Air Devices Inc.) two-phase permanent magnet stepper motor was used in the experiment. Mounted to the back of the motor is a 1024 line encoder which is used for position measurements. The 310SMB3 has a stall torque of $2.6N\cdot m$, a bus voltage of $80Vdc$, a rated phase current of $6A$, a full step angle of $1.8°$, a maximum radial load of $156N$, and a maximum thrust load of $267N$. A single link robotic load is designed as a metal bar link attached to the rotor shaft and a brass ball attached to the free end. The parameters M and N in the mechanical subsystem model of (3.1) are explicitly expressed as

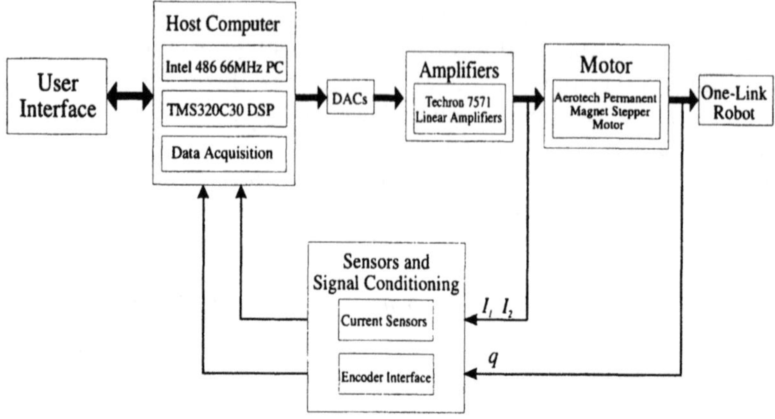

Figure 3.2: Block Diagram of PMS Motor Experimental Setup

$$M = \frac{J}{K_m} + \frac{m_1 l^2}{3K_m} + \frac{m_0 l^2}{K_m} + \frac{2m_0 r_0^2}{5K_m}, \quad N = \frac{m_1 lG}{2K_m} + \frac{m_0 lG}{K_m}, \text{ and } B = \frac{B_0}{K_m} \tag{3.76}$$

where J represents the motor rotor inertia, m_1 represents the link mass, m_0 represents the load mass, l represents the link length, r_0 represents the radius of the load, B_0 is the coefficient of viscous friction at the joint, and G is the coefficient of gravity. The electromechanical system parameters for the motor model in (3.1) through (3.3) were measured using standard test procedures and were determined to be

$J = 1.872 \cdot 10^{-4} kg \cdot m^2$	$m_1 = 0.4014kg$	$m_0 = 0.3742kg$
$B_0 = 0.002N \cdot m \cdot \sec/rad$	$l = 0.305m$	$r_0 = 0.017m$
$L = 0.7mH$	$R = 0.9\Omega$	$G = 9.81kg \cdot m/\sec^2$
$K_D = 0.176N \cdot m$	$K_m = 0.25N \cdot m/A$	$N_r = 50.$

The velocity information required by the nonlinear controllers was obtained through a backwards difference of the encoder position signal. This velocity obtained in this manner was filtered using a second-order, low-pass filter in order to remove the high frequency content that results from the backwards difference algorithm.

The desired position trajectory was selected as

$$q_d(t) = \frac{\pi}{2} \sin(2t) \left(1 - e^{-0.3t^3}\right) \; rad$$

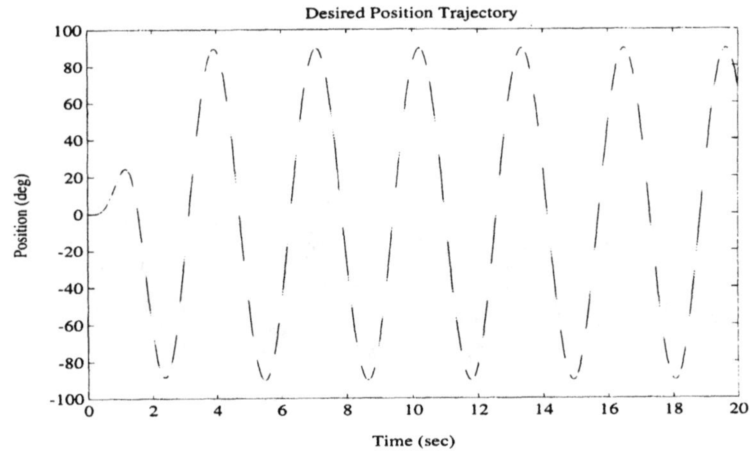

Figure 3.3: Desired Position Trajectory

which has the desirable property $q_d(0) = \dot{q}_d(0) = \ddot{q}_d(0) = \dddot{q}_d(0) = 0$. The above desired position trajectory is shown in Figure 3.3. The exact model knowledge controller and the adaptive controller with reduced overparameterization were implemented for tracking of the above desired position trajectory.

3.8.1 Exact Model Knowledge Control Experiment

The above electromechanical parameter values were used for the exact model knowledge controller of (3.19) and (3.28). The best tracking performance was found using the following control gains

$$\alpha = 100, \qquad k_s = 13, \quad \text{and} \quad k_1 = k_2 = 0.9.$$

The resulting position tracking error is given in Figure 3.4 which shows that the maximum position tracking error is approximately within ± 0.2 degrees. The corresponding motor per phase stator winding current and voltage control input are shown in Figure 3.5 through Figure 3.8, respectively.

3.8.2 Adaptive Control Experiment

The adaptive controller of (3.43), (3.67), (3.71), (3.72), and (3.70) was also implemented. The initial conditions for the parametric estimates were set to 50% of their nominal values given above. The best position tracking performance with the following controller gain values

$$\alpha = 110, \ \ k_s = 5.6, \ \ k_1 = k_2 = 0.7$$

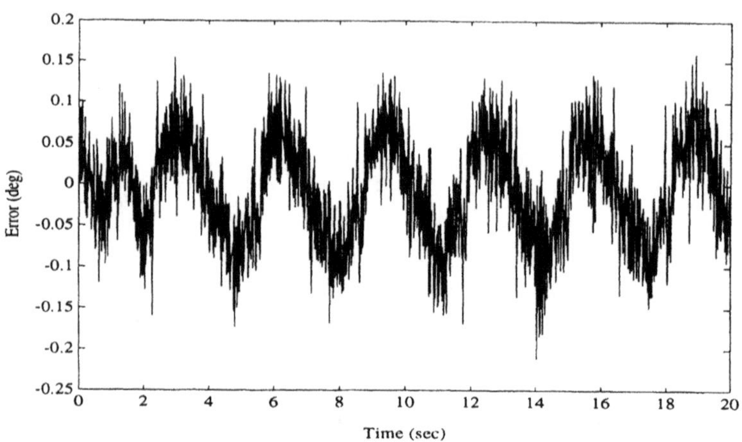

Figure 3.4: Position Tracking Error for Exact Knowledge Controller

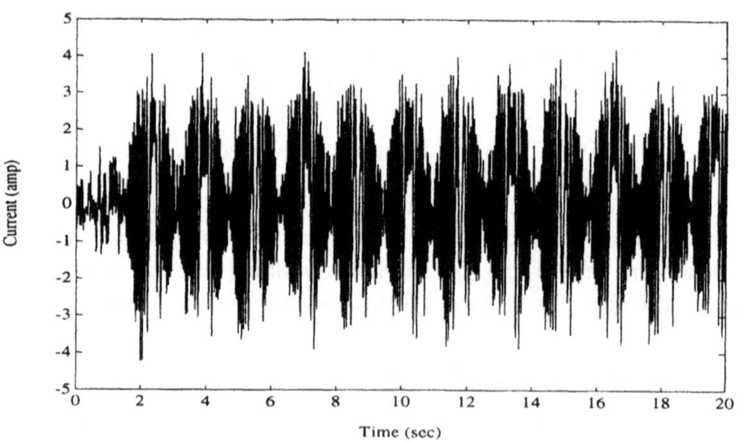

Figure 3.5: Phase One Current for Exact Knowledge Controller

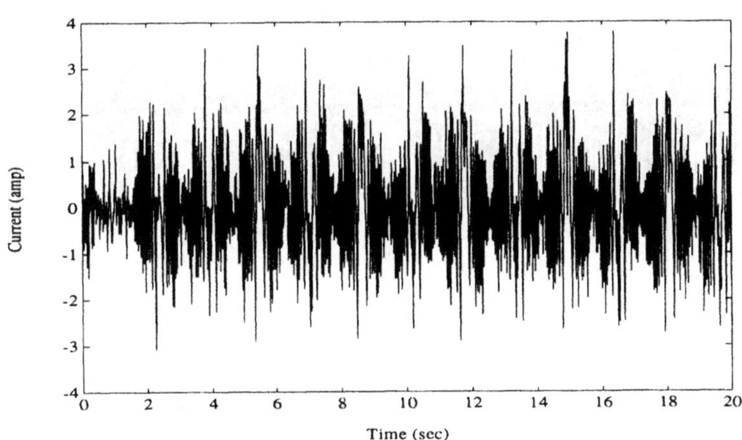

Figure 3.6: Phase Two Current for Exact Knowledge Controller

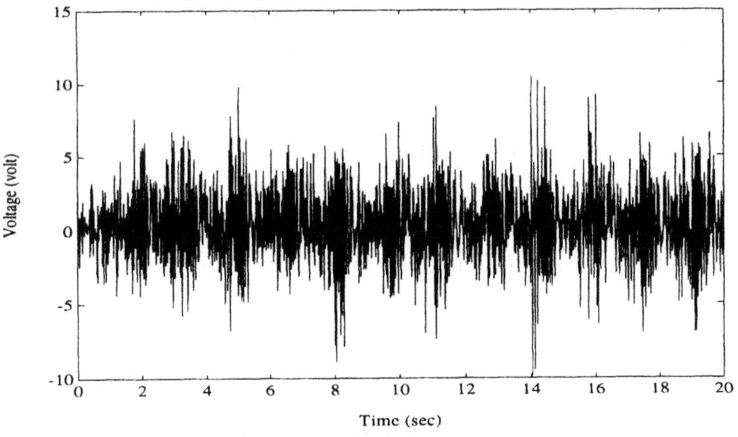

Figure 3.7: Phase One Voltage for Exact Knowledge Controller

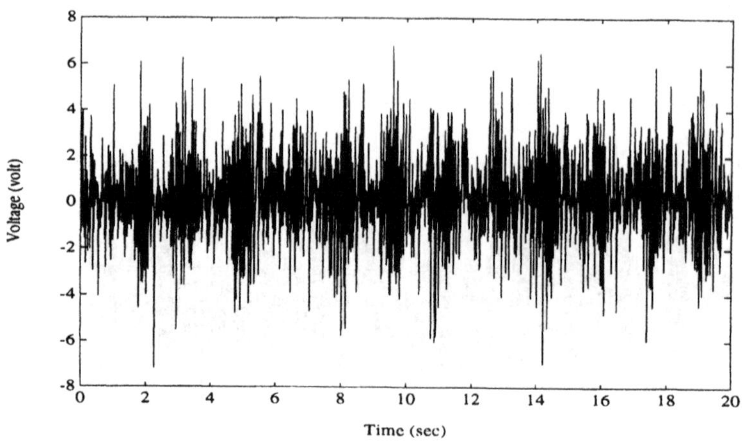

Figure 3.8: Phase Two Voltage for Exact Knowledge Controller

and

$$\Gamma_\tau = diag\{0.0001, 0.1, 0.1, 0.3\}, \quad \Gamma_b = 0.0001,$$
$$\Gamma = diag\{1, 1, 0.0001\}.$$

The adaptation update laws were implemented with a standard trapezoid-type integration rule. The resulting position tracking error is shown in Figure 3.9 which indicates that the maximum position tracking error finally settles to within ± 0.1 degrees. Note that due to measurement noise, quantization error in applying the control, and resolution of the encoder the tracking error does not approach zero as predicted by the theory but rather is driven to a small value. For brevity, only the per phase voltages are plotted in Figure 3.10 and Figure 3.11.

3.8.3 Linear Control Experiment

For comparison purposes, we also implemented a proportional-derivative outer loop controller in conjunction with a current feedback inner loop controller. That is, the input control voltages were taken to be

$$v_j = k_j \eta_j$$

while the desired torque trajectory is taken to be

$$\tau_d = k_v \dot{e} + k_p e.$$

The best performance was found by using the following control gains

$$k_p = 750, \qquad k_v = 2.5, \qquad \text{and} \qquad k_1 = k_2 = 1.0.$$

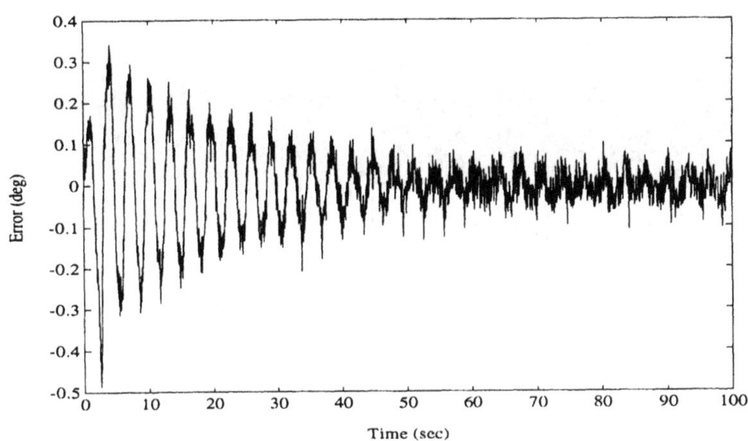

Figure 3.9: Position Tracking Error for Adaptive Controller

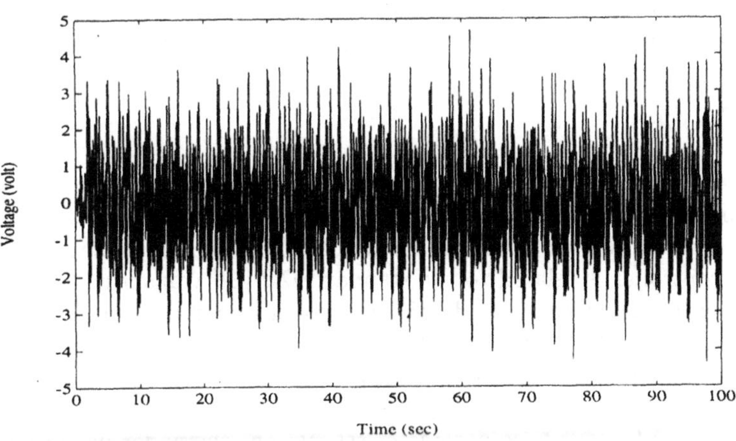

Figure 3.10: Phase One Voltage for Adaptive Controller

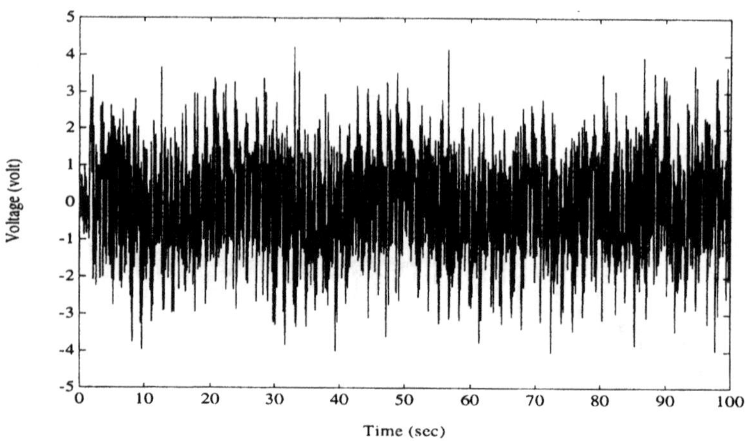

Figure 3.11: Phase Two Voltage for Adaptive Controller

The resulting position tracking error is given in Figure 3.12 which indicates that the best steady state tracking error is approximately within ±0.6 degrees. For brevity, only the phase one voltage is shown in Figure 3.13.

Remark 3.5

From the above experimental results, we can see that, with regard to the steady-state tracking error, the adaptive controller outperformed the linear controller by approximately a factor of six times and outperformed the exact model knowledge controller by approximately a factor of two.

3.9 Notes

A good deal of current literature relates the development of nonlinear controllers for stepper-type motors. For example, in [3], Zribi *et al.* design a feedback linearization controller for position tracking and illustrate how the proposed controller is related to the standard *DQ* [12] transformation. In [15], Bodson *et al.* develop an adaptive controller to compensate for parametric uncertainty. In [2], Bodson *et al.* develop a model-based control law using an exact linearization methodology while also considering practical issues such as speed estimation and voltage saturation. In [1], Blauch *et al.* present a continuous time parameter estimation scheme for use in high speed control applications. In [16] [17], Chen *et al.* develop an adaptive linearization scheme for torque ripple cancellation in hybrid stepper motors. To compensate for modeling uncertainty in position tracking applications, Speagle *et al.* present adaptive and robust full state feedback controllers

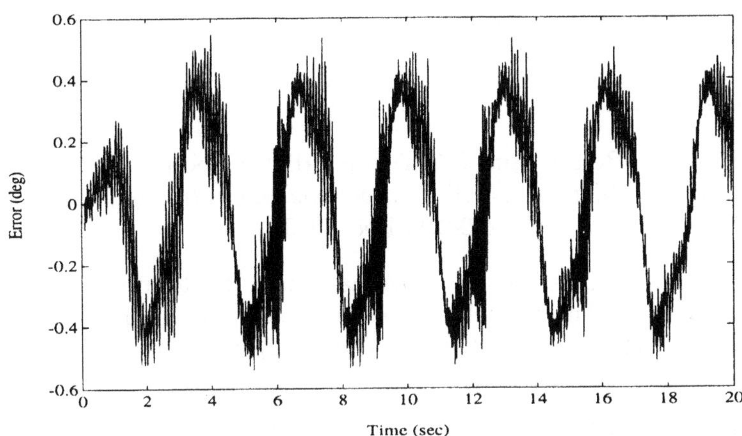

Figure 3.12: Position Tracking Error for a Linear Controller

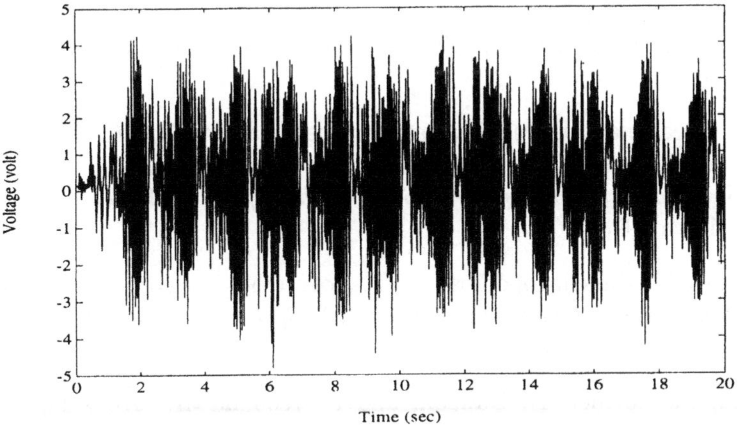

Figure 3.13: Phase One Voltage for a Linear Controller

in [7] and [18], respectively, and a robust partial state feedback controller in [19]. In [6], Marino *et al.* designed a nonlinear adaptive controller using an extended matching technique for the nonlinear PMS motor model. An adaptive tracking controller, which reduced the number of adaptive update laws, was developed by Hu *et al.* in [9].

Bibliography

[1] A. Blauch, M. Bodson, and J. Chiasson, "High-Speed Parameter Estimation of Step Motors", *IEEE Transactions on Control Systems Technology,* Vol. 1, No. 4, pp. 270-279, Dec. 1993.

[2] M. Bodson, J. Chiasson, R. Novotnak, and R. Rekowski, "High-Performance Nonlinear Feedback Control of a Permanent Magnet Stepper Motor", *IEEE Transactions on Control Systems Technology,* Vol. 1, No. 1, pp. 5-14, March 1993.

[3] M. Zribi and J. Chiasson, "Position Control of a PM Stepper Motor by Exact Linearization", *IEEE Transactions on Automatic Control,* Vol. 36, No. 5, pp. 620-625, May 1991.

[4] I. Kanellakopoulos, P. V. Kokotovic, and A. S. Morse, "Systematic Design of Adaptive Controllers for Feedback Linearizable Systems", *IEEE Transactions on Automatic Control,* Vol. 36, pp. 1241-1253, 1991.

[5] M. Krstic, I. Kanellakopoulos, and P. Kokotovic, *Nonlinear and Adaptive Control Design,* John Wiley & Sons, 1995.

[6] R. Marino, S. Peresada, and P. Tomei, "Nonlinear Adaptive Control of Permanent Magnet Step Motors", *Automatica,* Vol 31., No. 11, pp. 1595-1604.

[7] R. C. Speagle and D. M. Dawson, "Robust Tracking Control of a Permanent Magnet Stepper Motor Driving a Mechanical Load", *Proc. of the Southeastern Symposium on System Theory,* pp. 43-47, Tuscaloosa, AL, March 1993.

[8] M. Krstic, I. Kanellakopoulos, and P. Kokotovic, "Adaptive Nonlinear Control Without Overparameterization", *Systems & Control Letters,* Vol. 19, pp. 177-185, 1992.

[9] J. Hu, M. Leviner, D. Dawson, and P. Vedaghabar, "An Adaptive Tracking Controller for Multi-Phase Permanent Magnet Machines", *International Journal of Control,* Vol. 64, No. 6, 1996, pp. 997-1022.

[10] B. Kuo and J. Tal, *Incremental Motion Control, Step Motors and Control Systems,* Vol. II, SRL Publishing, Champaign, IL, 1979.

[11] M. Spong and M. Vidyasagar, *Robot Dynamics and Control*, New York: John Wiley and Sons, Inc., 1989.

[12] P. C. Krause and O. Wasynczuk, *Electromechanical Motion Devices*, New York: McGraw-Hill, 1989.

[13] J. Slotine and W. Li, *Applied Nonlinear Control*, Englewood Cliffs, NJ: Prentice Hall Co., 1991.

[14] S. Sastry and M. Bodson, *Adaptive Control: Stability, Convergence, and Robustness*, Englewood Cliffs, NJ: Prentice Hall Co., 1989.

[15] M. Bodson and J. Chiasson, "Application of Nonlinear Control Methods to the Positioning of a Permanent Magnet Stepper Motor", *Proc. of the IEEE Conference on Decision and Control*, Tampa, FL, pp. 531-532, Dec. 1989.

[16] D. Chen and B. Paden, "Nonlinear Adaptive Torque Ripple Cancellation for Step Motors", *Proc. of the IEEE Conference on Decision and Control*, Honolulu, HI, pp. 3319-3324, Dec. 1990.

[17] D. Chen and B. Paden, "Adaptive Linearization of Hybrid Step Motors: Stability Analysis", *IEEE Transactions on Automatic Control*, Vol. 38, No. 6, pp. 874-887, June 1993.

[18] R. C. Speagle and D. M. Dawson, "Adaptive Tracking Control of a Permanent Magnet Stepper Motor Driving a Mechanical Load", *Proc. of the IEEE Southeastcon 93*, Apr., 1993, Charlotte, NC.

[19] R. C. Speagle, J. Hu, D. M. Dawson, and Z. Qu, "Robust Control of a Permanent Magnet Stepper Motor without Current Measurements", *2nd IEEE Conference on Control Applications*, Vancouver, Canada, pp. 153-158, Sept. 1993.

Chapter 4

BLDC Motor (FSFB)

4.1 Introduction

Since the BLDC motor is basically an AC machine with a permanent magnet on the rotor, there are no mechanical brushes or commutators; therefore, expensive maintenance procedures which are utilized for the permanent magnet, brushed DC (BDC) motor are not required. In addition to a reduction in maintenance procedures, the utilization of BLDC motors for industrial applications yields the following advantages: high torque production, significant reduction in friction, small size, high reliability, and good heat dissipation characteristics [1] [2] [14]. However, all of the above advantages are purchased at the price of a more difficult position control problem than that of the BDC motor. That is, the precise position control of a BLDC motor is a more complicated problem due to the multi-input nature of the motor and the significant nonlinear coupling among its phase winding currents and the rotor velocity [3]. Specifically, the BLDC transformed fixed rotor frame electromechanical dynamics exhibit: i) a bilinear term, which is the product of the electrical currents, in the static torque transmission equation, and ii) bilinear terms, which are a products of rotor velocity and electrical current, in the electrical subsystem dynamics.

In this chapter, we apply the adaptive integrator backstepping tools to the design of a position tracking controller for the BLDC motor driving a mechanical load. Since the BLDC is typically a three phase machine, we utilize the standard, fixed rotor transformation to substantially simplify the nonlinear electromechanical dynamics. In spite of the simplification rendered by the transformation, the static torque transmission equation consists of linear and bilinear expressions in terms of the transformed electrical current; hence, a method for commutating[1] the motor must be fused

[1] The term commutation or commutating is often used to describe the method for developing the desired current signals which are based on the rotor position, the desired torque signal, and the static torque transmission model.

into the backstepping procedure. First, we view the motor as a torque source and thus design a desired torque signal to ensure that the load follows the desired position trajectory. Since the developed motor torque is a function of the electrical winding currents, we utilize a simple commutation strategy to restate the desired torque signal as a set of desired current trajectories. The voltage control inputs are then formulated to force the electrical winding currents to follow the desired current trajectories. That is, the electrical dynamics are taken into account through the current tracking objective, and hence the position tracking control objective is embedded inside the current tracking objective. Therefore, if the voltage control input can be designed to guarantee that the actual currents track the desired currents, then the load position will follow the desired position trajectory.

4.2 System Model

Under the assumption of a linear magnetic circuit, the dynamics of a Y-connected BLDC motor driving a single-link, direct-drive robot arm in the rotor-fixed reference frame are given by [3] [4] [5]

$$M\ddot{q} + B\dot{q} + N\sin(q) = (K_b I_b + 1) I_a, \tag{4.1}$$

$$L_a \dot{I}_a = -RI_a - n_p L_b I_b \dot{q} - K_{\tau 2}\dot{q} + V_a, \tag{4.2}$$

and

$$L_b \dot{I}_b = -RI_b + n_p L_a I_a \dot{q} + V_b \tag{4.3}$$

where K_b and $K_{\tau 2}$ are two positive torque transmission constants, M is a positive constant related to the mechanical inertia of the system (including rotor inertia), N is a positive constant related to the mass of the load and the coefficient of gravity, and B is the positive coefficient of viscous friction at the joint. The subscripts a and b refer to the quadrature and direct axes, respectively. The motor parameters and variables are defined as follows: L_a and L_b represent the positive transformed winding inductance constants, R is the positive winding resistance constant, n_p is the number of permanent magnet rotor pole pairs, $q(t)$, $\dot{q}(t)$, $\ddot{q}(t)$ represent the load position, velocity, and acceleration, respectively, $I_a(t)$, $I_b(t)$ represent the transformed winding currents, and $V_a(t)$, $V_b(t)$ represent the transformed winding voltages. Note that the parameters M, B, and N described in (4.1) are defined to include the effects of the torque coefficient constant $K_{\tau 2}$. That is, the original mechanical parameters in (4.1) have been divided by the constant $K_{\tau 2}$ (See Appendix A for details).

Remark 4.1

In Appendix A, we present the detailed development for the above transformed model. That is, we will give the transformation of the BLDC electrical dynamics and the torque transmission terms in the rotor-fixed reference frame. For control purposes, we assume the states $q(t)$, $\dot{q}(t)$, $I_a(t)$, and $I_b(t)$ are measurable. In addition, since the zero axis [5] does not affect mechanical system rotation, we can neglect the control consideration for the zero axis dynamics.

4.3 Control Objective

Given full state measurement (*i.e.*, $q(t)$, $\dot{q}(t)$, $I_a(t)$, and $I_b(t)$), the control objective is to develop load position tracking controllers for the electromechanical dynamics given by (4.1) through (4.3). To begin the development, we define the load position tracking error $e(t)$ as

$$e = q_d - q \tag{4.4}$$

where $q_d(t)$ represents the desired load position trajectory, and $q(t)$ was defined in (4.1). We will assume that $q_d(t)$ and its first, second, and third derivatives are all bounded functions of time. To simplify the control formulation and the stability analysis, we define the filtered link tracking error $r(t)$ [6] as

$$r = \dot{e} + \alpha e \tag{4.5}$$

where α is a positive constant control gain. As explained in Chapter 1, use of the filtered tracking error allows us to analyze the second-order dynamics of (4.1) as though it is a first-order system.

To form the open-loop filtered tracking error system, we differentiate (4.5) with respect to time and rearrange terms to yield

$$\dot{r} = (\ddot{q}_d + \alpha\dot{e}) - \ddot{q}. \tag{4.6}$$

Multiplying (4.6) by M and substituting the mechanical subsystem dynamics of (4.1) yields the filtered tracking error dynamics as shown

$$M\dot{r} = M(\ddot{q}_d + \alpha\dot{e}) + B\dot{q} + N\sin(q) - (K_b I_b + 1)\, I_a. \tag{4.7}$$

To reduce the notational burden, the right-hand side of (4.7) can be rewritten as

$$M\dot{r} = W_\tau \theta_\tau - (K_b I_b + 1)\, I_a \tag{4.8}$$

where the known regression matrix [7] $W_\tau(q, \dot{q}, t) \in \Re^{1\times 3}$ is given by

$$W_\tau = \begin{bmatrix} \ddot{q}_d + \alpha\dot{e} & \dot{q} & \sin(q) \end{bmatrix} \tag{4.9}$$

and the parameter vector $\theta_\tau \in \Re^3$ is given by

$$\theta_\tau = \begin{bmatrix} M & B & N \end{bmatrix}^T. \tag{4.10}$$

Considering the structure of the electromechanical systems given by (4.1), (4.2), and (4.3), we are only free to specify the motor voltages V_a

and V_b. In other words, the mechanical subsystem error dynamics lack a true current (torque) level control input. For this reason, we shall add and subtract the desired current trajectories I_{da} and I_{db} to the right-hand side of (4.8), as shown

$$M\dot{r} = W_\tau \theta_\tau - I_{da} + \eta_a - K_b I_a I_{db} + K_b I_a \eta_b \tag{4.11}$$

where η_a and η_b represent the current tracking error perturbations to the mechanical subsystem dynamics of the form

$$\eta_a = I_{da} - I_a \quad \text{and} \quad \eta_b = I_{db} - I_b. \tag{4.12}$$

If the current tracking error terms $\eta_a(t)$ and $\eta_b(t)$ in (4.11) were equal to zero, then $I_{da}(t)$ and $I_{db}(t)$ could be designed to achieve good load position tracking utilizing standard control techniques (*e.g.*, [6]) and appropriate commutation strategy. Since the current tracking error is not equal to zero in general, we must design the voltage control inputs which compensate for the effects of $\eta_a(t)$ and $\eta_b(t)$ in (4.11). To accomplish this control objective, the dynamics of the current tracking errors are needed. Taking the time derivative of the current tracking error $\eta_a(t)$ in (4.12) and then multiplying by L_a yields

$$L_a \dot{\eta}_a = L_a \dot{I}_{da} - L_a \dot{I}_a. \tag{4.13}$$

Substituting the right-hand side of (4.2) for $L_a \dot{I}_a(t)$ in (4.13) results in the current tracking error dynamics for $\eta_a(t)$ as shown

$$L_a \dot{\eta}_a = L_a \dot{I}_{da} + R I_a + n_p L_b I_b \dot{q} + K_{\tau 2} \dot{q} - V_a. \tag{4.14}$$

Following the same procedure for $\eta_b(t)$, we take the time derivative of the current tracking error $\eta_b(t)$ in (4.12) and then multiply by L_b to yield

$$L_b \dot{\eta}_b = L_b \dot{I}_{db} - L_b \dot{I}_b. \tag{4.15}$$

Substituting the right-hand side of (4.3) for $L_b \dot{I}_b(t)$ in (4.15) results in the current tracking error dynamics for $\eta_b(t)$ as shown

$$L_b \dot{\eta}_b = L_b \dot{I}_{db} + R I_b - n_p L_a I_a \dot{q} - V_b. \tag{4.16}$$

4.4 Exact Model Knowledge Controller

First, based on the exact model knowledge of the system and full state feedback, we design a position tracking controller for the open-loop dynamics of (4.11), (4.14), and (4.16). To facilitate the following stability analysis, the closed-loop electromechanical systems will also be formulated. As mentioned before, we first design the desired current trajectories (*i.e.*, desired torque) to force the load to follow the desired position trajectory. That is, we specify the desired torque, using I_{da} and I_{db}, to make r go to zero. To this end, we select I_{da} and I_{db} as

$$I_{da} = W_\tau \theta_\tau + k_s r \quad \text{and} \quad I_{db} = 0 \tag{4.17}$$

where W_τ and θ_τ were previously defined in (4.9) and (4.10), respectively, and k_s is a positive constant control gain. Substituting (4.17) into the open-loop dynamics of (4.11) yields the closed-loop filtered tracking error dynamics as shown

$$M\dot{r} = -k_s r + \eta_a + K_b I_a \eta_b. \tag{4.18}$$

From (4.18), we can see that if the current tracking errors $\eta_a(t)$ and $\eta_b(t)$ were equal to zero, then the filtered tracking error $r(t)$ will go to zero exponentially fast. However, in general, this is not true; hence, we must design the voltage control inputs to ensure $\eta_a(t)$ and $\eta_b(t)$ converge to zero. To this end, we need to first complete the open-loop system descriptions for the current tracking error dynamics for $\eta_a(t)$ and $\eta_b(t)$. Calculating the term $\dot{I}_{da}(t)$ in (4.14) by taking the time derivative of $I_{da}(t)$ in (4.17) yields

$$\dot{I}_{da} = \dot{W}_\tau \theta_\tau + k_s \dot{r}. \tag{4.19}$$

Substituting the time derivative of (4.9) and (4.6) into the right-hand side of (4.19) yields

$$\dot{I}_{da} = M\left(\dddot{q}_d + \alpha\left(\ddot{q}_d - \ddot{q}\right)\right) + B\ddot{q} + N\dot{q}\cos(q) + k_s\left(\ddot{q}_d - \ddot{q} + \alpha\dot{e}\right) \tag{4.20}$$

which contains measurable states (*i.e.*, q and $\dot{q}$), known functions, known constant parameters, and the unmeasurable load acceleration $\ddot{q}$. From (4.1), we can solve for $\ddot{q}$ in the following form

$$\ddot{q} = -\frac{B}{M}\dot{q} - \frac{N}{M}\sin(q) + \frac{1}{M}\left(K_b I_b + 1\right) I_a. \tag{4.21}$$

Substituting for $\ddot{q}$ from the right-hand side of (4.21) into (4.20), we can write $\dot{I}_{da}$ in terms of measurable states (*i.e.*, q, $\dot{q}$, I_a, and I_b), known functions, and known constant parameters in the following form

$$\begin{aligned}\dot{I}_{da} = {} & M\left(\dddot{q}_d + \alpha\ddot{q}_d\right) + N\dot{q}\cos(q) + \\ & k_s\left(\ddot{q}_d + \alpha\dot{e}\right) + \left(B - M\alpha - k_s\right)\left(-\frac{B}{M}\dot{q}\right. \\ & \left. -\frac{N}{M}\sin(q) + \frac{1}{M}\left(K_b I_b + 1\right) I_a\right).\end{aligned} \tag{4.22}$$

Substituting the expression for $\dot{I}_{da}$ of (4.22) into (4.14) yields the final open-loop dynamics for the current tracking error η_a in the form

$$L_a \dot{\eta}_a = w_a - V_a \tag{4.23}$$

where the auxiliary scalar variable $w_a(q, \dot{q}, I_a, I_b, t)$ is given by

$$
\begin{aligned}
w_a = \; & L_a \left(M \left(\dddot{q}_d + \alpha \ddot{q}_d \right) + N \dot{q} \cos(q) \right. \\
& \left. + k_s \left(\ddot{q}_d + \alpha \dot{e} \right) \right) + L_a \left(B - M\alpha \right. \\
& \left. - k_s \right) \left(-\frac{B}{M}\dot{q} - \frac{N}{M}\sin(q) + \frac{1}{M}\left(K_b I_b + 1 \right) I_a \right) \\
& + R I_a + n_p L_b I_b \dot{q} + K_{\tau 2} \dot{q}.
\end{aligned}
\tag{4.24}
$$

From (4.23), we can easily design the voltage control input V_a to force η_a to zero. That is, the voltage control input V_a can be specified as

$$V_a = w_a + k_1 \eta_a + r. \tag{4.25}$$

where k_1 is a positive control gain. Substituting V_a of (4.25) into (4.23) yields the closed-loop error system for the current tracking error η_a as

$$L_a \dot{\eta}_a = -k_1 \eta_a - r. \tag{4.26}$$

Similar to the above development for the current tracking error η_a, the dynamics for the current tracking error η_b can be obtained by substituting the time derivative of I_{db} in (4.17) into (4.16) to obtain

$$L_b \dot{\eta}_b = w_b - V_b \tag{4.27}$$

where

$$w_b = R I_b - n_p L_a I_a \dot{q}.$$

Based on (4.27), we now design the voltage control input V_b to force the current tracking error η_b to zero as

$$V_b = w_b + k_2 \eta_b + K_b I_a r \tag{4.28}$$

where k_2 is a positive control gain. Substituting the voltage control input of (4.28) into (4.27) yields the closed-loop error system for the current tracking error η_b as

$$L_b \dot{\eta}_b = -k_2 \eta_b - K_b I_a r. \tag{4.29}$$

Remark 4.2

In [8], the desired current I_{db} was designed to help rotate the mechanical system along the desired position trajectory. However, in the above controller, we set the desired current I_{db} equal to zero as delineated by (4.17). This zeroing out of the I_{db} was done to simplify the backstepping procedure. An additional motivation for setting I_{db} equal to zero is based on the practical consideration that the torque constant K_b in the bilinear torque transmission term (*i.e.*, $K_b I_a I_b$) of (4.1) is usually very small compared to *one*; therefore, very little torque can be produced from the bilinear torque transmission term (*i.e.*, $K_b I_b I_a$). For example, the Baldor BSM3R3-33 BLDC motor exhibits the following nominal values for the torque transmission constant: $K_b = 2 \times 10^{-3}$ N-m/A^2 which shows that very little additional torque will be available from the bilinear torque transmission term as compared to the linear torque transmission term (*i.e.*, I_a).

The dynamics given by (4.18), (4.26), and (4.29) represent the electromechanical closed-loop system for which the stability analysis is performed while the exact model knowledge controller given by (4.17), (4.25), and (4.28) represents the control input which is implemented at the voltage terminals of the motor. Note that the desired current trajectories I_{da} and I_{db} are embedded (in the guise of the variable η_a and η_b) inside of the voltage control inputs V_a and V_b. The theorem given below delineates the performance of the closed-loop system under the proposed control.

Theorem 4.1

The proposed exact knowledge controller ensures that the filtered tracking error goes to zero exponentially fast for the electromechanical dynamics of (4.1), (4.2), and (4.3) as shown

$$\|x(t)\| \leq \sqrt{\frac{\lambda_2}{\lambda_1}} \|x(0)\| \, \underline{e}^{-\gamma t} \qquad \forall t \in [0, \infty) \tag{4.30}$$

where

$$x = \begin{bmatrix} r & \eta_a & \eta_b \end{bmatrix}^T \in \Re^3, \tag{4.31}$$

$$\lambda_1 = \min\{M, L_a, L_b\}, \qquad \lambda_2 = \max\{M, L_a, L_b\}, \tag{4.32}$$

and

$$\gamma = \frac{\min\{k_s, k_1, k_2\}}{\max\{M, L_a, L_b\}}. \tag{4.33}$$

Proof. First, we define the following non-negative function

$$V(t) = \frac{1}{2} M r^2 + \frac{1}{2} L_a \eta_a^2 + \frac{1}{2} L_b \eta_b^2 = \frac{1}{2} x^T diag\{M, L_a, L_b\}\, x \tag{4.34}$$

where $x(t)$ was defined in (4.31), and $diag\{\cdot\}$ is used to denote the diagonal elements of a diagonal matrix. By applying Lemma 1.7 in Chapter 1 to the matrix term in (4.34), we can form upper and lower bounds on $V(t)$ as follows

$$\frac{1}{2}\lambda_1 \|x(t)\|^2 \leq V(t) \leq \frac{1}{2}\lambda_2 \|x(t)\|^2 \tag{4.35}$$

where λ_1 and λ_2 were defined in (4.32). Differentiating (4.34) with respect to time yields

$$\dot{V}(t) = r M \dot{r} + L_a \eta_a \dot{\eta}_a + L_b \eta_b \dot{\eta}_b. \tag{4.36}$$

Substituting (4.18), (4.26), and (4.29) into (4.36) yields

$$\dot{V}(t) = -k_s r^2 - k_1\eta_a^2 - k_2\eta_b^2 = -x^T diag\{k_s, k_1, k_2\}\, x. \tag{4.37}$$

After applying Lemma 1.7 in Chapter 1 to the matrix term in (4.37), $\dot{V}(t)$ in (4.37) can be upper bounded as

$$\dot{V}(t) \leq -\min\{k_s, k_1, k_2\} \|x(t)\|^2. \tag{4.38}$$

From (4.35), it easy to see that $\|x(t)\|^2 \geq 2V(t)/\lambda_2$; hence, $\dot{V}(t)$ in (4.38) can be further upper bounded as

$$\dot{V}(t) \leq -2\gamma V(t) \tag{4.39}$$

where γ was defined (4.33). Applying Lemma 1.1 in Chapter 1 to (4.39) yields

$$V(t) \leq V(0)\underline{e}^{-2\gamma t}. \tag{4.40}$$

From (4.35), we have that

$$\frac{1}{2}\lambda_1 \|x(t)\|^2 \leq V(t) \qquad \text{and} \qquad V(0) \leq \frac{1}{2}\lambda_2 \|x(0)\|^2. \tag{4.41}$$

Substituting (4.41) appropriately into the left-hand and right-hand sides of (4.40) allows us to form the following inequality

$$\frac{1}{2}\lambda_1 \|x(t)\|^2 \leq \frac{1}{2}\lambda_2 \|x(0)\|^2 \underline{e}^{-2\gamma t}. \tag{4.42}$$

Using (4.42), we can solve for $\|x(t)\|$ to obtain the result given in (4.30). □

Remark 4.3

Using the result of Theorem 4.1, the control structure, and the electromechanical model, it is straightforward to illustrate that all signals remain bounded during closed-loop operation. Specifically, from (4.30) and (4.31), we know that $r(t) \in L_\infty$, $\eta_a(t) \in L_\infty$, and $\eta_b(t) \in L_\infty$. From Lemma 1.4 in Chapter 1, we know that if $r(t) \in L_\infty$ then $e(t) \in L_\infty$ and $\dot{e}(t) \in L_\infty$. Since $q_d(t) \in L_\infty$ and $\dot{q}_d(t) \in L_\infty$ by assumption, then (4.4) can be utilized to show that $q(t) \in L_\infty$ and $\dot{q}(t) \in L_\infty$. From the definitions of $I_{da}(t)$ and $I_{db}(t)$ given in (4.17) and the fact that $\ddot{q}_d(t) \in L_\infty$ by assumption, it is easy to show that $I_{da}(t) \in L_\infty$ and $I_{db}(t) \in L_\infty$; hence, since $\eta_a(t) \in L_\infty$ and $\eta_b(t) \in L_\infty$, we can use (4.12) to state that $I_a(t) \in L_\infty$ and $I_b(t) \in L_\infty$. Using the structure of the voltage control input of (4.25), (4.28), the assumption of $\dddot{q}_d \in L_\infty$, and the above information, we now know that $V_a(t) \in L_\infty$ and $V_b(t) \in L_\infty$. Finally, the electromechanical dynamics of (4.1), (4.2), and (4.3) can be used to show that $\ddot{q}(t) \in L_\infty$, $\dot{I}_a(t) \in L_\infty$ and $\dot{I}_b(t) \in L_\infty$.

Remark 4.4

From the result given by (4.30), it is now straightforward to see that the premise of Lemma 1.5 in Chapter 1 is satisfied; hence, we know that the position tracking error (*i.e.*, $e(t)$) and the velocity tracking error (*i.e.*, $\dot{e}(t)$) both go to zero exponentially fast.

4.5 Adaptive Controller

Given the dynamics of (4.1), (4.2), and (4.3), we now design an adaptive position tracking controller under the constraint of parametric uncertainty. That is, without any apriori knowledge of the electromechanical system parameters we achieve our stated control objectives while measuring load position, load velocity, and the electrical winding currents. The advantage that this approach offers compared to typical controllers found in motor/robot controls literature is that this controller is designed to compensate for the nonlinear dynamics used to describe the fourth-order electromechanical system. Specifically, the well known nonlinear model [5] of the BLDC motor is used for the control design/stability analysis. Furthermore, the electrical dynamics are not simplified into a static set of equations as commonly done in more classical motor control approaches.

The first step in the procedure is to design the desired current trajectories such that the desired torque can be achieved to ensure load position tracking. Given the open-loop dynamics for the mechanical subsystem dynamics of (4.11), we define the adaptive desired current trajectories as [6]

$$I_{da} = W_\tau \hat{\theta}_\tau + k_s r, \quad \text{and} \qquad I_{db} = 0 \tag{4.43}$$

where $W_\tau \in \Re^{1\times 3}$ was defined in (4.9), $\hat{\theta}_\tau(t) \in \Re^3$ represents a dynamic estimate of the unknown parameter vector θ_τ defined in (4.10), and k_s is a positive, constant controller gain. The parameter estimate $\hat{\theta}_\tau$ defined in (4.43) is updated online according to the following adaptation law

$$\hat{\theta}_\tau = \int_0^t \Gamma_\tau W_\tau^T(\sigma) r(\sigma)\, d\sigma \tag{4.44}$$

where $\Gamma_\tau \in \Re^{3\times 3}$ is a constant, positive definite, diagonal adaptive gain matrix. If we define the mismatch between $\hat{\theta}_\tau$ and θ_τ as

$$\tilde{\theta}_\tau = \theta_\tau - \hat{\theta}_\tau, \tag{4.45}$$

then the adaptation law of (4.44) can be written in terms of the parameter error as

$$\dot{\tilde{\theta}}_\tau = -\Gamma_\tau W_\tau^T r = -\dot{\hat{\theta}}_\tau\ . \tag{4.46}$$

Substituting (4.43) into the open-loop dynamics of (4.11) yields the closed-loop filtered tracking error dynamics as shown

$$M\dot{r} = W_{\tau}\tilde{\theta}_{\tau} - k_s r + \eta_a + K_b I_a \eta_b. \tag{4.47}$$

Now that we have completed the control design for the mechanical subsystem dynamics, we can complete the open-loop system description for the current tracking errors η_a and η_b. We first calculate the term $\dot{I}_{da}$ in (4.14) by taking the time derivative of I_{da} in (4.43) to yield

$$\dot{I}_{da} = \dot{W}_{\tau}\hat{\theta}_{\tau} + W_{\tau}\dot{\hat{\theta}}_{\tau} + k_s\dot{r}. \tag{4.48}$$

Substituting the time derivative of (4.9) along with (4.6) and (4.46) into the right-hand side of (4.48) yields

$$\begin{aligned} \dot{I}_{da} = \; & \hat{M}\left(\dddot{q}_d + \alpha\left(\ddot{q}_d - \ddot{q}\right)\right) + \hat{B}\ddot{q} + \hat{N}\dot{q}\cos(q) \\ & + W_{\tau}\Gamma_{\tau}W_{\tau}^T r + k_s\left(\ddot{q}_d - \ddot{q} + \alpha\dot{e}\right) \end{aligned} \tag{4.49}$$

where $\hat{M}$, $\hat{B}$, and $\hat{N}$ denote the scalar components of the vector $\hat{\theta}_{\tau}$ (*i.e.*, $\hat{\theta}_{\tau} = \begin{bmatrix} \hat{M} & \hat{B} & \hat{N} \end{bmatrix}^T$). Note that $\dot{I}_{da}$ of (4.49) is written in terms of measurable states (*i.e.*, q and $\dot{q}$), known functions, and the unmeasurable quantity $\ddot{q}$. After substituting for $\ddot{q}$ from the right-hand side of (4.21) into (4.49), we can write $\dot{I}_{da}$ in terms of measurable states (*i.e.*, q, $\dot{q}$, I_a, and I_b), known functions, and unknown constant parameters. Substituting this expression for $\dot{I}_{da}$ into (4.14) yields the linear parameterized open-loop current tracking error dynamics for η_a in the form

$$L_a\dot{\eta}_a = W_a\theta_a - V_a \tag{4.50}$$

where the known regression matrix $W_a(q, \dot{q}, I_a, I_b, \hat{\theta}_{\tau}, t) \in \Re^{1\times 8}$ and the unknown constant parameter vector $\theta_a \in \Re^8$ are explicitly defined as follows

$$\theta_a = \begin{bmatrix} \frac{L_a}{M} & \frac{L_a B}{M} & \frac{L_a N}{M} & \frac{L_a K_b}{M} & R & K_{\tau 2} & L_b & L_a \end{bmatrix}^T, \tag{4.51}$$

$$W_a = \begin{bmatrix} W_{a1} & W_{a2} & W_{a3} & W_{a4} & W_{a5} & W_{a6} & W_{a7} & W_{a8} \end{bmatrix}, \tag{4.52}$$

with

$$W_{a1} = \left(\hat{B} - \alpha\hat{M} - k_s\right)I_a, \quad W_{a2} = -\left(\hat{B} - \alpha\hat{M} - k_s\right)\dot{q},$$

$$W_{a3} = -\left(\hat{B} - \alpha\hat{M} - k_s\right)\sin(q), \quad W_{a4} = \left(\hat{B} - \alpha\hat{M} - k_s\right)I_a I_b,$$

$$W_{a5} = I_a, \; W_{a6} = \dot{q}, \; W_{a7} = n_p I_b\dot{q}$$

and

$$W_{a8} = \hat{M}\left(\dddot{q}_d + \alpha\ddot{q}_d\right) + \hat{N}\dot{q}\cos(q) + k_s\left(\ddot{q}_d + \alpha\dot{e}\right) + W_\tau\Gamma_\tau W_\tau^T r.$$

Similar to the above procedure, we can use (4.16) to obtain the dynamics for the current tracking error η_b as

$$L_b\dot{\eta}_b = RI_b - n_pL_aI_a\dot{q} - V_b \tag{4.53}$$

which can be rewritten as

$$L_b\dot{\eta}_b = W_b\theta_b - V_b - K_bI_ar \tag{4.54}$$

where the term K_bI_ar has been added and subtracted and $W_b\left(q,\dot{q},I_a,I_b,t\right) \in \Re^{1\times 3}$ is a known regression matrix given by

$$W_b = \begin{bmatrix} W_{b1} & W_{b2} & W_{b3} \end{bmatrix} \in \Re^{1\times 3} \tag{4.55}$$

with

$$W_{b1} = I_b, \quad W_{b2} = -n_pI_a\dot{q}, \quad W_{b3} = I_ar,$$

and $\theta_b \in \Re^3$ is an unknown parameter vector given by

$$\theta_b = [R \;\; L_a \;\; K_b] \in \Re^3. \tag{4.56}$$

Remark 4.5

The parameter n_p (*i.e.*, the number of pole pairs) is assumed to be known *a prior* (Indeed, n_p is required in the rotor-fixed reference frame transformation).

The second step in the design procedure involves the design of the voltage control inputs $V_a(t)$ and $V_b(t)$ for the open-loop systems of (4.50), and (4.54). Given the structure of (4.47), (4.50), and (4.54), we define the voltage input controllers as

$$V_a = W_a\hat{\theta}_a + k_1\eta_a + r \quad \text{and} \quad V_b = W_b\hat{\theta}_b + k_2\eta_b \tag{4.57}$$

where k_1 and k_2 are positive control gains, $\hat{\theta}_a(t) \in \Re^8$ is a dynamic estimate of the unknown parameter vector θ_a, and $\hat{\theta}_b(t) \in \Re^3$ is a dynamic estimate of the unknown parameter vector θ_b. The parameter estimates $\hat{\theta}_a(t)$ and $\hat{\theta}_b(t)$ are updated on-line by the following adaptation laws

$$\hat{\theta}_a = \int_0^t \Gamma_a W_a^T(\sigma)\eta_a(\sigma)\,d\sigma \quad \text{and} \quad \hat{\theta}_b = \int_0^t \Gamma_b W_b^T(\sigma)\eta_b(\sigma)\,d\sigma \tag{4.58}$$

where $\Gamma_a \in \Re^{8\times 8}$, $\Gamma_b \in \Re^{3\times 3}$ are constant positive definite, diagonal adaptive gain matrices. If we define the parameter mismatch as follows

$$\tilde{\theta}_a = \theta_a - \hat{\theta}_a \qquad \text{and} \qquad \tilde{\theta}_b = \theta_b - \hat{\theta}_b, \tag{4.59}$$

then the adaptation laws of (4.58) can be written in terms of the parameter estimation errors as

$$\dot{\tilde{\theta}}_a = -\Gamma_a W_a^T \eta_a = -\dot{\hat{\theta}}_a \quad \text{and} \quad \dot{\tilde{\theta}}_b = -\Gamma_b W_b^T \eta_b = -\dot{\hat{\theta}}_b . \tag{4.60}$$

Substituting V_a and V_b of (4.57) into the open-loop dynamics of (4.50) and (4.54), respectively yields the closed-loop current tracking error dynamics for η_a and η_b in the form

$$L_a \dot{\eta}_a = W_a \tilde{\theta}_a - k_1 \eta_a - r \tag{4.61}$$

and

$$L_b \dot{\eta}_b = W_b \tilde{\theta}_b - k_2 \eta_b - K_b I_a r. \tag{4.62}$$

The dynamics given by (4.46), (4.47), (4.60), (4.61), and (4.62) represent the electromechanical closed-loop system for which the stability analysis is performed while the adaptive controller given by (4.43), (4.44), (4.57), and (4.58) represent the controller which is implemented at the voltage terminals of the motor. Similar to the exact model knowledge controller, the adaptive desired current trajectory I_{da} and I_{db} are embedded (in the guise of the variables η_a and η_b) inside of the voltage control input V_a and V_b. The theorem given below delineates the performance of the closed-loop system under the proposed control.

Theorem 4.2

The proposed adaptive controller ensures that the filtered tracking error goes to zero asymptotically for the electromechanical dynamics of (4.1), (4.2), and (4.3) as shown

$$\lim_{t \to \infty} r(t) = 0. \tag{4.63}$$

Proof. First, define the following non-negative function

$$\begin{aligned} V(t) = \; & \tfrac{1}{2} M r^2 + \tfrac{1}{2} L_a \eta_a^2 + \tfrac{1}{2} L_b \eta_b^2 + \\ & \tfrac{1}{2} \tilde{\theta}_\tau^T \Gamma_\tau^{-1} \tilde{\theta}_\tau + \tfrac{1}{2} \tilde{\theta}_a^T \Gamma_a^{-1} \tilde{\theta}_a + \tfrac{1}{2} \tilde{\theta}_b^T \Gamma_b^{-1} \tilde{\theta}_b . \end{aligned} \tag{4.64}$$

Taking the time derivative of (4.64) with respect to time yields

$$\dot{V}(t) = r M \dot{r} + \eta_a L_a \dot{\eta}_a + \eta_b L_b \dot{\eta}_b + \tilde{\theta}_\tau^T \Gamma_\tau^{-1} \dot{\tilde{\theta}}_\tau + \tilde{\theta}_a^T \Gamma_a^{-1} \dot{\tilde{\theta}}_a + \tilde{\theta}_b^T \Gamma_b^{-1} \dot{\tilde{\theta}}_b \tag{4.65}$$

where the facts that: i) scalars can be transposed and ii) Γ_τ, Γ_a, and Γ_b are diagonal matrices, have been used. Substituting the error dynamics of (4.46), (4.60), (4.47), (4.61), and (4.62) into (4.65) yields

$$\begin{aligned} \dot{V}(t) \; = \; & -k_s r^2 - k_1 \eta_a^2 - k_2 \eta_b^2 + \left(r W_\tau \tilde{\theta}_\tau - \tilde{\theta}_\tau^T W_\tau^T r \right) \\ & + \left(\eta_a W_a \tilde{\theta}_a - \tilde{\theta}_a^T W_a^T \eta_a \right) + \left(\eta_b W_b \tilde{\theta}_b - \tilde{\theta}_b^T W_b^T \eta_b \right) . \end{aligned} \tag{4.66}$$

Since any scalar quantity can be transposed, (4.66) can be simplified as follows

$$\dot{V}(t) = -k_s r^2 - k_1\eta_a^2 - k_2\eta_b^2. \tag{4.67}$$

From the form of (4.67), we can see that $\dot{V}(t)$ is negative or zero; hence, we know from calculus that $V(t)$ given in (4.64) is either decreasing or constant. Since $V(t)$ is nonnegative, it is lower bounded by zero; hence, from the form of $V(t)$, we know that $r(t) \in L_\infty$, $\tilde{\theta}_\tau(t) \in L_\infty^3$, $\eta_a(t) \in L_\infty$, $\eta_b(t) \in L_\infty$, $\tilde{\theta}_a(t) \in L_\infty^8$, and $\tilde{\theta}_b(t) \in L_\infty^3$. Since $r(t) \in L_\infty$, Lemma 1.4 in Chapter 1 can be used to show that $e(t) \in L_\infty$ and $\dot{e}(t) \in L_\infty$; therefore, since $q_d(t) \in L_\infty$ and $\dot{q}_d(t) \in L_\infty$ by assumption, (4.4) can be utilized to show that $q(t) \in L_\infty$ and $\dot{q}(t) \in L_\infty$. Since $\tilde{\theta}_\tau(t) \in L_\infty^3$, $\tilde{\theta}_a(t) \in L_\infty^8$, and $\tilde{\theta}_b(t) \in L_\infty^3$, we can use (4.45) and (4.59) to show that $\hat{\theta}_\tau(t) \in L_\infty^3$, $\hat{\theta}_a(t) \in L_\infty^8$, and $\hat{\theta}_b(t) \in L_\infty^3$. From the definition of $I_{da}(t)$ given in (4.43) and the fact that $\ddot{q}_d(t) \in L_\infty$ by assumption, it is easy to show that $I_{da}(t) \in L_\infty$; hence, since $\eta_a(t) \in L_\infty$, and $\eta_b \in L_\infty$, we can use (4.12) to state that $I_a(t) \in L_\infty$ and $I_b(t) \in L_\infty$. Using the structure of the voltage control input of (4.57), the assumption that $\dddot{q}_d$ is bounded, and the above information, we can now state that $V_a(t) \in L_\infty$ and $V_b(t) \in L_\infty$. The electromechanical dynamics of (4.1), (4.2), and (4.3) can be used to show that $\ddot{q}(t) \in L_\infty$, $\dot{I}_a(t) \in L_\infty$ and $\dot{I}_b(t) \in L_\infty$. Finally, we can use the above information and (4.47), (4.61), and (4.62) to illustrate that $\dot{r}(t) \in L_\infty$, $\dot{\eta}_a(t) \in L_\infty$, and $\dot{\eta}_b(t) \in L_\infty$.

We now illustrate how Corollary 1.1 in Chapter 1 can be used to show that the filtered tracking error goes to zero. First, we integrate both sides of (4.67) with respect to time to yield

$$\int_0^\infty \frac{dV(\sigma)}{d\sigma}\,d\sigma = -\int_0^\infty \left(k_s r^2(\sigma) + k_1\eta_a^2(\sigma) + k_2\eta_b^2(\sigma)\right)\,d\sigma. \tag{4.68}$$

If we integrate the left-hand side of (4.68), we obtain

$$\sqrt{V(0) - V(\infty)} = \sqrt{\int_0^\infty \left(k_s r^2(\sigma) + k_1\eta_a^2(\sigma) + k_2\eta_b^2(\sigma)\right)\,d\sigma} \tag{4.69}$$

after some minor algebraic operations. Since $\dot{V}(t) \le 0$ as illustrated by (4.67), $V(t)$ of (4.64) is decreasing or constant; hence, $V(0) \ge V(\infty) \ge 0$. We now use this information and (4.69) to obtain the following inequality

$$\begin{aligned} \sqrt{k_s \int_0^\infty r^2(\sigma)\,d\sigma} &\le \sqrt{\int_0^\infty \left(k_s r^2(\sigma) + k_1\eta_a^2(\sigma) + k_2\eta_b^2(\sigma)\right)\,d\sigma} \\ &\le \sqrt{V(0)} < \infty \end{aligned} \tag{4.70}$$

which indicates according to Definition 1.2 in Chapter 1 that $r(t) \in L_2$. Since $r(t) \in L_\infty$, $\dot{r}(t) \in L_\infty$, and $r(t) \in L_2$, we can invoke Corollary 1.1 in Chapter 1 to obtain the result given by (4.63). □

Remark 4.6

From the result given by (4.63) and the proof of Theorem 4.2, we know that $r(t) \in L_\infty$, $r(t) \in L_2$, and $\lim_{t\to\infty} r(t) = 0$; hence, we can use Lemma 1.6 in Chapter 1 to state that $\lim_{t\to\infty} e(t) = 0$ and $\lim_{t\to\infty} \dot{e}(t) = 0$. As noted by Remark 4.2, I_{db} can be selected as a positive function of time to help boost the motor torque production capability. However, a special update law based on the projection algorithm [10] must be used to avoid control singularities.

4.6 Reduction of Overparameterization

From the dynamics given by (4.1), (4.2), and (4.3), we can see that there are only eight unknown parameters (n_p is assumed known); however, in the adaptive controller presented above, there are fourteen adaptive update laws. That is, $\hat{\theta}_\tau(t) \in \Re^3$ of (4.44) contains three dynamics estimates while $\hat{\theta}_a(t) \in \Re^8$ and $\hat{\theta}_b(t) \in \Re^3$ of (4.58) contain eleven dynamic estimates. We now present a modification [19] of the adaptive controller which reduces this overparameterization problem. Note, during the development given below, all variables are as defined originally unless otherwise stated.

First, the adaptive desired current trajectories $I_{da}(t)$ and $I_{db}(t)$ defined in (4.43) are unchanged and hence the closed-loop filtered tracking error system for the mechanical system is the same as that given by (4.47). The adaptation law for θ_τ is now changed to

$$\dot{\hat{\theta}}_\tau = \Gamma_\tau \left(W_\tau^T r + Y_a^T \eta_a + Y_b^T \eta_b \right) = -\dot{\tilde{\theta}}_\tau \tag{4.71}$$

where η_a, η_b were defined in (4.12), Y_a, Y_b are defined as

$$Y_a = \begin{bmatrix} 0 & W_{a2} & W_{a3} \end{bmatrix} \in \Re^{1\times 3} \quad Y_b = \begin{bmatrix} W_{b2} & 0 & 0 \end{bmatrix} \in \Re^{1\times 3}, \tag{4.72}$$

where W_{a2}, W_{a3}, W_{b2} were defined in (4.52) and (4.55), respectively. Now proceeding as before, we parameterize the dynamics for η_a in the following way

$$L_a \dot{\eta}_a = W_{a1} b + Y_{e1}\theta_e + bW_{a4}\theta_k + bY_a\theta_\tau - V_a \tag{4.73}$$

where W_{a1} was defined in (4.52),

$$b = \frac{L_a}{M}, \; \theta_e = \begin{bmatrix} L_a & L_b & R & K_{\tau 2} \end{bmatrix}^T \in \Re^{4\times 1}, \; \theta_k = K_b,$$

$$Y_{e1} = \begin{bmatrix} W_{a8} & W_{a7} & W_{a5} & W_{a6} \end{bmatrix} \in \Re^{1\times 4},$$

in which W_{a8}, W_{a7}, W_{a5} , and W_{a6} were defined in (4.52). Note that the $W_\tau \Gamma_\tau W_\tau^T r$ term in W_{a8} will now change to $W_\tau \Gamma_\tau \left(W_\tau^T r + Y_a^T \eta_a + Y_b^T \eta_b \right)$

as a result of the adaptive update law modification of (4.71). Similar to η_a, we can rewrite the dynamics for the current tracking error η_b as

$$L_b\dot{\eta}_b = Y_{e2}\theta_e + bY_b\theta_\tau + bW_{b3}\theta_k - V_b - bK_bI_ar \tag{4.74}$$

where the term bK_bI_ar has been added and subtracted and W_{b3} is defined in (4.55) and

$$Y_{e2} = \begin{bmatrix} 0 & 0 & W_{b1} & 0 \end{bmatrix} \in \Re^{1\times 4}.$$

Based on the parameterized error systems given by (4.47), (4.73), and (4.74), the new voltage control inputs are given by

$$V_a = \hat{b}\left(Y_a\hat{\theta}_\tau + W_{a1} + r\right) + Y_{e1}\hat{\theta}_e + k_1\eta_a + \hat{b}W_{a4}\hat{\theta}_k \tag{4.75}$$

and

$$V_b = \hat{b}Y_b\hat{\theta}_\tau + Y_{e2}\hat{\theta}_e + k_2\eta_b + \hat{b}W_{b3}\hat{\theta}_k. \tag{4.76}$$

The adaptive update laws are now given by

$$\dot{\hat{b}} = \Gamma_b\left((W_{a1}+r)\,\eta_a + Y_a\hat{\theta}_\tau\eta_a + Y_b\hat{\theta}_\tau\eta_b + W_{a4}\hat{\theta}_k\eta_a + W_{b3}\hat{\theta}_k\eta_b\right) = -\dot{\tilde{b}} \tag{4.77}$$

$$\dot{\hat{\theta}}_e = \Gamma_e\left(Y_{e1}^T\eta_a + Y_{e2}^T\eta_b\right) = -\dot{\tilde{\theta}}_e \tag{4.78}$$

and

$$\dot{\hat{\theta}}_k = \Gamma_k\left(W_{a4}\eta_a + W_{b3}\eta_b\right) = -\dot{\tilde{\theta}}_k \tag{4.79}$$

where $\tilde{b} = b - \hat{b}$, $\tilde{\theta}_e = \theta_e - \hat{\theta}_e$, $\tilde{\theta}_k = \theta_k - \hat{\theta}_k$, Γ_b, Γ_k are positive adaptive gain constants, and $\Gamma_e \in \Re^{4\times 4}$ is a diagonal, positive definite, adaptive gain matrix. Note from (4.71), (4.77), (4.78), and (4.79), we can see that there are only nine, as opposed to fourteen, parameter update laws, and hence the overparameterization problem has been reduced.

After substituting (4.75) into (4.73) and multiplying the resulting equation by $\frac{M}{L_a}$, we can form the following closed-loop equation for η_a

$$\begin{aligned} M\dot{\eta}_a = {} & \frac{1}{b}\left(-k_1\eta_a + Y_{e1}\tilde{\theta}_e + (W_{a1}+r)\,\tilde{b} + \tilde{b}W_{a4}\hat{\theta}_k + \tilde{b}Y_a\hat{\theta}_\tau\right) \\ & + W_{a4}\tilde{\theta}_k + Y_a\tilde{\theta}_\tau - r. \end{aligned} \tag{4.80}$$

Similar to η_a, we substitute (4.76) into (4.74), and multiply the resulting expression by b^{-1} to obtain the following closed-loop equation for η_b as

$$\begin{aligned} b^{-1}L_b\dot{\eta}_b = {} & \tfrac{1}{b}\left(-k_2\eta_b + Y_{e2}\tilde{\theta}_e + \tilde{b}Y_b\hat{\theta}_\tau + \tilde{b}W_{b3}\hat{\theta}_k\right) \\ & + Y_b\tilde{\theta}_\tau + W_{b3}\tilde{\theta}_k - K_bI_ar. \end{aligned} \tag{4.81}$$

Now, to establish the stability result, we use the non-negative function

$$V = \frac{1}{2}Mr^2 + \frac{1}{2}M\eta_a^2 + \frac{1}{2}b^{-1}L_b\eta_b^2 + \frac{1}{2}\tilde{\theta}_\tau^T\Gamma_\tau^{-1}\tilde{\theta}_\tau + \frac{1}{2b}\tilde{\theta}_e^T\Gamma_e^{-1}\tilde{\theta}_e + \frac{1}{2b}\Gamma_b^{-1}\tilde{b}^2 + \frac{1}{2}\Gamma_k^{-1}\tilde{\theta}_k^2. \quad (4.82)$$

After differentiating (4.82) with respect to time and substituting (4.47), (4.80), (4.81), (4.71), (4.77), (4.78), and (4.79), we can simplify the resulting expression to obtain

$$\dot{V} = -k_s r^2 - \frac{1}{b}k_1\eta_a^2 - \frac{1}{b}k_2\eta_b^2. \quad (4.83)$$

From (4.82) and (4.83), we can use the same arguments used in the proof of Theorem 4.2 to state the same results as those claimed by Theorem 4.2.

4.7 Experimental Results

The proposed controllers in this chapter were successfully applied to a BLDC motor turning a position dependent mechanical load. The experimental setup is the same as that in Chapter 2 except the motor is replaced by a three-phase Baldor BSM3R3-33 BLDC motor with resolver for position measurements. An Analog Devices AD2S90 chip is used to convert the analog revolver signal to a 1024 count encoder signal. Since the motor has three phases, three Hall-effect current sensors and three Techron model 7571 linear power amplifiers are used in the experiments. The three-phase BLDC motor used for the experiments has 4 poles, a rated speed of 6000 RPM, a rated current of 10.4 A, a rated voltage of 100V, and a peak torque of 11.25 N-m. A single link robotic load is designed as a metal bar link attached to the rotor shaft and a brass ball attached to the free end (See Figure 4.1).

The parameters M and N in the mechanical subsystem model of (4.1) are therefore expressed as

$$M = \frac{J}{K_{\tau 2}} + \frac{m_1 l^2}{3K_{\tau 2}} + \frac{m_0 l^2}{K_{\tau 2}} + \frac{2m_0 r_0^2}{5K_{\tau 2}}, \quad N = \frac{m_1 lG}{2K_{\tau 2}} + \frac{m_0 lG}{K_{\tau 2}}, \quad \text{and} \quad B = \frac{B_0}{K_{\tau 2}} \quad (4.84)$$

where J represents the motor rotor inertia, m_1 represents the link mass, m_0 represents the load mass, l represents the link length, r_0 represents the radius of the load, B_0 is the coefficient of viscous friction at the joint, and G is the coefficient of gravity. The electromechanical system parameters for the motor model in (4.1) through (4.3) were measured using standard

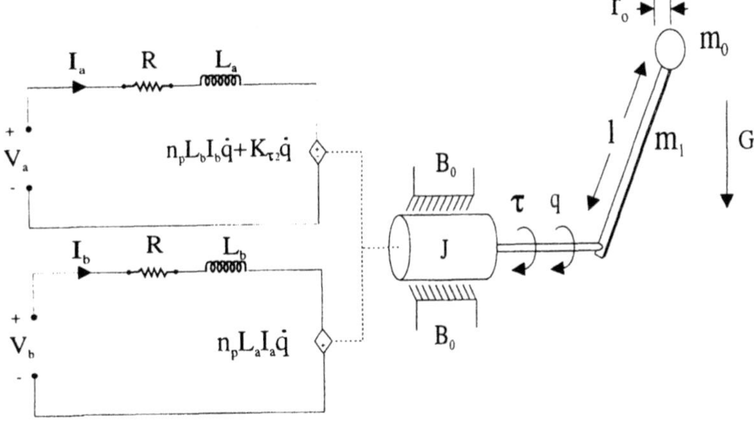

Figure 4.1: Schematic Diagram of a BLDC Motor/Load System

test procedures and were determined to be (before division by $K_{\tau 2}$)

$J = 2.99 \cdot 10^{-4} kg \cdot m^2$	$m_1 = 0.4014kg$	$m_0 = 0.3742kg$
$B_0 = 0.035N \cdot m \cdot \sec/rad$	$l = 0.305m$	$r_0 = 0.017m$
$K_{\tau 1} = 1.01 \cdot 10^{-3} N - m/A^2$	$L_b = 0.723$	$L_a = 0.216mH$
$G = 9.81kg \cdot m/\sec^2$	$R = 0.9\Omega$	$K_{\tau 2} = 0.506N - m/A.$

The parameter K_b in (4.1) is defined as $\dfrac{K_{\tau 1}}{K_{\tau 2}}$ with $K_{\tau 1}$ and $K_{\tau 2}$ given above. The above electromechanical parameter values were used for the exact model knowledge controller. For the adaptive controller, the initial conditions for the parametric estimates were set to zero. The velocity information required by the nonlinear controllers was obtained through a backwards difference of the encoder position signal. This velocity obtained in this manner was filtered using a second-order, low-pass filter in order to remove the high frequency content that results from the backwards difference algorithm.

The desired position trajectory was selected as

$$q_d(t) = \frac{\pi}{2}\sin(2t)\left(1 - \text{e}^{-0.3t^3}\right)\ rad$$

which has the desirable property $q_d(0) = \dot{q}_d(0) = \ddot{q}_d(0) = \dddot{q}_d(0) = 0$. The above desired position trajectory is shown in Figure 4.2. The exact model knowledge controller of (4.17), (4.25), and (4.28) and the adaptive controller with reduced overparameterization of (4.43), (4.75), (4.76), (4.71), (4.77),

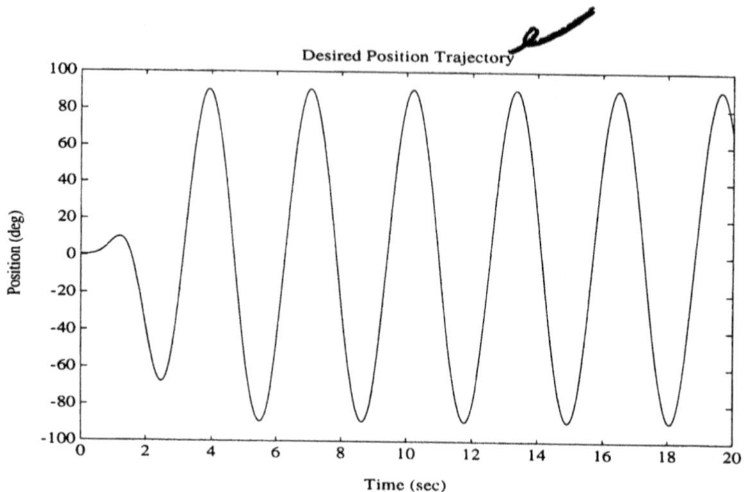

Figure 4.2: Desired Position Trajectory

(4.78), and (4.79) were implemented on the BLDC motor/load system for tracking the above desired position trajectory.

4.7.1 Exact Model Knowledge Control Experiment

For the exact model knowledge controller, the best tracking performance was found using the following control gains

$$\alpha = 100, \qquad k_s = 13, \quad \text{and} \quad k_1 = k_2 = 0.9.$$

The resulting position tracking error is given in Figure 4.3 which shows that the maximum position tracking error is approximately bounded by ± 0.15 degrees. The corresponding measured (*i.e.*, nontransformed) motor winding currents and voltages are shown in Figure 4.4 through Figure 4.9, respectively.

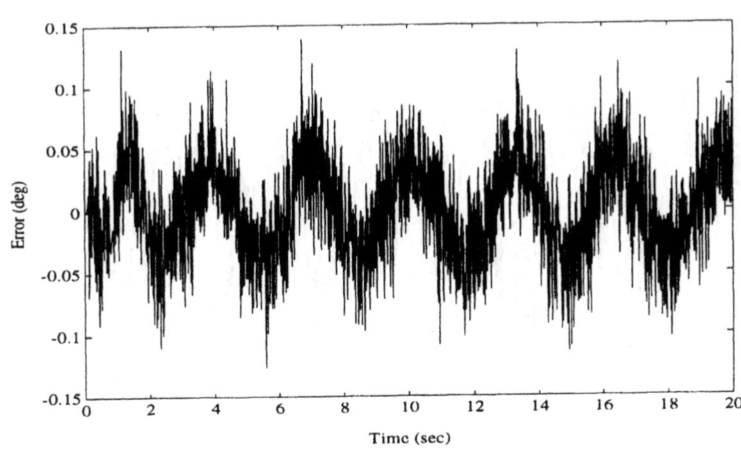

Figure 4.3: Position Tracking Error for Exact Knowledge Controller

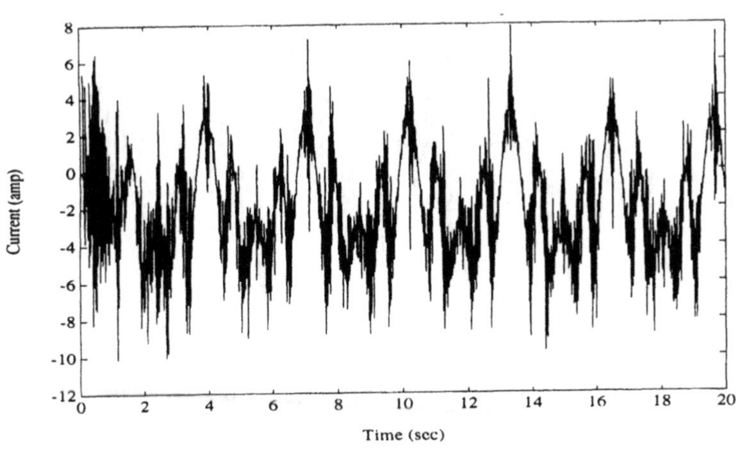

Figure 4.4: Phase One Current for Exact Knowledge Controller

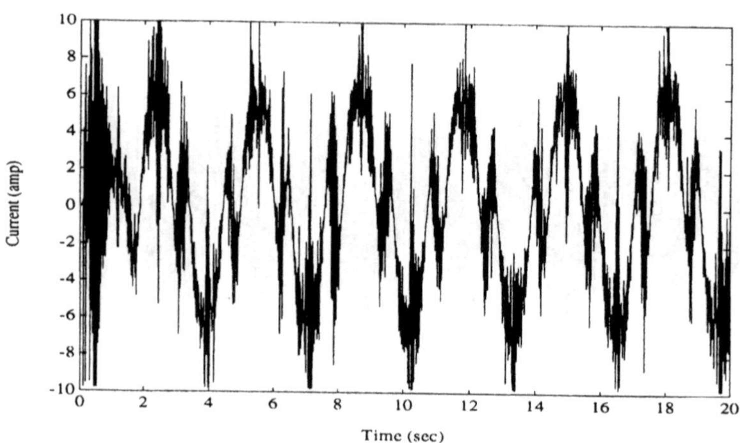

Figure 4.5: Phase Two Current for Exact Knowledge Controller

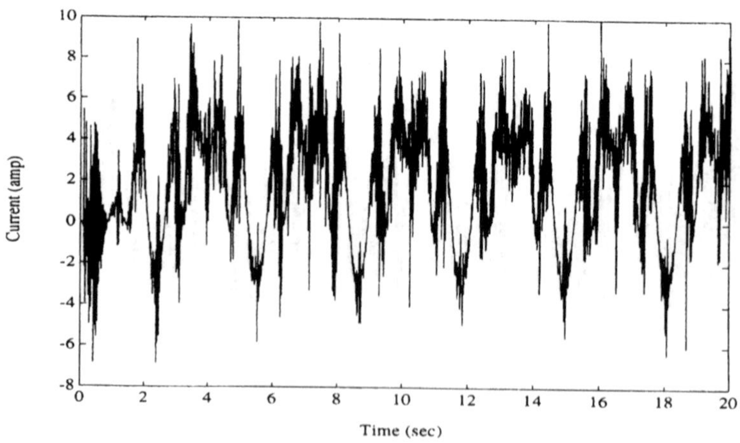

Figure 4.6: Phase Three Current for Exact Knowledge Controller

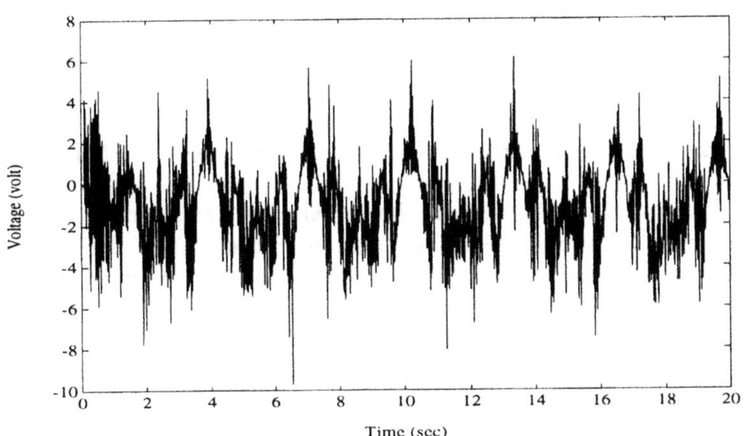

Figure 4.7: Phase One Voltage for Exact Knowledge Controller

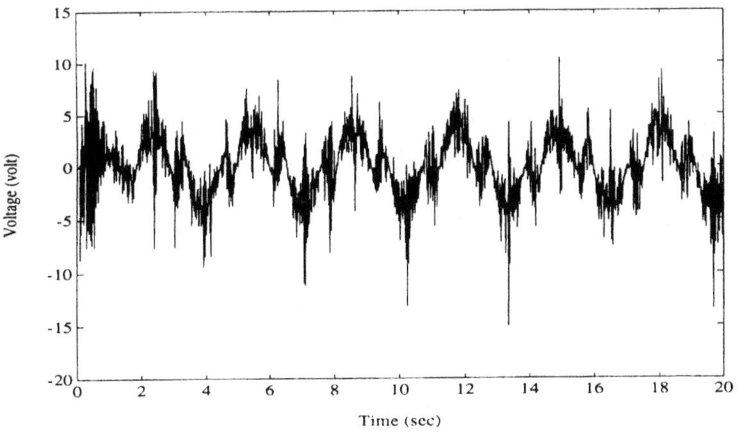

Figure 4.8: Phase Two Voltage for Exact Knowledge Controller

4.7.2 Adaptive Control Experiment

The adaptive controller was found to yield the best position tracking performance with the following controller gain values

$$\alpha = 80, \quad k_s = 7, \quad k_1 = 0.7, \quad k_2 = 2, \quad \Gamma_b = 0.001,$$

and

$$\Gamma_\tau = diag\{0.0001, 0.1, 0.001\}, \quad \Gamma_k = 0.1, \quad \Gamma_e = diag\{0.1, 0.5.0.2\}.$$

The adaptation update laws were implemented with a standard trapezoid-type integration rule. The resulting position tracking error is shown in Figure 4.10 which indicates that the maximum position tracking error finally settles to within ± 0.15 degrees. Note that due to measurement noise, quantization error in applying the control, and resolution of the encoder, the tracking error does not approach zero as predicted by the theory but rather is driven to a small value. For brevity, only the phase voltages are plotted in Figure 4.11, Figure 4.12, and Figure 4.13, respectively.

4.7.3 Linear Control Experiment

For comparison purposes, we also implemented a high gain, proportional-derivative outer loop controller in conjunction with a high gain current feedback inner loop controller. That is, the input control voltages were taken to be

$$V_a = k_a \eta_a \quad \text{and} \quad V_b = k_b \eta_b$$

where k_a and k_b are positive control gains, I_{da} and I_{db} are defined as

$$I_{da} = k_v \dot{e} + k_p e \quad \text{and} \quad I_{db} = 0$$

where k_v and k_p are positive control gains. The best performance was found by using the following control gain values

$$k_p = 460, \qquad k_v = 4, \qquad k_a = 1.5, \quad \text{and} \quad k_b = 2.$$

The resulting position tracking error is given in Figure 4.14 which indicates that the best steady state tracking error is approximately within ± 1.7 degrees. For brevity, we only plotted out the phase one voltage in Figure 4.15 for the linear control experiment.

Remark 4.7

From the above experimental results, we can see that, with regard to the steady-state tracking error, the nonlinear controllers outperformed the linear controller by approximately 15 times.

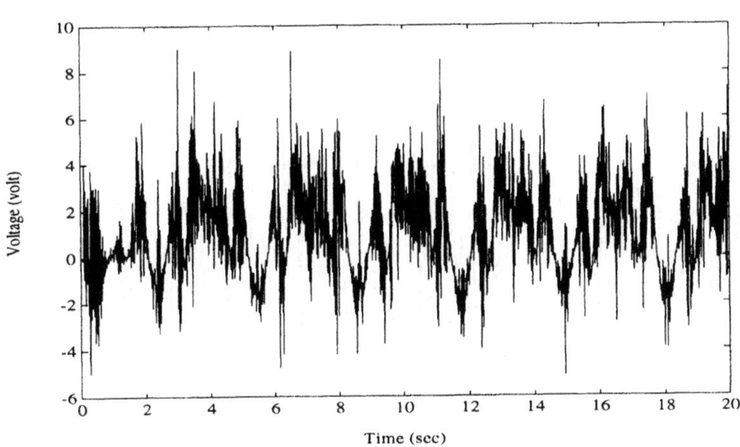

Figure 4.9: Phase Three Voltage for Exact Knowledge Controller

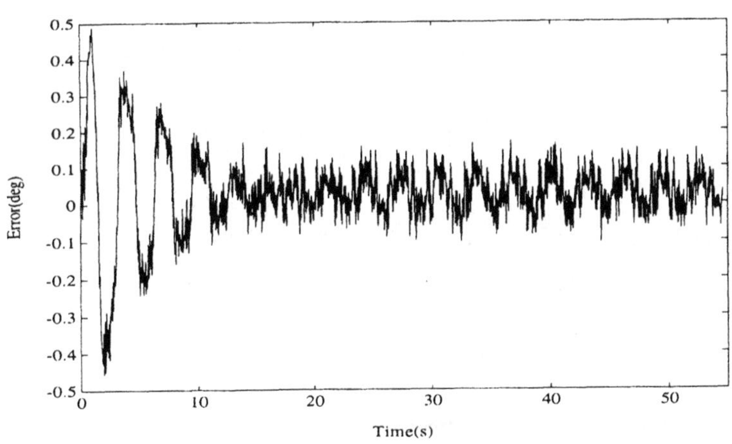

Figure 4.10: Position Tracking Error for Adaptive Controller

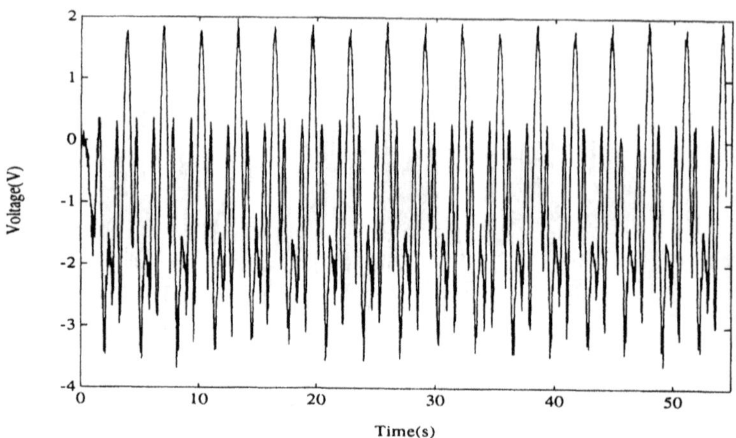

Figure 4.11: Phase One Voltage for Adaptive Controller

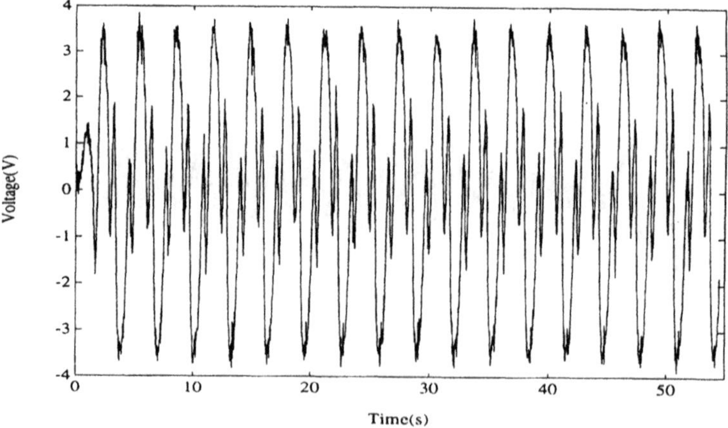

Figure 4.12: Phase Two Voltage for Adaptive Controller

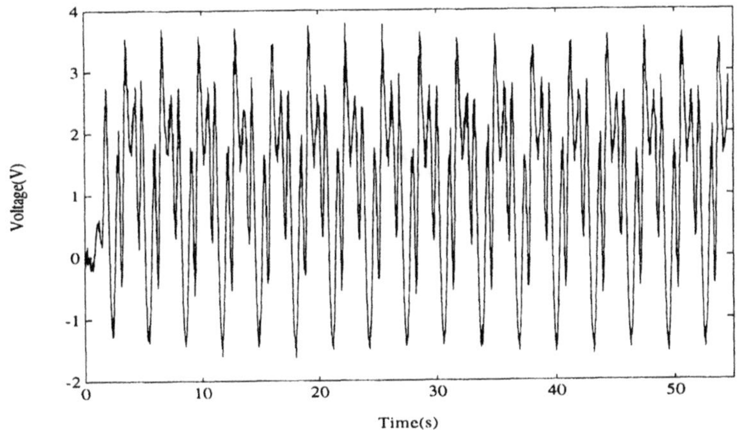

Figure 4.13: Phase Three Voltage for Adaptive Controller

4.8 Notes

Many recent publications discuss the development of nonlinear controllers for BLDC motors. Specifically, to compensate for system uncertainties, advanced nonlinear techniques such as adaptive control and robust control have been considered for the control of BLDC motors. For example, in [15], Hemati *et al.* developed a robust feedback linearizing controller for a single-link robot actuated by a BLDC motor. Under certain matching conditions, this controller is robust with respect to parametric uncertainty throughout the electromechanical model. In [16], Hemati also investigated both global and local characteristics of BLDC motors to provide guide-lines for designing controllers. A compact representation for the BLDC motor dynamic equations was also presented in [9]. In [8], Carroll *et al.* developed a robust tracking controller for a BLDC motor which achieved a globally uniform ultimate bounded (GUUB) result for the rotor position tracking error despite parametric uncertainties and additive bounded disturbances. Utilizing a projection type parameter update law, Hu *et al.* [10] developed an adaptive position tracking controller for a BLDC motor/load system which achieved a global asymptotic tracking performance. This controller was modified to reduce the overparameterization problem and was applied to a BLDC motor/load system in [11]. Several nonlinear tracking controllers and the corresponding experimental results were presented in [12] and [13].

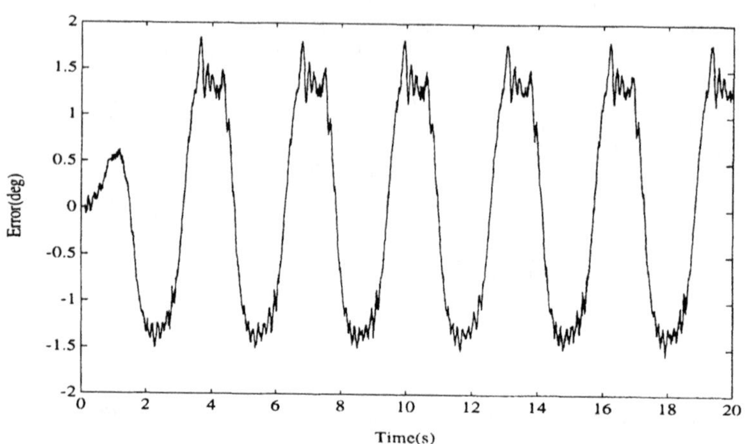

Figure 4.14: Position Tracking Error for Linear Controller

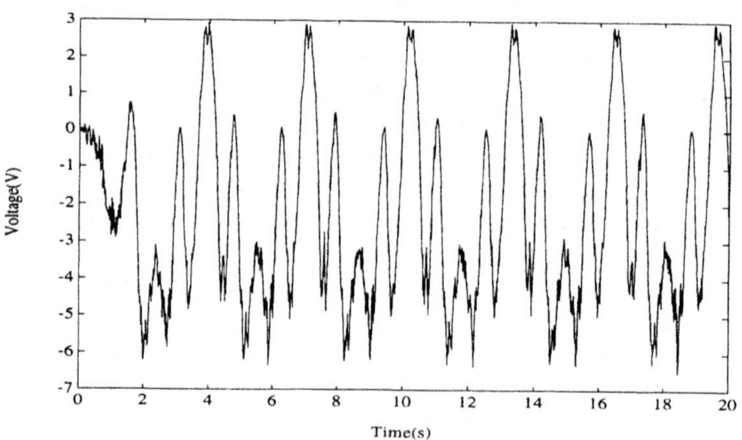

Figure 4.15: Phase One Voltage for Linear Controller

Bibliography

[1] H. Asada and K. Youcef-Toumi, *Direct Drive Robots: Theory and Practice*, Cambridge: MIT Press, 1987.

[2] T. J. E. Miller, *Brushless Permanent-Magnet and Reluctance Motor Drives*, Oxford University Press, 1989.

[3] N. Hemati and M. Leu, "A Complete Model Characterization of Brushless DC Motors", *IEEE Transactions on Industry Applications*, Vol. 28, No. 1, pp. 172-180, Jan/Feb., 1992.

[4] M. Spong and M. Vidyasagar, *Robot Dynamics and Control*, New York: John Wiley and Sons, Inc., 1989.

[5] P. C. Krause, *Analysis of Electric Machinery*, McGraw-Hill, Inc., 1986.

[6] J. Slotine and W. Li, *Applied Nonlinear Control*, Englewoods Cliff, NJ: Prentice Hall Co., 1991.

[7] S. Sastry, and M. Bodson, *Adaptive Control: Stability, Convergence, and Robustness*, Englewoods Cliff, NJ: Prentice Hall Co., 1989.

[8] J.J. Carroll, and D.M. Dawson, "Robust Tracking Control of a BLDC Motor with Application to Robotics", *Proc. of IEEE International Conference on Robotics and Automation*, Vol. 1, pp. 94-99, June, 1993.

[9] N. Hemati, "Dynamic Analysis of Brushless Motors Based on Compact Representations of Equations of Motion", *Proc. of IEEE Industrial Application Society Annual Meeting*, Vol. 1, pp. 51-58, September, 1993.

[10] J. Hu, J. J. Carroll, D.M. Dawson, and P. Vedagarbha, "An Adaptive Integrator Backstepping Tracking Controller for Brushless DC Motor/Robotic Load", *Proc. of the American Control Conference*, Vol. 2, pp. 1401-1405 Baltimore, MD, June, 1994.

[11] J. Hu, M. Leviner, D. Dawson, and P. Vedaghabar, "An Adaptive Tracking Controller for Multi-Phase Permanent Magnet Machines", *International Journal of Control*, Vol. 64, No. 6, 1996, pp. 997-1022.

[12] K. Anderson, *Nonlinear Tracking Controllers for a Brushless DC Motor*, M.S. Thesis, Clemson University, May, 1994.

[13] J. Hu, T. C. Burg, and D. M. Dawson, "Nonlinear Tracking Controllers for Brushless DC Motors", *Proc. of the 1994 IEEE Industrial Applications Society Annual Meeting*, Vol. 1, pp. 480-487, Denver, Colorado, October, 1994.

[14] A. Glumineau, M. Hamy, C. Lanier, and C. H. Moog, "Robust Control of a Brushless Servo Motor via Sliding Mode Techniques", *International Journal of Control,* Vol. 58, No. 5, pp. 979-990, 1993.

[15] N. Hemati, J. Thorp, and M. Leu, "Robust Nonlinear Control of Brushless DC Motors for Direct-Drive Robotic Applications", *IEEE Transactions on Industrial Electronics,* Vol. 37, No. 6, pp. 498-501, 1990.

[16] N. Hemati, "The Global and Local Dynamics of Direct-Drive Brushless DC Motors", *Proc. of the IEEE International Conference on Robotics and Automation,* Vol. 2, pp. 1858-1863, May, 1992.

[17] I. Kanellakopoulos, P.V. Kokotovic, and A.S. Morse, "Systematic Design of Adaptive Controllers for Feedback Linearizable Systems", *IEEE Transactions on Automatic Control*, Vol. 36, pp. 1241-1253, 1991.

[18] M. Krstic, I. Kanellakopoulos, and P. Kokotovic, *Nonlinear and Adaptive Control Design,* John Wiley & Sons, 1995.

[19] M. Krstic, I. Kanellakopoulos, and P. Kokotovic, "Adaptive Nonlinear Control Without Overparameterization", *Systems & Control Letters,* Vol. 19, pp. 177-185, 1992.

Chapter 5

SR Motor (FSFB)

5.1 Introduction

Since the switched reluctance (SR) motor is basically an AC machine with an iron core rotor, there are no mechanical brushes that require expensive maintenance as are required with the permanent magnet, brushed DC motor. In addition to a reduction in maintenance, there are several advantages to using SR motors as actuators in high performance motion control applications [1] [2]. These advantages include their low cost and high reliability through simple design and construction, increased capability to withstand high temperature environments, significant reduction in friction, and their ability to increase torque production through electromagnetic gearing. While the SR motor complicates the control problem by coupling multi-input nonlinear dynamics into the overall electromechanical system dynamics, it is possible to develop an accurate dynamic model. Unfortunately, as far as control design is concerned, the SR motor/load electromechanical dynamics exhibit: i) nonlinear terms, which are composed of the square of the electrical current multiplied by trigonometric terms, in the static torque transmission equation, and ii) nonlinear terms, which are products of the rotor velocity, electrical current, and trigonometric terms, in the electrical subsystem dynamics.

In this chapter, we apply the adaptive integrator backstepping tools [26] to the design of a position tracking controller for the SR motor driving a mechanical load. Since there is no standard rotor fixed frame transformation *(e.g.*, similar to the BLDC motor) for the SR motor, a closed form, differentiable commutation[1] strategy must be fused into the backstepping procedure. In contrast with the simple commutation strategy for the PMS motor given in Chapter 3, the SR motor's highly nonlinear, multi-phase

[1]The term commutation strategy is often used to describe the method for developing the desired current signals which are based on the rotor position, the desired torque signal, and the static torque transmission model.

(*e.g.*, four phases) torque transmission relationship demands a more sophisticated commutation strategy. In addition, the backstepping procedure places constraints on the commutation strategy that must be accounting for during its design. For example, the backstepping control design technique requires that the commutation strategy be differentiable and that its derivative be bounded if the desired torque signal is bounded. Indeed, it is the merging of the commutation architecture with the control design which makes the SR motor one of the more challenging electric machines to control precisely.

To facilitate the design of the controller, we first view the motor as a torque source and thus design a desired torque signal to ensure that the load follows the desired position trajectory. Since the developed motor torque is a function of rotor position and the electrical winding currents, we utilize a commutation strategy to restate the desired torque signal as a set of desired current trajectories. The voltage control inputs are then formulated to force the electrical winding currents to follow the desired current trajectories. That is, the electrical dynamics are taken into account through the current tracking objective, and hence the position tracking control objective is embedded inside the current tracking objective. Therefore, if the voltage control input can be designed to guarantee that the actual currents track the desired currents, then the load position will follow the desired position trajectory.

5.2 System Model

We shall consider the model of a SR motor turning a robotic load for control purposes. The dynamic model of the mechanical subsystem is taken to be a single-link, direct drive robot [3] of the form

$$M\ddot{q} + B\dot{q} + N\sin(q) = \tau \tag{5.1}$$

where the constant parameter M denotes the mechanical inertia of the motor rotor and the connected load, B is the viscous friction coefficient, N denotes the constant lumped load term, $q(t)$, $\dot{q}(t)$, and $\ddot{q}(t)$ represent the load position, velocity, and acceleration, respectively, and $\tau(t)$ represents the electromechanical coupling torque.

In this chapter, we will limit the discussion to control design for a four-phase SR motor. To characterize the electrical subsystem dynamics, N_r will be used to represent the total number of rotor saliencies (*i.e.*, poles) and the subscript j will be used to refer to the individual motor phases, for $j = 1, \ldots, 4$. For the four-phase SR motor, the number of control inputs (*i.e.*, voltages) and states (*i.e.*, load position, load velocity, and winding currents) are 4 and 6, respectively. The expression relating the per phase winding voltage and current for the SR motor with a linear magnetic circuit

[4] is given by

$$L_j(q)\,\dot{\gamma}_j = -R_j(q,\dot{q},\gamma_j) + V_j, \tag{5.2}$$

where

$$L_j(q) = L_0 - L_1\cos(x_j) > 0, \tag{5.3}$$

$$R_j(q,\dot{q},\gamma_j) = R_0\gamma_j + N_r L_1 \sin(x_j)\gamma_j\dot{q}, \tag{5.4}$$

and

$$x_j = N_r q - \frac{\pi(j-1)}{2}. \tag{5.5}$$

The constant parameters L_0, L_1, and R_0 represent the coefficients of static winding inductance, dynamic winding inductance, and winding resistance, respectively, $\gamma_j(t)$ denotes the per phase winding currents, and $V_j(t)$ denotes the per phase winding input voltages. The electromechanical torque coupling between the electrical and mechanical subsystems is given by [4]

$$\tau = \sum_{j=1}^{4} \sin(x_j)\gamma_j^2. \tag{5.6}$$

Note that the parameters M, B, and N introduced in (5.1) are defined to include the effects of the torque coefficient constant. That is, the original mechanical parameters have been divided by the constant $\frac{N_r L_1}{2}$. The schematic diagram of a SR motor turning a single-link robotic load is shown in Figure 5.1.

5.3 Control Objective

Given full state measurements (*i.e.*, q, $\dot{q}$, γ_j), the objective of the control design procedure is to develop load position tracking controllers for the electromechanical dynamics of (5.1) through (5.6). To begin the development, we define the load position tracking error $e(t)$ as

$$e = q_d - q \tag{5.7}$$

where $q_d(t)$ represents the desired load position trajectory, and $q(t)$ was defined in (5.1). We will assume that $q_d(t)$ and its first, second, and third time derivatives are all bounded functions of time. To simplify the control formulation and the stability analysis, we define the filtered position tracking error $r(t)$ [5] as

$$r = \dot{e} + \alpha e \tag{5.8}$$

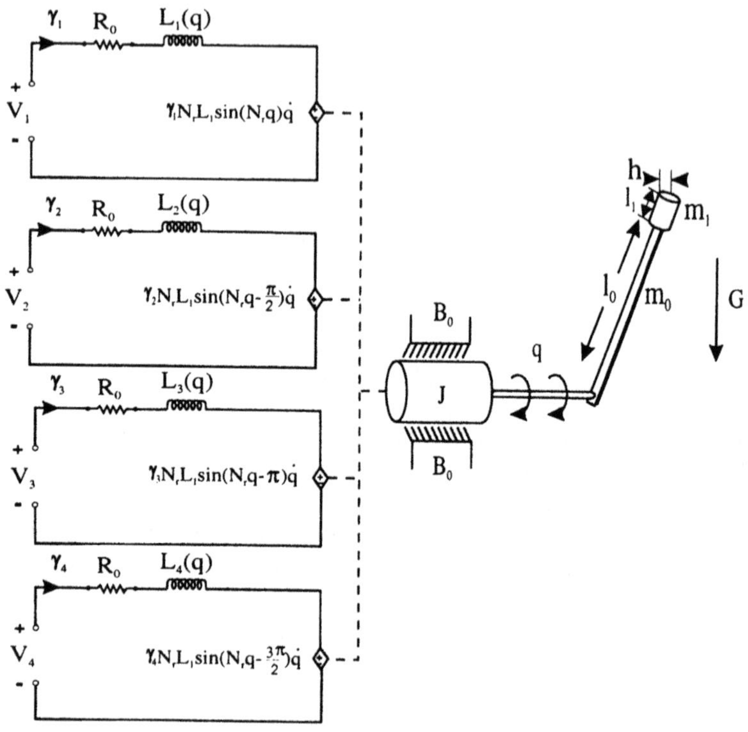

Figure 5.1: Schematic Diagram of the SR Motor/Load System

where α is a positive constant control gain. As explained in Chapter 1, use of the filtered tracking error allows us to analyze the second-order dynamics of (5.1) as though it is a first-order system.

To form the open-loop filtered tracking error dynamics, we differentiate (5.8) with respect to time and rearrange terms to yield

$$\dot{r} = (\ddot{q}_d + \alpha\dot{e}) - \ddot{q}. \tag{5.9}$$

Multiplying (5.9) by M and substituting the mechanical subsystem dynamics of (5.1), and the coupling torque of (5.6) yields the filtered tracking error dynamics as shown

$$M\dot{r} = M(\ddot{q}_d + \alpha\dot{e}) + B\dot{q} + N\sin(q) - \sum_{j=1}^{4}\sin(x_j)\gamma_j^2. \tag{5.10}$$

To prepare for the controller developments, the right-hand side of (5.10) can be rewritten as

$$M\dot{r} = W_{\tau}\theta_{\tau} - \sum_{j=1}^{4} \sin(x_j)\gamma_j^2 \tag{5.11}$$

where the known regression matrix [6] $W_{\tau}(q, \dot{q}, t) \in \Re^{1\times 3}$ is given by

$$W_{\tau} = \begin{bmatrix} \ddot{q}_d + \alpha\dot{e} & \dot{q} & \sin(q) \end{bmatrix} \tag{5.12}$$

and the parameter vector $\theta_{\tau} \in \Re^3$ is given by

$$\theta_{\tau} = \begin{bmatrix} M & B & N \end{bmatrix}^T. \tag{5.13}$$

Due to the structure of the electromechanical systems given by (5.1) through (5.6), we are only free to specify the per phase voltages, $V_1(t)$ through $V_4(t)$. In other words, the mechanical subsystem error dynamics lack true current (torque) level control inputs. For this reason, we shall add and subtract the desired current trajectories, γ_{dj}, to the right-hand side of (5.11), to yield

$$M\dot{r} = W_{\tau}\theta_{\tau} - \sum_{j=1}^{4} \sin(x_j)\gamma_{dj}^2 + \sum_{j=1}^{4} \sin(x_j)\left(\gamma_{dj} + \gamma_j\right)\eta_j \tag{5.14}$$

where $\eta_j(t)$ represents the current tracking error perturbations to the mechanical subsystem dynamics of the form

$$\eta_j = \gamma_{dj} - \gamma_j. \tag{5.15}$$

If the current tracking error terms $\eta_j(t)$ in (5.14) were equal to zero, then $\gamma_{dj}(t)$ could be designed to achieve good load position tracking by utilizing standard control techniques (*e.g.*, [5]) and an appropriate commutation strategy. Since the current tracking errors are not equal to zero in general, we must design voltage control inputs at V_j that will compensate for the effects of η_j in (5.14). To accomplish this control objective, the dynamics of the current tracking errors are needed. Taking the time derivative of the current tracking error in (5.15) and then multiplying the result by $L_j(q)$ yields

$$L_j(q)\dot{\eta}_j = L_j(q)\dot{\gamma}_{dj} - L_j(q)\dot{\gamma}_j. \tag{5.16}$$

Substituting the right-hand side of (5.2) for $L_j(q)\dot{\gamma}_j$ into (5.16) results in the open-loop current tracking error dynamics as shown

$$L_j(q)\dot{\eta}_j = L_j(q)\dot{\gamma}_{dj} + R_j(q, \dot{q}, \gamma_j) - V_j. \tag{5.17}$$

5.4 Commutation Strategy

An inherent feature of the SR motor is that it must be electronically commutated to produce motion. That is, based on the motor construction, the individual phases are energized as a function of the rotor position (or load position) in order to turn the rotor. The commutation can be achieved through dedicated electronics or integrated into the controller design via a mathematical commutation strategy. Various mathematical commutation strategies have been proposed for the SR motor in the literature ([1], [2], [7], [8], [9], [4], [10]) corresponding to different control objectives and methodologies. In [12], a new commutation strategy was introduced to allow voltage input controllers to be designed for robots actuated by SR motors. Utilizing this commutation strategy, we design the desired per phase winding currents to "share" the control responsibilities in a continuously differentiable fashion. In other words each individual phase will be assigned the responsibility of producing some fraction of the desired torque based on the position of the rotor. At certain rotor positions this fraction is small while at other rotor positions it may be nearly 100% of the desired torque.

The commanded per phase winding current, $\gamma_{dj}(t)$, is generated from the desired torque signal $\tau_d(t)$, where $\tau_d(t)$ is designed to force the mechanical load track the desired position trajectory [12], according to

$$\gamma_{dj} = \sqrt{\frac{\tau_d \sin(x_j) S\left(\sin(x_j)\tau_d\right)}{S_T} + \gamma_{c0}^2}\,. \tag{5.18}$$

The positive scalar design parameter γ_{c0} of (5.18) is used to set the desired per phase threshold winding current. The term $S(z)$ of (5.18) is defined to be the following differentiable function for arbitrary scalars z

$$S(z) = \begin{cases} 0 & , \ z \leq 0 \\ 1 - \underline{e}^{-\epsilon_0 z^2} & , \ z > 0 \end{cases}, \tag{5.19}$$

where ϵ_0 is a positive scalar design parameter which determines how closely $S(z)$ approximates the unit-step function. The term S_T of (5.18) is also defined to be a differentiable function as shown

$$S_T = \sum_{j=1}^{4} \sin^2(x_j) S\left(\sin(x_j)\tau_d\right). \tag{5.20}$$

Given the definitions of $S(\cdot)$ and S_T in (5.19) and (5.20), respectively, it is easy to show that

$$\frac{\tau_d \sin(x_j) S\left(\sin(x_j)\tau_d\right)}{S_T} \geq 0. \tag{5.21}$$

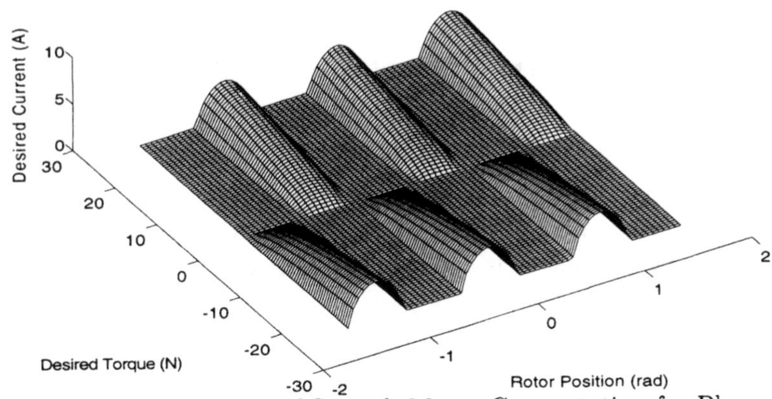

Figure 5.2: Example Plot of Smooth Motor Commutation for Phase One of the Four-Phase SR Motor

However, at first glance, it seems that (5.18) is undefined for $S_T = 0$. Given the definition of $S(\cdot)$ and S_T, it is easy to see that $S_T = 0$ when $\tau_d = 0$. Evaluating the left-hand side of (5.21) in the limit as $\tau_d(t)$ approaches zero yields [13]

$$\lim_{\tau_d \to 0} \left[\frac{\tau_d \sin(x_j) S\left(\sin(x_j)\tau_d\right)}{S_T} \right] = 0 \tag{5.22}$$

for all q; hence, from (5.18) and (5.22), we can state that

$$\lim_{\tau_d \to 0} \gamma_{dj} = \gamma_{c0}. \tag{5.23}$$

Therefore, given (5.18) through (5.23), we can state that γ_{dj} exists and is bounded for all bounded desired torque signals and rotor positions.

To graphically illustrate the commutation strategy for a 4-phase SR motor with $N_r = 6$, the first phase desired current is plotted as a function of desired torque τ_d and rotor position q in Figure 5.2 with $\gamma_{c0} = \epsilon_0 = 1.0$.

From Figure 5.2, we can see that the proposed commutation strategy is a smooth function of τ_d and q. In Figure 5.2, it is insightful to consider a constant desired torque and observe how the desired current signal changes as a function of the rotor position (*i.e.*, how the motor is commutated).

To illustrate the motivation for the structure of (5.18), suppose we set $\gamma_j = \gamma_{dj}$ so that we can rewrite (5.1) as

$$M\ddot{q} + B\dot{q} + N\sin(q) = \sum_{j=1}^{4} \sin(x_j)\gamma_{dj}^2. \tag{5.24}$$

After substituting the proposed commutation strategy of (5.18) into (5.24), we have

$$M\ddot{q}+B\dot{q}+N\sin(q)=\sum_{j=1}^{4}\sin(x_j)\left(\frac{\tau_d\sin(x_j)S\left(\sin(x_j)\tau_d\right)}{S_T}+\gamma_{c0}^2\right). \quad (5.25)$$

After substituting (5.20) into (5.25) and simplifying the result, we have

$$M\ddot{q}+B\dot{q}+N\sin(q)=\tau_d+\gamma_{c0}^2\sum_{j=1}^{4}\sin(x_j). \quad (5.26)$$

Using the definition of x_j in (5.5), the summation in (5.26) is expanded, and a trigonometric identity is applied to show that

$$\begin{aligned}\sum_{j=1}^{4}\sin(x_j) &= \sin\left(N_r q\right)+\sin\left(N_r q-\frac{\pi}{2}\right) \\ &\quad +\sin\left(N_r q-\pi\right)+\sin\left(N_r q-\frac{3\pi}{2}\right)=0.\end{aligned} \quad (5.27)$$

Therefore, (5.26) can be simplified to

$$M\ddot{q}+B\dot{q}+N\sin(q)=\tau_d. \quad (5.28)$$

From (5.28), we can see that the proposed commutation strategy has been developed such that the desired torque trajectory inherently becomes the control input to the mechanical subsystem and hence can be designed to ensure that the rotor (and hence the load) follows the desired position trajectory. Of course due to the electrical dynamics, we can not guarantee that $\gamma_j=\gamma_{dj}$; hence, we substitute (5.18) into (5.14) to obtain the open-loop dynamics for the filtered tracking error r as

$$M\dot{r}=W_\tau\theta_\tau-\tau_d+\sum_{j=1}^{4}\sin(x_j)\left(\gamma_{dj}+\gamma_j\right)\eta_j. \quad (5.29)$$

Based on the structure of (5.29), we are motivated to design the voltage control inputs to force the actual currents to track the desired per phase winding currents, that is, to force the current tracking errors $\eta_j(t)$ of (5.15) to zero.

5.5 Exact Model Knowledge Controller

Based on exact knowledge of the electromechanical system model and full state feedback, we now design a load position tracking controller for the open-loop dynamics of (5.29) and (5.17). As stated in the introduction, we first view the motor as a torque source and design the desired torque τ_d to ensure that the load tracks the desired position trajectory. Next, through the commutation strategy of (5.18), the desired torque is transformed into four desired current trajectories. Finally, we specify the voltage control inputs to force the stator winding currents to track the desired current trajectories.

As mentioned above, we first design the desired torque signal to force the load along the desired position trajectory. That is, we specify the desired torque $\tau_d(t)$ to drive $r(t)$ to *zero*. To this end, we select $\tau_d(t)$ as

$$\tau_d = W_\tau \theta_\tau + k_s r \tag{5.30}$$

where W_τ and θ_τ were previously defined in (5.12) and (5.13), respectively, and k_s is a positive constant control gain. Substituting (5.30) into the open-loop dynamics of (5.29) yields the closed-loop filtered tracking error dynamics as shown

$$M\dot{r} = -k_s r + \sum_{j=1}^{4} \sin(x_j) \left(\gamma_{dj} + \gamma_j\right) \eta_j. \tag{5.31}$$

Given τ_d of (5.30), the desired phase currents, γ_{dj}, can be directly calculated from the commutation strategy in (5.18). To complete the open-loop system description for the current tracking error dynamics, we now calculate the term $\dot{\gamma}_{dj}$ required in (5.17) by taking the time derivative of (5.18) to yield

$$\dot{\gamma}_{dj} = \Sigma_j \dot{x}_j + \Pi_j \dot{\tau}_d \tag{5.32}$$

where Σ_j and Π_j are partial derivatives of γ_{dj} with respect to x_j and τ_d, respectively (see Appendix B for explicit definitions of Σ_j and Π_j). Equation (5.32) now requires $\dot{\tau}_d$ which can be found from (5.30) as

$$\dot{\tau}_d = \dot{W}_\tau \theta_\tau + k_s \dot{r}. \tag{5.33}$$

Substituting the time derivatives of (5.8) and (5.12) into the right-hand side of (5.33) yields

$$\begin{aligned} \dot{\tau}_d \;=\; & M\left(\dddot{q}_d + \alpha\left(\ddot{q}_d - \ddot{q}\right)\right) + B\ddot{q} + N\dot{q}\cos(q) \\ & + k_s\left(\ddot{q}_d - \ddot{q} + \alpha\dot{e}\right) \end{aligned} \tag{5.34}$$

which expresses $\dot{\tau}_d$ in terms of measurable states (*i.e.*, q and $\dot{q}$), known functions, known constant parameters, and the unmeasurable load acceleration $\ddot{q}$. From (5.1) and (5.6), we can solve for $\ddot{q}$ in the following form

$$\ddot{q} = -\frac{B}{M}\dot{q} - \frac{N}{M}\sin(q) + \frac{1}{M}\sum_{j=1}^{4} \sin(x_j)\gamma_j^2. \tag{5.35}$$

Substituting for $\ddot{q}$ from the right-hand side of (5.35) into (5.34), we can write $\dot{\gamma}_{dj}$ in terms of measurable states (*i.e.*, q, $\dot{q}$, γ_j), known functions, and known constant parameters in the following form

$$\begin{aligned}\dot{\gamma}_{dj} = \; & \Sigma_j N_r \dot{q} + \Pi_j \left(M \left(\dddot{q}_d + \alpha \ddot{q}_d \right) + N \dot{q} \cos(q) \right. \\ & + k_s \left(\ddot{q}_d + \alpha \dot{e} \right)) + \Pi_j \left(B - M\alpha - k_s \right) \left(-\frac{B}{M} \dot{q} \right. \\ & \left. - \frac{N}{M} \sin(q) + \frac{1}{M} \sum_{j=1}^{4} \sin(x_j) \gamma_j^2 \right)\end{aligned} \tag{5.36}$$

Substituting the expression for $\dot{\gamma}_{dj}$ of (5.36) into (5.17) yields the final open-loop model for the current tracking error in the form

$$L_j(q)\dot{\eta}_j = w_j - V_j - \frac{1}{2}\dot{L}_j(q)\eta_j \tag{5.37}$$

where the auxiliary scalar variable $w_j(q, \dot{q}, \gamma_j, t)$ is given by

$$\begin{aligned}w_j = \; & L_j(q)\Sigma_j N_r \dot{q} + L_j(q)\Pi_j \left(B - M\alpha - k_s \right) \\ & \cdot \left(-\frac{B}{M}\dot{q} - \frac{N}{M}\sin(q) + \frac{1}{M}\sum_{j=1}^{4}\sin(x_j)\gamma_j^2 \right) \\ & + L_j(q)\Pi_j \left(M \left(\dddot{q}_d + \alpha \ddot{q}_d \right) + N\dot{q}\cos(q) + k_s \left(\ddot{q}_d + \alpha \dot{e} \right) \right) \\ & + R_0 \gamma_j + N_r L_1 \sin(x_j)\gamma_j \dot{q} + \frac{1}{2} L_1 \sin(x_j) N_r \dot{q} \eta_j .\end{aligned} \tag{5.38}$$

Remark 5.1

Note that the term $\frac{1}{2}\dot{L}_j(q)\eta_j$ was added and subtracted in (5.37) to facilitate the stability analysis. Also from (5.32), we see that the first order time derivative of the commutation strategy proposed in (5.18) is required for the voltage control input design. That is, the commutation strategy presented in (5.18) must be first order differentiable.

We now design the voltage control inputs, V_j, for the open-loop system of (5.37). Given the structure of (5.37) and (5.31), we define the voltage control inputs as

$$V_j = w_j + k_j \eta_j + \sin(x_j) \left(\gamma_{dj} + \gamma_j \right) r \tag{5.39}$$

where k_j are positive control gains. Substituting (5.39) into the open-loop dynamics of (5.37) yields the closed-loop current tracking error dynamics in the form

$$L_j(q)\dot{\eta}_j = -k_j\eta_j - \sin(x_j)\left(\gamma_{dj} + \gamma_j\right)r - \frac{1}{2}\dot{L}_j(q)\eta_j. \tag{5.40}$$

The dynamics given by (5.31) and (5.40) represent the electromechanical closed-loop system for which the stability analysis is performed while the exact model knowledge controller given by (5.30) and (5.39) represents the control inputs which are implemented at the voltage terminals of the motor. Note that the desired current trajectories, γ_{dj} of (5.18), are embedded (in the guise of the variable η_j) inside of the voltage control inputs, V_j. The theorem given below delineates the performance of the closed-loop system under the proposed control.

Theorem 5.1

The proposed exact knowledge controller for the electromechanical dynamics of (5.1) through (5.6) ensures that the filtered tracking error goes to zero exponentially fast as shown

$$\|x(t)\| \leq \sqrt{\frac{\lambda_2}{\lambda_1}}\,\|x(0)\|\,\underline{e}^{-\gamma t} \qquad \forall t \in [0, \infty) \tag{5.41}$$

where

$$x = \begin{bmatrix} r & \eta_1 & \eta_2 & \eta_3 & \eta_4 \end{bmatrix}^T \in \Re^5, \tag{5.42}$$

$$\lambda_1 = \min_j\{M, L_j(q)\} \quad \text{and} \quad \lambda_2 = \max_j\{M, L_j(q)\} \quad \text{for } j = 1, \cdots, 4 \tag{5.43}$$

and

$$\gamma = \frac{\min_j\{k_s,\, k_j\}}{\max_j\{M, L_j(q)\}} \quad \text{for } j = 1, \cdots, 4. \tag{5.44}$$

Proof. First, we define the following non-negative function

$$\begin{aligned} V(t) &= \tfrac{1}{2}Mr^2 + \tfrac{1}{2}\textstyle\sum_{j=1}^4 L_j(q)\eta_j^2 \\ &= \tfrac{1}{2}x^T diag\left\{M, L_1(q), L_2(q), L_3(q), L_4(q)\right\}x \end{aligned} \tag{5.45}$$

where $x(t)$ was defined in (5.42), and $diag\{\cdot\}$ represents a diagonal matrix in which the braced terms are the diagonal elements. Application of Lemma 1.7 in Chapter 1 to the matrix term in (5.45), provides the following upper and lower bounds on $V(t)$ as

$$\frac{1}{2}\lambda_1\|x(t)\|^2 \leq V(t) \leq \frac{1}{2}\lambda_2\|x(t)\|^2 \tag{5.46}$$

where λ_1 and λ_2 were defined in (5.43) (Note due to the form of (5.3), the inductance can be upper and lower bounded by positive constants). Differentiating (5.45) with respect to time yields

$$\dot{V}(t) = rM\dot{r} + \sum_{j=1}^{4} L_j(q)\eta_j\dot{\eta}_j + \frac{1}{2}\sum_{j=1}^{4} \dot{L}_j(q)\eta_j^2. \tag{5.47}$$

Substituting the closed-loop dynamics from (5.31) and (5.40) into (5.47) yields

$$\dot{V}(t) = -k_s r^2 - \sum_{j=1}^{4} k_j\eta_j^2 = -x^T diag\left\{k_s, k_1, k_2, k_3, k_4\right\} x \tag{5.48}$$

where $x(t)$ was defined in (5.42). The derivative of V in (5.48) can be upper bounded by applying Lemma 1.7 in Chapter 1 to the matrix term in (5.48) as follows

$$\dot{V}(t) \leq -\min_{j}\{k_s, k_j\} \left\|x(t)\right\|^2 \quad \text{for} \quad j = 1, \cdots, 4. \tag{5.49}$$

From (5.46), it is easy to see that $\left\|x(t)\right\|^2 \geq 2V(t)/\lambda_2$; hence, $\dot{V}(t)$ in (5.49) can be further upper bounded as

$$\dot{V}(t) \leq -2\gamma V(t) \tag{5.50}$$

where γ was defined (5.44). Applying Lemma 1.1 in Chapter 1 to (5.50) and (5.45) yields

$$V(t) \leq V(0)\underline{e}^{-2\gamma t}. \tag{5.51}$$

From (5.46), we have that

$$\frac{1}{2}\lambda_1 \left\|x(t)\right\|^2 \leq V(t) \qquad \text{and} \qquad V(0) \leq \frac{1}{2}\lambda_2 \left\|x(0)\right\|^2. \tag{5.52}$$

Substituting (5.52) appropriately into the left-hand and right-hand sides of (5.51) allows us to form the following inequality

$$\frac{1}{2}\lambda_1 \left\|x(t)\right\|^2 \leq \frac{1}{2}\lambda_2 \left\|x(0)\right\|^2 \underline{e}^{-2\gamma t}. \tag{5.53}$$

Using (5.53), we can solve for $\left\|x(t)\right\|$ to obtain the result given in (5.41). □

Remark 5.2

Using the result of Theorem 5.1, the control structure, and the electromechanical model, it is straightforward to illustrate that all signals remain bounded during closed-loop operation. Specifically, from (5.41) and (5.42), we know that $r(t) \in L_\infty$ and $\eta_j(t) \in L_\infty$. From Lemma 1.4 in Chapter 1, we know that if $r(t) \in L_\infty$ then $e(t) \in L_\infty$ and $\dot{e}(t) \in L_\infty$. Since

$q_d(t) \in L_\infty$ and $\dot{q}_d(t) \in L_\infty$ by assumption, then (5.7) can be utilized to show that $q(t) \in L_\infty$ and $\dot{q}(t) \in L_\infty$. From the definition of $\tau_d(t)$ given in (5.30) and the fact that $\ddot{q}_d(t) \in L_\infty$ by assumption, it is now easy to show that $\tau_d(t) \in L_\infty$ and therefore $\gamma_{dj}(t) \in L_\infty$ from (5.18); hence, since $\eta_j(t) \in L_\infty$, we can use (5.15) to state that $\gamma_j(t) \in L_\infty$. Using the structure of the voltage control inputs of (5.39), the information in Appendix B, the above information, and the assumption that $\dddot{q}_d \in L_\infty$, we now know that $V_j(t) \in L_\infty$. Finally, the electromechanical dynamics of (5.1) and (5.2) can be used to show that $\ddot{q}(t) \in L_\infty$ and $\dot{\gamma}_j(t) \in L_\infty$.

From the result given by (5.41), it is now straightforward to see that the premise of Lemma 1.5 in Chapter 1 is satisfied; hence, we know that position tracking error (*i.e.*, $e(t)$) and the velocity tracking error (*i.e.*, $\dot{e}(t)$) both go to zero exponentially fast.

5.6 Adaptive Controller

Given the dynamics of (5.1) through (5.6), we now design an adaptive position tracking controller under the constraint of parametric uncertainty. That is, without any apriori knowledge of the electromechanical system parameters we achieve our stated control objectives based on measurement of load position, load velocity, and the per phase winding currents. The steps followed to design the adaptive controller are the same as those used in the development of the exact model knowledge controller coupled with the introduction of adaptation mechanisms for the uncertain system parameters (for a general overview of adaptive control design the reader is referred to [11]).

The first step in the design procedure is to develop an adaptive desired torque signal τ_d for the mechanical dynamics of (5.29) as [5]

$$\tau_d = W_\tau \hat{\theta}_\tau + k_s r \tag{5.54}$$

where $W_\tau \in \Re^{1\times 3}$ was defined in (5.12), $\hat{\theta}_\tau(t) \in \Re^3$ represents a dynamic estimate of the unknown parameter vector θ_τ defined in (5.13), and k_s is a positive constant controller gain. The parameter estimate $\hat{\theta}_\tau$ defined in (5.54) is updated online according to the following adaptation law

$$\hat{\theta}_\tau = \int_0^t \Gamma_\tau W_\tau^T(\sigma) r(\sigma)\, d\sigma \tag{5.55}$$

where $\Gamma_\tau \in \Re^{3\times 3}$ is a constant, positive definite, diagonal adaptive gain matrix. If we define the mismatch between $\hat{\theta}_\tau$ and θ_τ as

$$\tilde{\theta}_\tau = \theta_\tau - \hat{\theta}_\tau, \tag{5.56}$$

then the adaptation law of (5.55) can be written in terms of the parameter error as

$$\dot{\tilde{\theta}}_{\tau} = -\Gamma_{\tau} W_{\tau}^{T} r. \tag{5.57}$$

Substituting (5.54) into the open-loop dynamics of (5.29) yields the closed-loop filtered tracking error dynamics, as shown

$$M\dot{r} = W_{\tau}\tilde{\theta}_{\tau} - k_s r + \sum_{j=1}^{4} \sin(x_j) \left(\gamma_{dj} + \gamma_j\right) \eta_j. \tag{5.58}$$

The desired current trajectories can be found by direct application of the commutation strategy of (5.18) and the proposed desired torque signal in (5.54). Hence, we can now complete the open-loop system description for the current tracking error dynamics. We first calculate the term $\dot{\gamma}_{dj}$ in (5.17) by taking the time derivative of (5.18) to yield

$$\dot{\gamma}_{dj} = \Sigma_j \dot{x}_j + \Pi_j \dot{\tau}_d \tag{5.59}$$

where Σ_j and Π_j are partial derivatives given in Appendix B. Given (5.54), $\dot{\tau}_d$ can be calculated as

$$\dot{\tau}_d = \dot{W}_{\tau}\hat{\theta}_{\tau} + W_{\tau} \dot{\hat{\theta}}_{\tau} + k_s \dot{r}. \tag{5.60}$$

Substituting the time derivatives of (5.8), (5.12), and (5.55) into the right-hand side of (5.60) yields

$$\begin{aligned} \dot{\tau}_d = \; & \hat{M}\left(\dddot{q}_d + \alpha\left(\ddot{q}_d - \ddot{q}\right)\right) + \hat{B}\ddot{q} + \hat{N}\dot{q}\cos(q) \\ & + k_s\left(\ddot{q}_d - \ddot{q} + \alpha\dot{e}\right) + W_{\tau}\Gamma_{\tau}W_{\tau}^{T} r \end{aligned} \tag{5.61}$$

where $\hat{M}$, $\hat{B}$, and $\hat{N}$ denote the scalar components of the vector $\hat{\theta}_{\tau}$ (*i.e.*, $\hat{\theta}_{\tau} = \begin{bmatrix} \hat{M} & \hat{B} & \hat{N} \end{bmatrix}^T$). Note that $\dot{\tau}_d$ of (5.61) is expressed in terms of measurable states (*i.e.*, q and $\dot{q}$), known functions, and the unmeasurable state $\ddot{q}$. Substituting for $\ddot{q}$ from the right-hand side of (5.35) into (5.61), we can write $\dot{\tau}_d$ in terms of measurable states (*i.e.*, q, $\dot{q}$, γ_j), known functions, and unknown constant parameters. Substituting this expression for $\dot{\tau}_d$ into (5.59) yields

$$\begin{aligned} \dot{\gamma}_{dj} = \; & \Pi_j \left(\hat{M}\left(\dddot{q}_d + \alpha\ddot{q}_d\right) + \hat{N}\dot{q}\cos(q) + k_s\left(\ddot{q}_d + \alpha\dot{e}\right)\right. \\ & \left. + W_{\tau}\Gamma_{\tau}W_{\tau}^{T} r\right) + \Sigma_j N_r \dot{q} + \Pi_j \left(\hat{B} - \hat{M}\alpha - k_s\right) \\ & \cdot \left(-\frac{B}{M}\dot{q} - \frac{N}{M}\sin(q) + \frac{1}{M}\sum_{j=1}^{4}\sin(x_j)\gamma_j^2\right) \end{aligned} \tag{5.62}$$

Finally, substituting equation (5.62) into (5.17) and adding and subtracting $\frac{1}{2}\dot{L}_j(q)\eta_j$, yields the following linear parameterized open-loop dynamics for the current error as shown

$$L_j(q)\dot{\eta}_j = W_j\theta - V_j - \frac{1}{2}\dot{L}_j(q)\eta_j \tag{5.63}$$

where the known regression matrix $W_j(q,\dot{q},\gamma_j,\hat{\theta}_\tau,t) \in \Re^{1\times 9}$ and the unknown constant parameter vector $\theta \in \Re^9$ are explicitly defined as follows

$$\theta = \begin{bmatrix} L_0 & L_1 & \frac{L_0 B}{M} & \frac{L_0 N}{M} & \frac{L_0}{M} & \frac{L_1 B}{M} & \frac{L_1 N}{M} & \frac{L_1}{M} & R_0 \end{bmatrix}^T \tag{5.64}$$

and

$$W_j = \begin{bmatrix} W_{j1} & W_{j2} & W_{j3} & W_{j4} & W_{j5} & W_{j6} & W_{j7} & W_{j8} & W_{j9} \end{bmatrix} \tag{5.65}$$

with

$$\begin{aligned} W_{j1} = {} & \Sigma_j N_r\dot{q} + \Pi_j\left[\hat{M}\left(\dddot{q}_d + \alpha\ddot{q}_d\right)\right. \\ & \left. + \hat{N}\dot{q}\cos(q) + k_s\left(\ddot{q}_d + \alpha\dot{e}\right) + W_\tau\Gamma_\tau W_\tau^T r\right], \end{aligned}$$

$$\begin{aligned} W_{j2} = {} & -\cos(x_j)\,\Sigma_j N_r\dot{q} \\ & -\cos(x_j)\,\Pi_j\left(\hat{M}\left(\dddot{q}_d + \alpha\ddot{q}_d\right) + \hat{N}\dot{q}\cos(q) + k_s\left(\ddot{q}_d + \alpha\dot{e}\right)\right) \\ & -\cos(x_j)\,\Pi_j W_\tau\Gamma_\tau W_\tau^T r + N_r\sin(x_j)\,\gamma_j\dot{q} + \tfrac{1}{2}\sin(x_j)\,N_r\dot{q}\eta_j \end{aligned}$$

$$W_{j3} = -\Pi_j\left(\hat{B} - \alpha\hat{M} - k_s\right)\dot{q},$$

$$W_{j4} = -\Pi_j\left(\hat{B} - \alpha\hat{M} - k_s\right)\sin(q),$$

$$W_{j5} = \Pi_j\left(\hat{B} - \alpha\hat{M} - k_s\right)\sum_{j=1}^{4}\sin(x_j)\gamma_j^2,$$

$$W_{j6} = \cos(x_j)\,\Pi_j\left(\hat{B} - \alpha\hat{M} - k_s\right)\dot{q},$$

$$W_{j7} = \cos(x_j)\,\Pi_j\left(\hat{B} - \alpha\hat{M} - k_s\right)\sin(q),$$

$$W_{j8} = -\cos(x_j)\,\Pi_j\left(\hat{B} - \alpha\hat{M} - k_s\right)\sum_{j=1}^{4}\sin(x_j)\gamma_j^2, \text{ and } W_{j9} = \gamma_j.$$

The second step in the design procedure involves the design of the voltage control inputs V_j for the open-loop system of (5.63). Given the structure of (5.58) and (5.63), we define the voltage-level control inputs as

$$V_j = W_j\hat{\theta} + k_j\eta_j + \sin(x_j)\left(\gamma_{dj} + \gamma_j\right)r \tag{5.66}$$

where k_j are positive control gains, and $\hat{\theta}(t) \in \Re^9$ is a dynamic estimate of the unknown parameter vector θ. The parameter estimates are updated online by the following adaptation law

$$\hat{\theta} = \int_0^t \Gamma \sum_{j=1}^{4} \left(W_j^T(\sigma)\eta_j(\sigma)\right) d\sigma \tag{5.67}$$

where $\Gamma \in \Re^{9\times 9}$ is a constant positive definite, diagonal adaptive gain matrix. If we define the mismatch between θ and $\hat{\theta}$ as

$$\tilde{\theta} = \theta - \hat{\theta}, \tag{5.68}$$

then the adaptation law of (5.67) can be written in terms of the parameter error as

$$\dot{\tilde{\theta}} = -\Gamma \sum_{j=1}^{4} W_j^T\eta_j. \tag{5.69}$$

Substituting (5.66) into the open-loop dynamics of (5.63) yields the closed-loop current tracking error dynamics in the form

$$L_j(q)\dot{\eta}_j = W_j\tilde{\theta} - k_j\eta_j - \sin(x_j)\left(\gamma_{dj} + \gamma_j\right)r - \frac{1}{2}\dot{L}_j(q)\eta_j. \tag{5.70}$$

The dynamics given by (5.57), (5.58), (5.69), and (5.70) represent the electromechanical closed-loop system for which the stability analysis is performed while the adaptive controller given by (5.54), (5.55), (5.66), and (5.67) represents the controller which is implemented at the voltage terminals of the motor. Similar to the exact model knowledge controller, the adaptive desired current trajectories γ_{dj} are embedded (in the guise of the variable η_j) inside of the voltage control inputs V_j. The theorem given below delineates the performance of the closed-loop system under the proposed control.

Theorem 5.2

The proposed adaptive controller for the electromechanical dynamics of (5.1) through (5.6) ensures that the filtered tracking error goes to zero asymptotically as shown

$$\lim_{t\to\infty} r(t) = 0. \tag{5.71}$$

Proof. First, we define the following non-negative function

$$V=\frac{1}{2}Mr^2+\frac{1}{2}\sum_{j=1}^{4}L_j(q)\eta_j^2+\frac{1}{2}\tilde{\theta}_\tau^T\Gamma_\tau^{-1}\tilde{\theta}_\tau+\frac{1}{2}\tilde{\theta}^T\Gamma^{-1}\tilde{\theta}. \tag{5.72}$$

The derivative of (5.72) with respect to time is given by

$$\dot{V}=rM\dot{r}+\frac{1}{2}\sum_{j=1}^{4}\dot{L}_j(q)\eta_j^2+\sum_{j=1}^{4}\eta_jL_j(q)\dot{\eta}_j+\tilde{\theta}_\tau^T\Gamma_\tau^{-1}\dot{\tilde{\theta}}_\tau+\tilde{\theta}^T\Gamma^{-1}\dot{\tilde{\theta}} \tag{5.73}$$

where the facts that: i) scalars can be transposed and ii) Γ_τ, Γ are diagonal matrices, have been used. Substituting the error dynamics of (5.57), (5.69), (5.58), and (5.70) into (5.73) yields

$$\dot{V}=-k_sr^2-\sum_{j=1}^{4}k_j\eta_j^2+rW_\tau\tilde{\theta}_\tau-\tilde{\theta}_\tau^TW_\tau^Tr+\sum_{j=1}^{4}\left(\eta_jW_j\tilde{\theta}-\tilde{\theta}^TW_j^T\eta_j\right). \tag{5.74}$$

Since any scalar quantity can be transposed, (5.74) can be simplified to yield

$$\dot{V}=-k_sr^2-\sum_{j=1}^{4}k_j\eta_j^2. \tag{5.75}$$

From the form of (5.75), we can see that $\dot{V}(t)$ is negative or *zero*; hence, we know from calculus that $V(t)$ given in (5.72) is either decreasing or constant. Since $V(t)$ is non-negative, it is lower bounded by *zero*; hence, from the form of $V(t)$, we know that $r(t)\in L_\infty$, $\tilde{\theta}_\tau(t)\in L_\infty^3$, $\eta_j(t)\in L_\infty$, and $\tilde{\theta}(t)\in L_\infty^9$. Since $r(t)\in L_\infty$, Lemma 1.4 in Chapter 1 can be used to show that $e(t)\in L_\infty$ and $\dot{e}(t)\in L_\infty$; therefore, since $q_d(t)\in L_\infty$ and $\dot{q}_d(t)\in L_\infty$, (5.7) can be utilized to show that $q(t)\in L_\infty$ and $\dot{q}(t)\in L_\infty$. Since $\tilde{\theta}_\tau(t)\in L_\infty^3$ and $\tilde{\theta}(t)\in L_\infty^9$, we can use (5.56) and (5.68) to show that $\hat{\theta}_\tau(t)\in L_\infty^3$ and $\hat{\theta}(t)\in L_\infty^9$. From the definition of $\gamma_{dj}(t)$ given in (5.18) and the fact that $\ddot{q}_d(t)\in L_\infty$ by assumption, it is now easy to show that $\gamma_{dj}(t)\in L_\infty$; hence, since $\eta_j(t)\in L_\infty$, we can use (5.15) to state that $\gamma_j(t)\in L_\infty$. Using the structure of the voltage control input of (5.66), the above information, the information in Appendix B, and the assumption $\dddot{q}_d\in L_\infty$, we can now show that $V_j(t)\in L_\infty$. The electromechanical dynamics of (5.1) and (5.2) can be used to show that $\ddot{q}(t)\in L_\infty$ and $\dot{\gamma}_j(t)\in L_\infty$. Finally, we can use the above information and (5.58) and (5.70) to illustrate that $\dot{r}(t)\in L_\infty$ and $\dot{\eta}_j(t)\in L_\infty$.

We now illustrate how Corollary 1.1 in Chapter 1 can be used to show that the filtered tracking error goes to zero. First, we integrate both sides of (5.75) with respect to time to yield

$$\int_0^\infty\frac{dV(\sigma)}{d\sigma}\,d\sigma=-\int_0^\infty\left(k_sr^2(\sigma)+\sum_{j=1}^{4}k_j\eta_j^2(\sigma)\right)\,d\sigma. \tag{5.76}$$

If we perform the integration on the left-hand side of (5.76), we obtain

$$\sqrt{V(0) - V(\infty)} = \sqrt{\int_0^{\infty} \left(k_s r^2(\sigma) + \sum_{j=1}^{4} k_j \eta_j^2(\sigma) \right) d\sigma} \tag{5.77}$$

after some minor algebraic manipulation. Since $\dot{V}(t) \leq 0$ as illustrated by (5.75), $V(t)$ of (5.72) is decreasing or constant; hence, $V(0) \geq V(\infty) \geq 0$. We now use this information and (5.77) to obtain the following inequality

$$\begin{aligned} \sqrt{k_s \int_0^{\infty} r^2(\sigma)\, d\sigma} &\leq \sqrt{\int_0^{\infty} \left(k_s r^2(\sigma) + \sum_{j=1}^{4} k_j \eta_j^2(\sigma) \right) d\sigma} \\ &\leq \sqrt{V(0)} < \infty \end{aligned} \tag{5.78}$$

which indicates according to Definition 1.2 in Chapter 1 that $r(t) \in L_2$. Since $r(t) \in L_\infty$, $\dot{r}(t) \in L_\infty$, and $r(t) \in L_2$, we can invoke Corollary 1.1 in Chapter 1 to obtain the result given by (5.71). □

Remark 5.4

From the result given by (5.71) and the proof of Theorem 5.2, we know that $r(t) \in L_\infty$, $r(t) \in L_2$, and $\lim_{t\to\infty} r(t) = 0$; hence, we can use Lemma 1.6 in Chapter 1 to state that $\lim_{t\to\infty} e(t) = 0$ and $\lim_{t\to\infty} \dot{e}(t) = 0$.

5.7 Torque Ripple

Over the last decade, there has been much attention focused on the subject of *torque ripple* in the context of motor control ([27], [14], [15], [20], [16], [17], [18]). We believe that the torque ripple phenomenon is often associated with the SR motor because of the inherent difficulty of using this machine in a closed-loop position tracking control application. That is, the dynamics of other electric machines, such as the brushless DC motor, can be greatly simplified through the use of the rotor fixed reference frame transformation; hence, a closed-loop control strategy is much easier to design.

For purposes of discussion, we can think of torque ripple as an unwanted signal added to the dynamics of the mechanical sub-system actuated by a SR motor. This torque disturbance or ripple can be attributed to three major sources: i) inadequate torque transmission modeling, ii) non-zero values of the current tracking errors, and iii) improper commutation of the motor. To explain this concept, let the mechanical dynamics of a SR motor actuated system be given by

$$M\ddot{q} + B\dot{q} + N\sin(q) = \tau\left(q, \gamma_j\right) \tag{5.79}$$

where q denotes the rotor position, γ_j denotes the per-phase winding current, M, B, N are mechanical subsystem parameters, and $\tau\left(q,\gamma_j\right)$ is the unknown torque transmission model. To develop a controller for the SR motor, we must develop a model for $\tau\left(q,\gamma_j\right)$ to aid in the development of the requisite commutation strategy. If we denote this model by $\tau_m\left(q,\gamma_j\right)$, we can add and subtract $\tau_m\left(q,\gamma_j\right)$ to the right-hand side of (5.79) to yield

$$M\ddot{q}+B\dot{q}+N\sin(q)=\tau_m\left(q,\gamma_j\right)+\tau\left(q,\gamma_j\right)-\tau_m\left(q,\gamma_j\right). \tag{5.80}$$

which can be rewritten as

$$M\ddot{q}+B\dot{q}+N\sin(q)=\tau_m\left(q,\gamma_j\right)+\tau_{me}\left(q,\gamma_j\right) \tag{5.81}$$

where $\tau_{me}\left(q,\gamma_j\right)$ is the torque ripple caused by inexact torque transmission modeling and is explicitly given by

$$\tau_{me}=\tau\left(q,\gamma_j\right)-\tau_m\left(q,\gamma_j\right). \tag{5.82}$$

Remark 5.5

Note that during the development of the controller presented in this chapter, we assumed that τ_{me} was equal to zero. That is, since it is extremely difficult to deal with a unknown term such as τ_{me}, we used the standard linear magnetic circuit model as though it precisely modeled the actual torque transmission term τ. However, we can indeed see from (5.81) that τ_{me} is an unwanted signal added into the mechanical dynamics; hence, this modeling error will be a possible source of torque ripple. In [19] and [20], Chen and Paden used a Fourier series representation to more accurately model the torque transmission term in an effort to reduce the torque ripple contributed by τ_{me}.

Continuing with our investigation of torque ripple sources, we note that every motor controller which has been presented thus far has involved the use of a current tracking error control objective to facilitate the design of the voltage input controller. Hence, we add and subtract the model (*i.e.*, denoted by $\tau_m\left(q,\gamma_{dj}\right)$) for the torque produced by the desired current γ_{dj} to the right-hand side of (5.81) to yield

$$M\ddot{q}+B\dot{q}+N\sin(q)=\tau_m\left(q,\gamma_{dj}\right)-\eta_\tau+\tau_{me} \tag{5.83}$$

where η_τ is the torque ripple caused by the torque/current tracking error and is explicitly given by

$$\eta_\tau=\tau_m\left(q,\gamma_j\right)-\tau_m\left(q,\gamma_{dj}\right). \tag{5.84}$$

Remark 5.6

From (5.83), we can see that η_τ is an unwanted signal added into the mechanical dynamics; hence, the torque tracking error will also be a source of torque ripple. To counter this torque ripple source, the voltage input controller is usually designed to ensure that $\gamma_{dj} - \gamma_j$ goes to zero; hence, one may then be able to use the structure of τ_m to show that the torque tracking error η_τ goes to zero. Note that during the development of the controller presented in this chapter, we use the specific structure of τ_m to reformulate the torque tracking control problem (*e.g.*, which involves forcing η_τ to zero) into four current tracking control problems (*e.g.*, which involves forcing $\gamma_{dj} - \gamma_j$ to zero). Hence, if the controller is theoretically designed to provide exponential or asymptotic current tracking (and hence torque tracking), η_τ will only be a transient source of torque ripple. However, due to implementation constraints (*e.g.*, discretization, quantization error, measurement noise, *etc.*), the current tracking error, and hence the torque tracking error, will never be driven identically to zero; therefore, the current tracking error represents the second possible source of torque ripple.

Concluding with our investigation of torque ripple sources, we can see from (5.83) that the desired current trajectories, denoted by γ_{dj}, are usually designed using the structure of τ_m to deliver the desired torque, denoted by $\tau_d(q, \dot{q})$, to the mechanical subsystem. Hence, we add and subtract $\tau_d(q, \dot{q})$ to the right-hand side of (5.83) to yield

$$M\ddot{q} + B\dot{q} + N\sin(q) = \tau_d(q, \dot{q}) - \tau_{ce} - \eta_\tau + \tau_{me} \tag{5.85}$$

where τ_{ce} is the torque ripple caused by inappropriate commutation and is explicitly given by

$$\tau_{ce} = \tau_m(q, \gamma_{dj}) - \tau_d(q, \dot{q}). \tag{5.86}$$

Note τ_d is usually designed according to the position or velocity control objectives for the second-order mechanical subsystem.

Remark 5.7

From (5.85), we can see that τ_{ce} is an unwanted signal added into the mechanical dynamics; hence, this commutation error will also be a source of torque ripple. To counter this torque ripple source, the commutation strategy is usually designed to ensure that τ_{ce} is small in magnitude. Note that during the development of the controller presented in this chapter, we use the specific structure of τ_m to formulate a differentiable commutation strategy which ensured that τ_{ce} was equal to zero for all time. That is, the desired current trajectory $\gamma_{dj}(q, \tau_d)$ was designed to ensure that

$$\tau_m(q, \gamma_{dj}(q, \tau_d)) = \tau_d(q, \dot{q}); \tag{5.87}$$

hence, $\tau_{ce} = 0$. Depending on the structure of τ_m, one may not be able to find a $\gamma_{dj}(q, \tau_d)$ which satisfies (5.87); therefore, the commutation error τ_{ce} represents the third possible source of torque ripple.

5.8 Experimental Results

The basic hardware setup was described in the experimental section of Chapter 2. To accommodate the 4-phase SR motor, three additional linear amplifiers and current sensors were added to the basic setup. The software was modified to accommodate the new control algorithms and the sampling time was set to $500\mu\,\text{sec}$.

The Magna Physics model 90 four-phase, 6:8 switched reluctance motor (*i.e.*, $N_r = 6$) was used in the experiment. Mounted to the back of the motor is a resolver which is used for position measurements. The resolver signal is converted to a 1024 line encoder signal, through use of the Analog Devices AD2S90, which can then be decoded by the DS-2 I/O board. The model 90 SR motor has a peak torque of $2.64N{\cdot}m$, a rated speed of $600rpm$, a bus voltage of $100Vdc$, and a peak current of $10A$. A single link robotic load is designed as a metal bar attached to the rotor shaft with a cylindrical load mass attached to the free end so that the parameters for the mechanical subsystem dynamics of (5.1) are calculated according to

$$M = \frac{2}{N_r L_1}\left[J + \frac{1}{12}(m_0 + m_1) l_0^2 + (m_0 + m_1) l_{cm}^2 ,\right.$$

$$\left. + \frac{1}{12} m_1 \left(l_1^2 + h^2\right) + m_1 l_0^2\right]$$

$$N = \frac{2}{N_r L_1}\left[\frac{m_0 l_0 G}{2} + m_1 l_0 G\right], \qquad \text{and} \qquad B = \frac{2B_0}{N_r L_1},$$

where J represents the motor rotor inertia, m_0 represents the link mass, m_1 represents the load mass, l_0 represents the link length, l_{cm} represents the length to the center of mass for the link, l_1 and h represent the dimensions of a cylindrical load, G is the coefficient of gravity, and B_0 represents the coefficient of viscous load dampening. The electromechanical system parameters for the motor model in (5.1) through (5.6) were measured using standard test procedures and were determined to be

$$L_0 = 6.52mH, \quad L_1 = 3.42mH, \quad R_0 = 1.1\Omega, \; l_{cm} = 0.095m,$$

$$m_0 = 0.506kg, \quad m_1 = 0.0795kg, \quad l_0 = 0.305m,$$

$$l_1 = 0.0234m, \quad J = 5.0 \times 10^{-4} kg \cdot m^2, \quad h = 0.013m,$$

$$B_0 = 0.001N \cdot m \cdot \sec/rad, \text{ and } \quad G = 9.81kg \cdot m/\sec^2 .$$

The velocity information required by the nonlinear controllers was obtained through a backwards difference of the position signal. The velocity signal obtained in this manner was then filtered using a second-order, low-pass filter in order to remove the high frequency content that results from the backwards difference algorithm.

The desired position trajectory was selected as

$$q_d(t) = \frac{\pi}{2}\sin(2t)\left(1 - \underline{e}^{-0.3t^3}\right)\ rad$$

which has the desirable property $q_d(0) = \dot{q}_d(0) = \ddot{q}_d(0) = \dddot{q}_d(0) = 0$. The above desired position trajectory is shown in Figure 5.3.

The exact model knowledge controller and the adaptive controller were implemented on the SR motor for tracking the above desired position trajectory.

5.8.1 Exact Model Knowledge Control Experiment

The above electromechanical parameter values were used for the exact model knowledge controller of (5.30) and (5.39). The best tracking performance was found using the following control gains

$$\alpha = 160, \quad k_s = 9, \quad \gamma_{co} = 0.05, \ \epsilon_0 = 0.01,$$

$$\text{and} \quad k_1 = k_2 = k_3 = k_4 = 1.$$

The resulting position tracking error is given in Figure 5.4 which shows that the maximum position tracking error is approximately within ± 0.35 degrees of zero. The corresponding motor per phase voltage control inputs are shown in Figure 5.5 through Figure 5.8, respectively.

5.8.2 Adaptive Control Experiment

The adaptive controller of (5.54), (5.55), (5.66), and (5.67) was also implemented. The initial conditions for the parametric estimates were set to 50% of their nominal values given above. The best position tracking performance was achieved with the following controller gain values

$$\alpha = 65, \ k_s = 0.8, \ k_1 = k_2 = k_3 = k_4 = 2.9, \ \gamma_{co} = 0.1, \ \epsilon_0 = 0.01$$

and

$$\Gamma_\tau = diag\{0.25, 3, 2\},$$

$$\Gamma_e = diag\left\{10^{-6}, 10^{-4}, 10^{-4}, 10^{-5}, 10^{-5}, 10^{-5}, 10^{-5}, 10^{-5}, 0.8\right\}.$$

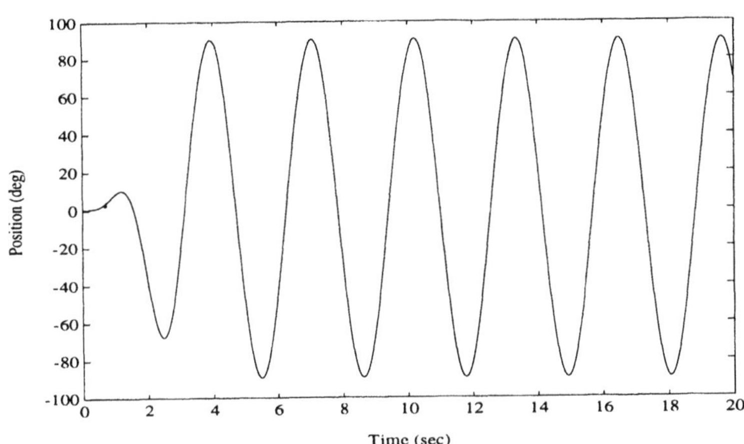

Figure 5.3: Desired Position Trajectory

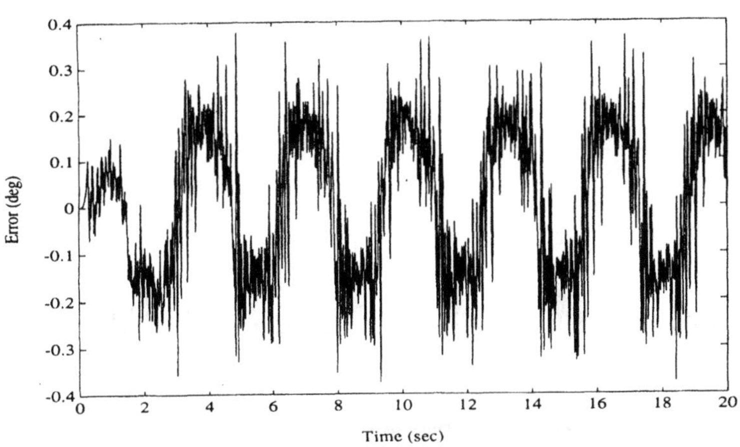

Figure 5.4: Position Tracking Error for Exact Knowledge Controller

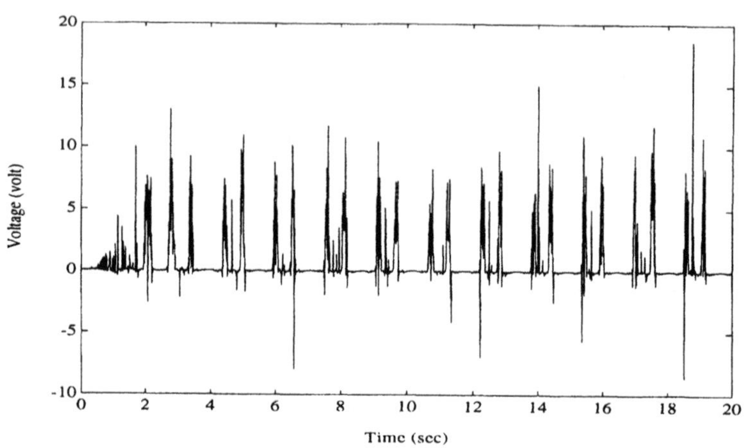

Figure 5.5: Phase One Voltage for Exact Knowledge Controller

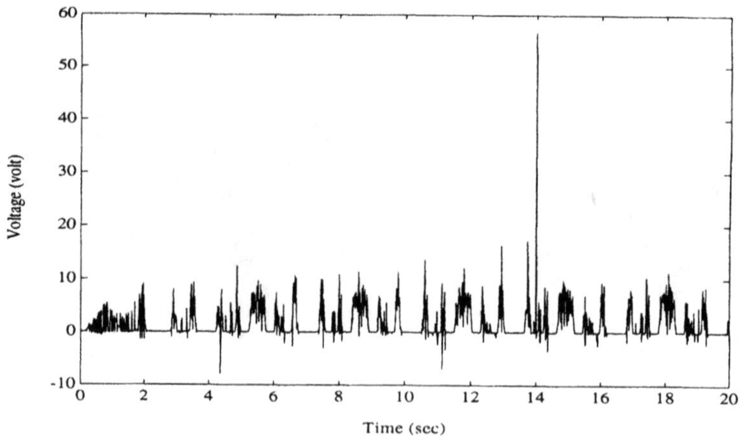

Figure 5.6: Phase Two Voltage for Exact Knowledge Controller

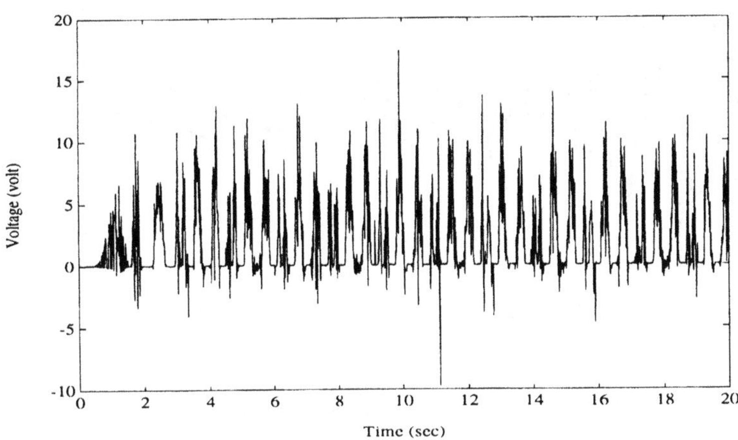

Figure 5.7: Phase Three Voltage for Exact Knowledge Controller

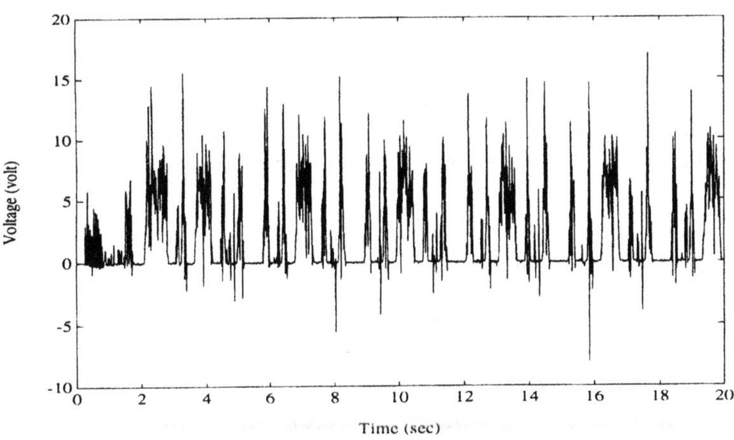

Figure 5.8: Phase Four Voltage for Exact Knowledge Controller

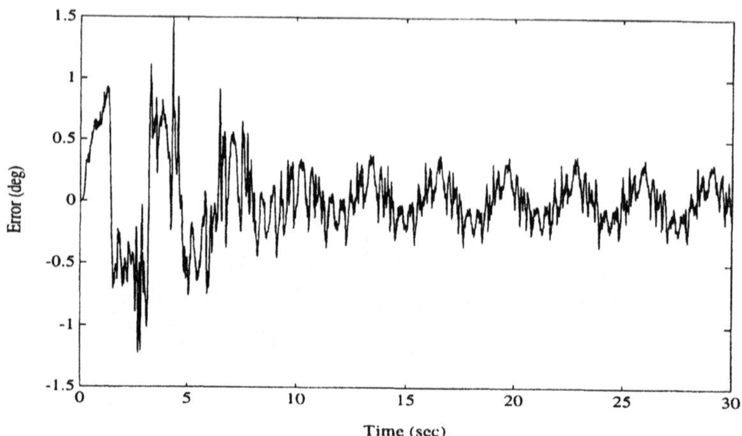

Figure 5.9: Position Tracking Error for Adaptive Controller

The adaptation update laws were implemented using a standard trapezoid-type integration rule. The resulting position tracking error is shown in Figure 5.9 and indicates that the maximum position tracking error finally settles to within ± 0.35 degrees of zero. Note that due to measurement noise, quantization error in applying the control, and resolution of the encoder the tracking error does not approach zero as predicted by the theory but rather is driven to a small value. The per phase voltages are plotted in Figure 5.10 through Figure 5.13.

5.8.3 Linear Control Experiment

For comparison purposes, we also implemented a proportional-derivative outer loop controller in conjunction with a current feedback inner loop controller. That is, the input control voltages are designed as

$$V_j = k_j \eta_j$$

and the desired torque trajectory is taken to be

$$\tau_d = k_v \dot{e} + k_p e.$$

The best performance was found by using the following control gains

$$k_p = 4800, \quad k_v = 32, \quad \gamma_{co} = 0.01, \quad \epsilon_0 = 0.01,$$

$$\text{and} \quad k_1 = k_2 = k_3 = k_4 = 4.5.$$

The resulting position tracking error is given in Figure 5.14 which indicates that the best steady state tracking error is approximately within ± 0.6

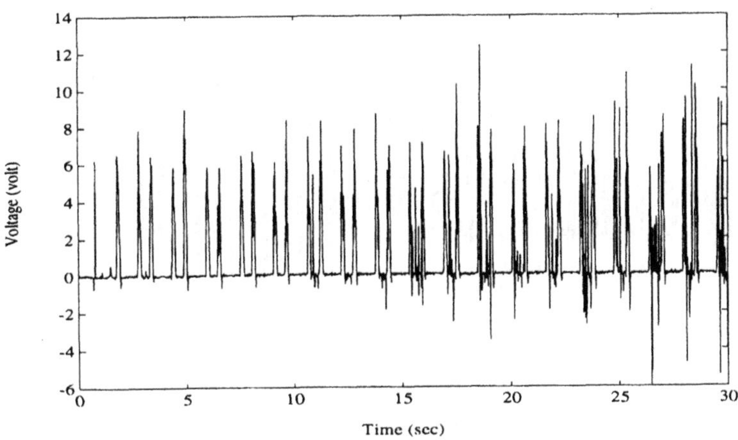

Figure 5.10: Phase One Voltage for Adaptive Controller

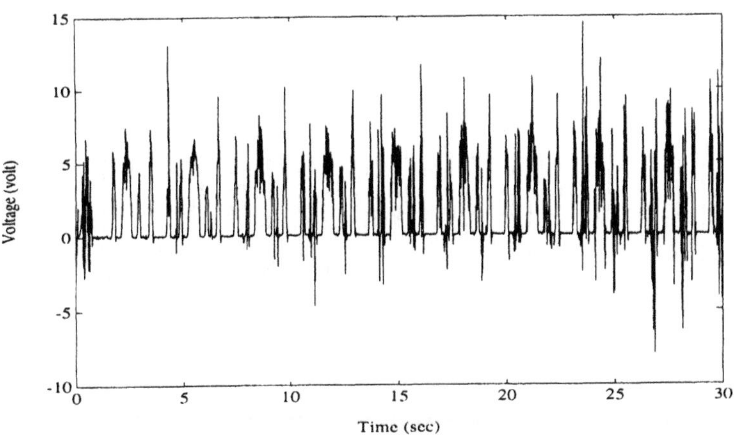

Figure 5.11: Phase Two Voltage for Adaptive Controller

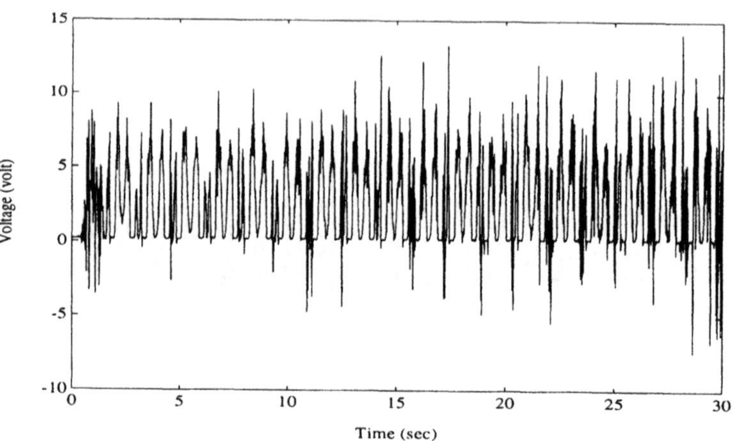

Figure 5.12: Phase Three Voltage for Adaptive Controller

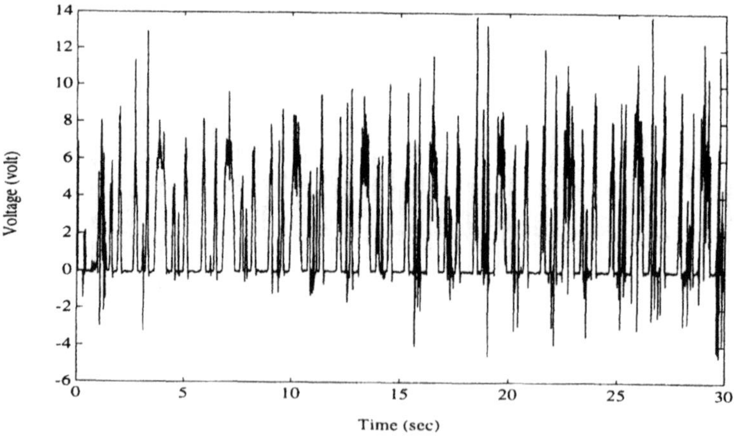

Figure 5.13: Phase Four Voltage for Adaptive Controller

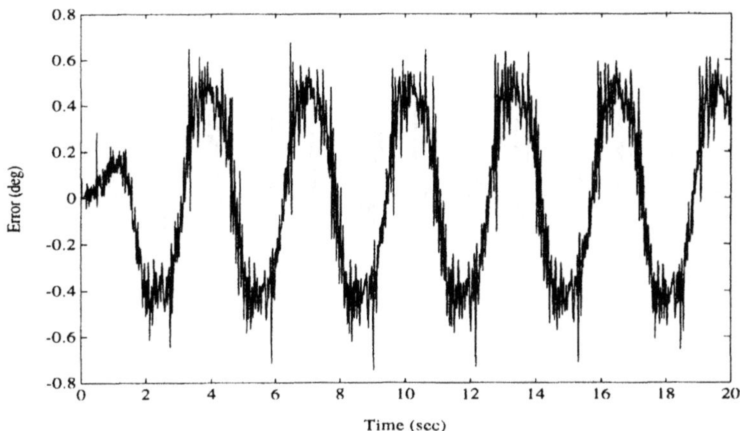

Figure 5.14: Position Tracking Error for a Linear Controller

degrees of zero. For brevity, only the phase one voltage is given in Figure 5.15 for the linear control experiment.

Remark 5.8

From the above experimental results, we can see that, with regard to the steady-state tracking error, the adaptive controller outperformed the linear controller by approximately two times and is about the same as the exact model knowledge controller.

5.9 Notes

A good deal of current literature relates the development of nonlinear controllers for switch reluctance motors. For example, in [7], Ilic'-Spong *et al.* introduced a detailed nonlinear model and an electronic commutation strategy for the SR motor and applied a state feedback control algorithm which compensates for the nonlinearities in the system. Given the assumption of constant rotor velocity, Ilic'-Spong *et al.* [8] also introduced an instantaneous torque control for a SR motor driving an inertial load. The work in [7] was then generalized to a direct-drive manipulator with SR actuation in [9]. A composite control based on a singularly perturbed model of a n-link direct-drive manipulator with SR actuator dynamics was then developed in [1]. A similar strategy using reduced-order feedback linearization techniques is presented by Taylor in [2] to reduce torque ripple in the actuation of an experimental load. A torque ripple study of SR motors was pursued in

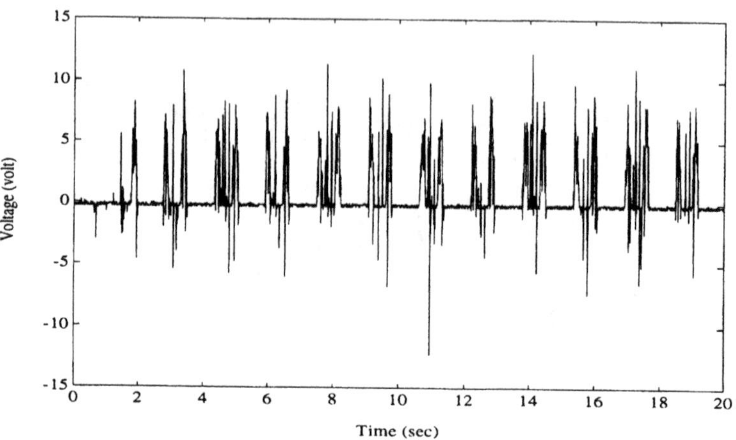

Figure 5.15: Phase One Voltage for a Linear Controller

[27] where Wallace *et al.* investigated the performance of four related motors using finite-element analysis with a commutation algorithm for torque control. Wallace *et al.* also presented a new method of computing the reference currents utilizing a balanced commutator for current tracking feedback control in [14]. In [25], Carroll *et al.* use a backstepping technique to develop a singularity free adaptive trajectory tracking controller for the SR motor driving an inertial load. The controller is capable of compensating for parametric uncertainty throughout the electromechanical dynamics, given full state feedback. In [4], Taylor uses a certainty equivalence argument to develop an adaptive feedback linearizing control for a single-link robot with a SR actuator. This control yields asymptotic link tracking despite uncertainty in the mechanical parameters and selected electrical parameters. In [21], Amor utilizes an instantaneous torque measurement method based on flux observations to develop an adaptive feedback linearizing torque controller for a three-phase SR motor. In [12], Carroll *et al.* designed a robust controller for a n-link robotic manipulator actuated by SR motors. This controller was shown to be robust to parametric uncertainty and additive bounded disturbances while delivering a global uniform ultimately bounded stability result for the link position tracking error. In [24], Kohan *et al* developed an adaptive control for SR motors which utilized B-spline functions to model the torque transmission relationship. In [22], Kim *et al* designed a commutation strategy for a sophisticated model of the SR motor. In [23], Vedagarbha *et al* utilized a general, nonlinear model of the switched reluctance (SR) motor to design an adaptive controller which accounted for magnetic saturation associated with a proposed flux linkage model.

Bibliography

[1] D. G. Taylor, "Composite Control of Direct-Drive Robots", *Proc. 28th IEEE Conference on Decision and Control*, pp. 1670-1675, Dec. 1989.

[2] D. G. Taylor, "An Experimental Study on Composite Control of Switched Reluctance Motors", *IEEE Control Systems Magazine*, pp. 31-36, Feb. 1991.

[3] M. Spong and M. Vidyasagar, *Robot Dynamics and Control*, New York: John Wiley and Sons, Inc., 1989.

[4] D. G. Taylor, "Adaptive Control Design for a Class of Doubly-Salient Motors", *Proc. 30th IEEE Conference on Decision and Control*, pp. 2903-2908, Dec. 1991.

[5] J. Slotine and W. Li, *Applied Nonlinear Control*, Englewood Cliffs, NJ: Prentice Hall Co., 1991.

[6] S. Sastry and M. Bodson, *Adaptive Control: Stability, Convergence, and Robustness*, Englewood Cliffs, NJ: Prentice Hall Co., 1989.

[7] M. Ilic'-Spong, R. Marino, S. M. Peresada, and D. G. Taylor, "Feedback Linearizing Control of Switched Reluctance Motors", *IEEE Transactions on Automatic Control*, Vol. AC-32, No. 5, pp. 371-379, May 1987.

[8] M. Ilic'-Spong, T. J. Miller, S. R. Macminn, and J. S. Thorp, "Instantaneous Torque Control of Electric Motor Drives", *IEEE Transactions on Power Electronics*, Vol. PE-2, No. 1, pp. 55-61, Jan. 1987.

[9] D. G. Taylor, M. Ilic'-Spong, R. Marino, and S. Peresada, "A Feedback Linearizing Control for Direct-Drive Robots with Switched Reluctance Motors", *Proc. 25th IEEE Conference on Decision and Control*, pp. 388-396, Dec. 1986.

[10] G. S. Buja and M. I. Valla, "Control Characteristics of the SRM Drives-Part I: Operation in the Linear Region", *IEEE Transactions on Industrial Electronics*, Vol. 38, No. 5, pp. 313-321, Oct. 1991.

[11] I. Kanellakopoulos, P. V. Kokotovic, and A. S. Morse, "Systematic Design of Adaptive Controllers for Feedback Linearizable Systems", *IEEE Transactions on Automatic Control*, Vol. 36, pp. 1241-1253, 1991.

[12] J. J. Carroll, D. M. Dawson, Z. Qu, and M. D. Leviner, "Robust Tracking Control of Rigid-Link Electrically Driven Robot Robots Actuated by Switched Reluctance Motors", *Proc. ASME 1992 Annual Winter Meeting*, DSC-Vol. 43, pp. 97-103, March 1992.

[13] J. J. Carroll, *Nonlinear Control of Electric Machines*, Ph.D. Dissertation, Clemson University, August 1993.

[14] R. S. Wallace and D. G. Taylor, "A Balanced Commutator for Switched Reluctance Motors to Reduce Torque Ripple", *IEEE Transactions on Power Electronics*, Vol. 7, No. 4, pp. 617-626, Oct. 1992.

[15] H. Nagase, T. Okuyama, J. Takahashi, and K. Saitoh, "A Method for Suppressing torque Ripple of AC Motor by Current Amplitude Control", *IEEE Transactions on Industrial Electronics*, Vol. 36, No. 4, pp. 504-510, 1989.

[16] H. Le-Huy, R. Perret, and R. Feuillet, "Minimization of Torque Ripple in Brushless DC Motor Drives", *IEEE Transactions on Industry Applications*, Vol. IA-22, No. 4, pp. 748-755, July/August, 1986.

[17] P. Pillay, R. Krishnan, "Application Characteristics of Permanent Magnet Synchronous and Brushless DC Motors for Servo Drives", *IEEE Transactions on Industry Applications,* Vol. 27, No. 5, pp. 986-996, September/October, 1991.

[18] T. S. Low, T. H. Lee, K. J. Tseng, and K. S. Lock, "Servo Performance of a Brushless Drive with Instantaneous Torque Control", *IEEE Transactions on Industry Applications,* Vol. 28, No. 2, pp. 455-462, March/April, 1992.

[19] D. Chen and B. Paden, "Nonlinear Adaptive Torque Ripple Cancellation for Step Motors", *Proc. of the IEEE Conference on Decision and Control*, Honolulu, HI, pp. 3319-3324, Dec. 1990.

[20] D. Chen and B. Paden, "Adaptive Linearization of Hybrid Step Motors: Stability Analysis", *IEEE Transactions on Automatic Control*, Vol. 38, No. 6, pp. 874-887, June 1993.

[21] L. B. Amor, O. Akhrif, L. A. Dessaint, and G. Oliver, "Adaptive Nonlinear Torque Control of a Switched Reluctance Motor", *Proc. American Control Conference*, pp. 2831-2836, June 1993.

[22] C. Kim, I. Ha, H. Huh, and M. Ko, "A New Approach to Feedback Linearizing Control of Variable Reluctance Motors for Direct-Drive Applications", *Proc. of the IECON*, 1994.

[23] P. Vedagarbha, D. M. Dawson, and W. Rhodes, "An Adaptive Control for General Class of Switch Reluctance Motor Models", *Automatica*, October, 1997, to appear.

[24] R. Kohan and S. Bortoff, "Adaptive Control of Variable Reluctance Motors Using Spline Functions, *IEEE Conference on Decision and Control*, Lake Buena Vista, FL, pp. 1694-1699, Dec., 1994.

[25] J. J. Carroll, D. M. Dawson, and M. D. Leviner, "Adaptive Tracking Control of a Switched Reluctance Motor Turning an Inertial Load", *Proc. American Control Conference*, pp. 670-674, June 1993.

[26] M. Krstic, I. Kanellakopoulos, and P. Kokotovic, *Nonlinear and Adaptive Control Design*, John Wiley & Sons, 1995.

[27] R. S. Wallace and D. G. Taylor, "Low-Torque-Ripple Switched Reluctance Motors for Direct-Drive Robotics", *IEEE Transactions on Robotics and Automation*, Vol. 7, No. 6, pp. 733-742, Dec. 1991.

Chapter 6

Induction Motor (FSFB)

6.1 Introduction

Since the induction motor is one of the most widely used actuator for industrial applications due to its reliability, ruggedness and relatively low cost, the use of this motor in position/velocity tracking applications is expected to increase in the near future. While the induction motor complicates the control problem by coupling multi-input nonlinear dynamics into the overall electromechanical system dynamics, it is possible to develop an accurate dynamic model. Unfortunately, as far as control design is concerned, the stator fixed transformed induction motor/load electromechanical dynamics exhibit: i) bilinear terms, which are products of the stator electrical current and the rotor flux, in the static torque transmission equation, and ii) bilinear terms, which are products of the rotor velocity and the stator electrical current (or the rotor flux), in the electrical subsystem dynamics. In addition, in contrast with all of the previous electric machine models, the induction motor possesses an additional set of rotor flux dynamic equations which must be accounted for during the control design procedure.

In this chapter, we apply the adaptive integrator backstepping tools [25] to the design of a position tracking controller for the induction motor driving a mechanical load. Since the motor is typically a three phase machine, we utilize the standard, stator fixed transformation to substantially simplify the nonlinear electromechanical dynamics. In spite of the simplification rendered by the transformation, the static torque transmission equation consists of bilinear expressions in terms of the transformed stator current and transformed rotor flux; hence, a method for delivering the required torque to the mechanical system must be devised. In contrast with the commutation strategies developed for the PMS, the BLDC, and the SR motor given in the previous chapters, the unique dynamic characteristics of the induction motor demand a different control solution. That is, due to the additional set of dynamic equations (*i.e.*, the rotor flux electrical dynamics), we do not separate the torque tracking objective into several, distinct electrical state variable tracking problems. That is, since there are two stator current variables and two rotor flux variables, it is not obvious how to use the two voltage control inputs to force these four independent variables to track four independent signals.

To facilitate the design of the controller, we first view the motor as a torque source and thus design a desired torque signal to ensure that the load follows the desired position trajectory. We then develop one voltage input control relationship[1] to force the entire torque transmission term to track the desired torque. That is, the electrical dynamics are taken into account through the torque tracking objective, and hence the position tracking control objective is embedded inside the torque tracking objective. Therefore, if the voltage input control relationship can be designed to guarantee that the actual torque tracks the desired torque signal, then the load position will follow the desired position trajectory. In addition, to ensure that all of the electromechanical state variables remain bounded during closed-loop operation, we formulate a desired rotor flux tracking control objective. Roughly speaking, this secondary control objective requires the magnitude of the rotor flux to track a positive function; hence, the second voltage input control relationship is designed to guarantee that the rotor flux tracking goes to zero.

6.2 System Model

The reader is referred to [1] [2] for the general theory of induction motor operation and related motor control problems. We summarize the dynamic behavior of a balanced three-phase induction motor in the electromechanical model given below. Under the assumption of a linear magnetic circuit, the dynamics of an induction motor driving a position-dependent load in a stator-fixed reference frame are given by (see Appendix C)

$$M\ddot{q} + B\dot{q} + N\sin(q) = \psi_a I_b - \psi_b I_a, \tag{6.1}$$

$$L_I \dot{I}_a = -R_I I_a + \alpha_1 \psi_a + \alpha_2 \psi_b \dot{q} + V_a, \tag{6.2}$$

$$L_I \dot{I}_b = -R_I I_b + \alpha_1 \psi_b - \alpha_2 \psi_a \dot{q} + V_b, \tag{6.3}$$

$$L_r \dot{\psi}_a = -R_r \psi_a - \alpha_3 \dot{q} \psi_b + K_I I_a, \tag{6.4}$$

and

$$L_r \dot{\psi}_b = -R_r \psi_b + \alpha_3 \dot{q} \psi_a + K_I I_b, \tag{6.5}$$

where $q(t)$, $\dot{q}(t)$, and $\ddot{q}(t)$ represent the load position, velocity, and acceleration, respectively, $I_a(t)$, $I_b(t)$ represent the transformed stator current,

[1] The terminology voltage input control relationship is used to highlight the fact that the control input is a nonlinear combination of the transformed stator input voltages.

$\psi_a(t)$, $\psi_b(t)$ denote the transformed rotor flux, and $V_a(t)$, $V_b(t)$ represent the transformed stator input voltage. The positive constants L_I, R_I, K_I, α_1, α_2, and α_3 in the electrical subsystems are related to the electrical parameters by the following relationships

$$L_I = L_s - M_e^2/L_r, \quad K_I = R_r M_e, \quad R_I = \left(M_e^2 R_r + L_r^2 R_s\right)/L_r^2,$$

$$\alpha_1 = M_e R_r / L_r^2, \quad \alpha_2 = n_p M_e / L_r, \quad \text{and} \quad \alpha_3 = n_p L_r.$$

In the mechanical subsystem, M is a positive constant related to the mechanical inertia of the system (including rotor inertia), N is a positive constant related to the mass of the load and the coefficient of gravity, and B is the positive coefficient of viscous friction. In the electrical subsystem, R_s, R_r, n_p, L_s, L_r, and M_e are positive constants representing the stator resistance, rotor resistance, number of pole pairs, stator inductance, rotor inductance, and mutual inductance, respectively. Note that the parameters M, B, and N described in (6.1) are defined to include the effects of the torque coefficient constant α_2. That is, equation (6.1) has been divided by the constant α_2.

Remark 6.1

The detailed development for the above stator-fixed reference frame transformed model is presented in Appendix C. The mechanical load specified in (6.1) is a one-link robot [3] which is directly coupled to the induction motor. We selected a robotic type load based on the belief that there is an increasing possibility of utilizing induction motors for robotic applications.

6.3 Control Objective

Given full state measurements (*i.e.*, $q(t)$, $\dot{q}(t)$, $I_a(t)$, $I_b(t)$, $\psi_a(t)$, and $\psi_b(t)$), the control objective is to develop load position tracking controllers for the electromechanical dynamics given by (6.1) through (6.5).

6.3.1 Position/Velocity Tracking Objective

To begin, we define the load position tracking error, $e(t)$, to be

$$e = q_d - q \tag{6.6}$$

where $q_d(t)$ represents the desired load position. We will assume that $q_d(t)$ and its first, second, and third derivatives are all bounded functions of time. In addition, we define the filtered tracking error, $r(t)$, [4] to be

$$r = \dot{e} + \alpha e \tag{6.7}$$

where α is a positive scalar constant. Note that the definition of the filtered tracking error allows us to analyze the second-order mechanical subsystem dynamics of (6.1) as a first-order system and thus simplifies the controller development. The overall tracking control objective will be met if $r(t)$, and hence $e(t)$, is driven to zero.

To form the open-loop filtered tracking error dynamics, we differentiate (6.7) with respect to time and rearrange terms to yield

$$\dot{r} = (\ddot{q}_d + \alpha\dot{e}) - \ddot{q}. \tag{6.8}$$

Multiplying (6.8) by M and substituting the mechanical subsystem dynamics of (6.1) yields the filtered tracking error dynamics as shown

$$M\dot{r} = M(\ddot{q}_d + \alpha\dot{e}) + B\dot{q} + N\sin(q) - (\psi_a I_b - \psi_b I_a). \tag{6.9}$$

To reduce the algebraic overhead, the right-hand side of (6.9) can be rewritten as

$$M\dot{r} = W_\tau \theta_\tau - (\psi_a I_b - \psi_b I_a) \tag{6.10}$$

where the known regression matrix [5] $W_\tau(q, \dot{q}, t) \in \Re^{1\times 3}$ is given by

$$W_\tau = \begin{bmatrix} \ddot{q}_d + \alpha\dot{e} & \dot{q} & \sin(q) \end{bmatrix} \tag{6.11}$$

and the parameter vector $\theta_\tau \in \Re^3$ is given by

$$\theta_\tau = \begin{bmatrix} M & B & N \end{bmatrix}^T. \tag{6.12}$$

Due to the structure of the electromechanical system given by (6.1) through (6.5), we are only free to specify the input motor voltages $V_a(t)$ and $V_b(t)$. In other words, the mechanical subsystem error dynamics lack a true torque level control input. For this reason, we shall add and subtract a desired torque signal, $\tau_d(t)$, to the right-hand side of (6.10) to yield

$$M\dot{r} = W_\tau \theta_\tau - \tau_d + \eta_\tau \tag{6.13}$$

where $\eta_\tau(t)$ is used to represent the torque tracking error and is explicitly defined by

$$\eta_\tau = \tau_d - (\psi_a I_b - \psi_b I_a). \tag{6.14}$$

If the torque tracking error term $\eta_\tau(t)$ in (6.13) were equal to zero, then $\tau_d(t)$ could be designed to achieve good load position tracking utilizing standard control techniques (*e.g.*, [7]). Since the torque tracking error is not equal to zero in general, we must design the voltage control inputs to compensate for the effects of $\eta_\tau(t)$ in (6.13). To accomplish this additional control objective, the dynamics of the torque tracking error are needed. Taking the time derivative of the torque tracking error $\eta_\tau(t)$ in (6.14) and then multiplying the result by L_I yields

$$L_I\dot{\eta}_\tau = L_I\dot{\tau}_d - L_I\left(\dot{\psi}_a I_b + \psi_a \dot{I}_b\right) + L_I\left(\dot{\psi}_b I_a + \psi_b \dot{I}_a\right). \tag{6.15}$$

Substituting the right-hand sides of (6.2) through (6.5) into (6.15) results in the open-loop torque tracking error dynamics as shown

$$\begin{aligned} L_I\dot{\eta}_\tau = {} & L_I\dot{\tau}_d - L_I L_r^{-1} I_b\left(-R_r\psi_a - \alpha_3\dot{q}\psi_b\right) \\ & + L_I L_r^{-1} I_a\left(-R_r\psi_b + \alpha_3\dot{q}\psi_a\right) - \psi_a\left(-R_I I_b - \alpha_2\psi_a\dot{q}\right) \\ & + \psi_b\left(-R_I I_a + \alpha_2\psi_b\dot{q}\right) + \psi_b V_a - \psi_a V_b. \end{aligned} \quad (6.16)$$

We can see that the voltage control inputs (*i.e.*, $V_a(t)$ and $V_b(t)$) have appeared on the right-hand side of (6.16). Later, we will design the voltage control inputs to force the torque tracking error, $\eta_\tau(t)$, to zero.

6.3.2 Flux Tracking Objective

An important subtask in meeting the position control objective is to ensure that all of the system signals stay bounded during closed-loop operation. Typically the boundness of all system signals can be ensured based on a stability analysis that includes the filtered tracking error and the current/torque tracking error. However, due to the special nature of the motor dynamics in (6.1) through (6.5), this standard approach allows for the possibility of unbounded rotor fluxes. As a means of preventing unbounded rotor fluxes, the rotor fluxes are forced to track a bounded signal. The closeness of the actual flux to this bounded signal is then incorporated into the stability analysis. To quantify this flux tracking objective, we define the flux tracking error, $\eta_\psi(t)$, as

$$\eta_\psi = \psi_d - \frac{1}{2}(\psi_a^2 + \psi_b^2) \quad (6.17)$$

where $\psi_d(t)$ is the "pseudo-magnitude" of the desired flux [6] which is assumed to be second-order differentiable with respect to time. Differentiating (6.17) with respect to time and multiplying the resulting expression by L_r yields

$$L_r\dot{\eta}_\psi = L_r\dot{\psi}_d - L_r(\psi_a\dot{\psi}_a + \psi_b\dot{\psi}_b). \quad (6.18)$$

Substituting for $L_r\dot{\psi}_a$ and $L_r\dot{\psi}_b$ from (6.4) and (6.5) into (6.18), respectively, the dynamics of the flux tracking error can be written in the following form

$$L_r\dot{\eta}_\psi = L_r\dot{\psi}_d + R_r\left(\psi_a^2 + \psi_b^2\right) - K_I\left(\psi_a I_a + \psi_b I_b\right). \quad (6.19)$$

To facilitate the stability analysis, we divide both sides of equation (6.19) by K_I to obtain

$$\bar{L}_r\dot{\eta}_\psi = Y_\psi\theta_\psi - \left(\psi_a I_a + \psi_b I_b\right) \quad (6.20)$$

where $Y_\psi\left(\psi_a, \psi_b, t\right) \in \Re^{1\times 2}$ and $\theta_\psi \in \Re^{2\times 1}$ are given by

$$Y_\psi = [\dot{\psi}_d \;\; \psi_a^2 + \psi_b^2] \quad \text{and} \quad \theta_\psi = [\bar{L}_r \;\; \bar{R}_r]^T \quad (6.21)$$

with

$$\bar{L}_r = \frac{L_r}{K_I} \quad \text{and} \quad \bar{R}_r = \frac{R_r}{K_I}.$$

From (6.20), we can see that the flux tracking error dynamics lack a control input which could be used to drive $\eta_\psi(t)$ to zero. Similar to the position tracking objective, we shall use the integrator backstepping approach to add and subtract a fictitious flux controller, $u_I(t)$, in (6.20) to obtain

$$\bar{L}_r \dot{\eta}_\psi = Y_\psi \theta_\psi - u_I + \eta_I \tag{6.22}$$

where the auxiliary tracking variable $\eta_I(t)$ is defined by

$$\eta_I = u_I - (\psi_a I_a + \psi_b I_b). \tag{6.23}$$

From (6.22), we can see that if the auxiliary variable $\eta_I(t)$ were zero, then the fictitious flux controller $u_I(t)$ could be easily designed to force $\eta_\psi(t)$ to zero. In order to ensure that $\eta_I(t)$ goes to zero, we will construct the open-loop dynamics for $\eta_I(t)$ and thereby introduce the voltage control inputs into the flux tracking control problem. The open-loop dynamics for $\eta_I(t)$ can be obtained by taking the time derivative of (6.23) and multiplying both sides of the resulting equation by L_I to yield

$$L_I \dot{\eta}_I = L_I \dot{u}_I - L_I \left(\dot{\psi}_a I_a + \dot{\psi}_b I_b \right) - L_I \left(\psi_a \dot{I}_a + \psi_b \dot{I}_b \right). \tag{6.24}$$

Substituting the right-hand sides of (6.2) through (6.5) for $L_I \dot{I}_a$, $L_I \dot{I}_b$, $\dot{\psi}_a$, and $\dot{\psi}_b$, respectively, into (6.24) yields

$$\begin{aligned} L_I \dot{\eta}_I \; = \; & L_I \dot{u}_I - L_I L_r^{-1} I_a \left(-R_r \psi_a - \alpha_3 \dot{q} \psi_b + K_I I_a \right) \\ & - L_I L_r^{-1} I_b \left(-R_r \psi_b + \alpha_3 \dot{q} \psi_a + K_I I_b \right) - \psi_a \left(-R_I I_a + \alpha_1 \psi_a \right) \\ & - \psi_b \left(-R_I I_b + \alpha_1 \psi_b \right) - \psi_a V_a - \psi_b V_b. \end{aligned} \tag{6.25}$$

We now see that the voltage control inputs have appeared in the right-hand side of (6.25). Later, we will design the voltage control inputs $V_a(t)$ and $V_b(t)$ to force $\eta_I(t)$ to zero.

Remark 6.2

It is important to note that we have two tracking objectives that must be met, namely, we must drive $\eta_I(t)$ in (6.23) and $\eta_\tau(t)$ in (6.14) to zero. Since $V_a(t)$ and $V_b(t)$ both appear in the dynamics of $\eta_\tau(t)$ and $\eta_I(t)$, it is not possible to independently assign one voltage control input to achieve one tracking objective and the second control input to achieve the other tracking objective. However, we can use the dynamic equation of (6.16) to design the quantity $\psi_b V_a - \psi_a V_b$ such that $\eta_\tau(t)$ is driven to zero. Likewise, we can use the dynamic equation of (6.25) to design the quantity $\psi_a V_a + \psi_b V_b$ such that $\eta_I(t)$ is driven to zero. After these two design procedures are completed, we can then solve for the transformed voltages $V_a(t)$ and $V_b(t)$.

6.4 Exact Model Knowledge Controller

Based on exact model knowledge of the system and full state feedback, we now design a position tracking controller for the open-loop dynamics of (6.13), (6.16), (6.22), and (6.25). To facilitate the following stability analysis, the closed-loop electromechanical error systems will also be formulated. As mentioned before, we first design the desired torque signal to force the load along the desired position trajectory. Based on the structure of (6.13) and the subsequent stability analysis, we specify $\tau_d(t)$ as

$$\tau_d = W_\tau \theta_\tau + k_s r \tag{6.26}$$

where $W_\tau(\cdot)$ and θ_τ were previously defined in (6.11) and (6.12), respectively, and k_s is a positive constant control gain. Substituting (6.26) into the open-loop dynamics of (6.13) yields the closed-loop filtered tracking error dynamics as shown

$$M\dot{r} = -k_s r + \eta_\tau . \tag{6.27}$$

From (6.27), we can see that if the torque tracking error $\eta_\tau(t)$ was equal to zero, then the filtered tracking error $r(t)$ will go to zero exponentially fast. However, in general, this is not true; hence, we must design the voltage control inputs to ensure that $\eta_\tau(t)$ converges to zero. To this end, we need to complete the open-loop system description for the dynamics of $\eta_\tau(t)$. Calculating the term $\dot{\tau}_d(t)$ in (6.16) by taking the time derivative of $\tau_d(t)$ in (6.26) yields

$$\dot{\tau}_d = \dot{W}_\tau \theta_\tau + k_s \dot{r}. \tag{6.28}$$

After substituting the time derivative of (6.11) and (6.8) into the right-hand side of (6.28) yields

$$\dot{\tau}_d = M\left(\dddot{q}_d + \alpha\left(\ddot{q}_d - \ddot{q}\right)\right) + B\ddot{q} + N\dot{q}\cos(q) + k_s\left(\ddot{q}_d - \ddot{q} + \alpha\dot{e}\right) \tag{6.29}$$

which contains measurable states (*i.e.*, $q(t)$ and $\dot{q}(t)$), known functions, known constant parameters, and the unmeasurable load acceleration $\ddot{q}(t)$. From (6.1), we can solve for $\ddot{q}(t)$ in the following form

$$\ddot{q} = -\frac{B}{M}\dot{q} - \frac{N}{M}\sin(q) + \frac{1}{M}\left(\psi_a I_b - \psi_b I_a\right). \tag{6.30}$$

After substituting for $\ddot{q}(t)$ from the right-hand side of (6.30) into (6.29), we can write $\dot{\tau}_d(t)$ in terms of measurable states (*i.e.*, $q(t)$, $\dot{q}(t)$, $I_a(t)$, $I_b(t)$, $\psi_a(t)$, and $\psi_b(t)$), known functions, and known constant parameters in the following form

$$\begin{aligned}\dot{\tau}_d = \; & M\left(\dddot{q}_d + \alpha\ddot{q}_d\right) + N\dot{q}\cos(q) + k_s\left(\ddot{q}_d + \alpha\dot{e}\right) \\ & + (B - M\alpha - k_s)\left(-\frac{B}{M}\dot{q} - \frac{N}{M}\sin(q)\right. \\ & \left. + \frac{1}{M}\left(\psi_a I_b - \psi_b I_a\right)\right).\end{aligned} \tag{6.31}$$

After substituting the expression for $\dot{\tau}_d(t)$ of (6.31) into (6.16) yields the final open-loop dynamics for the torque tracking error $\eta_\tau(t)$ in the form

$$L_I\dot{\eta}_\tau = w_a - (\psi_a V_b - \psi_b V_a) \tag{6.32}$$

where the auxiliary scalar variable $w_a(q, \dot{q}, I_a, I_b, \psi_a, \psi_b, t)$ is given by

$$\begin{aligned}w_a = \; & L_I\left(M\left(\dddot{q}_d + \alpha\ddot{q}_d\right) + N\dot{q}\cos(q) + k_s\left(\ddot{q}_d + \alpha\dot{e}\right)\right) \\ & + L_I(B - M\alpha - k_s)\left(-\frac{B}{M}\dot{q} - \frac{N}{M}\sin(q)\right. \\ & \left. + \frac{1}{M}\left(\psi_a I_b - \psi_b I_a\right)\right) - L_I L_r^{-1} I_b\left(-R_r\psi_a - \alpha_3\dot{q}\psi_b\right) \\ & + L_I L_r^{-1} I_a\left(-R_r\psi_b + \alpha_3\dot{q}\psi_a\right) \\ & - \psi_a\left(-R_I I_b - \alpha_2\psi_a\dot{q}\right) + \psi_b\left(-R_I I_a + \alpha_2\psi_b\dot{q}\right).\end{aligned} \tag{6.33}$$

From (6.32), we can easily design a voltage level control input to force the torque tracking error $\eta_\tau(t)$ to zero. That is, we specify the following voltage input control relationship as

$$\psi_a V_b - \psi_b V_a = w_a + k_1\eta_\tau + r \tag{6.34}$$

where k_1 is a positive control gain. Substituting the right-hand side of (6.34) into (6.32) yields the closed-loop error system for the torque tracking error $\eta_\tau(t)$ as

$$L_I\dot{\eta}_\tau = -k_1\eta_\tau - r. \tag{6.35}$$

As stated before, a secondary control objective is to force the flux tracking error $\eta_\psi(t)$ to zero. Based on the structure of the open-loop dynamics for the flux tracking error $\eta_\psi(t)$ of (6.22), we design the fictitious flux controller $u_I(t)$ as

$$u_I = Y_\psi\theta_\psi + k_2\eta_\psi \tag{6.36}$$

where k_2 is a positive control gain, $Y_\psi(\cdot)$, and θ_ψ are defined (6.21). Substituting $u_I(t)$ of (6.36) into (6.22), we obtain the closed-loop error dynamics for the flux tracking error $\eta_\psi(t)$ as

$$\bar{L}_r\dot{\eta}_\psi = -k_2\eta_\psi + \eta_I \tag{6.37}$$

where $\eta_I(t)$ was defined in (6.23). From (6.37), we can see that if the auxiliary tracking error variable $\eta_I(t)$ was equal to zero, then the flux tracking error $\eta_\psi(t)$ would converge to zero exponentially fast. However, to ensure $\eta_I(t)$ goes to zero, a voltage-level controller must be designed. To this end, we complete the open-loop description for the dynamics of $\eta_I(t)$ in (6.25). From (6.36), the time derivative of $u_I(t)$ can be obtained as

$$\dot{u}_I = \dot{Y}_\psi\theta_\psi + k_2\dot{\eta}_\psi. \tag{6.38}$$

After differentiating (6.21), (6.38) can be written as

$$\dot{u}_I = \bar{L}_r\ddot{\psi}_d + 2\bar{R}_r\left(\dot{\psi}_a\psi_a + \dot{\psi}_b\psi_b\right) + k_2\dot{\eta}_\psi. \tag{6.39}$$

Multiplying (6.39) by L_I and substituting (6.4) and (6.5) for $\dot{\psi}_a$ and $\dot{\psi}_b$, respectively, yields

$$\begin{aligned} L_I\dot{u}_I \;=\;& \bar{L}_rL_I\ddot{\psi}_d + 2L_IL_r^{-1}\bar{R}_r\psi_a\left(-R_r\psi_a - \alpha_3\dot{q}\psi_b + K_II_a\right) \\ & +2L_IL_r^{-1}\bar{R}_r\psi_b\left(-R_r\psi_b + \alpha_3\dot{q}\psi_a + K_II_b\right) + k_2L_I\dot{\eta}_\psi. \end{aligned} \tag{6.40}$$

After substituting the right-hand of (6.20) into (6.40) for $\dot{\eta}_\psi(t)$, we have

$$\begin{aligned} L_I\dot{u}_I \;=\;& \bar{L}_rL_I\ddot{\psi}_d + 2L_IL_r^{-1}\bar{R}_r\psi_a\left(-R_r\psi_a - \alpha_3\dot{q}\psi_b + K_II_a\right) \\ & +2L_IL_r^{-1}\bar{R}_r\psi_b\left(-R_r\psi_b + \alpha_3\dot{q}\psi_a + K_II_b\right) \\ & +k_2L_I\bar{L}_r^{-1}\left(Y_\psi\theta_\psi - \left(\psi_aI_a + \psi_bI_b\right)\right). \end{aligned} \tag{6.41}$$

After substituting (6.41) into (6.25), we complete the open-loop dynamics for $\eta_I(t)$ as follows

$$L_I\dot{\eta}_I = w_b - \left(\psi_aV_a + \psi_bV_b\right) \tag{6.42}$$

where the auxiliary scalar $w_b(\dot{q}, I_a, I_b, \psi_a, \psi_b, t)$ is given by

$$\begin{aligned} w_b =\;& \bar{L}_rL_I\ddot{\psi}_d + L_r^{-1}\left(2L_I\bar{R}_r\psi_a - L_II_a\right)\left(-R_r\psi_a - \alpha_3\dot{q}\psi_b\right. \\ & \left.+K_II_a\right) + k_2L_I\bar{L}_r^{-1}\left(Y_\psi\theta_\psi - \left(\psi_aI_a + \psi_bI_b\right)\right) \\ & +L_r^{-1}\left(2L_I\bar{R}_r\psi_b - L_II_b\right)\left(-R_r\psi_b + \alpha_3\dot{q}\psi_a + K_II_b\right) \\ & -\psi_a\left(-R_II_a + \alpha_1\psi_a\right) - \psi_b\left(-R_II_b + \alpha_1\psi_b\right). \end{aligned} \tag{6.43}$$

From the structure of (6.42), we propose the following voltage input control relationship to drive $\eta_I(t)$ to zero

$$\psi_a V_a + \psi_b V_b = w_b + k_3\eta_I + \eta_\psi \tag{6.44}$$

where k_3 is a positive control gain. After substituting the right-hand side of (6.44) into (6.42), we obtain the closed-loop description for $\eta_I(t)$ as

$$L_I\dot{\eta}_I = -k_3\eta_I - \eta_\psi. \tag{6.45}$$

Remark 6.3

Given (6.34) and (6.44), we can solve for the transformed voltage control inputs V_a and V_b as follows error dynamics

$$\begin{bmatrix} V_a \\ V_b \end{bmatrix} = C^{-1}\begin{bmatrix} w_a + k_1\eta_\tau + r \\ w_b + k_3\eta_I + \eta_\psi \end{bmatrix} \tag{6.46}$$

where C is given by

$$C = \begin{bmatrix} -\psi_b & \psi_a \\ \psi_a & \psi_b \end{bmatrix} \in \Re^{2\times 2}. \tag{6.47}$$

The determinant of matrix C in (6.47) is given by

$$\det(C) = -\left(\psi_a^2 + \psi_b^2\right). \tag{6.48}$$

From (6.48), we can see that the determinant of matrix C equals zero whenever $\psi_a = \psi_b = 0$. That is, the controller in (6.46) is not well-defined for $\psi_a = \psi_b = 0$. Roughly speaking, this rotor flux singularity is of primary concern during motor start-up because it is precisely at this point in time that the rotor flux is equal to zero. Indeed, it is this rotor flux singularity which motivates us to design the desired pseudo magnitude (*i.e.*, $\psi_d(t)$) of the rotor flux as a positive scalar function. That is, if the rotor flux tracking error defined in (6.17) is "small" and $\psi_d(t)$ is "large", we can be assured that the operating condition of $\psi_a = \psi_b = 0$ is always avoided. However, it is obvious at motor startup, we must employ some adhoc method of ensuring that the rotor flux singularity is avoided. It is also interesting to note that the same singularity exists for field oriented control as discussed in [6]. Later in Chapter 11, we will present a partial state feedback controller which estimates rotor flux measurements while simultaneously eliminating this control singularity problem.

The dynamics given by (6.27), (6.35), (6.37), and (6.45) represent the electromechanical closed-loop system for which the stability analysis is performed while the exact model knowledge controllers given by (6.26), (6.36), (6.34), (6.44), and (6.46) represent the control inputs which are implemented at the voltage terminals of the motor. Note that the desired torque signal $\tau_d(t)$ and the fictitious flux controller $u_I(t)$ are embedded (in the guise of the variables $\eta_\tau(t)$ and $\eta_I(t)$) inside of the voltage control inputs $V_a(t)$ and $V_b(t)$. The theorem given below delineates the performance of the closed-loop system under the proposed control.

Theorem 6.1

The proposed exact model knowledge controller ensures that the filtered tracking error goes to zero exponentially fast for the electromechanical dynamics of (6.1), (6.2), (6.3), (6.4), and (6.5) as shown

$$\|x(t)\| \leq \sqrt{\frac{\lambda_2}{\lambda_1}} \|x(0)\| \underline{e}^{-\gamma t} \qquad \forall t \in [0, \infty) \tag{6.49}$$

where

$$x = \begin{bmatrix} r & \eta_\tau & \eta_\psi & \eta_I \end{bmatrix}^T \in \Re^4, \tag{6.50}$$

$$\lambda_1 = \min\{M, \bar{L}_r, L_I\}, \qquad \lambda_2 = \max\{M, \bar{L}_r, L_I\}, \tag{6.51}$$

and

$$\gamma = \frac{\min\{k_s, k_1, k_2, k_3\}}{\max\{M, \bar{L}_r, L_I\}}. \tag{6.52}$$

Proof. We define the following non-negative function

$$V(t) = \frac{1}{2}Mr^2 + \frac{1}{2}L_I\eta_\tau^2 + \frac{1}{2}\bar{L}_r\eta_\psi^2 + \frac{1}{2}L_I\eta_I^2 = \frac{1}{2}x^T diag\{M, L_I, \bar{L}_r, L_I\}\, x \tag{6.53}$$

where $x(t)$ was defined in (6.50), and $diag\{\cdot\}$ is used to denote the on-diagonal elements of a diagonal matrix. By applying Lemma 1.7 in Chapter 1 to the matrix term in (6.53), we can form upper and lower bounds on $V(t)$ as follows

$$\frac{1}{2}\lambda_1 \|x(t)\|^2 \leq V(t) \leq \frac{1}{2}\lambda_2 \|x(t)\|^2 \tag{6.54}$$

where λ_1 and λ_2 were defined in (6.51). Differentiating (6.53) with respect to time yields

$$\dot{V}(t) = Mr\dot{r} + L_I\eta_\tau\dot{\eta}_\tau + \bar{L}_r\eta_\psi\dot{\eta}_\psi + L_I\eta_I\dot{\eta}_I. \tag{6.55}$$

Substituting (6.27), (6.35), (6.37), and (6.45) into (6.55) yields

$$\dot{V}(t) = -k_s r^2 - k_1\eta_\tau^2 - k_2\eta_\psi^2 - k_3\eta_I^2 = -x^T diag\{k_s, k_1, k_2, k_3\}\, x. \tag{6.56}$$

By application of Lemma 1.7 in Chapter 1 to the matrix term in (6.56), $\dot{V}(t)$ in (6.56) can be upper bounded as

$$\dot{V}(t) \leq -\min\{k_s, k_1, k_2, k_3\} \|x(t)\|^2. \tag{6.57}$$

From (6.54), it is easy to see that $\|x(t)\|^2 \geq 2V(t)/\lambda_2$; hence, $\dot{V}(t)$ in (6.57) can be further upper bounded as

$$\dot{V}(t) \leq -2\gamma V(t) \tag{6.58}$$

where γ was defined (6.52). Applying Lemma 1.1 in Chapter 1 to (6.58) yields

$$V(t) \leq V(0)\underline{e}^{-2\gamma t}. \tag{6.59}$$

From (6.54), we have that

$$\frac{1}{2}\lambda_1 \|x(t)\|^2 \leq V(t) \quad \text{and} \quad V(0) \leq \frac{1}{2}\lambda_2 \|x(0)\|^2. \tag{6.60}$$

Substituting (6.60) appropriately into the left-hand and right-hand sides of (6.59) allows us to form the following inequality

$$\frac{1}{2}\lambda_1 \|x(t)\|^2 \leq \frac{1}{2}\lambda_2 \|x(0)\|^2 \underline{e}^{-2\gamma t}. \tag{6.61}$$

Using (6.61), we can solve for $\|x(t)\|$ to obtain the result given in (6.49). □

Remark 6.4

Using the result of Theorem 6.1, the control structure, and the electromechanical model, it is straightforward to illustrate that all signals remain bounded during closed-loop operation. Specifically, from (6.49) and (6.50), we know that $r(t) \in L_\infty$, $\eta_\tau(t) \in L_\infty$, $\eta_\psi \in L_\infty$, and $\eta_I(t) \in L_\infty$. From Lemma 1.4 in Chapter 1, we know that if $r(t) \in L_\infty$ then $e(t) \in L_\infty$ and $\dot{e}(t) \in L_\infty$. Since $q_d(t) \in L_\infty$ and $\dot{q}_d(t) \in L_\infty$ by assumption, then (6.6) can be utilized to show that $q(t) \in L_\infty$ and $\dot{q}(t) \in L_\infty$. From (6.27), we have $\dot{r}(t) \in L_\infty$ and since $e(t)$, $\dot{e}(t)$, $\ddot{q}_d(t)$ are all bounded, we know from (6.8) that $\ddot{q}(t) \in L_\infty$. Given $\eta_\psi(t) \in L_\infty$ and the assumption that $\psi_d(t) \in L_\infty$, we can show from (6.17) that the quantity $(\psi_a^2 + \psi_b^2) \in L_\infty$ and hence $\psi_a(t) \in L_\infty$, $\psi_b(t) \in L_\infty$. Then from (6.26) and (6.36), it is easy to show $\tau_d(t) \in L_\infty$ and $u_I(t) \in L_\infty$. Since $\eta_\tau(t)$, $\eta_I(t)$, $\tau_d(t)$, and $u_I(t)$ are all bounded, we can state that the quantities $(\psi_a I_b - \psi_b I_a)$ and $(\psi_a I_a + \psi_b I_b)$ are all bounded. After multiplying $(\psi_a I_b - \psi_b I_a)$ by $\psi_a(t)$ and $(\psi_a I_a + \psi_b I_b)$ by $\psi_b(t)$ and adding the result, we obtain the bounded expression $I_b(\psi_a^2 + \psi_b^2)$ which shows that $I_b(t)$ (and similarly $I_a(t)$) is bounded for all time. Using the structure of the voltage control input of (6.34), (6.44), the assumption of $\dddot{q}_d(t) \in L_\infty$, and the above information, we now know that $V_a(t) \in L_\infty$ and $V_b(t) \in L_\infty$. Finally, the electrical dynamics of (6.2), (6.3), (6.4), and (6.5) can be used to show that $\dot{I}_a(t) \in L_\infty$, $\dot{I}_b(t) \in L_\infty$, $\dot{\psi}_a(t) \in L_\infty$, and $\dot{\psi}_b(t) \in L_\infty$. In order to show that $I_a(t) \in L_\infty$ and $I_b(t) \in L_\infty$, we have assumed the control singularity $\psi_a = \psi_b = 0$ never occurs during the closed-loop operation (See Remark 6.3). In addition, this same assumption is required to show that $V_a(t) \in L_\infty$ and $V_b(t) \in L_\infty$.

Remark 6.5

From the result given by (6.49), it is now straightforward to see that the premise of Lemma 1.5 in Chapter 1 is satisfied; hence, we know that the position tracking error (*i.e.*, $e(t)$) and the velocity tracking error (*i.e.*, $\dot{e}(t)$) both go to zero exponentially fast. As an added bonus, we can also use the result given by (6.49) to form the following upper bounds on the rotor flux tracking error term as

$$\left|\eta_\psi(t)\right| \leq \sqrt{\frac{\lambda_2}{\lambda_1}}\, \|x(0)\|\, \mathrm{e}^{-\gamma t} \quad \forall t \in [0, \infty). \tag{6.62}$$

6.5 Adaptive Controller

Given the dynamics of (6.1), (6.2), (6.3), (6.4), and (6.5), we now design an adaptive position tracking controller under the constraint of parametric uncertainty. That is, without any apriori knowledge of the electromechanical system parameters we achieve our stated control objectives while measuring load position, load velocity, the stator winding current, and the rotor flux. The advantage that this approach offers compared to typical controllers found in motor/robot controls literature is that this controller is designed to compensate for the nonlinear dynamics used to describe the full order electromechanical system.

The first step in the procedure is to design the desired torque signal τ_d such that load position tracking can be ensured. Given the open-loop dynamics for the mechanical subsystem dynamics of (6.13), we define the desired torque signal τ_d as [7]

$$\tau_d = W_\tau \hat{\theta}_\tau + k_s r, \tag{6.63}$$

where $W_\tau \in \Re^{1\times 3}$ was defined in (6.11), $\hat{\theta}_\tau(t) \in \Re^3$ represents a dynamic estimate for the unknown parameter vector θ_τ defined in (6.12), and k_s is a positive constant controller gain. The parameter estimate $\hat{\theta}_\tau(t)$ defined in (6.63) is updated online according to the following adaptation law

$$\hat{\theta}_\tau = \int_0^t \Gamma_\tau W_\tau^T(\sigma) r(\sigma) d\sigma \tag{6.64}$$

where $\Gamma_\tau \in \Re^{3\times 3}$ is a constant, positive definite, diagonal adaptive gain matrix. If we define the mismatch between $\hat{\theta}_\tau(t)$ and θ_τ as

$$\tilde{\theta}_\tau = \theta_\tau - \hat{\theta}_\tau, \tag{6.65}$$

then the adaptation law of (6.64) can be written in terms of the parameter estimation error as

$$\dot{\tilde{\theta}}_\tau = -\Gamma_\tau W_\tau^T r = -\dot{\hat{\theta}}_\tau\,. \tag{6.66}$$

Substituting (6.63) into the open-loop dynamics of (6.13) yields the closed-loop filtered tracking error dynamics, as shown

$$M\dot{r} = W_\tau \tilde{\theta}_\tau - k_s r + \eta_\tau. \tag{6.67}$$

Now that we have completed the control design for the mechanical subsystem dynamics, a voltage-level controller must be designed to force the torque tracking error, $\eta_\tau(t)$, to zero. To this end, we first need to complete the open-loop system description for $\eta_\tau(t)$ in (6.16). Given the specification of $\tau_d(t)$ in (6.63), we can calculate $\dot{\tau}_d(t)$ as follows

$$\dot{\tau}_d = \dot{W}_\tau \hat{\theta}_\tau + W_\tau \dot{\hat{\theta}}_\tau + k_s \dot{r}. \tag{6.68}$$

Substituting the time derivatives of (6.11) and (6.64) along with (6.8) into the right-hand side of (6.68) yields

$$\begin{aligned} \dot{\tau}_d &= \hat{M}\left(\dddot{q}_d + \alpha\left(\ddot{q}_d - \ddot{q}\right)\right) + \hat{B}\ddot{q} + \hat{N}\dot{q}\cos(q) \\ &\quad + W_\tau \Gamma_\tau W_\tau^T r + k_s\left(\ddot{q}_d - \ddot{q} + \alpha\dot{e}\right) \end{aligned} \tag{6.69}$$

where $\hat{M}(t)$, $\hat{B}(t)$, and $\hat{N}(t)$ denote the scalar components of the vector $\hat{\theta}_\tau$ (*i.e.*, $\hat{\theta}_\tau = \begin{bmatrix} \hat{M} & \hat{B} & \hat{N} \end{bmatrix}^T$). Note that $\dot{\tau}_d(t)$ of (6.69) is expressed in terms of measurable states (*i.e.*, $q(t)$ and $\dot{q}(t)$), known functions, known constant parameters, and the unmeasurable load acceleration $\ddot{q}(t)$. Substituting for $\ddot{q}(t)$ from the right-hand side of (6.30) into (6.69), we can write $\dot{\tau}_d(t)$ in terms of measurable states (*i.e.*, $q(t)$, $\dot{q}(t)$, $I_a(t)$, $I_b(t)$, $\psi_a(t)$, and $\psi_b(t)$), known functions, and unknown constant parameters in the following form

$$\begin{aligned} \dot{\tau}_d &= \hat{M}\left(\dddot{q}_d + \alpha\ddot{q}_d\right) + \hat{N}\dot{q}\cos(q) + \\ &\quad k_s\left(\ddot{q}_d + \alpha\dot{e}\right) + W_\tau \Gamma_\tau W_\tau^T r \\ &\quad + \left(\hat{B} - \hat{M}\alpha - k_s\right)\left(-\frac{B}{M}\dot{q} - \frac{N}{M}\sin(q) + \frac{1}{M}\left(\psi_a I_b - \psi_b I_a\right)\right). \end{aligned} \tag{6.70}$$

Substituting the expression for $\dot{\tau}_d(t)$ of (6.70) into (6.16) yields the final open-loop dynamics for the torque tracking error in the form

$$L_I \dot{\eta}_\tau = Y_1 \theta_1 - \left(\psi_a V_b - \psi_b V_a\right) \tag{6.71}$$

where the known regression vector $Y_1\left(q, \dot{q}, \psi_a, \psi_b, I_a, I_b, \hat{\theta}_\tau, t\right) \in \Re^{1\times 7}$ and the unknown constant parameter vector $\theta_1 \in \Re^7$ are explicitly defined as follows

$$\theta_1 = \begin{bmatrix} \frac{L_I}{M} & \frac{L_I B}{M} & \frac{L_I N}{M} & L_I & \frac{L_I R_r}{L_r} + R_I & \frac{L_I \alpha_3}{L_r} & \alpha_2 \end{bmatrix}^T, \tag{6.72}$$

$$Y_1 = \left[\begin{array}{ccccccc} Y_{11} & Y_{12} & Y_{13} & Y_{14} & Y_{15} & Y_{16} & Y_{17} \end{array} \right], \tag{6.73}$$

with

$$Y_{11} = \left(\hat{B} - \hat{M}\alpha - k_s\right)(\psi_a I_b - \psi_b I_a),$$

$$Y_{12} = -\left(\hat{B} - \hat{M}\alpha - k_s\right)\dot{q}, \; Y_{13} = -\left(\hat{B} - \hat{M}\alpha - k_s\right)\sin(q),$$

$$Y_{14} = \hat{M}\left(\dddot{q}_d + \alpha\ddot{q}_d\right) + \hat{N}\dot{q}\cos(q) + k_s(\ddot{q}_d + \alpha\dot{e}) + W_\tau\Gamma_\tau W_\tau^T r,$$

$$Y_{15} = \psi_a I_b - \psi_b I_a, \; Y_{16} = \dot{q}(\psi_a I_a + \psi_b I_b),$$

$$\text{and} \quad Y_{17} = \dot{q}\left(\psi_a^2 + \psi_b^2\right).$$

Given the open-loop dynamics for $\eta_\tau(t)$ in (6.71), we can specify the following voltage control input relationship to force $\eta_\tau(t)$ to zero

$$-\psi_b V_a + \psi_a V_b = Y_1\hat{\theta}_1 + k_1\eta_\tau + r \tag{6.74}$$

where k_1 is a positive control gain, and $\hat{\theta}_1(t) \in \Re^7$ is a dynamic estimate of the parameter vector θ_1 defined in (6.72). The dynamic estimate $\hat{\theta}_1(t)$ is updated according to the following update law

$$\hat{\theta}_1 = \int_0^t \Gamma_1 Y_1^T(\sigma)\eta_\tau(\sigma)\,d\sigma \tag{6.75}$$

where $\Gamma_1 \in \Re^{7\times 7}$ is a constant, positive definite, diagonal adaptive gain matrix. If we define the mismatch between $\hat{\theta}_1(t)$ and θ_1 as

$$\tilde{\theta}_1 = \theta_1 - \hat{\theta}_1, \tag{6.76}$$

then the adaptation law of (6.75) can be written in terms of the estimation error as

$$\dot{\tilde{\theta}}_1 = -\Gamma_1 Y_1^T\eta_\tau = -\dot{\hat{\theta}}_1. \tag{6.77}$$

After substituting the right-hand side of (6.74) into (6.71), we obtain the closed-loop dynamics for η_τ as

$$L_I\dot{\eta}_\tau = Y_1\tilde{\theta}_1 - k_1\eta_\tau - r. \tag{6.78}$$

Now that we have completed the adaptive controller design for the position tracking objective, we must also ensure that all signals stay bounded during closed-loop operation. That is, a voltage control input must be designed to force the rotor fluxes to track the desired pseudo-magnitude flux trajectory. Given the open-loop dynamics for the flux tracking error $\eta_\psi(t)$ of (6.22), we can specify the fictitious flux controller $u_I(t)$ as

$$u_I = Y_\psi\hat{\theta}_\psi + k_2\eta_\psi \tag{6.79}$$

where k_2 is positive control gain, $Y_\psi(\cdot)$ was defined in (6.21), and $\hat{\theta}_\psi(t) \in \Re^2$ represents a dynamic estimate for the unknown parameter vector θ_ψ defined in (6.21). The parameter estimate $\hat{\theta}_\psi(t)$ defined in (6.79) is updated online according to the following adaptation law

$$\hat{\theta}_\psi = \int_0^t \Gamma_\psi Y_\psi^T(\sigma)\eta_\psi(\sigma)d\sigma \tag{6.80}$$

where $\Gamma_\psi \in \Re^{2\times 2}$ is a constant, positive definite, diagonal adaptive gain matrix. If we define the mismatch between $\hat{\theta}_\psi(t)$ and θ_ψ as

$$\tilde{\theta}_\psi = \theta_\psi - \hat{\theta}_\psi, \tag{6.81}$$

then the adaptation law of (6.80) can be written in terms of the parameter error as

$$\dot{\tilde{\theta}}_\psi = -\Gamma_\psi Y_\psi^T \eta_\psi = -\dot{\hat{\theta}}_\psi\,. \tag{6.82}$$

After substituting the fictitious flux controller $u_I(t)$ into the open-loop dynamics of (6.22), we obtain the closed-loop error system for the flux tracking error $\eta_\psi(t)$ as

$$\bar{L}_r\dot{\eta}_\psi = Y_\psi\tilde{\theta}_\psi - k_2\eta_\psi + \eta_I. \tag{6.83}$$

From (6.83), it is easy to see that if the auxiliary variable $\eta_I(t)$ was equal to zero, we could obtain an asymptotic flux tracking result. Since $\eta_I(t)$ is not zero in general, a voltage control input must be designed to force $\eta_I(t)$ to zero. To complete the open-loop dynamic description for $\eta_I(t)$ of (6.25), we first calculate the time derivative of the fictitious flux controller $u_I(t)$ of (6.79) as

$$\dot{u}_I = \dot{Y}_\psi\hat{\theta}_\psi + Y_\psi\dot{\hat{\theta}}_\psi + k_2\dot{\eta}_\psi. \tag{6.84}$$

Substituting the time derivative of Y_ψ in (6.21) into (6.84), we have

$$\dot{u}_I = \hat{\bar{L}}_r\ddot{\psi}_d + 2\hat{\bar{R}}_r\left(\dot{\psi}_a\psi_a + \dot{\psi}_b\psi_b\right) + Y_\psi\dot{\hat{\theta}}_\psi + k_2\dot{\eta}_\psi \tag{6.85}$$

where $\hat{\bar{L}}_r(t)$ and $\hat{\bar{R}}_r(t)$ denote the scalar components of the vector $\hat{\theta}_\psi(t)$ (*i.e.*, $\hat{\theta}_\psi = \begin{bmatrix}\hat{\bar{L}}_r & \hat{\bar{R}}_r\end{bmatrix}^T$). Multiplying (6.85) by L_I and substituting (6.4) and (6.5) for $\dot{\psi}_a(t)$ and $\dot{\psi}_b(t)$, respectively, yields

$$\begin{aligned} L_I\dot{u}_I = {} & \hat{\bar{L}}_r L_I\ddot{\psi}_d + 2L_I L_r^{-1}\hat{\bar{R}}_r\psi_a\left(-R_r\psi_a + K_I I_a\right) \\ & +2L_I L_r^{-1}\hat{\bar{R}}_r\psi_b\left(-R_r\psi_b + K_I I_b\right) \\ & +L_I Y_\psi\dot{\hat{\theta}}_\psi + k_2 L_I\dot{\eta}_\psi. \end{aligned} \tag{6.86}$$

Substituting the right-hand of (6.19) for $\dot{\eta}_\psi(t)$ and (6.82) for $\dot{\hat{\theta}}_\psi\ (t)$ into (6.86), we have

$$\begin{aligned} L_I\dot{u}_I = \; & \widehat{\bar{L}}_r L_I \ddot{\psi}_d + 2L_I L_r^{-1}\widehat{\bar{R}}_r\psi_a\left(-R_r\psi_a + K_I I_a\right) \\ & +2L_I L_r^{-1}\widehat{\bar{R}}_r\psi_b\left(-R_r\psi_b + K_I I_b\right) \\ & +L_I Y_\psi \Gamma_\psi Y_\psi^T \eta_\psi + k_2 L_I L_r^{-1}\Big(L_r\dot{\psi}_d + \\ & R_r\left(\psi_a^2 + \psi_b^2\right) - K_I(\psi_a I_a + \psi_b I_b)\Big). \end{aligned} \tag{6.87}$$

We can complete the open-loop dynamics for $\eta_I(t)$ by substituting (6.87) into (6.25) to yield

$$L_I\dot{\eta}_I = Y_2\theta_2 - \left(\psi_a V_a + \psi_b V_b\right) \tag{6.88}$$

where the known regression vector $Y_2\left(q, \dot{q}, I_a, I_b, \psi_a, \psi_b, \hat{\theta}_\psi, t\right) \in \Re^{1\times 6}$ and the unknown constant parameter vector $\theta_2 \in \Re^6$ are explicitly defined as follows

$$\theta_2 = \begin{bmatrix} \dfrac{L_I R_r}{L_r} & \dfrac{L_I\alpha_3}{L_r} & \dfrac{L_I K_I}{L_r} & L_I & R_I & \alpha_1 \end{bmatrix}^T, \tag{6.89}$$

$$Y_2 = \begin{bmatrix} Y_{21} & Y_{22} & Y_{23} & Y_{24} & Y_{25} & Y_{26} \end{bmatrix}, \tag{6.90}$$

with

$$Y_{21} = \psi_a I_a + \psi_b I_b + \left(k_2 - 2\widehat{\bar{R}}_r\right)\left(\psi_a^2 + \psi_b^2\right),$$

$$Y_{22} = -\dot{q}\left(\psi_a I_b - \psi_b I_a\right),$$

$$Y_{23} = -I_a^2 - I_b^2 + \left(2\widehat{\bar{R}}_r - k_2\right)\left(\psi_a I_a + \psi_b I_b\right),$$

$$Y_{24} = \widehat{\bar{L}}_r\ddot{\psi}_d + k_2\dot{\psi}_d + Y_\psi\Gamma_\psi Y_\psi^T\eta_\psi,$$

$$Y_{25} = \psi_a I_a + \psi_b I_b, \text{ and } Y_{26} = -\left(\psi_a^2 + \psi_b^2\right).$$

Based on the open-loop dynamics for $\eta_I(t)$ in (6.88) and the subsequent stability analysis, we propose the following voltage control input relationship to force $\eta_I(t)$ to zero

$$\psi_a V_a + \psi_b V_b = Y_2\hat{\theta}_2 + k_3\eta_I + \eta_\psi \tag{6.91}$$

where k_3 is a positive control gain, and $\hat{\theta}_2(t) \in \Re^6$ is the estimate of the vector θ_2 defined in (6.89). The dynamic estimate $\hat{\theta}_2(t)$ is updated according to the following update law

$$\hat{\theta}_2 = \int_0^t \Gamma_2 Y_2^T(\sigma)\eta_I(\sigma)\, d\sigma \tag{6.92}$$

where $\Gamma_2 \in \Re^{6\times 6}$ is a constant, positive definite, diagonal adaptive gain matrix. If we define the mismatch between $\hat{\theta}_2(t)$ and θ_2 as

$$\tilde{\theta}_2 = \theta_2 - \hat{\theta}_2, \tag{6.93}$$

then the adaptation law of (6.92) can be written in terms of the parameter error as

$$\dot{\tilde{\theta}}_2 = -\Gamma_2 Y_2^T \eta_I = -\dot{\hat{\theta}}_2 . \tag{6.94}$$

After substituting the right-hand side of (6.91) into (6.88), we obtain the closed-loop dynamics for $\eta_I(t)$ as

$$L_I \dot{\eta}_I = Y_2 \tilde{\theta}_2 - k_3 \eta_I - \eta_\psi . \tag{6.95}$$

Given (6.74) and (6.91), we can now solve for the transformed voltage control inputs $V_a(t)$ and $V_b(t)$ as

$$\begin{bmatrix} V_a \\ V_b \end{bmatrix} = C^{-1} \begin{bmatrix} Y_1 \hat{\theta}_1 + k_1 \eta_\tau + r \\ Y_2 \hat{\theta}_2 + k_3 \eta_I + \eta_\psi \end{bmatrix} \tag{6.96}$$

where C is given by

$$C = \begin{bmatrix} -\psi_b & \psi_a \\ \psi_a & \psi_b \end{bmatrix} . \tag{6.97}$$

As stated earlier, field oriented control has become widely used for the speed control of induction motors. Roughly speaking, the basic objective of field oriented control is to make the induction motor behave like a DC motor. Some earlier research results on the field-oriented control of AC machines were reported in [8]. From the control point of view, as shown in [6], field oriented control is achieved through a partial-state feedback linearizing transformation. However, the proposed controller in this chapter does not rely on a feedback linearization strategy. Rather, the integrator backstepping approach and a Lyapunov-like analysis are used to design an adaptive controller which can compensate for parametric uncertainty.

The dynamics given by (6.66), (6.67), (6.77), (6.78), (6.82), (6.94), (6.83), and (6.95) represent the electromechanical closed-loop system for which the stability analysis is performed while the adaptive controller given by (6.63), (6.64), (6.75), (6.80), (6.92), (6.79), (6.74), and (6.91) represent the controller which is implemented at the voltage terminals of motor. Similar to the exact model knowledge controller, the adaptive desired torque input $\tau_d(t)$ and the fictitious controller $u_I(t)$ are embedded (in the guise of the variables $\eta_\tau(t)$ and $\eta_I(t)$) inside of the voltage control inputs $V_a(t)$ and $V_b(t)$. The theorem given below delineates the performance of the closed-loop system under the proposed control.

Theorem 6.2

The proposed adaptive controller ensures that the filtered tracking error goes to zero asymptotically for the electromechanical dynamics of (6.1), (6.2), (6.3), (6.4), and (6.5) as shown

$$\lim_{t\to\infty} r(t) = 0. \tag{6.98}$$

Proof. First, we define the following non-negative function

$$\begin{aligned} V &= \frac{1}{2}Mr^2 + \frac{1}{2}L_I\eta_\tau^2 + \frac{1}{2}\bar{L}_r\eta_\psi^2 + \frac{1}{2}L_I\eta_I^2 \\ &+\frac{1}{2}\tilde{\theta}_\tau^T\Gamma_\tau^{-1}\tilde{\theta}_\tau + \frac{1}{2}\tilde{\theta}_1^T\Gamma_1^{-1}\tilde{\theta}_1 + \frac{1}{2}\tilde{\theta}_\psi^T\Gamma_\psi^{-1}\tilde{\theta}_\psi + \frac{1}{2}\tilde{\theta}_2^T\Gamma_2^{-1}\tilde{\theta}_2. \end{aligned} \tag{6.99}$$

Taking the time derivative of (6.99) with respect to time yields

$$\begin{aligned} \dot{V} &= rM\dot{r} + \eta_\tau L_I\dot{\eta}_\tau + \eta_\psi\bar{L}_r\dot{\eta}_\psi + \eta_I L_I\dot{\eta}_I \\ &+\tilde{\theta}_\tau^T\Gamma_\tau^{-1}\dot{\tilde{\theta}}_\tau +\tilde{\theta}_1^T\Gamma_1^{-1}\dot{\tilde{\theta}}_1 +\tilde{\theta}_\psi^T\Gamma_\psi^{-1}\dot{\tilde{\theta}}_\psi +\tilde{\theta}_2^T\Gamma_2^{-1}\dot{\tilde{\theta}}_2 \end{aligned} \tag{6.100}$$

where the facts that: i) scalars can be transposed and ii) Γ_τ, Γ_1, Γ_ψ and Γ_2 are diagonal matrices, have been used. Substituting the closed-loop error dynamics of (6.66), (6.67), (6.77), (6.78), (6.82), (6.83), (6.94), and (6.95) into (6.100) yields

$$\begin{aligned} \dot{V} &= -k_s r^2 - k_1\eta_\tau^2 - k_2\eta_\psi^2 - k_3\eta_I^2 + \left(rW_\tau\tilde{\theta}_\tau - \tilde{\theta}_\tau^T W_\tau^T r\right) \\ &+\left(\eta_\tau Y_1\tilde{\theta}_1 - \tilde{\theta}_1^T Y_1^T\eta_\tau\right) + \left(\eta_\psi Y_\psi\tilde{\theta}_\psi - \tilde{\theta}_\psi^T Y_\psi^T\eta_\psi\right) \\ &+\left(\eta_I Y_2\tilde{\theta}_2 - \tilde{\theta}_2^T Y_2^T\eta_I\right). \end{aligned} \tag{6.101}$$

Since any scalar quantity can be transposed, (6.101) can be simplified to yield

$$\dot{V} = -k_s r^2 - k_1\eta_\tau^2 - k_2\eta_\psi^2 - k_3\eta_I^2. \tag{6.102}$$

From the form of (6.102), we can see that $\dot{V}(t)$ is negative or zero; hence, we know from calculus that $V(t)$ given in (6.99) is either decreasing or constant. Since $V(t)$ is non-negative, it is lower bounded by zero; hence, from the form of $V(t)$, we know that $r(t) \in L_\infty$, $\eta_\tau(t) \in L_\infty$, $\eta_\psi(t) \in L_\infty$, $\eta_I \in L_\infty$, $\tilde{\theta}_\tau(t) \in L_\infty^3$, $\tilde{\theta}_1(t) \in L_\infty^7$, $\tilde{\theta}_\psi \in L_\infty^2$, and $\tilde{\theta}_2(t) \in L_\infty^6$. Since $r(t) \in L_\infty$, Lemma 1.4 in Chapter 1 can be used to show that $e(t) \in L_\infty$ and $\dot{e}(t) \in L_\infty$; therefore, since $q_d(t) \in L_\infty$ and $\dot{q}_d(t) \in L_\infty$ by assumption, (6.6) can be utilized to show that $q(t) \in L_\infty$ and $\dot{q}(t) \in L_\infty$. Since

$\tilde{\theta}_\tau(t) \in L^3_\infty$, $\tilde{\theta}_1(t) \in L^7_\infty$, $\tilde{\theta}_\psi(t) \in L^2_\infty$ and $\tilde{\theta}_2(t) \in L^6_\infty$, we can use (6.65), (6.81), (6.76), and (6.93) to show that $\hat{\theta}_\tau(t) \in L^3_\infty$, $\hat{\theta}_1(t) \in L^7_\infty$, $\hat{\theta}_\psi(t) \in L^2_\infty$, and $\hat{\theta}_2(t) \in L^6_\infty$. From (6.67), we have $\dot{r}(t) \in L_\infty$ and since $e(t)$, $\dot{e}(t)$, $\ddot{q}_d(t)$ are all bounded, we know that $\ddot{q}(t) \in L_\infty$. Given $\eta_\psi(t) \in L_\infty$ and the assumption of $\psi_d(t) \in L_\infty$, we can show that the quantity $(\psi_a^2 + \psi_b^2) \in L_\infty$ and hence $\psi_a(t) \in L_\infty$, $\psi_b(t) \in L_\infty$. Then from (6.63) and (6.79), it is easy to show $\tau_d(t) \in L_\infty$ and $u_I(t) \in L_\infty$. Since $\eta_\tau(t)$, $\eta_I(t)$, $\tau_d(t)$, and $u_I(t)$ are all bounded, we can state the quantities $(\psi_a I_b - \psi_b I_a)$ and $(\psi_a I_a + \psi_b I_b)$ are all bounded. After multiplying $(\psi_a I_b - \psi_b I_a)$ by $\psi_a(t)$ and $(\psi_a I_a + \psi_b I_b)$ by $\psi_b(t)$ and adding the result, we obtain the bounded expression $I_b\left(\psi_a^2 + \psi_b^2\right)$ which shows that $I_b(t)$ (and similarly $I_a(t)$) is bounded for all time. Using the structure of the voltage control input of (6.34), (6.44), the assumption of $\dddot{q}_d(t) \in L_\infty$ and the above information, we now know that $V_a(t) \in L_\infty$ and $V_b(t) \in L_\infty$. In addition, from the closed-loop error systems (6.78), (6.83), and (6.95), we can see that $\dot{\eta}_\tau(t) \in L_\infty$, $\dot{\eta}_\psi(t) \in L_\infty$, and $\dot{\eta}_I(t) \in L_\infty$. Finally, the electrical dynamics of (6.2), (6.3), (6.4), and (6.5) can be used to show that $\dot{I}_a(t) \in L_\infty$, $\dot{I}_b(t) \in L_\infty$, $\dot{\psi}_a(t) \in L_\infty$, and $\dot{\psi}_b(t) \in L_\infty$.

We now illustrate how Corollary 1.1 in Chapter 1 can be used to show that the filtered tracking error goes to zero. First, we integrate both sides of (6.102) with respect to time to yield

$$\int_0^\infty \frac{dV(\sigma)}{d\sigma}\, d\sigma = -\int_0^\infty \left(k_s r^2(\sigma) + k_1\eta_\tau^2(\sigma) + k_2\eta_\psi^2(\sigma) + k_3\eta_I^2(\sigma)\right) d\sigma. \tag{6.103}$$

If we evaluate the left-hand side of (6.103), we obtain

$$\sqrt{V(0) - V(\infty)} = \sqrt{\int_0^\infty \left(k_s r^2(\sigma) + k_1\eta_\tau^2(\sigma) + k_2\eta_\psi^2(\sigma) + k_3\eta_I^2(\sigma)\right) d\sigma} \tag{6.104}$$

after some minor algebraic operations. Since $\dot{V}(t) \leq 0$ as illustrated by (6.102), $V(t)$ of (6.99) is decreasing or constant; hence, $V(0) \geq V(\infty) \geq 0$. We now use this information and (6.104) to obtain the following inequality

$$\begin{aligned} &\sqrt{k_s \int_0^\infty r^2(\sigma)\, d\sigma} \\ &\leq \sqrt{\int_0^\infty \left(k_s r^2(\sigma) + k_1\eta_\tau^2(\sigma) + k_2\eta_\psi^2(\sigma) + k_3\eta_I^2(\sigma)\right) d\sigma} \\ &\leq \sqrt{V(0)} < \infty \end{aligned} \tag{6.105}$$

which indicates according to Definition 1.2 in Chapter 1 that $r(t) \in L_2$. Since $r(t) \in L_\infty$, $\dot{r}(t) \in L_\infty$, and $r(t) \in L_2$, we can invoke Corollary 1.1

in Chapter 1 to obtain the result given by (6.98). As done for exact model knowledge controller, in order to show that $I_a(t) \in L_\infty$ and $I_b(t) \in L_\infty$, we have assumed the control singularity $\psi_a(t) = \psi_b(t) = 0$ never occurs during the closed-loop operation In addition, this same assumption is required to show that $V_a(t) \in L_\infty$ and $V_b(t) \in L_\infty$.

Remark 6.7

From the result given by (6.98) and the proof of Theorem 6.2, we know that $r(t) \in L_\infty$, $r(t) \in L_2$, and $\lim_{t\to\infty} r(t) = 0$; hence, we can use Lemma 1.6 in Chapter 1 to state that $\lim_{t\to\infty} e(t) = 0$ and $\lim_{t\to\infty} \dot{e}(t) = 0$. As an added bonus, we can also use the result given by (6.105) to state that

$$\sqrt{\int_0^\infty k_2\eta_\psi^2(\sigma)\, d\sigma} \leq \sqrt{V(0)} < \infty \tag{6.106}$$

which indicates that $\eta_\psi(t) \in L_2$. Since $\eta_\psi(t) \in L_\infty$, $\dot{\eta}_\psi(t) \in L_\infty$, and $\eta_\psi(t) \in L_2$, we can invoke Corollary 1.1 in Chapter 1 to state $\lim_{t\to\infty} \eta_\psi(t) = 0$.

6.6 Simulation

In this section, we present the result of a computer simulation of the theoretical developments presented in Section 5 (*i.e.*, the adaptive controller). A Baldor model M3541 induction motor is used for the simulation which has 2-poles, a rated speed of 3450 rpm, and a rated voltage of 230V. The mechanical load is selected as a single-link robot which is designed as a metal bar connected at one end to the motor shaft. Therefore, the parameters M and N of (6.1) can be expressed as

$$M = \frac{J}{\alpha_2} + \frac{mL_0^2}{3\alpha_2}, \quad N = \frac{mGL_0}{2\alpha_2}, \quad \text{and} \quad B = \frac{B_0}{\alpha_2}$$

where J is the rotor inertia, m is the link mass, L_0 is the link length, B_0 is the coefficient of viscous friction at the joint, and G is the gravity coefficient. The schematic diagram of the induction motor/load system is shown in Figure 6.1.

The values of electromechanical system parameters described in (6.1) through (6.5) are given by

$$\begin{aligned} &R_s = 3.05\Omega, \quad R_r = 2.12\Omega, \quad L_s = 0.243H, \\ &L_r = 0.306H, \quad M_e = 0.225H, \end{aligned} \tag{6.107}$$

$$J = 2.1 \times 10^{-4} Kg - m^2, \quad n_p = 1, \quad L_0 = 0.305m, \quad m = 0.401Kg,$$

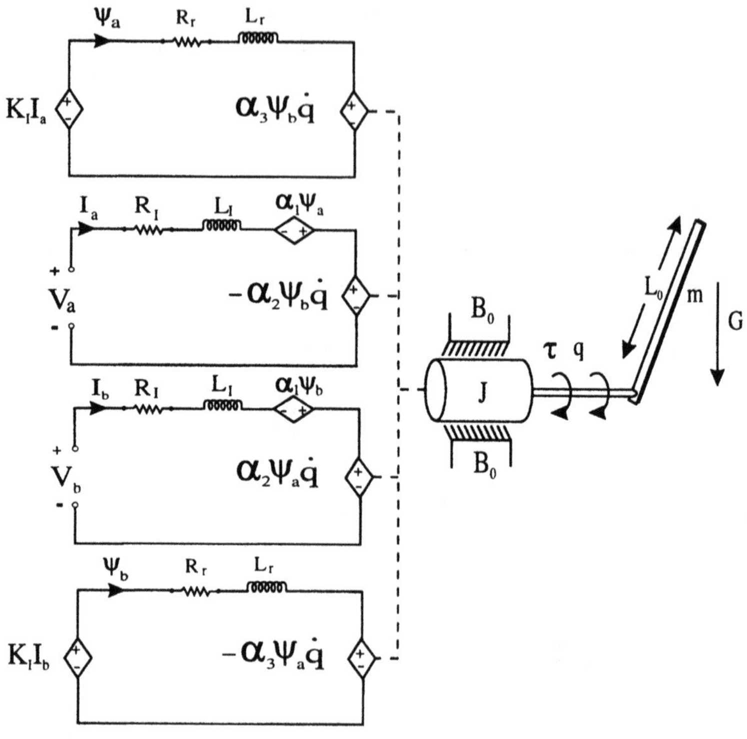

Figure 6.1: Schematic Diagram of the Induction Motor/Load System

and

$$B_0 = 0.015Nm - \sec/rad, \text{ and } G = 9.81Kg - m/sec^2.$$

The desired load position trajectory is assumed to be

$$q_d = \frac{\pi}{2}\sin(5t)\left(1 - \underline{e}^{-0.1t^3}\right) rad \tag{6.108}$$

which has the desirable property $q_d(0) = \dot{q}_d(0) = \ddot{q}_d(0) = \dddot{q}_d(0) = 0$. The desired position trajectory is plotted in Figure 6.2.The desired pseudo-magnitude flux trajectory is selected to be

$$\psi_d = 2\left(1 - \underline{e}^{-t^2}\right) + 1\ Wb \cdot Wb. \tag{6.109}$$

The initial conditions for the system states are assumed as: position is $0.1rad$, velocity is zero, stator currents are zero and the rotor fluxes are both $1\,Wb$ (Note the singularity problem referred to in Remark 6.3 is

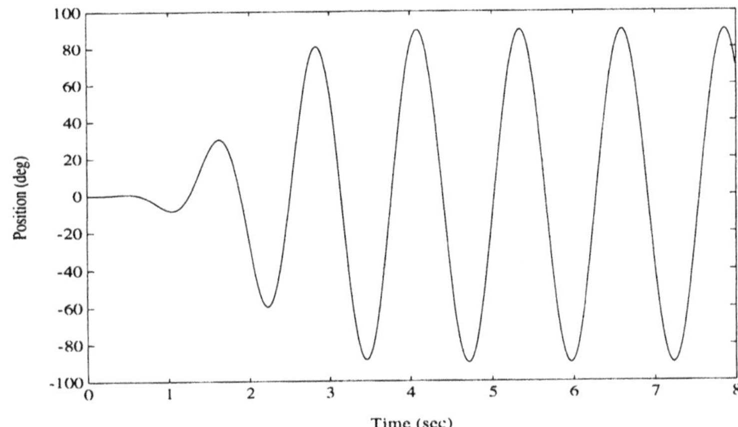

Figure 6.2: Desired Position Trajectory

initally avoided by setting the initial rotor flux to some initial value). The initial conditions on the adaptive parameters are assumed to differ from their actual values by +50%. The control gains are set to be

$$\Gamma = I_3,\ \Gamma_\psi = I_2,\ \Gamma_1 = I_7,\ \Gamma_2 = I_6,\ k_s = k_1 = 20,\ k_2 = k_3 = 20$$

where I_2, I_3, I_7, and I_6 are the 2×2, 3×3, 7×7, and 6×6 identity matrices, respectively.

The resulting load position tracking error is shown in Figure 6.3, the input control voltages V_a, V_b and the stator currents I_a, I_b are shown in Figures 6.4 through 6.7, respectively, the rotor fluxes ψ_a, ψ_b are shown in Figures 6.8 and Figure 6.9, respectively. From these figures, we can see that the position tracking error converges to zero, the fluxes remain bounded, and input voltages stay in the rated range.

Remark 6.8

The full state feedback controller was demonstrated using a computer simulation due to the difficulty associated with measuring the rotor flux. Measurement of the flux must take place inside of the motor and hence would require nontrivial modifications to our off-the-shelf motor. The full state feedback controller is provided more as a prelude to the partial state feedback controller presented in Chapter 11 than as an implementable control scheme. In Chapter 11, the rotor flux will be estimated in order to design a more practical controller.

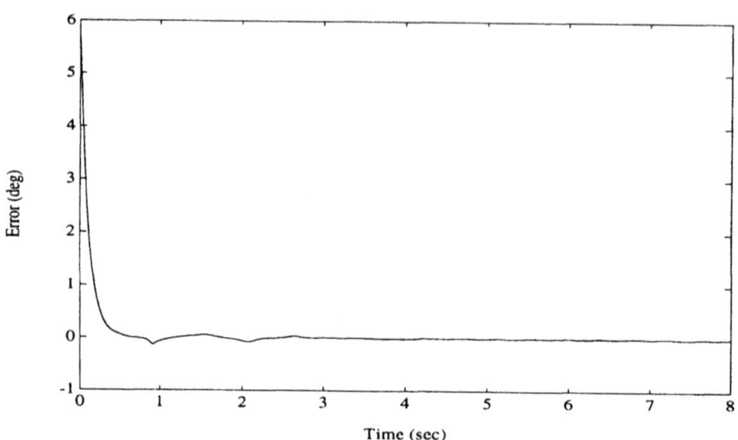

Figure 6.3: Position Tracking Error for Adaptive Controller

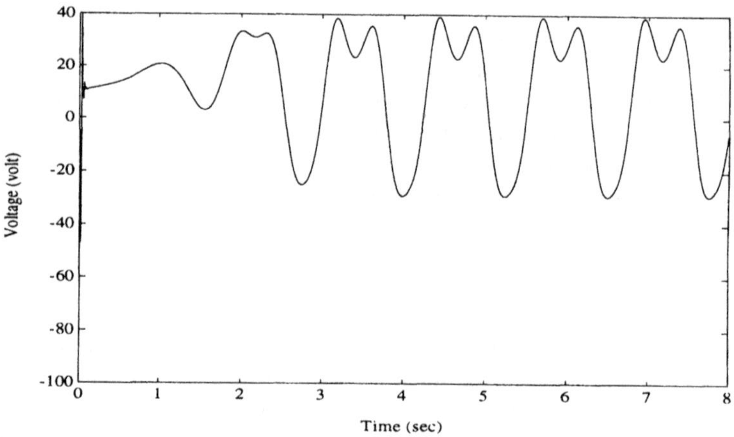

Figure 6.4: Transformed Phase One Voltage for Adpative Controller

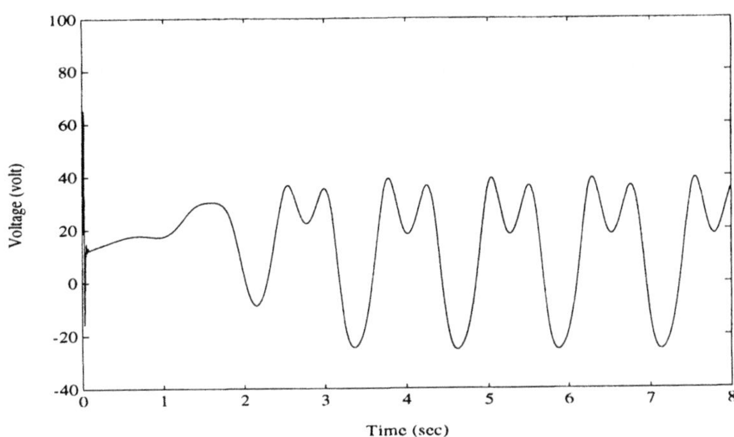

Figure 6.5: Transformed Phase Two Voltage for Adaptive Controller

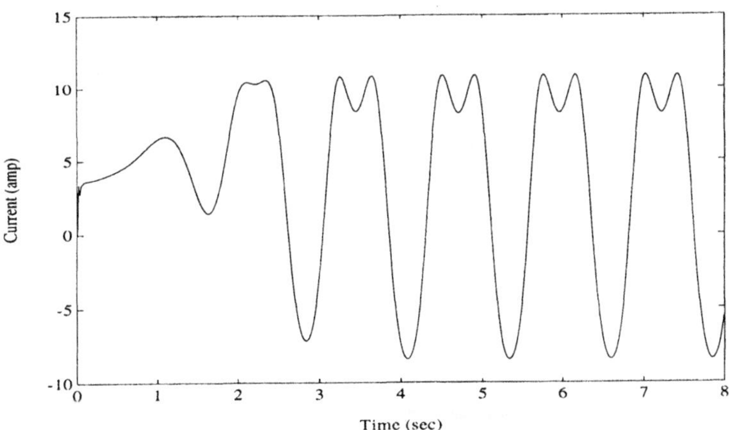

Figure 6.6: Transformed Phase One Current for Adaptive Controller

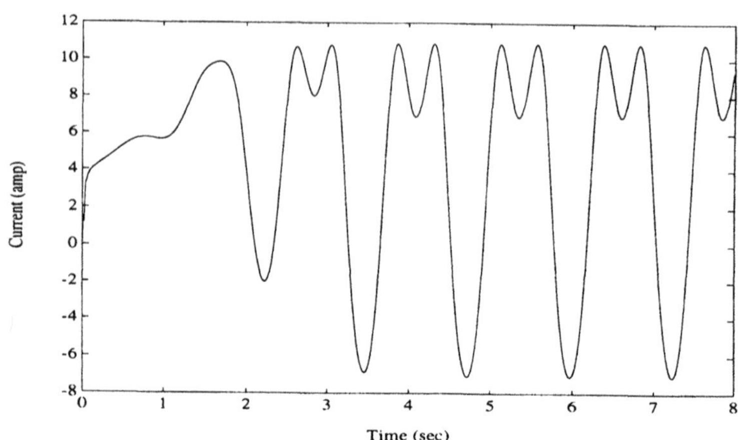

Figure 6.7: Transformed Phase Two Current for Adaptive Controller

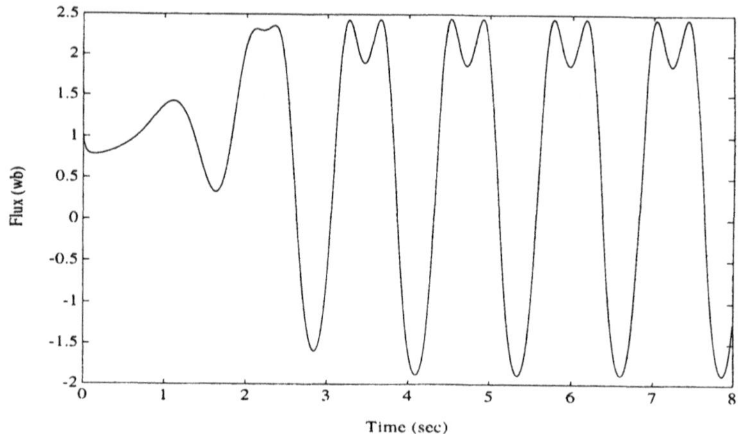

Figure 6.8: Transformed Phase One Flux for Adaptive Controller

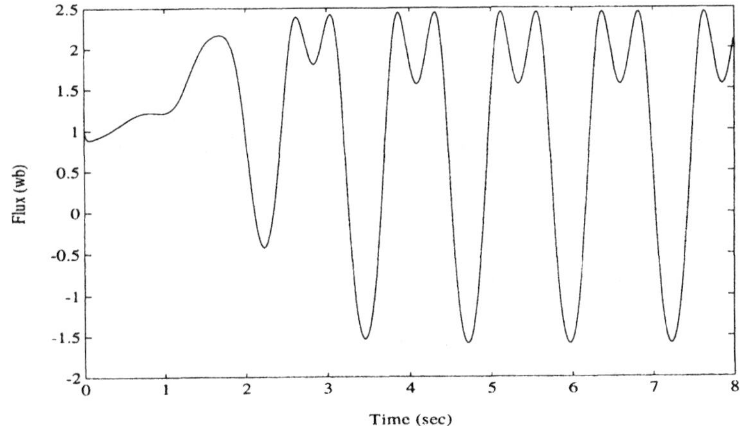

Figure 6.9: Transformed Phase Two Flux for Adaptive Controller

6.7 Notes

A good deal of current literature relates the development of nonlinear controllers for induction motors. Early work regarding induction motor control was primarily focused on the field-oriented vector control technique described in [9]. This technique uses nonlinear feedback to obtain approximate stator-rotor flux decoupling. An overview of the implementation of this control was presented in [10] by De Doncker. As pointed out in [6], a disadvantage of field-oriented control is that the rotor speed is only asymptotically decoupled from the rotor flux. In [11], Krzeminski introduced a new multi-scalar model of the induction motor and applied a nonlinear controller to this system to avoid the instabilities which may occur for certain conditions during field-oriented control. In [12] [13], Liaw *et al.* developed model reference adaptive controllers for velocity control of induction motors. In [14], Ho *et al.* proposed three different schemes for decoupling control of an induction motor based on the stator flux, airgap flux, and rotor flux field regulation. Given the assumption of slowly varying speed, Georgiou in [15] presented an adaptive feedback linearization controller for an induction motor. Based on suboptimal control theory, Sharkawi *et al.* developed an optimal quadratic tracking controller for an induction motor in [16] which required both exact knowledge of the motor parameters and full state measurements. Using a differential geometric approach, De Luca *et al.* [17] developed an exact model knowledge based nonlinear controller for induction motors with inputs taken to be the stator voltages and the slip frequency while the rotor speed was considered as a slowly varying parameter in relation to the electrical variables. Using a feedback linearization approach with input-output decoupling, Kim *et al.* [18] presented a controller for an induction motor which achieved both high dynamic performance and maximal power efficiency. In [6], Marino *et al.* developed an adaptive feedback linearization control which is capable of compensating for unknown load torque and rotor winding resistance. Chiasson *et al.* in [19] showed that if an integrator is added to one of the inputs to the induction motor, the resulting sixth-order system is feedback linearizable. In [20], Kadiyala presented an indirect adaptive nonlinear controller for an induction motor which achieved an asymptotic tracking result. Recently, in [21], Hu *et al.* developed an adaptive tracking controller for an induction motor driving a mechanical load which achieved asymptotic rotor position tracking. In [22], Dawson *et al.* developed a robust tracking controller for an induction motor driving a load which can compensate for parametric uncertainties and additive bounded disturbances throughout the entire electromechanical system. Bodson *et al.* [24] presented a motion controller for an induction motor through input-output linearization. In [23], Raumer *et al.* proposed and implemented a nonlinear controller for the induction motor using a digital signal processing.

Bibliography

[1] P. Krause, *Analysis of Electric Machinery*, McGraw Hill, 1986.

[2] W. Leonhard, *Control of Electrical Drives*, Springer-Verlag, 1985.

[3] M. Spong and M. Vidyasagar, *Robot Dynamics and Control*, New York: John Wiley and Sons, Inc., 1989.

[4] J. Slotine, "Putting Physics in Control — The example of Robotics", *Control System Magazine*, Vol. 8, pp. 12-17. Dec. 1988.

[5] S. Sastry and M. Bodson, *Adaptive Control: Stability, Convergence, and Robustness*, Englewoods Cliff, NJ: Prentice Hall Co., 1989.

[6] R. Marino, S. Peresada, and P. Valigi, "Adaptive Input-Output Linearizing Control of Induction Motors", *IEEE Transactions on Automatic Control*, Vol. 38, No. 2, Feb., 1993, pp. 208-221.

[7] J. Slotine and W. Li, *Applied Nonlinear Control*, Englewood Cliff, NJ: Prentice Hall Co., 1991.

[8] P. Vas, *Vector Control of AC Machines,* Oxford Science Publication, Clarendon Press, 1990.

[9] F. Blaschke, "The Principle of Field Orientation Applied to the New Transvector Closed-Loop Control System for Rotation Field Machines", *Siemens Review*, Vol. 39, pp. 217-220, 1972.

[10] R.W. De Doncker, D.W. Novotny, "The Universal Field Oriented Controller", *Conf. Rec., IEEE Industry Applications Society Annual Meeting,* 1988, pp. 450-456.

[11] Z. Krzeminski, "Nonlinear Control of the Induction Motor", *10th IFAC World Congress,* Munich, 1987, pp. 349-354.

[12] C.M. Liaw, C.T. Pan, and Y.C. Chen, "Design and Implementation of an Adaptive Controller for Current-Fed Induction Motor", *IEEE Transactions on Industrial Electronics.* Vol. 35, No. 3, pp. 393-401, 1988.

[13] F. Lin, and C.M. Liaw, "Reference Model Selection and Adaptive Control for Induction Motors Drives", *IEEE Transactions on Automatic Control,* Vol. 38, No. 10, pp. 1594-1600, 1993.

[14] E.Y. Ho, and P.C. Sen, "Decoupling Control of Induction Motor Drives", *IEEE Transactions on Industrial Electronics.* Vol. 35, No. 2, pp. 253-262, May, 1988.

[15] G. Georgiou, "Adaptive Feedback Linearization and Tracking for Induction Motors", *IFAC Evaluation of Adaptive Control Strategies,* Tbilisi, USSR, pp. 255-260, 1989.

[16] M.A. El-Sharkawi, and M. Akherraz, "Tracking Control Technique for Induction Motors", *IEEE Transactions on Energy Conversion,* Vol.4, No. 1, pp. 81-87, March 1989.

[17] A. De Luca, and G. Ulivi, "Design of Exact Nonlinear Controller for Induction Motors", *IEEE Transactions on Automatic Control,* Vol. 34, No. 12, pp. 1304-1307, 1989.

[18] D.I. Kim, I.J. Ha and M.S. Ko, "Control of Induction Motors via Feedback Linearization with Input-Output Decoupling", *International Journal of Control,* Vol. 51, No. 4, pp. 863-883, 1990.

[19] J. Chiasson, "Dynamic Feedback Linearization of the Induction Motor", *IEEE Transactions on Automatic Control,* Vol. 38, No. 10, pp. 1588-1594, October, 1993.

[20] R. Kadiyala, "Indirect Adaptive Nonlinear Control of Induction Motors", *Memorandum No. UCB/ERL M92/93,* Electronics Research Laboratory, College of Engineering, University of California, Berkeley.

[21] J. Hu, D. M. Dawson and Z. Qu, "Adaptive Tracking Control of an Induction Motor with Robustness to Parametric Uncertainty", *IEE Proceedings-B: Electric Power Applications,* Vol. 141, No. 2, pp. 85-94, March, 1994.

[22] J. Hu, D. Dawson, and Z. Qu, "Robust Tracking Control of an Induction Motor", *International Journal of Robust and Nonlinear Control*, Vol. 6, pp. 201-219, 1996.

[23] T. Raumer, J. Dion, and L. Dugard, "Applied Nonlinear Control of an Induction Motor using Digital Signal Processing", *IEEE Transactions on Control Systems Technology,* Vol. 2, No. 4, pp. 327-335, Dec. 1994.

[24] M. Bodson, J. Chiasson, and R. Novotnak, "High-Performance Induction Motor Control via Input-Output Linearization", *IEEE Control Systems Magazine*, Vol. 14, No. 4, pp. 25-33, Aug. 1994.

[25] M. Krstic, I. Kanellakopoulos, and P. Kokotovic, *Nonlinear and Adaptive Control Design,* John Wiley & Sons, 1995.

Chapter 7

BDC Motor (OFB)

7.1 Introduction

All of the controllers presented in the preceding chapters require full state feedback; therefore, for control implementation, we must measure the electrical current, the load velocity, and the load position. Since a reduction in the sensor count reduces the cost of the overall control system, we are compelled to investigate the construction of observers for estimating electrical current and load velocity for use in a closed-loop, load position tracking controller (*i.e.*, an output feedback (OFB) controller). In addition to eliminating the load velocity sensor for cost considerations, the use of a tachometer for measurement of load velocity typically renders a noisy signal. While the form of an exact-model knowledge based observer for load velocity and electrical current is motivated by the structure of the electromechanical model, the controller design is complicated by the fact that the use of the observer signals in the closed-loop control strategy could result in unsatisfactory tracking performance.

In this chapter, we revisit the problem of developing a controller for a brush dc motor (BDC). Specifically, we develop a load position tracking controller for the electromechanical model of a BDC motor driving a position-dependent load under the assumptions that only load position measurements are available and that an exact model for the electromechanical system dynamics can be determined. Specifically, by utilizing the structure of the electromechanical system, we design nonlinear observers to estimate load velocity and electrical current. A Lyapunov-like argument is then used to prove that the open-loop observers ensure that the load velocity and electrical current observation error converges to zero exponentially fast. We then illustrate that all of the internal signals inherent to the observer remain bounded as long as the controller is designed properly. That is, any proposed controller which uses the estimates must ensure that all of the signals in the actual electromechanical systems (*e.g.*, load position, load velocity, electrical current, and input voltage) remain bounded during closed-loop operation. Based on the structure of these observers, we then use an observed-integrator backstepping procedure and nonlinear damping tools [1] to design a load position tracking controller. A Lyapunov-like argument is then used to prove that all of the signals in the actual electromechanical systems and the observer remain bounded during closed-loop operation; furthermore, the controller ensures that the load position tracks the desired trajectory exponentially fast.

The reader should be aware that the observed integrator backstepping and nonlinear damping tools are the key to the design of the nonlinear observer-controller algorithm which ensures the boundedness of the internal signals inherent to the overall closed-loop system. That is, the separation principle [2] used in the design of linear feedback control systems utilizing state observers does not hold, in general, for nonlinear systems. Thus, one might question the validity of designing a controller using full state feedback and a state observer, and then simply using the observed states in place of the actual states. Rather, in this chapter, a more complex methodology motivated by the Lyapunov-like analysis is presented whereby the observer is designed in parallel with the controller. Specifically, the design of the composite observer-controller is based on a set of equations which describe the position tracking error dynamics, the observation error dynamics, and the corresponding Lyapunov-like functions. Moreover, the role of the observers is not limited to the usual function of providing an accurate estimate of a particular quantity but is expanded to include its contribution to the performance of the closed-loop system.

Due to the special structure of the BDC electromechanical model [3], the adaptive output feedback control structures of [4], [5], and [6] can be used to design a controller which only requires measurements of load position and compensates for parametric uncertainty throughout the entire systems. However, at the present time, it is not clear how to extend these techniques to the design of adaptive OFB controllers for multi-phase machines (*e.g.*, the switched reluctance motor). Hence, one should view the material in this chapter as introductory material for illustrating the design of exact model knowledge OFB and partial state feedback controllers for multi-phase machines.

7.2 System Model

The mechanical subsystem dynamics for a position-dependent load actuated by a permanent magnet brush dc motor are assumed to be of the form [7]

$$M\ddot{q} + B\dot{q} + N\sin(q) = I \tag{7.1}$$

where M denotes the lumped inertia, B denotes the friction coefficient, N denotes the lumped load term, $q(t)$ is the angular load position (and hence the position of the motor rotor), $\dot{q}(t)$ is the angular load velocity, $\ddot{q}(t)$ is the angular load acceleration, and $I(t)$ is the motor armature current (Note that the constant parameters M, B, and N described in (7.1) are defined to include the effects of the torque coefficient constant which characterizes the electromechanical conversion of armature current to torque). The electrical subsystem dynamics for the permanent magnet brush dc motor are assumed to be

$$L\dot{I} = v - RI - K_B\dot{q} \tag{7.2}$$

where L is the constant armature inductance, R is the constant armature resistance, K_B is the constant back-emf coefficient, and $v(t)$ is the input control voltage.

7.3 Control Objective

Given measurement of load position only (*i.e.*, $q(t)$), the control objective is to develop a load position tracking controller for the electromechanical dynamics of (7.1) and (7.2). To facilitate the tracking control formulation, we define the load position tracking error $e(t)$ as

$$e = q_d - q \tag{7.3}$$

where $q_d(t)$ represents the desired load position trajectory, and $q(t)$ was defined in (7.1). We will assume that $q_d(t)$ and its first, second, and third time derivatives are all bounded functions of time.

Since the control problem is constrained by the fact that the only measurable state is the load position, the remaining state variables (*i.e.*, the electrical current $I(t)$ and load velocity $\dot{q}(t)$) will be observed. As the term observation implies there will be some variation between the physical and observed quantities. This variation, called observation error, is defined for the velocity and current observers as

$$\dot{\tilde{q}} = \dot{q} - \dot{\hat{q}} \tag{7.4}$$

and

$$\tilde{I} = I - \hat{I} \tag{7.5}$$

where $\dot{\hat{q}}\,(t)$ denotes the estimate of the load velocity, and $\hat{I}(t)$ denotes the estimate of the electrical current. A necessary requirement in the overall control strategy will be to ensure that the observation errors $\dot{\tilde{q}}\,(t)$ and $\tilde{I}(t)$ are driven to zero.

7.4 Observer Formulation

Based on the structure of the mechanical subsystem dynamics given by (7.1) and the subsequent stability analysis, the following load velocity observer is proposed

$$\dot{\hat{q}} = y + \frac{1}{M}\left(k_{n1}\left(1 - K_B\right)^2 + k_{n2} + k_s\right)\tilde{q} \tag{7.6}$$

where the position observation error is defined as

$$\tilde{q} = q - \hat{q}, \tag{7.7}$$

the auxiliary variable $y(t)$ is updated according to

$$\dot{y} = \frac{1}{M}\left(\hat{I} - B\,\dot{\hat{q}} - N\sin(q)\right), \tag{7.8}$$

and k_{n1}, k_{n2}, k_s are positive, constant controller gains. The electrical current observer is designed to mimic the electrical subsystem in (7.2) where the estimates for electrical current and load velocity replace the corresponding physical system variables. The proposed electrical current observer is given by

$$\dot{\hat{I}} = \frac{1}{L}\left(v - R\hat{I} - K_B\,\dot{\hat{q}}\right). \tag{7.9}$$

Remark 7.1

After taking the derivative of (7.6) with respect to time, we have

$$\ddot{\hat{q}} = \dot{y} + \frac{1}{M}\left(k_{n1}\left(1 - K_B\right)^2 + k_{n2} + k_s\right)\dot{\tilde{q}}. \tag{7.10}$$

Substituting for $\dot{y}(t)$ from (7.8) into (7.10) then multiplying both sides of the resulting expression by M yields the following second-order expression for the load velocity observer

$$M\,\ddot{\hat{q}} + B\,\dot{\hat{q}} + N\sin(q) - \left(k_{n1}\left(1 - K_B\right)^2 + k_{n2} + k_s\right)\dot{\tilde{q}} = \hat{I}. \tag{7.11}$$

During the subsequent analysis, we will use the load velocity observer in the form given by (7.11) to develop the load velocity observation error system. However, we should note that (7.11) can only be utilized for analysis purposes since the load velocity observer in the above second-order form contains the unmeasurable quantity $\dot{\tilde{q}}\,(t)$. Hence, the velocity observer expressions given by (7.6) and (7.8) are utilized for actual implementation.

Remark 7.2

While the appearance of most of the terms in the above observer dynamics are due to the structure of the corresponding actual electromechanical dynamics, we note that there are several additional terms which are used for the compensation of other terms during the stability analysis. For example, some of the terms multiplied by $\tilde{q}(t)$ in (7.6) (and hence the same terms multiplying $\dot{\tilde{q}}\,(t)$ in (7.11)) are nonlinear damping terms which are utilized to "damp-out" other terms associated with the unmeasurable quantity $\dot{\tilde{q}}\,(t)$. This nonlinear damping action, which is directly related to Lemma 1.8 in Chapter 1, will be made clear during the subsequent stability analysis of the observation error dynamics and the position tracking error dynamics.

7.4.1 Observation Error Dynamics

In this section, we formulate the observation error dynamics for the load velocity and current observers. A Lyapunov-like function is then used to illustrate that the observation errors go to zero exponentially fast when used in an open-loop fashion (*i.e.*, the estimates, $\dot{\hat{q}}(t)$ and $\hat{I}(t)$, have not yet been utilized in a voltage control input algorithm). In the next section, we will design a voltage input controller which utilizes these estimates in a stable, closed-loop fashion.

The velocity observation error dynamics are derived from the definition of the velocity observation error. Specifically, the velocity observation error of (7.4) is differentiated with respect to time and multiplied by M to yield

$$M\ddot{\tilde{q}} = M\ddot{q} - M\ddot{\hat{q}}. \tag{7.12}$$

Substitution of the mechanical dynamics given in (7.1) for $M\ddot{q}$ and the second-order velocity observer given in (7.11) for $M\ddot{\hat{q}}$ into the right-hand side of (7.12) yields

$$M\ddot{\tilde{q}} = -B\dot{\tilde{q}} - \left(k_{n1}\left(1 - K_B\right)^2 + k_{n2} + k_s\right)\dot{\tilde{q}} + \tilde{I} \tag{7.13}$$

where (7.4) and (7.5) have been utilized. Roughly speaking, the form of (7.13) motivates one to classify the observer given by (7.6) and (7.8) as a closed-loop observer because a controller gain (*e.g.*, k_s) can be increased to cause $\dot{\tilde{q}}(t)$ to go to zero faster (Note, for this statement to be true in the strictest sense, $\tilde{I}(t)$ must be assumed to be zero).

The current observation error dynamics are derived from the definition of the current observation error. Specifically, the current observation error of (7.5) is differentiated with respect to time and multiplied by L to yield

$$L\dot{\tilde{I}} = L\dot{I} - L\dot{\hat{I}}. \tag{7.14}$$

Substitution of the electrical dynamics given in (7.2) for $L\dot{I}$ and the current observer given in (7.9) for $L\dot{\hat{I}}$ into the right-hand side of (7.14) yields

$$L\dot{\tilde{I}} = -R\tilde{I} - K_B\dot{\tilde{q}} \tag{7.15}$$

where (7.4) and (7.5) have been utilized. Roughly speaking, the form of (7.15) motivates one to classify the observer given by (7.9) as an open-loop observer because one can not increase a controller gain to cause $\tilde{I}(t)$ to go to zero faster (Note, for this statement to have meaning in the strictest sense, $\dot{\tilde{q}}(t)$ must be assumed to be zero).

7.4.2 Stability of the Observation Error Systems

To simplify the subsequent stability analysis of the voltage control input which will be designed in the next section and to provide insight into the construction of the observers, we now illustrate that the observers of (7.6), (7.8), and (7.9) provide for exponentially stable observation of load velocity and electrical current. The theorem given below delineates the performance of the observers.

Theorem 7.1

The observation error systems of (7.13) and (7.15) are exponentially stable in the sense that

$$\|z(t)\| \leq \sqrt{\frac{\lambda_b}{\lambda_a}}\,\|z(0)\|\,e^{-\gamma_c t} \qquad \forall t \in [0, \infty) \tag{7.16}$$

where

$$z = \begin{bmatrix} \dot{\tilde{q}} & \tilde{I} \end{bmatrix}^T \in \Re^2, \tag{7.17}$$

$$\lambda_a = \min\{M, L\}, \qquad \lambda_b = \max\{M, L\}, \tag{7.18}$$

and

$$\gamma_c = \frac{\min\left\{k_s + k_{n2}, \left(R - \frac{1}{k_{n1}}\right)\right\}}{\max\{M, L\}}. \tag{7.19}$$

Note that the controller gain k_{n1} must be selected to satisfy

$$k_{n1} > \frac{1}{R} \tag{7.20}$$

to guarantee the stability result of (7.16) (*i.e.*, to ensure that γ_c of (7.19) is positive).

Proof. First, we define the following non-negative function

$$V_o(t) = \frac{1}{2}M\,\dot{\tilde{q}}^2 + \frac{1}{2}L\tilde{I}^2 = \frac{1}{2}z^T diag\{M, L\}z \tag{7.21}$$

where $z(t)$ was defined in (7.17), and $diag\{M, L\} \in \Re^{2\times 2}$ is used to denote a diagonal 2×2 matrix with M and L arranged along the main diagonal, and with zeros inserted for all the other elements. From the matrix form of $V_o(t)$ given on the right-hand side of (7.21), Lemma 1.7 in Chapter 1 can be used to form the following upper and lower bounds for $V_o(t)$ as

$$\frac{1}{2}\lambda_a\,\|z(t)\|^2 \leq V_o(t) \leq \frac{1}{2}\lambda_b\,\|z(t)\|^2 \tag{7.22}$$

where λ_a and λ_b are defined in (7.18).

Differentiating (7.21) with respect to time yields

$$V_o(t) = \dot{\tilde{q}}\, M\, \ddot{\tilde{q}} + \tilde{I} L\, \dot{\tilde{I}}\,. \tag{7.23}$$

Substituting the right-hand sides of (7.13) for $M\,\ddot{\tilde{q}}$ and of (7.15) for $L\,\dot{\tilde{I}}$ into (7.23) yields

$$\begin{aligned} \dot{V}_o(t) &= -(B+k_s)\,\dot{\tilde{q}}^2 - R\tilde{I}^2 - k_{n2}\,\dot{\tilde{q}}^2 + \\ &\quad (1-K_B)\,\tilde{I}\,\dot{\tilde{q}} - k_{n1}\,(1-K_B)^2\,\dot{\tilde{q}}^2 \end{aligned} \tag{7.24}$$

after collecting some of the common terms. From (7.24), the following upper on $\dot{V}_o(t)$ can be formulated

$$\begin{aligned} \dot{V}_o(t) &\leq -k_s\,\dot{\tilde{q}}^2 - R\tilde{I}^2 - k_{n2}\,\dot{\tilde{q}}^2 + \\ &\quad \left[|1-K_B|\left|\tilde{I}\right|\left|\dot{\tilde{q}}\right| - k_{n1}\,(1-K_B)^2\,\dot{\tilde{q}}^2\right]. \end{aligned} \tag{7.25}$$

The bracketed term in (7.25) is a nonlinear damping pair; hence, we can use Lemma 1.8 in Chapter 1 to further upper bound $\dot{V}_o(t)$ as

$$\begin{aligned} \dot{V}_o(t) &\leq -(k_s+k_{n2})\,\dot{\tilde{q}}^2 - \left(R - \frac{1}{k_{n1}}\right)\tilde{I}^2 \\ &= -z^T diag\left\{k_s + k_{n2}, \left(R - \frac{1}{k_{n1}}\right)\right\} z. \end{aligned} \tag{7.26}$$

where $z(t)$ was defined in (7.17). Since the controller gain k_{n1} is selected to satisfy (7.20), Lemma 1.7 in Chapter 1 can be utilized to upper bound $\dot{V}_o(t)$ in (7.26) as

$$\dot{V}_o(t) \leq -\min\left\{k_s + k_{n2}, \left(R - \frac{1}{k_{n1}}\right)\right\} \|z(t)\|^2 . \tag{7.27}$$

From (7.22), it easy to see that $\|z(t)\|^2 \geq 2V_o(t)/\lambda_b$; hence, $\dot{V}_o(t)$ in (7.27) can be further upper bounded as

$$\dot{V}_o(t) \leq -2\gamma_c V_o(t) \tag{7.28}$$

where γ_c was defined in (7.19). Applying Lemma 1.1 in Chapter 1 to (7.28) yields

$$V_o(t) \leq V_o(0) e^{-2\gamma_c t}. \tag{7.29}$$

From (7.22), we have that

$$\frac{1}{2}\lambda_a \|z(t)\|^2 \leq V_o(t) \quad \text{and} \quad V_o(0) \leq \frac{1}{2}\lambda_b \|z(0)\|^2 . \tag{7.30}$$

Substituting (7.30) appropriately into the left-hand and right-hand sides of (7.29) allows us to form the following inequality

$$\frac{1}{2}\lambda_a \|z(t)\|^2 \leq \frac{1}{2}\lambda_b \|z(0)\|^2 \mathrm{e}^{-2\gamma_c t}. \tag{7.31}$$

Using (7.31), we can solve for $\|z(t)\|$ to obtain the result given in (7.16). □

Remark 7.3

From the result given by (7.16) and (7.17), we form the following upper bounds for $\dot{\tilde{q}}(t)$ and $\tilde{I}(t)$

$$\left|\dot{\tilde{q}}(t)\right| \leq \sqrt{\frac{\lambda_b}{\lambda_a}} \|z(0)\| \mathrm{e}^{-\gamma_c t} \quad \text{and} \quad \left|\tilde{I}(t)\right| \leq \sqrt{\frac{\lambda_b}{\lambda_a}} \|z(0)\| \mathrm{e}^{-\gamma_c t} \tag{7.32}$$

$\forall t \in [0,\infty)$; hence, the velocity observation error (*i.e.*, $\dot{\tilde{q}}(t)$) and the current observation error (*i.e.*, $\tilde{I}(t)$) both go to zero exponentially fast.

Remark 7.4

Using the result of Theorem 7.1, the structure of the observers, and the electromechanical model, it is straightforward to illustrate that all signals remain bounded during open-loop operation provided that $q(t) \in L_\infty$, $\dot{q}(t) \in L_\infty$, $\ddot{q}(t) \in L_\infty$, $I(t) \in L_\infty$, and $v(t) \in L_\infty$. Specifically, from (7.16) and (7.17), we know that $\dot{\tilde{q}}(t) \in L_\infty$ and $\tilde{I}(t) \in L_\infty$; hence, we can use (7.4) and (7.5) to show that $\dot{\hat{q}}(t) \in L_\infty$ and $\hat{I}(t) \in L_\infty$. In addition since $\dot{\tilde{q}}(t)$ satisfies (7.32), Lemma 1.9 in Chapter 1 can be used to show that $\tilde{q}(t) \in L_\infty$; hence, (7.7) can be used to show that $\hat{q}(t) \in L_\infty$. Since all the terms on the right-hand side of (7.13) are bounded, we now know that $\ddot{\tilde{q}}(t) \in L_\infty$; hence, (7.12) can be used to show that $\ddot{\hat{q}}(t) \in L_\infty$. Based on the above information, (7.6) and (7.8) can be utilized to show that $y(t) \in L_\infty$ and $\dot{y}(t) \in L_\infty$.

Remark 7.5

As illustrated by (7.6) and (7.10), there are two nonlinear damping gains k_{n1} and k_{n2} (See Lemma 1.8 in Chapter 1) in the velocity observer. It can easily be established that the result of Theorem 7.1 is still valid even if k_{n2} is set to zero. However, as we will see later, the controller gain k_{n2} is used to "damp-out" a velocity observation error term during the composite controller-observer analysis presented in a subsequent section.

7.5 Voltage Control Input Design

In this section, we design a voltage control input which drives the position tracking error defined in (7.3) to zero. As mentioned in the introduction, the control design is constrained by the fact that only measurements of load position (*i.e.*, $q(t)$) are assumed to be available (*i.e.*, output feedback (OFB)). Since measurements of load velocity and electrical current are not available, the observed load velocity and the observed electrical current are utilized in the formulation of the voltage control input. In many ways the design of the output feedback controller is similar to the design of the full state feedback controller (FSFB), exact model knowledge controller presented in Chapter 2. For example, the FSFB controller utilized the following filtered tracking error definition

$$r = \dot{e} + \alpha e = \dot{q}_d - \dot{q} + \alpha e \tag{7.33}$$

where α is a positive control constant. Note that the above definition of the filtered tracking error given by (7.33) depends on measurements of load velocity (*i.e.*, $\dot{q}(t)$) and load position. Since the OFB controller can only use load position measurements, we define the observed filtered tracking error as

$$\hat{r} = \dot{q}_d - \dot{\hat{q}} + \alpha e. \tag{7.34}$$

Note that the above definition of the observed filtered tracking error given by (7.34) only depends on the measurement of load position.

A second similarity between the FSFB controller and the OFB controller is the notion of current tracking error. That is, in Chapter 2, the current tracking error was defined as

$$\eta_I = I_d - I \tag{7.35}$$

where $I_d(t)$ was used to denote the desired current trajectory. Note that the above definition of the current tracking error given by (7.35) depends on measurements of the electrical current (*i.e.*, $I(t)$). Since the OFB controller can only use load position measurements, we define the observed current tracking error as

$$\hat{\eta}_I = I_d - \hat{I} \tag{7.36}$$

where $I_d(t)$ is again used to denote the desired current trajectory which must be designed to move the mechanical load along a desired position trajectory. Note that the above definition of the observed current tracking error given by (7.36) does not depend on measurements of electrical current.

Before we begin the design of the voltage control input, the overall voltage input control strategy can be summarized, in a heuristic manner, using the above auxiliary definitions. That is, the role of the voltage control input will be to force $\hat{I}(t)$ to track $I_d(t)$ in the electrical subsystem while the combined action of the observer and controller will cause $I(t)$ to track $\hat{I}(t)$; hence, the overall effect is that $I(t)$ will be forced to track $I_d(t)$. The result of the actual current tracking the desired current trajectory is that the desired torque will be effectively applied to the mechanical subsystem, which the will cause $q(t)$ to track $q_d(t)$ and $\dot{\hat{q}}(t)$ to track $\dot{q}_d(t)+\alpha e(t)$ in the mechanical subsystem. The combined action of the velocity observer and the desired current trajectory will then force $\dot{q}(t)$ to track $\dot{\hat{q}}(t)$. In short, the combined function of the composite observer-controller is to force $I(t)$ to track $I_d(t)$ and as a result force $q(t)$ to track $q_d(t)$ and force $\dot{q}(t)$ to track $\dot{q}_d(t) + \alpha e(t)$. Thus the observers and the voltage control input work together to cause tracking of the internal system states to the designed trajectories and thereby achieve tracking in the output state $q(t)$ (*i.e.*, the load position trajectory).

7.5.1 Position Tracking Error Dynamics

In this section, we develop the position tracking error dynamics which are utilized in the subsequent stability proof to show that the position tracking error goes to zero exponentially fast. During the development of the error dynamics, we formulate the desired current trajectory and the voltage control input. First, the position tracking of (7.3) can be differentiated with respect to time to yield

$$\dot{e} = \dot{q}_d - \dot{q}. \tag{7.37}$$

Since there is no control input on the right-hand side of (7.37), we add and subtract the observed filtered tracking error of (7.34) to yield

$$\dot{e} = \dot{q}_d - \hat{r} + \hat{r} - \dot{q}. \tag{7.38}$$

After substituting the right-hand side of the (7.34) for the first occurrence of $\hat{r}(t)$ only, we obtain the closed-loop position tracking error dynamics in the form

$$\dot{e} = -\alpha e + \hat{r} - \dot{\tilde{q}} \tag{7.39}$$

where (7.4) has been utilized. From (7.39), we can see that if $\hat{r}(t)$ and $\dot{\tilde{q}}(t)$ were both equal to zero, then the position tracking error would converge to zero exponentially fast. From our previous analysis of the observation error systems analysis, we have a good feeling about our ability to drive $\dot{\tilde{q}}(t)$ to zero. Hence, motivated by the above discussion and the form of (7.39), our controller must ensure that $\hat{r}(t)$ is driven to zero.

To accomplish this new control objective, we need to construct the open-loop dynamics for $\hat{r}(t)$. To this end, we multiply both sides of (7.34) by M and then take the time derivative to yield

$$M \dot{\hat{r}} = M\left(\ddot{q}_d + \alpha\dot{e}\right) - M \ddot{\hat{q}}. \tag{7.40}$$

Substituting the second-order load velocity observer dynamics of (7.11) into (7.40) and arranging the resulting expression in an advantageous manner, we obtain

$$M\dot{\hat{r}} = \Omega_1 + \Omega_2\dot{\tilde{q}} - \hat{I} \tag{7.41}$$

where (7.4) has been utilized, and $\Omega_1(t)$, Ω_2 are *measurable* auxiliary variables defined as

$$\Omega_1 = M\left(\ddot{q}_d + \alpha\dot{q}_d\right) + N\sin(q) + (B - M\alpha)\,\dot{\hat{q}} \tag{7.42}$$

and

$$\Omega_2 = -\left(k_{n1}\left(1 - K_B\right)^2 + k_{n2} + k_s + M\alpha\right). \tag{7.43}$$

Since there is no control input on the right-hand side of (7.41), we add and subtract a desired current trajectory $I_d(t)$ to yield

$$M\dot{\hat{r}} = \Omega_1 + \Omega_2\dot{\tilde{q}} - I_d + \hat{\eta}_I \tag{7.44}$$

where $\hat{\eta}_I(t)$ is used to represent the observed current tracking error and was explicitly defined in (7.36). The desired current trajectory $I_d(t)$ in (7.44) will now be designed to force the observed filtered tracking error $\hat{r}(t)$ to zero. Specifically, we select $I_d(t)$ as follows

$$I_d = \Omega_1 + k_{n3}\Omega_2^2\hat{r} + k_{n4}\hat{r} + k_s\hat{r} \tag{7.45}$$

where k_{n3}, k_{n4}, and k_s are positive control gains. Substituting $I_d(t)$ of (7.45) into (7.44) yields the closed-loop dynamics for $\hat{r}(t)$ as follows

$$M\dot{\hat{r}} = -k_s\hat{r} + \left[\Omega_2\dot{\tilde{q}} - k_{n3}\Omega_2^2\hat{r}\right] - k_{n4}\hat{r} + \hat{\eta}_I. \tag{7.46}$$

Remark 7.6

To explain the origins of some of the terms in (7.46), we note that the auxiliary control term $k_{n4}\hat{r}$ will be used to damp-out the $\hat{r}$ term in (7.39) during the subsequent position tracking analysis. The bracketed term in (7.46) is a nonlinear damping pair (See Lemma 1.8 in Chapter 1), and hence, since the previous observation error stability analysis indicates that $\dot{\tilde{q}}(t)$ can be driven to zero, we are motivated by the form of (7.46) to drive $\hat{\eta}_I(t)$ to zero. Based on the above observations, we can see that if $\hat{\eta}_I(t)$ is zero then $\hat{r}(t)$ will go to zero.

To ensure that the observed current tracking error goes to zero, we first construct the open-loop dynamics for $\hat{\eta}_I(t)$. To this end, we multiply both sides of (7.36) by L and take the time derivative to obtain the following expression

$$L\dot{\hat{\eta}}_I = L\dot{I}_d - L\dot{\hat{I}}. \tag{7.47}$$

After taking the time derivative of (7.45) for $\dot{I}_d$ and substituting the resulting expression into (7.47), we have

$$L\,\dot{\hat{\eta}}_I = L\left(\dot{\Omega}_1 + \left(k_{n3}\Omega_2^2 + k_{n4} + k_s\right)\dot{\hat{r}}\right) - L\,\dot{\hat{I}}\,. \tag{7.48}$$

Continuing with the construction of the open-loop, observed current tracking error dynamics, we substitute the time derivative of (7.42) for $\dot{\Omega}_1(t)$ into (7.48) to obtain

$$\begin{aligned} L\,\dot{\hat{\eta}}_I &= L\left(M\left(\dddot{q}_d + \alpha\ddot{q}_d\right) + N\cos(q)\dot{q} + (B - M\alpha)\,\ddot{\hat{q}}\right) \\ &\quad + L\left(k_{n3}\Omega_2^2 + k_{n4} + k_s\right)\dot{\hat{r}} - L\,\dot{\hat{I}}\,. \end{aligned} \tag{7.49}$$

Finally to complete the description, we substitute the velocity observer dynamics of (7.11) for $\ddot{\hat{q}}\,(t)$, the observed filtered tracking error dynamics of (7.46) for $\dot{\hat{r}}\,(t)$, the current observer dynamics of (7.9) for $L\,\dot{\hat{I}}$, and group the measurable and unmeasurable terms in an advantageous manner to obtain the following expression

$$L\,\dot{\hat{\eta}}_I = \Omega_3 + \Omega_4\dot{q} + \Omega_5\,\ddot{\tilde{q}} - v \tag{7.50}$$

where $\Omega_3(t)$, $\Omega_4(t)$, and $\Omega_5(t)$ are measurable auxiliary variables defined as

$$\begin{aligned} \Omega_3 = {} & \frac{L}{M}(-B + M\alpha)\left(B\,\dot{\hat{q}} + N\sin(q) - \hat{I}\right) \\ & + LM\left(\dddot{q}_d + \alpha\ddot{q}_d\right) + \frac{L}{M}\left(k_{n3}\Omega_2^2 + k_{n4} + k_s\right) \\ & \cdot\left(-k_s\hat{r} - k_{n3}\Omega_2^2\hat{r} - k_{n4}\hat{r} + \hat{\eta}_I\right) + R\hat{I} + K_B\,\dot{\hat{q}}, \end{aligned} \tag{7.51}$$

$$\Omega_4 = LN\cos(q), \tag{7.52}$$

and

$$\begin{aligned} \Omega_5 &= \frac{L}{M}(B - M\alpha)\left(k_{n1}\left(1 - K_B\right)^2 + k_{n2} + k_s\right) \\ &\quad + \frac{L}{M}\left(k_{n3}\Omega_2^2 + k_{n4} + k_s\right)\Omega_2. \end{aligned} \tag{7.53}$$

Based on the structure of (7.50) and the subsequent stability analysis, we propose the following voltage control input to drive the observed current tracking error to zero

$$v = \Omega_3 + \Omega_4\,\dot{\hat{q}} + k_{n5}\left(\Omega_4 + \Omega_5\right)^2\hat{\eta}_I + k_s\hat{\eta}_I + k_{n6}\hat{\eta}_I \tag{7.54}$$

where k_s, k_{n5}, and k_{n6} are positive control gains. Substituting the voltage control of (7.54) into (7.50) yields the closed-loop dynamics for $\hat{\eta}_I(t)$ in the form

$$L\dot{\hat{\eta}}_I = -k_s\hat{\eta}_I - k_{n6}\hat{\eta}_I + \left[(\Omega_4+\Omega_5)\dot{\tilde{q}} - k_{n5}(\Omega_4+\Omega_5)^2\hat{\eta}_I\right]. \qquad (7.55)$$

Remark 7.7

To explain the origins of some of the terms in (7.55), we note that the auxiliary control term $k_{n6}\hat{\eta}_I$ will be used to damp-out the $\hat{\eta}_I$ term in (7.46) during the subsequent position tracking analysis. The bracketed term in (7.55) is a nonlinear damping pair, and hence, since the previous observation error stability analysis indicates that $\dot{\tilde{q}}(t)$ can be driven to zero, we now see that the form of (7.55) has been specifically constructed to ensure that $\hat{\eta}_I(t)$ is driven to zero.

Remark 7.8

The dynamics given by (7.39), (7.46), and (7.55), represent the closed-loop position tracking error system for which the stability analysis is performed while the controller given by (7.54) represents the controller which is implemented at the voltage terminals of the motor. Note that the desired current trajectory $I_d(t)$ of (7.45) is embedded (in the guise of the variable $\hat{\eta}_I(t)$) inside of the voltage control input $v(t)$.

7.5.2 Stability of the Position Tracking Error System

In order to analyze the stability of the closed-loop position tracking error system, we define the following non-negative function

$$V_p(t) = \frac{1}{2}e^2 + \frac{1}{2}M\hat{r}^2 + \frac{1}{2}L\hat{\eta}_I^2. \qquad (7.56)$$

The time derivative of (7.56) is given by

$$\dot{V}_p(t) = e\dot{e} + \hat{r}M\dot{\hat{r}} + \hat{\eta}_I L\dot{\hat{\eta}}_I. \qquad (7.57)$$

Substituting (7.39), (7.46), and (7.55) into (7.57) for $\dot{e}$, $M\dot{\hat{r}}$, and $L\dot{\hat{\eta}}_I$, respectively, yields

$$\begin{aligned} \dot{V}_p(t) = \; & -\alpha e^2 + e\hat{r} - \dot{\tilde{q}}\,e - k_s\hat{r}^2 + \left[\Omega_2\dot{\tilde{q}}\hat{r} - k_{n3}\Omega_2^2\hat{r}^2\right] \\ & -k_{n4}\hat{r}^2 + \hat{\eta}_I\hat{r} - k_s\hat{\eta}_I^2 - k_{n6}\hat{\eta}_I^2 \\ & + \left[(\Omega_4+\Omega_5)\dot{\tilde{q}}\hat{\eta}_I - k_{n5}(\Omega_4+\Omega_5)^2\hat{\eta}_I^2\right]. \end{aligned} \qquad (7.58)$$

Utilizing (7.58), $\dot{V}_p(t)$ can be upper bounded as follows

$$\begin{aligned}\dot{V}_p(t) \leq\ & -\alpha e^2 - k_s\hat{r}^2 - k_s\hat{\eta}_I^2 + \left|\dot{\tilde{q}}\right| |e| + \left[|e|\,|\hat{r}| - k_{n4}\hat{r}^2\right] \\ & + \left[|\hat{\eta}_I|\,|\hat{r}| - k_{n6}\hat{\eta}_I^2\right] + \left[|\Omega_2| \left|\dot{\tilde{q}}\right| |\hat{r}| - k_{n3}\Omega_2^2\hat{r}^2\right] \\ & + \left[|\Omega_4 + \Omega_5| \left|\dot{\tilde{q}}\right| |\hat{\eta}_I| - k_{n5}(\Omega_4 + \Omega_5)^2\hat{\eta}_I^2\right].\end{aligned} \tag{7.59}$$

We now apply Lemma 1.8 in Chapter 1 to the four bracketed terms on the right-hand side of (7.59) to form the following upper bound for $\dot{V}_p(t)$ as

$$\begin{aligned}\dot{V}_p(t) \leq\ & -\left(\alpha - \frac{1}{k_{n4}}\right)e^2 - \left(k_s - \frac{1}{k_{n6}}\right)\hat{r}^2 - k_s\hat{\eta}_I^2 \\ & + \left[\left|\dot{\tilde{q}}\right| |e|\right] + \left[\left(\frac{1}{k_{n3}} + \frac{1}{k_{n5}}\right)\dot{\tilde{q}}^2\right].\end{aligned} \tag{7.60}$$

Remark 7.9

From the form of (7.60), we can see that if $\dot{\tilde{q}}\ (t)$ was equal to zero for all time, and the controller gains were selected to satisfy the inequalities $\alpha > \dfrac{1}{k_{n4}}$ and $k_s > \dfrac{1}{k_{n6}}$, then we could utilize Lemma 1.1 in Chapter 1 to guarantee that $e(t)$, $\hat{r}(t)$, and $\hat{\eta}_I(t)$ would all go to zero. As we will see later during the analysis of the composite observer-controller error systems, the last two bracketed terms in (7.60) can be combined with other terms in (7.26) to obtain the desired stability result. For example, it is easy to see that the last bracketed term in (7.60) can be combined with a similar term in (7.26). Therefore if the controller gains are selected appropriately, these two terms can be made to produce a non-positive term when the time derivatives of the two Lyapunov-like functions (*i.e.*, $V_o(t)$ of (7.21) and $V_p(t)$ of (7.56)) are added together.

7.6 Stability of the Composite Error Systems

In this section, we will analyze the performance of the composite observer-controller error systems. The performance of these closed-loop error systems is illustrated by the following theorem.

Theorem 7.2

Given the nonlinear observers of (7.6), (7.8), and (7.9) and the voltage input control of (7.54), the load position tracking error is exponentially stable in the sense that

$$\|x(t)\| \leq \sqrt{\frac{\lambda_2}{\lambda_1}}\, \|x(0)\|\, \underline{e}^{-\gamma t} \qquad \forall t \in [0, \infty) \tag{7.61}$$

where

$$x = \begin{bmatrix} \dot{\tilde{q}} & \tilde{I} & e & \hat{r} & \hat{\eta}_I \end{bmatrix}^T \in \Re^5, \tag{7.62}$$

$$\lambda_1 = \min\{M,\ L,\ 1\}, \qquad \lambda_2 = \max\{M,\ L,\ 1\}, \tag{7.63}$$

and

$$\gamma = \frac{\min\left\{\left(R - \frac{1}{k_{n1}}\right), \left(\alpha - \frac{1}{k_{n2}} - \frac{1}{k_{n4}}\right), \left(k_s - \frac{1}{k_{n6}}\right), \left(k_s - \frac{1}{k_{n3}} - \frac{1}{k_{n5}}\right)\right\}}{\max\{M,\ L,\ 1\}}. \tag{7.64}$$

To ensure the above stability result, the controller gains must be adjusted to satisfy the following sufficient condition

$$k_{ni} > \max\left\{\frac{1}{R}, \frac{2}{k_s}, \frac{2}{\alpha}\right\} \quad \text{where } i = 1,\ \cdots,\ 6. \tag{7.65}$$

Proof. First, we define the following non-negative function

$$V(t) = V_o(t) + V_p(t) = \frac{1}{2}x^T diag\{M, L, 1, M, L\}x \tag{7.66}$$

where $x(t)$ was defined in (7.62), $V_o(t)$ was defined in (7.21), $V_p(t)$ was defined in (7.56), and $diag\{M, L, 1, M, L\} \in \Re^{5\times5}$. From the matrix form of $V(t)$ given on the right-hand side of (7.66), Lemma 1.7 in Chapter 1 can be used to form the following upper and lower bounds for $V(t)$

$$\frac{1}{2}\lambda_1 \|x(t)\|^2 \leq V(t) \leq \frac{1}{2}\lambda_2 \|x(t)\|^2 \tag{7.67}$$

where λ_1 and λ_2 are defined in (7.63).

From (7.26) and (7.60), an upper bound on the time derivative of $V(t)$ defined in (7.66) is given by

$$\begin{aligned} \dot{V}(t) \leq & -\left(\alpha - \frac{1}{k_{n4}}\right)e^2 - \left(k_s - \frac{1}{k_{n6}}\right)\hat{r}^2 + \left[\left|\dot{\tilde{q}}\right||e| - k_{n2}\,\dot{\tilde{q}}^2\right] \\ & -k_s\hat{\eta}_I^2 - \left(k_s - \frac{1}{k_{n3}} - \frac{1}{k_{n5}}\right)\dot{\tilde{q}}^2 - \left(R - \frac{1}{k_{n1}}\right)\tilde{I}^2. \end{aligned} \tag{7.68}$$

After applying Lemma 1.8 in Chapter 1 to the bracketed term in (7.68), Lemma 1.7 in Chapter 1 can be utilized to upper bound $\dot{V}(t)$ in (7.68) as

$$\begin{aligned} \dot{V}(t) \leq & -\min\left\{\left(R - \frac{1}{k_{n1}}\right), \left(\alpha - \frac{1}{k_{n2}} - \frac{1}{k_{n4}}\right),\right. \\ & \left.\left(k_s - \frac{1}{k_{n6}}\right), \left(k_s - \frac{1}{k_{n3}} - \frac{1}{k_{n5}}\right)\right\}\|x(t)\|^2 \end{aligned} \tag{7.69}$$

where $x(t)$ was defined in (7.62). From (7.67), it easy to see that $\|x(t)\|^2 \geq 2V(t)/\lambda_2$; hence, $\dot{V}(t)$ in (7.69) can be further upper bounded as

$$\dot{V}(t) \leq -2\gamma V(t) \tag{7.70}$$

where γ was defined in (7.64). Applying Lemma 1.1 in Chapter 1 to (7.70) yields

$$V(t) \leq V(0)\underline{e}^{-2\gamma t}. \tag{7.71}$$

From (7.67), we have that

$$\frac{1}{2}\lambda_1 \|x(t)\|^2 \leq V(t) \qquad \text{and} \qquad V(0) \leq \frac{1}{2}\lambda_2 \|x(0)\|^2. \tag{7.72}$$

Substituting (7.72) appropriately into the left-hand and right-hand sides of (7.71) allows us to form the following inequality

$$\frac{1}{2}\lambda_1 \|x(t)\|^2 \leq \frac{1}{2}\lambda_2 \|x(0)\|^2 \underline{e}^{-2\gamma t}. \tag{7.73}$$

Using (7.73), we can solve for $\|x(t)\|$ to obtain the result given in (7.61). □

Remark 7.10

From the result given by (7.61) and (7.62), we can form the following upper bound on the load position tracking error

$$|e(t)| \leq \sqrt{\frac{\lambda_2}{\lambda_1}} \|x(0)\| \underline{e}^{-\gamma t} \quad \forall t \in [0, \infty);$$

hence, the load position tracking error goes to zero exponentially fast. In a similar manner to the above inequality, it is easy to show that the velocity observation error (*i.e.*, $\dot{\tilde{q}}\,(t)$) and the current observation error (*i.e.*, $\tilde{I}(t)$) both go to zero exponentially fast.

Remark 7.11

An important necessary property of the output feedback controller is that all system quantities remain bounded. The result of the above analysis, as summarized in (7.61), is that $x(t) \in L_\infty^5$ and therefore the variables $\dot{\tilde{q}}\,(t) \in L_\infty$, $\tilde{I}(t) \in L_\infty$, $e(t) \in L_\infty$, $\hat{r}(t) \in L_\infty$, and $\hat{\eta}_I(t) \in L_\infty$. A direct result of the definition of $\dot{\tilde{q}}\,(t)$, $e(t)$, and $\hat{r}(t)$ given in (7.4), (7.3), and (7.34) is that $q(t) \in L_\infty$, $\dot{q}(t) \in L_\infty$, and $\dot{\hat{q}}\,(t) \in L_\infty$. Equation (7.45) can be written to show the dependence of $I_d(t)$ on bounded variables as

$$I_d = g_1\left(q_d, \dot{q}_d, \ddot{q}_d, q, \dot{\hat{q}}\right)$$

where $g_1(\cdot) \in L_\infty$ due to the structure of the desired current trajectory and the fact that it is only dependent on bounded quantities. Thus, we can conclude that $I_d(t) \in L_\infty$. Given that $I_d(t) \in L_\infty$, the definitions of $\hat{\eta}_I(t)$ and $\tilde{I}(t)$ in (7.36) and (7.5) can be used to show that $I(t) \in L_\infty$ and $\hat{I}(t) \in L_\infty$. Equation (7.54) can now be written to show the dependence of $v(t)$ on bounded variables as

$$v = g_2\left(q_d, \dot{q}_d, \ddot{q}_d, \dddot{q}_d, q, \dot{\hat{q}}, \hat{I}\right)$$

where $g_2(\cdot) \in L_\infty$ due to the structure of the voltage input controller and the fact that it is only dependent on bounded quantities. Thus we can conclude that $v(t) \in L_\infty$. With regard to the observer which is actually implemented, we know that $q(t) \in L_\infty$, $\dot{\hat{q}}\ (t) \in L_\infty$, and $\hat{I}(t) \in L_\infty$; therefore, (7.8) implies that $\dot{y}(t) \in L_\infty$. Now using the fact that $\dot{\tilde{q}}\ (t)$ is exponentially stable as described in (7.61), it can be concluded from Lemma 1.9 in Chapter 1 that $\tilde{q}(t) \in L_\infty$. Hence, (7.7) can be utilized to show that $\hat{q}(t) \in L_\infty$. Since of $\tilde{q}(t) \in L_\infty$ and $\dot{\hat{q}}\ (t) \in L_\infty$, (7.6) may be used to conclude that $y(t) \in L_\infty$. Finally, the above information, (7.1), and (7.2) imply that $\ddot{q}(t) \in L_\infty$ and $\dot{I}(t) \in L_\infty$. It has now been shown that all signals in the system remain bounded during closed-loop operation.

Remark 7.12

Theorem 7.2 can be extended to provide a statement on load velocity tracking. That is, from the position tracking error dynamics developed in (7.39), the following bound can be written for the velocity tracking error

$$|\dot{e}(t)| = |\dot{q}_d - \dot{q}| = \left|-\alpha e + \hat{r} - \dot{\tilde{q}}\right| \leq \alpha\,|e| + |\hat{r}| + \left|\dot{\tilde{q}}\right|.$$

The implication of the above inequality is that, based on the exponential tracking of each term on the right-hand side of the inequality (See (7.61)), exponential load velocity tracking is also achieved.

7.7 Experimental Results

The hardware used to implement the output feedback controller presented in this chapter is exactly the same as that described in Chapter 2; however, only the encoder is used to provide sensory information to the input voltage control algorithm. The desired load position trajectory was selected as

$$q_d(t) = \frac{\pi}{2}\left(1 - \underline{e}^{-0.1t^3}\right)\sin\left(\frac{8\pi}{5}t\right)$$

which has the advantageous property that

$$q_d(0) = \dot{q}_d(0) = \ddot{q}_d(0) = \dddot{q}_d\ (0) = 0.$$

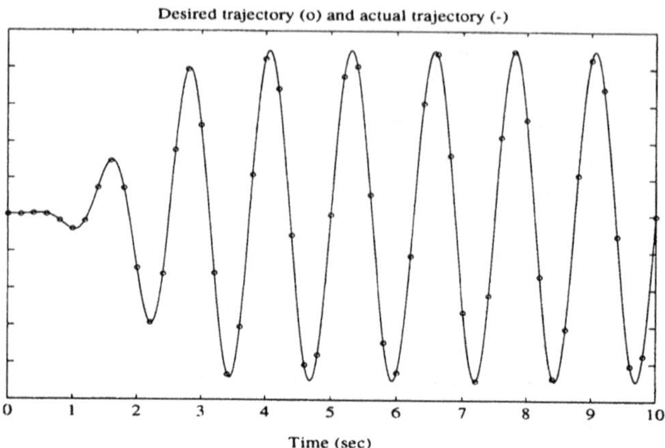

Figure 7.1: Desired (o) and Actual (-) Rotor Position Trajectories

The desired position trajectory and the actual position trajectory are shown in Figure 7.1.

The output feedback controller was found to yield good load position tracking error performance with the following controller gain values

$$k_s = 0.1, \quad k_{n1} = k_{n2} = k_{n6} = 0.7, \quad \text{and} \quad k_{n3} = k_{n4} = k_{n5} = 1.4.$$

The resulting position tracking error is given in Figure 7.2. Note that the maximum, steady-state position tracking error is approximately within ± 0.5 degrees. The corresponding rotor current and voltage are shown in Figures 7.3 and 7.4, respectively.

Remark 7.13

From the above experimental results, we can make the following observations. First, we can see from Figure 7.2, that the output feedback controller exhibited a relatively large, transient tracking error (*i.e.*, approximately ± 6 degrees) during the first three seconds of the experiment. This occurrence may be due to the transient time involved with the observers attempting to converge to the actual quantities. Second, it is interesting to note that the frequency content of the input voltage of Figure 7.4 is much lower than that of the exact model knowledge controller presented in Chapter 2. One possible reason for this phenomenon is that noise on the state measurements (especially the tachometer's velocity signal) are directly incorporated into the exact model knowledge controller. From this perspective, the behavior of the output feedback controller is more desirable since chattering voltages and currents can ruin the brushes and commutators and hence shorten the longevity of the motor.

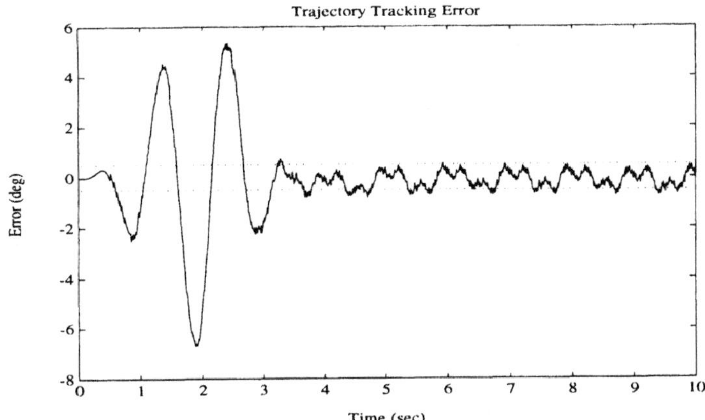

Figure 7.2: Position Tracking Error for Output Feedback Controller

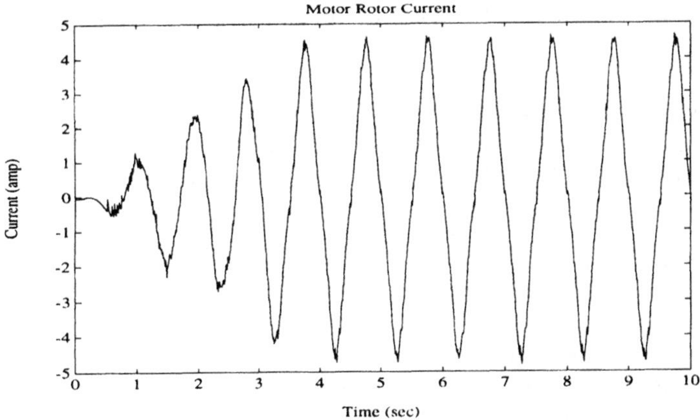

Figure 7.3: Rotor Current for Output Feedback Controller

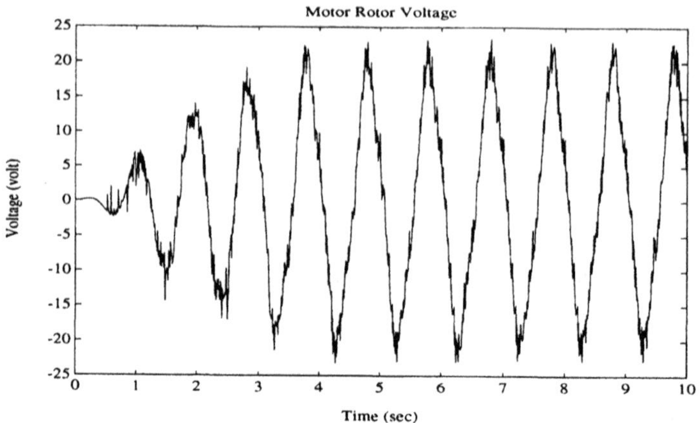

Figure 7.4: Rotor Voltage for Output Feedback Controller

Bibliography

[1] M. Krstic, I. Kanellakopoulos, and P. Kokotovic, *Nonlinear and Adaptive Control Design,* John Wiley & Sons, 1995.

[2] T. Kailaith, *Linear Systems*, Englewood Cliffs, NJ, Prentice Hall, 1980.

[3] J.J. Carroll and D.M. Dawson, "Integrator Backstepping Techniques for the Tracking Control of Permanent Magnet Brush DC Motors", *IEEE Transactions on Industrial Applications*, Vol. 31, No. 2, Mar./Apr., 1995, pp. 248-255.

[4] I. Kanellakopoulos, P.V. Kokotovic, and A.S. Morse, "Adaptive Output-Feedback Control of Systems with Output Nonlinearities," *IEEE Transactions on Automatic Control*, Vol. 37, pp. 1266-1282, 1992.

[5] R. Marino and P. Tomei, "Global Adaptive Output-Feedback Control of Nonlinear Systems, Part I: Linear Parameterization", *IEEE Transactions on Automatic Control*, Vol. 38, No.1, pp. 17-31, Jan., 1993.

[6] M. Krstic, I. Kanellakopoulos, and P. Kokotovic, "Nonlinear Design of Adaptive Controllers for Linear Systems", *IEEE Transactions on Automatic Control,* Vol. 39, No. 4, pp. 738-752, Apr., 1994.

[7] M. Spong and M. Vidyasagar, *Robot Dynamics and Control*, New York: John Wiley and Sons, Inc., 1989.

Chapter 8

PMS MOTOR (OFB)

8.1 Introduction

In this chapter, we use the framework developed in Chapter 7 to develop a load position tracking controller for the electromechanical model of a PMS motor driving a position-dependent load under the assumptions that only load position measurements are available and that an exact model for the electromechanical system dynamics can be determined. While the BDC OFB motor controller provides a starting point for the design of a PMS OFB motor controller, the PMS motor control design is more complicated due to the fact that the multi-phase PMS motor/load electromechanical dynamics exhibit: i) nonlinear terms, which are products of the electrical current and trigonometric terms, in the static torque transmission equation, and ii) nonlinear terms, which are products of rotor velocity and trigonometric terms, in the electrical subsystem dynamics. Despite these differences between the BDC and the PMS electromechanical dynamics, the observed-integrator backstepping procedure [1] used in Chapter 7 can be utilized to design a PMS OFB controller. Specifically, by utilizing the structure of the PMS electromechanical system, we design nonlinear observers to estimate the load velocity and the stator current. A Lyapunov-like argument is then used to prove that the open-loop observers ensure that the load velocity and stator current observation errors converge to zero exponentially fast. We then illustrate that all of the internal signals inherent to the observer remain bounded as long as the controller is designed properly[1]. Based on the structure of these observers, we then use an observed-integrator backstepping procedure, nonlinear damping tools, and the commutation strategy given in Chapter 3 to design a load position tracking controller. A Lyapunov-like argument is then used to prove that all of the signals in the actual electromechanical systems and the observer

[1]Specifically, the controller must be designed to ensure that the load position, load velocity, stator current, and input voltage remain bounded during closed-loop operation.

remain bounded during closed-loop operation; furthermore, the controller ensures that the load position tracks the desired trajectory exponentially fast. Finally, experimental results are utilized to illustrate the performance of the proposed controller.

8.2 System Model

The load position dynamics, $q(t)$, for a two-phase permanent magnet stepper motor can be described by a set of differential equations as in [2] or [3]. The dynamics of the mechanical subsystem for a position-dependent load actuated by a permanent magnet stepper motor are assumed to be of the form [4] [5]

$$M\ddot{q} + B\dot{q} + N\sin(q) + K_D\sin(4N_rq) = \sum_{j=1}^{2} -K_m\sin(x_j)I_j \tag{8.1}$$

where $q(t)$, $\dot{q}(t)$, and $\ddot{q}(t)$ represent the load position, velocity, and acceleration, respectively. The constant parameter M denotes the mechanical inertia of the motor rotor and the connected load, B represents the viscous friction coefficient, and N denotes the constant lumped load term. The term $K_D\sin(4N_rq)$ is used to model the detente torque, and K_D is usually referred to as the detente torque constant. The term $\sum_{j=1}^{2} -K_m\sin(x_j)I_j$ can be considered as the torque input originating in the electrical subsystems. In (8.1), $I_j(t)$ denotes a particular phase current, and $x_j(t)$ is given by

$$x_j = N_rq - (j-1)\frac{\pi}{2} \tag{8.2}$$

in which N_r accounts for the number of teeth on the rotor. The current dynamics in the two electrical subsystems are described by [6]

$$L\dot{I}_j = v_j - RI_j + K_m\dot{q}\sin(x_j) \quad \text{for} \quad j = 1,2 \tag{8.3}$$

where $v_j(t)$ is the voltage input to a particular phase. The electrical parameters R and L describe the winding resistance and inductance constants, respectively. The back-emf term $K_m\dot{q}\sin(x_j)$ can be considered as inherent feedback from the mechanical subsystem. The electrical subsystem parameters R, L, and K_m are assumed to be the same for each of the two phases.

Remark 8.1

Note that in this chapter we do not divide the parameters M, B, N, and K_D in (8.1) by the torque coefficient constant K_m. This helps to reduce the complexity of the controller by allowing for exact cancellation of certain terms during the subsequent stability analysis. If we did divide (8.1) by K_m, then the controller would require the injection of an additional nonlinear damping term to compensate for the resulting difference between the torque transmission terms and the back-emf terms.

8.3 Control Objective

The goal of the controller is to force the load position $q(t)$ in the electromechanical dynamics of (8.1) through (8.3) to follow a desired position trajectory $q_d(t)$ given only load position measurements (*i.e.*, $q(t)$). To facilitate the tracking control formulation, we define the load position tracking error $e(t)$ as

$$e = q_d - q. \tag{8.4}$$

We will assume that $q_d(t)$ and its first, second, and third time derivatives are all bounded functions of time.

Since the problem is constrained by the fact that the only measurable state is the load position, the remaining state variables (*i.e.*, the electrical phase currents $I_j(t)$ and the load velocity $\dot{q}(t)$) will be observed. As the term observation implies, there will be some variation between the physical and observed quantities. This variation, called observation error, is defined for the velocity and phase current observers as

$$\dot{\tilde{q}} = \dot{q} - \dot{\hat{q}} \tag{8.5}$$

and

$$\tilde{I}_j = I_j - \hat{I}_j \tag{8.6}$$

where $\dot{\hat{q}}\,(t)$ denotes the observed load velocity, and $\hat{I}_j(t)$ denotes the jth observed phase current. A necessary requirement in the overall control strategy will be to ensure that the observation errors $\dot{\tilde{q}}\,(t)$ and $\tilde{I}_j(t)$ are both driven to zero.

8.4 Observer Formulation

Based on the structure of the mechanical subsystem dynamics given by (8.1) and the subsequent stability analysis, the following load velocity observer is proposed

$$\dot{\hat{q}} = y + \frac{1}{M}\left(k_{s2} + k_{n1}\right)\tilde{q} \tag{8.7}$$

where the position observation error is defined as

$$\tilde{q} = q - \hat{q}, \tag{8.8}$$

the auxiliary variable $y(t)$ is updated according to

$$\dot{y} = \frac{1}{M}\left(-B\,\dot{\hat{q}} - N\sin(q) - K_D\sin(4N_r q) - \sum_{j=1}^{2} K_m \sin(x_j)\hat{I}_j\right), \quad (8.9)$$

and k_{s2}, k_{n1} are positive controller gains. The electrical current observers are designed to mimic the electrical subsystems in (8.3) where the observed phase current and load velocity replace the corresponding physical system variables. Specifically, the proposed electrical phase current observers are given by

$$\dot{\hat{I}}_j = \frac{1}{L}\left(v_j - R\hat{I}_j + K_m\,\dot{\hat{q}}\sin(x_j)\right) \quad \text{for} \quad j = 1, 2. \quad (8.10)$$

Remark 8.2

The time derivative of the velocity observer in (8.7) is given by

$$\ddot{\hat{q}} = \dot{y} + \frac{1}{M}\left(k_{s2} + k_{n1}\right)\dot{\tilde{q}}\,. \quad (8.11)$$

The definition of $\dot{y}(t)$ in (8.9) is substituted into (8.11) and the result is multiplied by M to produce

$$M\,\ddot{\hat{q}} = -B\,\dot{\hat{q}} - N\sin(q) - K_D\sin(4N_r q) - \sum_{j=1}^{2} K_m\sin(x_j)\hat{I}_j + \left(k_{s2} + k_{n1}\right)\dot{\tilde{q}}\,. \quad (8.12)$$

8.4.1 Observation Error Dynamics

In this section, we formulate the observation error dynamics for the load velocity and current observers. A Lyapunov-like function is then used to illustrate that the observation errors go to zero exponentially fast when used in open-loop operation (*i.e.*, the observed quantities, $\dot{\hat{q}}(t)$ and $\hat{I}_j(t)$, have not yet been utilized in a voltage control input algorithm). In the next section, we will design a voltage input controller which utilizes these observed quantities in a stable, closed-loop fashion.

The velocity observation error dynamics are derived from the definition of the velocity observation error in (8.5). The development begins by differentiating the velocity observation error of (8.5) with respect to time and then multiplying by M to yield

$$M\,\ddot{\tilde{q}} = M\ddot{q} - M\,\ddot{\hat{q}}\,. \quad (8.13)$$

Substitution of the mechanical dynamics given in (8.1) for $M\ddot{q}$ and the second-order velocity observer given in (8.12) for $M\ \ddot{\hat{q}}$ into the right-hand side of (8.13) yields

$$M\ \ddot{\tilde{q}}= -B\ \dot{\tilde{q}} - (k_{s2} + k_{n1})\ \dot{\tilde{q}} - \sum_{j=1}^{2} K_m \sin(x_j)\tilde{I}_j \tag{8.14}$$

where (8.5) and (8.6) have been utilized. Roughly speaking, the form of (8.14) motivates one to classify the observer given by (8.7) and (8.9) as a closed-loop observer because a controller gain (*e.g.*, k_{s2}) can be increased to cause $\dot{\tilde{q}}\ (t)$ to go to zero faster (Note, for this statement to be true in the strictest sense, $\tilde{I}_j(t)$ must be assumed to be zero).

The current observation error dynamics are derived from the definition of the current observation error. The development begins by differentiating the phase current observation error of (8.6) with respect to time and then multiplying by L to yield

$$L\ \dot{\tilde{I}}_j= L\dot{I}_j - L\ \dot{\hat{I}}_j\ . \tag{8.15}$$

Substitution of the electrical dynamics given in (8.3) for $L\dot{I}_j$ and the phase current observer given in (8.10) for $L\ \dot{\hat{I}}_j$ into the right-hand side of (8.15) yields

$$L\ \dot{\tilde{I}}_j= -R\tilde{I}_j + K_m\ \dot{\tilde{q}}\sin(x_j) \tag{8.16}$$

where (8.5) and (8.6) have been utilized. Roughly speaking, the form of (8.16) motivates one to classify the observer given by (8.10) as an open-loop observer because one can not increase a controller gain to cause $\tilde{I}_j(t)$ to go to zero faster (Note, for this statement to have meaning in the strictest sense, $\dot{\tilde{q}}\ (t)$ must be assumed to be zero).

8.4.2 Stability of the Observation Error Systems

To simplify the subsequent stability analysis of the voltage control input which will be designed in the next section and to provide insight into the construction of the observers, we now illustrate that the observers of (8.7), (8.9), and (8.10) provide for exponentially stable observation of the load velocity and the electrical phase currents. The theorem given below delineates the performance of the open-loop observers.

Theorem 8.1

The observation error systems of (8.14) and (8.16) are exponentially stable in the sense that

$$\|z(t)\| \leq \sqrt{\frac{\lambda_b}{\lambda_a}}\, \|z(0)\|\, \underline{e}^{-\gamma_c t} \qquad \forall t \in [0, \infty) \tag{8.17}$$

where

$$z = \begin{bmatrix} \dot{\tilde{q}} & \tilde{I}_1 & \tilde{I}_2 \end{bmatrix}^T \in \Re^3, \tag{8.18}$$

$$\lambda_a = \min\{M, L\}, \qquad \lambda_b = \max\{M, L\}, \tag{8.19}$$

and

$$\gamma_c = \frac{\min\{(k_{s2} + k_{n1}), R\}}{\max\{M, L\}}. \tag{8.20}$$

Proof. First, we define the following non-negative function

$$V_o = \frac{1}{2} M \dot{\tilde{q}}^2 + \frac{1}{2} \sum_{j=1}^{2} L \tilde{I}_j^2 = \frac{1}{2} z^T diag\{M, L, L\} z \tag{8.21}$$

where $z(t)$ was defined in (8.18), and $diag\{M, L, L\} \in \Re^{3\times3}$ is used to denote a diagonal 3×3 matrix with M and L arranged along the main diagonal, and with zeros inserted for all the other elements. From the matrix form of $V_o(t)$ given on the right-hand side of (8.21), Lemma 1.7 in Chapter 1 can be use to form the following upper and lower bounds for $V_o(t)$ as

$$\frac{1}{2}\lambda_a \|z(t)\|^2 \leq V_o(t) \leq \frac{1}{2}\lambda_b \|z(t)\|^2 \tag{8.22}$$

where λ_a and λ_b are defined in (8.19).

Differentiating (8.21) with respect to time yields

$$\dot{V}_o = \dot{\tilde{q}} \left(M \ddot{\tilde{q}} \right) + \sum_{j=1}^{2} \tilde{I}_j \left(L \dot{\tilde{I}}_j \right). \tag{8.23}$$

Substitutions from the right-hand sides of (8.14) for $M \ddot{\tilde{q}}$ and (8.16) for $L \dot{\tilde{I}}_j$ into (8.23) yields the following upper bound

$$\dot{V}_o \leq -(k_{s2} + k_{n1}) \dot{\tilde{q}}^2 - \sum_{j=1}^{2} R \tilde{I}_j^2 \tag{8.24}$$

after collecting common terms and neglecting the term $-B \dot{\tilde{q}}^2$. Equation (8.24) can be rewritten using matrix notation as

$$\dot{V}_o \leq -z^T diag\{(k_{s2} + k_{n1}), R, R\} z \tag{8.25}$$

where $z(t)$ was defined in (8.18). An upper bound for $\dot{V}_o(t)$ in (8.25) is found using Lemma 1.7 in Chapter 1 as

$$\dot{V}_o(t) \leq -\min\{(k_{s2} + k_{n1}), R\} \|z(t)\|^2. \tag{8.26}$$

From (8.22), it easy to see that $\|z(t)\|^2 \geq 2V_o(t)/\lambda_b$; hence, $\dot{V}_o(t)$ in (8.26) can be further upper bounded as

$$\dot{V}_o(t) \leq -2\gamma_c V_o(t) \tag{8.27}$$

where γ_c was defined in (8.20). Applying Lemma 1.1 in Chapter 1 to (8.27) yields

$$V_o(t) \leq V_o(0)\mathrm{e}^{-2\gamma_c t}. \tag{8.28}$$

From (8.22), we have that

$$\frac{1}{2}\lambda_a \|z(t)\|^2 \leq V_o(t) \quad \text{and} \quad V_o(0) \leq \frac{1}{2}\lambda_b \|z(0)\|^2. \tag{8.29}$$

Substituting (8.29) appropriately into the left-hand and right-hand sides of (8.28) allows us to form the following inequality

$$\frac{1}{2}\lambda_a \|z(t)\|^2 \leq \frac{1}{2}\lambda_b \|z(0)\|^2 \mathrm{e}^{-2\gamma_c t}. \tag{8.30}$$

Using (8.30), we can solve for $\|z(t)\|$ to obtain the result given in (8.17). □

Remark 8.3

Using the result of Theorem 8.1, the structure of the observers, and the electromechanical model, it is straightforward to illustrate that all signals remain bounded during open-loop operation provided that $q(t) \in L_\infty$, $\dot{q}(t) \in L_\infty$, $\ddot{q}(t) \in L_\infty$, $I_1 \in L_\infty$, $I_2 \in L_\infty$, and $v(t) \in L_\infty$. Specifically, from (8.17) and (8.18), we know that $\dot{\tilde{q}}\ (t) \in L_\infty$, $\tilde{I}_1(t) \in L_\infty$ and $\tilde{I}_2(t) \in L_\infty$; hence, we can use (8.5) and (8.6) to show that $\dot{\hat{q}}\ (t) \in L_\infty$, $\hat{I}_1(t) \in L_\infty$, and $\hat{I}_2(t) \in L_\infty$. In addition since $\dot{\tilde{q}}\ (t)$ satisfies (??), Lemma 1.9 in Chapter 1 can be used to show that $\tilde{q}(t) \in L_\infty$; hence, (8.5) can be used to show that $\hat{q}(t) \in L_\infty$. Since all the terms on the right-hand side of (8.14) are bounded, we now know that $\ddot{\tilde{q}}\ (t) \in L_\infty$; hence, (8.13) can be used to show that $\ddot{\hat{q}}\ (t) \in L_\infty$. Based on the above information, (8.7) and (8.9) can be utilized to show that $y(t) \in L_\infty$ and $\dot{y}(t) \in L_\infty$.

8.5 Voltage Control Inputs Design

In this section, we design the voltage control input which drives the position tracking error defined in (8.4) to zero. As mentioned in the introduction, the control design is constrained by the fact that only measurements of load position (*i.e.*, $q(t)$) are assumed to be available (*i.e.*, output feedback (OFB)). Since measurements of load velocity and electrical phase currents are not available, the observed load velocity and the observed electrical phase currents are utilized in the formulation of the voltage control inputs. In many ways the design of the output feedback controller is similar to the design of the full state feedback (FSFB), exact model knowledge controller presented in Chapter 3. For example, the FSFB controller utilized the following filtered tracking error definition

$$r = \dot{e} + \alpha e = \dot{q}_d - \dot{q} + \alpha e \tag{8.31}$$

where α is a positive control constant. Note that the above definition of the filtered tracking error given by (8.31) depends on measurements of load velocity (*i.e.*, $\dot{q}(t)$) and load position. Since the OFB controller can only use load position measurements, we define the observed filtered tracking error as

$$\hat{r} = \dot{q}_d - \dot{\hat{q}} + \alpha e. \tag{8.32}$$

Note that the above definition of the observed filtered tracking error given by (8.32) only depends on the measurement of load position.

A second similarity between the FSFB controller and the OFB controller is the notion of current tracking error. That is, in Chapter 3, the current tracking error was defined as

$$\eta_{Ij} = I_{dj} - I_j \tag{8.33}$$

where $I_{dj}(t)$ was used to denote the desired current trajectory. Note that the above definition of the current tracking error given by (8.33) depends on measurements of the electrical current (*i.e.*, $I_j(t)$). Since the OFB controller can only use load position measurements, we define the observed current tracking as

$$\hat{\eta}_{Ij} = I_{dj} - \hat{I}_j \tag{8.34}$$

where $I_{dj}(t)$ is again used to denote the desired phase current trajectory which must be designed to move the mechanical load along a desired position trajectory. Note that the above definition of the observed current tracking error given by (8.34) does not depend on measurements of electrical current.

Before we begin the design of the voltage control input, the overall voltage input control strategy can be summarized, in a heuristic manner, using the above auxiliary definitions. That is, the role of the voltage control input will be to force $\hat{I}_j(t)$ to track $I_{dj}(t)$ in the electrical subsystem while the combined action of the observer and controller will cause $I_j(t)$ to track $\hat{I}_j(t)$; hence, the overall effect is that $I_j(t)$ will be forced to track $I_{dj}(t)$. The result of the actual current tracking the desired current trajectory is that the desired torque will in effect be applied to the mechanical subsystem, which then will cause $q(t)$ to track $q_d(t)$ and $\dot{\hat{q}}(t)$ to track $\dot{q}_d(t)+\alpha e(t)$ in the mechanical subsystem. The combined action of the velocity observer and the desired current trajectory will then force $\dot{q}(t)$ to track $\dot{\hat{q}}\ (t)$. In short, the combined function of the composite observer-controller is to force $I_j(t)$ to track $I_{dj}(t)$ and as a result force $q(t)$ to track $q_d(t)$ and force $\dot{q}(t)$ to track $\dot{q}_d(t) + \alpha e(t)$. Thus the observers and the voltage control inputs work together to cause tracking of the internal system states to the designed trajectories and thereby achieve tracking in the output state $q(t)$ (*i.e.*, the load position trajectory).

8.5.1 Position Tracking Error Dynamics

In this section, we develop the position tracking error dynamics which are utilized in the subsequent stability proof to show that the position tracking error goes to zero exponentially fast. During the development of the error dynamics, we formulate the desired current trajectory and the voltage control input. First, the position tracking of (8.4) can be differentiated with respect to time to yield

$$\dot{e} = \dot{q}_d - \dot{q}. \tag{8.35}$$

Since there is no control input on the right-hand side of (8.35), we add and subtract the observed filtered tracking error of (8.32) to yield

$$\dot{e} = \dot{q}_d - \hat{r} + \hat{r} - \dot{q}. \tag{8.36}$$

After substituting the right-hand side of (8.32) for the first occurrence of $\hat{r}(t)$ only, we obtain the closed-loop position tracking error dynamics in the form

$$\dot{e} = -\alpha e + \hat{r} - \dot{\tilde{q}} \tag{8.37}$$

where (8.5) has been utilized. From (8.37), we can see that if $\hat{r}(t)$ and $\dot{\tilde{q}}\ (t)$ were both equal to zero, then the position tracking error would converge to zero exponentially fast. From our previous analysis of the observation error systems analysis, we have a good feeling about our ability to drive $\dot{\tilde{q}}\ (t)$ to zero. Hence, motivated by the above discussion and the form of (8.37), our controller must ensure that $\hat{r}(t)$ is driven to zero.

To accomplish this new control objective, we need to construct the open-loop dynamics for $\hat{r}(t)$. To this end, we multiply both sides of (8.32) by M and then take the time derivative to yield

$$M\,\dot{\hat{r}} = M\left(\ddot{q}_d + \alpha\dot{e}\right) - M\,\ddot{\hat{q}}\,. \tag{8.38}$$

The second-order load velocity observer dynamics are substituted from (8.12) for $M\,\ddot{\hat{q}}$, $\dot{e}$ is substituted from (8.35), and the resulting expression is arranged in an advantageous manner to produce

$$M\,\dot{\hat{r}} = \Omega_1 + \Omega_2\dot{q} + \Omega_3\,\dot{\tilde{q}} + \sum_{j=1}^{2} K_m \sin(x_j)\hat{I}_j \tag{8.39}$$

where $\Omega_1(q,\hat{q},t)$, Ω_2, and Ω_3 are measurable auxiliary variables defined as

$$\Omega_1 = M\ddot{q}_d + M\alpha\dot{q}_d + N\sin(q) + K_D\sin(4N_r q) + B\,\dot{\hat{q}}\,, \tag{8.40}$$

$$\Omega_2 = -M\alpha\,, \tag{8.41}$$

and

$$\Omega_3 = -k_{s2} - k_{n1}. \tag{8.42}$$

Since there is no control input on the right-hand side of (8.39), we add and subtract two desired current trajectories in the form $\sum_{j=1}^{2} K_m\sin(x_j)I_{dj}$ to yield

$$M\,\dot{\hat{r}} = \Omega_1 + \Omega_2\dot{q} + \Omega_3\,\dot{\tilde{q}} + \sum_{j=1}^{2} K_m\sin(x_j)I_{dj} - \sum_{j=1}^{2} K_m\sin(x_j)\hat{\eta}_{Ij} \tag{8.43}$$

where $\hat{\eta}_{Ij}(t)$ is used to represent the observed current tracking error and was explicitly defined in (8.34). We can see from the form of (8.43) that the desired phase current trajectories should be designed to force the observed filtered tracking error $\hat{r}(t)$ to zero.

8.5.2 Commutation Strategy

In order to generate the desired torque to force $\hat{r}(t)$ in (8.43) to zero, we propose the following continuous, differentiable commutation strategy for the PMS motor

$$I_{dj} = -\tau_d\sin(x_j) \quad \text{for} \quad j = 1,\ 2 \tag{8.44}$$

where $\tau_d(t)$ denotes the desired torque. The commutation strategy in (8.44) can be thought of as the means by which the task of moving the load is shared between the two electrical phases. The commutation strategy of (8.44) is substituted for $I_{dj}(t)$ in the observed filtered tracking error dynamics of (8.43) to yield

$$M\,\dot{\hat{r}} = \Omega_1 + \Omega_2\dot{q} + \Omega_3\,\dot{\tilde{q}} - K_m\tau_d - \sum_{j=1}^{2} K_m\sin(x_j)\hat{\eta}_{Ij} \tag{8.45}$$

where the fact that $\sum_{j=1}^{2}\sin^2(x_j) = 1$ has been used to obtain (8.45).

From (8.45), we can see that the proposed commutation strategy has been developed such that the desired torque trajectory inherently becomes the control input to the mechanical subsystem and hence can be designed to ensure that the load follows the desired position trajectory. Of course due to the electrical dynamics, we can not guarantee that $\hat{I}_j = I_{dj}$; therefore, we are motivated to design the voltage control inputs to force the observed phase current to track the desired per phase current, that is, to force the observed phase current tracking error $\hat{\eta}_{Ij}$ of (8.34) to zero.

8.5.3 Voltage Input Controller

Based on the structure of (8.45), the desired torque trajectory will now be designed to force the observed filtered tracking error $\hat{r}(t)$ to zero. Specifically, we select $\tau_d(t)$ as follows

$$\tau_d = \frac{1}{K_m}\left(\Omega_1 + \Omega_2\,\dot{\hat{q}} + k_{n2}\left(\Omega_2+\Omega_3\right)^2\hat{r} + k_{n3}\hat{r} + k_{s1}\hat{r}\right) \tag{8.46}$$

where k_{n2}, k_{n3}, and k_{s1} are positive control gains. Substituting $\tau_d(t)$ of (8.46) into (8.45) yields the closed-loop dynamics for $\hat{r}(t)$ as

$$\begin{aligned} M\,\dot{\hat{r}} = & -\left(k_{s1}+k_{n3}\right)\hat{r} - \sum_{j=1}^{2} K_m \sin(x_j)\hat{\eta}_{Ij} \\ & + \left[\left(\Omega_2+\Omega_3\right)\dot{\tilde{q}} - k_{n2}\left(\Omega_2+\Omega_3\right)^2\hat{r}\right]. \end{aligned} \tag{8.47}$$

To ensure that the observed phase current tracking error goes to zero, we first construct the open-loop dynamics for $\hat{\eta}_{Ij}(t)$. To this end, we multiply both sides of (8.34) by L and take the time derivative to obtain the following expression

$$L\,\dot{\hat{\eta}}_{Ij} = L\dot{I}_{dj} - L\,\dot{\hat{I}}_j\,. \tag{8.48}$$

The time derivative of (8.44) for $\dot{I}_{dj}(t)$ is given by

$$\dot{I}_{dj} = -\dot{\tau}_d\sin(x_j) - \tau_d\cos(x_j)\dot{x}_j \tag{8.49}$$

where $\dot{x}_j(t)$ is found directly from differentiation of (8.2) to be

$$\dot{x}_j = N_r\dot{q}, \tag{8.50}$$

$\dot{\tau}_d(t)$ is found from differentiation of (8.46) to be

$$\dot{\tau}_d = \frac{1}{K_m}\left(\dot{\Omega}_1 + \Omega_2\,\ddot{\hat{q}} + k_{n2}\left(\Omega_2+\Omega_3\right)^2\dot{\hat{r}} + k_{n3}\,\dot{\hat{r}} + k_{s1}\,\dot{\hat{r}}\right), \tag{8.51}$$

$\dot{\Omega}_1(\cdot)$ is found by differentiating (8.40) and substituting for $\ddot{q}\,(t)$ from (8.12) to be

$$\begin{aligned} \dot{\Omega}_1 = \; & M\,\dddot{q}_d + M\alpha\ddot{q}_d + N\cos(q)\dot{q} + 4N_rK_D\cos(4N_rq)\dot{q} \\ & + \frac{B}{M}\Big\{-B\,\dot{\hat{q}} + \left(k_{s2}+k_{n1}\right)\dot{\tilde{q}} - N\sin(q) - K_D\sin(4N_rq) \\ & - \sum_{j=1}^{2} K_m\sin(x_j)\hat{I}_j\Big\}, \end{aligned} \tag{8.52}$$

and the additional time derivatives for $\ddot{\hat{q}}(t)$, and $\dot{\hat{r}}(t)$ in (8.51) were previously given in (8.12) and (8.39), respectively. The next step is to substitute all of the time derivatives given in (8.49) through (8.52), the expression for $\ddot{\hat{q}}(t)$ in (8.12), the expression for $\dot{\hat{r}}(t)$ in (8.39), and the right-hand side of (8.10) for $\dot{\hat{I}}_j(t)$ into (8.48). The result of these substitutions along with some algebraic manipulation yields

$$L\,\dot{\hat{\eta}}_{Ij} = \Omega_{4j} + \Omega_{5j}\dot{q} + \Omega_{6j}\,\dot{\tilde{q}} - v_j \tag{8.53}$$

where $\Omega_{4j}(t)$, $\Omega_{5j}(t)$, and $\Omega_{6j}(t)$ are *measurable* auxiliary variables defined as

$$\begin{aligned}
\Omega_{4j} = \;& R\hat{I}_j - K_m\,\dot{\hat{q}}\sin(x_j) - \frac{L\sin(x_j)}{K_m}\left(M\,\dddot{q}_d + M\alpha\ddot{q}_d\right) \\
& -\frac{L\sin(x_j)}{K_m M}\left(k_{n2}\left(\Omega_2+\Omega_3\right)^2 + k_{n3} + k_{s1}\right) \\
& \cdot\left(\Omega_1 + \sum_{j=1}^{2} K_m\hat{I}_j\sin(x_j)\right) + \frac{BL\sin(x_j)}{K_m M}\left(B\,\dot{\hat{q}}\right. \\
& \left. + N\sin(q) + K_D\sin(4N_r q) + \sum_{j=1}^{2} K_m\hat{I}_j\sin(x_j)\right) \\
& + \frac{\Omega_2 L\sin(x_j)}{K_m M}\cdot\left(B\,\dot{\hat{q}} + N\sin(q)\right. \\
& \left. + K_D\sin(4N_r q) + \sum_{j=1}^{2} K_m\hat{I}_j\sin(x_j)\right)
\end{aligned} \tag{8.54}$$

$$\begin{aligned}
\Omega_{5j} = \;& -\frac{L}{K_m M}\Omega_2\sin(x_j)\left(k_{n2}\left(\Omega_2+\Omega_3\right)^2 + k_{n3} + k_{s1}\right) \\
& -\frac{L}{K_m}\sin(x_j)\left(N\cos(q) + 4K_D N_r\cos(4N_r q)\right) \\
& -L\tau_d N_r\cos(x_j)\,,
\end{aligned} \tag{8.55}$$

and

$$\begin{aligned}
\Omega_{6j} = \;& -\frac{L}{M}\sin(x_j)\left\{\left(B+\Omega_2\right)\left(k_{s2}+k_{n1}\right)\right. \\
& \left. + \frac{\Omega_3}{M}\left(k_{n2}\left(\Omega_2+\Omega_3\right)^2 + k_{n3} + k_{s1}\right)\right\}.
\end{aligned} \tag{8.56}$$

Based on the structure of (8.53) and the subsequent stability analysis, we propose the following phase voltage control input to drive the observed current tracking error to zero

$$v_j = \Omega_{4j} + \Omega_{5j}\ \dot{\hat{q}} + k_{s3}\hat{\eta}_{Ij} + k_{n5}K_m^2\hat{\eta}_{Ij} + k_{n4}\left(\Omega_{5j} + \Omega_{6j}\right)^2\hat{\eta}_{Ij}, \qquad (8.57)$$

where k_{s3}, k_{n4}, and k_{n5} are positive control gains. Substituting the voltage control of (8.57) into (8.53) yields the closed-loop dynamics for $\hat{\eta}_{Ij}(t)$ in the form

$$\begin{aligned} L\,\dot{\hat{\eta}}_{Ij} = & -k_{s3}\hat{\eta}_{Ij} - k_{n5}K_m^2\hat{\eta}_{Ij} \\ & + \left[\left(\Omega_{5j} + \Omega_{6j}\right)\dot{\tilde{q}} - k_{n4}\left(\Omega_{5j} + \Omega_{6j}\right)^2\hat{\eta}_{Ij}\right]. \end{aligned} \qquad (8.58)$$

Remark 8.4

The dynamics given by (8.37), (8.47), and (8.58), represent the closed-loop position tracking error system for which the stability analysis is performed while (8.57) and (8.46) represent the controller which is implemented at the voltage terminals of the motor. Note that the desired phase current trajectory $I_{dj}(t)$ of (8.44) is embedded (in the guise of the variable $\hat{\eta}_{Ij}(t)$) inside of the voltage control input $v_j(t)$.

8.5.4 Stability of the Position Tracking Error System

In order to analyze the stability of the closed-loop position tracking error system, we define the following non-negative function

$$V_p = \frac{1}{2}e^2 + \frac{1}{2}M\hat{r}^2 + \frac{1}{2}\sum_{j=1}^{2} L\hat{\eta}_{Ij}^2. \qquad (8.59)$$

The time derivative of (8.59) is given by

$$\dot{V}_p = e\dot{e} + \hat{r}\left(M\,\dot{\hat{r}}\right) + \sum_{j=1}^{2}\hat{\eta}_{Ij}\left(L\,\dot{\hat{\eta}}_{Ij}\right). \qquad (8.60)$$

Substituting (8.37), (8.47), and (8.58) into (8.60) for $\dot{e}$, $M\ \dot{\hat{r}}$, and $L\ \dot{\hat{\eta}}_{Ij}$, respectively, yields

$$\begin{aligned}\dot{V}_p = \ & -\alpha e^2 - k_{s1}\hat{r}^2 - \sum_{j=1}^{2} k_{s3}\hat{\eta}_{Ij}^2 - e\,\dot{\tilde{q}} \\ & + \left[(\Omega_2+\Omega_3)\,\dot{\tilde{q}}\,\hat{r} - k_{n2}\,(\Omega_2+\Omega_3)^2\,\hat{r}^2\right] + \left[e\hat{r} - k_{n3}\hat{r}^2\right] \\ & + \sum_{j=1}^{2}\left[(\Omega_{5j}+\Omega_{6j})\,\dot{\tilde{q}}\,\hat{\eta}_{Ij} - k_{n4}\,(\Omega_{5j}+\Omega_{6j})^2\,\hat{\eta}_{Ij}^2\right] \\ & + \sum_{j=1}^{2}\left[-K_m\sin(x_j)\hat{\eta}_{Ij}\hat{r} - k_{n5}K_m^2\hat{\eta}_{Ij}^2\right].\end{aligned} \tag{8.61}$$

A new upper bound for $\dot{V}_p(t)$ can be found from (8.61) as

$$\begin{aligned}\dot{V}_p \leq \ & -\alpha e^2 - k_{s1}\hat{r}^2 - \sum_{j=1}^{2} k_{s3}\hat{\eta}_{Ij}^2 + |e|\left|\dot{\tilde{q}}\right| \\ & + \left[|\Omega_2+\Omega_3|\left|\dot{\tilde{q}}\right||\hat{r}| - k_{n2}\,(\Omega_2+\Omega_3)^2\,\hat{r}^2\right] \\ & + \left[|e|\,|\hat{r}| - k_{n3}\hat{r}^2\right] \\ & + \sum_{j=1}^{2}\left[|\Omega_{5j}+\Omega_{6j}|\left|\dot{\tilde{q}}\right|\left|\hat{\eta}_{Ij}\right| - k_{n4}\,(\Omega_{5j}+\Omega_{6j})^2\,\hat{\eta}_{Ij}^2\right] \\ & + \sum_{j=1}^{2}\left[K_m\left|\hat{\eta}_{Ij}\right||\hat{r}| - k_{n5}K_m^2\hat{\eta}_{Ij}^2\right].\end{aligned} \tag{8.62}$$

We now apply Lemma 1.8 in Chapter 1 to the four bracketed terms on the right-hand side of (8.62) to form the following upper bound for $\dot{V}_p(t)$

$$\begin{aligned}\dot{V}_p \leq \ & -\alpha e^2 - k_{s1}\hat{r}^2 - \sum_{j=1}^{2} k_{s3}\hat{\eta}_{Ij}^2 + |e|\left|\dot{\tilde{q}}\right| \\ & + \frac{\dot{\tilde{q}}^2}{k_{n2}} + \frac{e^2}{k_{n3}} + \sum_{j=1}^{2}\frac{\dot{\tilde{q}}^2}{k_{n4}} + \sum_{j=1}^{2}\frac{\hat{r}^2}{k_{n5}}\end{aligned} \tag{8.63}$$

which can be rewritten as

$$\begin{aligned}\dot{V}_p \leq & -\left(\alpha - \frac{1}{k_{n3}}\right)e^2 - \left(k_{s1} - \frac{2}{k_{n5}}\right)\hat{r}^2 \\ & -\sum_{j=1}^{2} k_{s3}\hat{\eta}_{Ij}^2 + \left[\|e\|\left|\dot{\tilde{q}}\right|\right] + \left[\left(\frac{1}{k_{n2}} + \frac{2}{k_{n4}}\right)\dot{\tilde{q}}^2\right].\end{aligned} \tag{8.64}$$

8.6 Stability of the Composite Error System

In this section, we will analyze the performance of the composite observer-controller error systems. The performance of these closed-loop error systems is illustrated by the following theorem.

Theorem 8.2

Given the nonlinear observers of (8.7), (8.9), and (8.10) and the voltage input control of (8.57), the load position tracking error is exponentially stable in the sense that

$$\|x(t)\| \leq \sqrt{\frac{\lambda_2}{\lambda_1}}\|x(0)\| e^{-\gamma t} \qquad \forall t \in [0, \infty) \tag{8.65}$$

where

$$x(t) = \begin{bmatrix} \dot{\tilde{q}} & \tilde{I}_1 & \tilde{I}_2 & e & \hat{r} & \hat{\eta}_{I1} & \hat{\eta}_{I2} \end{bmatrix}^T \in \Re^7, \tag{8.66}$$

$$\lambda_1 = \min\{M,\ L,\ 1\}, \qquad \lambda_2 = \max\{M,\ L,\ 1\}, \tag{8.67}$$

and

$$\gamma = \frac{\min\left\{\left(\alpha - \frac{1}{k_{n1}} - \frac{1}{k_{n3}}\right), \left(k_{s1} - \frac{2}{k_{n5}}\right), \left(k_{s2} - \frac{2}{k_{n4}} - \frac{1}{k_{n2}}\right), R, k_{s3}\right\}}{\max\{M,\ L,\ 1\}}. \tag{8.68}$$

To ensure the above stability result, the controller gains must be adjusted to satisfy the following sufficient condition

$$k_{ni} > \max\left\{\frac{2}{k_{s1}}, \frac{3}{k_{s2}}, \frac{2}{\alpha}\right\} \quad \text{where } i = 1, \ldots, 5. \tag{8.69}$$

Proof. First, we define the following non-negative function

$$V(t) = V_o(t) + V_p(t) = \frac{1}{2}x^T diag\{M, L, L, 1, M, L, L\}x \tag{8.70}$$

where $x(t)$ was defined in (8.66), $V_o(t)$ was defined in (8.21), $V_p(t)$ was defined in (8.59), and $diag\{M, L, L, 1, M, L, L\} \in \Re^{7\times7}$ is used to denote a diagonal matrix. From the matrix form of $V(t)$ given on the right-hand side of (8.70), Lemma 1.7 in Chapter 1 can be used to form the following upper and lower bounds for $V(t)$

$$\frac{1}{2}\lambda_1 \|x(t)\|^2 \leq V(t) \leq \frac{1}{2}\lambda_2 \|x(t)\|^2 \tag{8.71}$$

where λ_1 and λ_2 are defined in (8.67).

From (8.25) and (8.64), an upper bound on the time derivative of $V(t)$ defined in (8.70) is given by

$$\begin{aligned} \dot{V} \leq \; & -\left(k_{s2} - \frac{1}{k_{n2}} - \frac{2}{k_{n4}}\right)\dot{\tilde{q}}^2 - \sum_{j=1}^{2} R\tilde{I}_j^2 \\ & -\left(\alpha - \frac{1}{k_{n3}}\right)e^2 - \left(k_{s1} - \frac{2}{k_{n5}}\right)\hat{r}^2 - \sum_{j=1}^{2} k_{s3}\hat{\eta}_{Ij}^2 \\ & + \left[\|e\|\left|\dot{\tilde{q}}\right| - k_{n1}\dot{\tilde{q}}^2\right]. \end{aligned} \tag{8.72}$$

After applying Lemma 1.8 in Chapter 1 to the bracketed term in (8.72), Lemma 1.7 in Chapter 1 can be utilized to upper bound $\dot{V}(t)$ in (8.72) as

$$\begin{aligned} \dot{V}(t) \leq \; & -\min\left\{\left(k_{s2} - \frac{1}{k_{n2}} - \frac{2}{k_{n4}}\right), \left(k_{s1} - \frac{2}{k_{n5}}\right), \right. \\ & \left.\left(\alpha - \frac{1}{k_{n1}} - \frac{1}{k_{n3}}\right), R, \; k_{s3}\right\}\|x(t)\|^2 \end{aligned} \tag{8.73}$$

where $x(t)$ was defined in (8.66). From (8.71), it easy to see that $\|x(t)\|^2 \geq 2V(t)/\lambda_2$; hence, $\dot{V}(t)$ in (8.73) can be further upper bounded as

$$\dot{V}(t) \leq -2\gamma V(t) \tag{8.74}$$

where γ was defined in (8.68). Applying Lemma 1.1 in Chapter 1 to (8.74) yields

$$V(t) \leq V(0)\underline{e}^{-2\gamma t}. \tag{8.75}$$

From (8.71), we have that

$$\frac{1}{2}\lambda_1 \|x(t)\|^2 \leq V(t) \quad \text{and} \quad V(0) \leq \frac{1}{2}\lambda_2 \|x(0)\|^2. \tag{8.76}$$

Substituting (8.76) appropriately into the left-hand and right-hand sides of (8.75) allows us to form the following inequality

$$\frac{1}{2}\lambda_1 \|x(t)\|^2 \leq \frac{1}{2}\lambda_2 \|x(0)\|^2 \underline{e}^{-2\gamma t}. \tag{8.77}$$

Using (8.77), we can solve for $\|x(t)\|$ to obtain the result given in (8.65). □

Remark 8.5

An important property of the output feedback controller is that all system quantities remain bounded. The result of the above analysis, as summarized in (8.65), is that $x(t) \in L_\infty^7$ and therefore the variables $\dot{\tilde{q}}(t) \in L_\infty$, $\tilde{I}_1(t) \in L_\infty$, $\tilde{I}_2(t) \in L_\infty$, $e(t) \in L_\infty$, $\hat{r}(t) \in L_\infty$, $\hat{\eta}_{I1}(t) \in L_\infty$, and $\hat{\eta}_{I2}(t) \in L_\infty$. A direct result of the definition of $\dot{\tilde{q}}(t)$, $e(t)$, and $\hat{r}(t)$ given in (8.5), (8.4), and (8.32) is that $q(t) \in L_\infty$, $\dot{q}(t) \in L_\infty$, and $\dot{\hat{q}}(t) \in L_\infty$. Equation (8.46) can be written to show the dependence of $\tau_d(t)$ on bounded variables as

$$\tau_d = g_1\left(q_d, \dot{q}_d, \ddot{q}_d, q, \dot{\hat{q}}\right)$$

where $g_1(\cdot) \in L_\infty$ due to the structure of the desired torque trajectory and the fact that it is only dependent on bounded quantities. Thus, we can conclude that $\tau_d(t) \in L_\infty$. Given that $\tau_d(t) \in L_\infty$, the commutation strategy in (8.44) can be used to show that $I_{d1}(t) \in L_\infty$ and $I_{d2}(t) \in L_\infty$. The definitions of $\hat{\eta}_{Ij}(t)$ and $\tilde{I}_j(t)$ in (8.34) and (8.6) can be used to show that $I_1(t) \in L_\infty$, $I_2(t) \in L_\infty$, $\hat{I}_1(t) \in L_\infty$ and $\hat{I}_2(t) \in L_\infty$. Equation (8.57) can now be written to show the dependence of $v_j(t)$ on bounded variables as

$$v_j = g_2\left(q_d, \dot{q}_d, \ddot{q}_d, \dddot{q}_d, q, \dot{\hat{q}}, \hat{I}_j\right)$$

where $g_2(\cdot) \in L_\infty$ due to the structure of the voltage input controller and the fact that it is only dependent on bounded quantities. Thus, we can conclude that $v_1(t) \in L_\infty$ and $v_2(t) \in L_\infty$. With regard to the observer which is actually implemented, we know that $q(t) \in L_\infty$, $\dot{\hat{q}}(t) \in L_\infty$, $\hat{I}_1(t) \in L_\infty$, and $\hat{I}_2(t) \in L_\infty$; therefore, (8.9) implies that $\dot{y}(t) \in L_\infty$. Now using the fact that $\dot{\tilde{q}}(t)$ is exponentially stable as described in (8.65), it can be concluded from Lemma 1.9 in Chapter 1 that $\tilde{q}(t) \in L_\infty$. Hence, (8.5) can be utilized to show that $\hat{q}(t) \in L_\infty$. Since $\tilde{q}(t) \in L_\infty$ and $\dot{\hat{q}}(t) \in L_\infty$, (8.7) may be used to conclude that $y(t) \in L_\infty$. Finally, the above information, (8.1), and (8.3) imply that $\ddot{q}(t) \in L_\infty$, $\dot{I}_1(t) \in L_\infty$, and $\dot{I}_2(t) \in L_\infty$. It has now been shown that all signals in the system and the controller remain bounded during closed-loop operation.

Remark 8.6

It should be noted that it is the use of the standard PMS motor model [6] in conjunction with the commutation strategy of (8.44) as opposed to the *DQ* PMS motor model which allows us to obtain a global stability result. Specifically, the *DQ* PMS motor model has a nonlinear term, which is a product of load velocity and the stator current, while the standard PMS motor model has a nonlinear term, which is product of load velocity and sinusoidal functions of the load position. Even though this difference in the model may sound trivial on the surface, it is these differences which allow us to design an observed backstepping, output feedback controller that yields global exponential load position tracking and is free of control singularities.

8.7 Experimental Results

The hardware used to implement the output feedback controller presented in this chapter is exactly the same as that described in Chapter 3; however, only the encoder is used to provide sensory information to the input voltage control algorithm. The desired trajectory for the load position was selected as

$$q_d(t) = \frac{\pi}{2}\sin(2t)\left(1 - \underline{e}^{-0.3t^3}\right)$$

which has the desirable property

$$q_d(0) = \dot{q}_d(0) = \ddot{q}_d(0) = \dddot{q}_d(0) = 0.$$

The output feedback controller was found to yield good load position tracking error performance with the following controller gain values

$$k_{n1} = 5 \times 10^{-4},\ k_{n2} = 10^{-6},\ k_{n3} = 0.5,$$

$$k_{n4} = 5 \times 10^{-4},\ k_{n5} = 5 \times 10^{-4},$$

$$\alpha = 100,\ k_{s1} = 4.5,\ k_{s2} = 2.5,\ \text{and}\ k_{s3} = 0.8.$$

The results are summarized in Figure 8.1 through Figure 8.3. The plot in Figure 8.1 provides the basis for concluding the control system is functioning as desired (*i.e.*, the position tracking error has become small). Note that due to measurement noise, quantization error in applying the control, and resolution of the encoder the tracking error does not approach zero as predicted by the theory but rather is driven to a small value. Figure 8.2 and Figure 8.3 demonstrate that the input voltages stay within the operating range of the motor.

Remark 8.7

It is interesting to note that the price we pay for the elimination of current and velocity measurements is a degradation in tracking performance. The exact model knowledge, full-state controller in Chapter 3 had less than half of the tracking error of the output feedback controller presented here. As both controllers were applied to the same motor and used the same electromechanical model, this decrease in tracking performance can likely be attributed to the fact that the velocity and current observation errors are not zero.

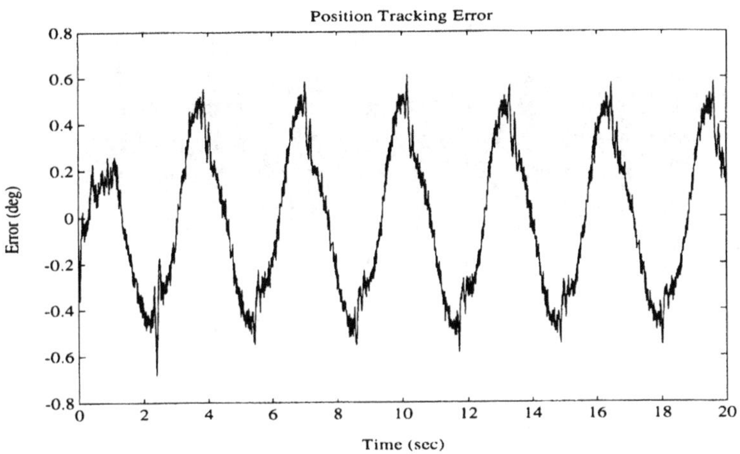

Figure 8.1: Position Tracking Error

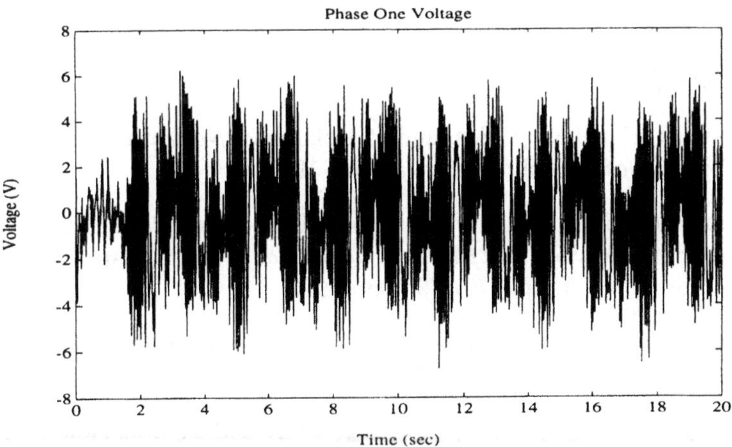

Figure 8.2: Input Voltage - Phase 1

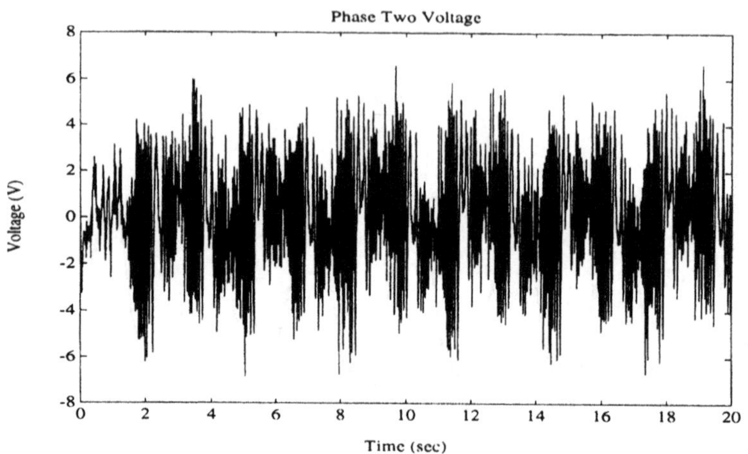

Figure 8.3: Input Voltage - Phase 2

8.8 Notes

In [7], Chiasson and Novotnak construct a nonlinear full order speed observer for a permanent magnet stepper motor which is validated with experimental results. In [8], Burg *et al.* developed an output feedback tracking controller which eliminated both velocity and current measurements while achieving global exponential position tracking performance (this work formed the basis for this chapter).

Bibliography

[1] M. Krstic, I. Kanellakopoulos, and P. Kokotovic, *Nonlinear and Adaptive Control Design,* John Wiley & Sons, 1995.

[2] A. Blauch, M. Bodson, and J. Chiasson, "High-Speed Parameter Estimation of Step Motors", *IEEE Transactions on Control Systems Technology,* Vol. 1, No. 4, pp. 270-279, Dec. 1993.

[3] B. Kuo and J. Tal, *Incremental Motion Control, Step Motors and Control Systems,* Vol. II, SRL Publishing, Champaign, IL, 1979.

[4] M. Spong and M. Vidyasagar, *Robot Dynamics and Control,* New York: John Wiley and Sons, Inc., 1989.

[5] M. Zribi and J. Chiasson, "Position Control of a PM Stepper Motor by Exact Linearization", *IEEE Transactions on Automatic Control,* Vol. 36, No.5, pp. 620-625, May 1991.

[6] P. C. Krause and O. Wasynczuk, *Electromechanical Motion Devices*, New York: McGraw-Hill, 1989.

[7] J. Chiasson and R. Novotnak, "Nonlinear Speed Observer for the PM Stepper Motor", *IEEE Transactions on Automatic Control,* Vol. 38, No. 10, pp. 1584-1588, October, 1993.

[8] T. Burg, D. Dawson, J. Hu, and P. Vedagarbha, "Position Tracking Control of a Stepper Motor via Output Feedback", *IEEE Conference on Control Applications*, Glasgow, Scotland, Aug., 1994, pp. 213-218.

Chapter 9

BLDC Motor (OFB)

9.1 Introduction

In this chapter, we use the framework developed in Chapter 7 to develop a load position tracking controller for the electromechanical model of a BLDC motor driving a position-dependent load under the assumptions that only load position measurements are available and that an exact model for the electromechanical system dynamics can be determined. While the BDC OFB motor controller provides a starting point for the design of a BLDC OFB motor controller, the BLDC motor control design is more complicated due to the fact that the multi-phase BLDC motor/load electromechanical dynamics exhibit: i) a bilinear term, which is the product of the electrical currents, in the static torque transmission equation, and ii) bilinear terms, which are a products of rotor velocity and electrical current, in the electrical subsystem dynamics. Despite these differences between the BDC and the BLDC electromechanical dynamics, the observed-integrator backstepping procedure [1] used in Chapter 7 can be utilized to design an BLDC OFB controller. Specifically, by utilizing the structure of the BLDC electromechanical system, we design nonlinear observers to estimate the load velocity and the transformed stator current. A Lyapunov-like argument is then used to motivate the construction of the observers; however in contrast with the BDC and PMS OFB control approaches, we can not explicitly show that the open-loop observation error goes to zero exponentially fast. Specifically, a general statement on the open-loop observation error is prevented by the fact that the observation error dynamics contain higher-order, nonlinear terms which can only be written in terms of the position tracking control objective. Notwithstanding this difference from previous chapters, we then use the structure of these observers, the observed-integrator backstepping procedure, nonlinear damping tools, and a commutation strategy different from that used in Chapter 4 to design a load position tracking controller. A Lyapunov-like argument is then used to prove that all of the signals in the actual electromechanical systems and the observer remain bounded during closed-loop operation; furthermore, the controller ensures that the load position tracks the desired trajectory exponentially fast. Finally, simulation results are utilized to illustrate the performance of the proposed controller.

9.2 System Model

To facilitate the controller development, the system model presented in Chapter 4 for a BLDC motor driving a position-dependent mechanical load is adopted for use in this chapter. Under the assumption of a linear magnetic circuit, the dynamics of a Y-connected BLDC motor driving a nonlinear position-dependent load in the rotor-fixed reference frame (See Appendix A) are given by [2] [3] [4]

$$M\ddot{q} + B\dot{q} + f(q) = \left(K_{\tau 1} I_b + K_{\tau 2}\right) I_a \tag{9.1}$$

where $q(t)$, $\dot{q}(t)$, and $\ddot{q}(t)$ represent the load position, velocity, and acceleration, respectively, M and B are the lumped rotor-load inertia and friction coefficients, respectively, and $f(q)$ is a nonlinear, differentiable function used to model the load (Note that we assume that $f(q) \in L_\infty$ if $q(t) \in L_\infty$ and $\dot{f}(q) \in L_\infty$ if $q(t), \dot{q}(t) \in L_\infty$). $I_a(t)$ and $I_b(t)$ are the transformed stator winding currents, $K_{\tau 1}$ and $K_{\tau 2}$ are positive constants related to the electrical parameters and are explicitly given by

$$K_{\tau 1} = n_p \left(L_b - L_a\right) \qquad \text{and} \qquad K_{\tau 2} = \sqrt{\frac{3}{2}} n_p K_B, \tag{9.2}$$

where L_a and L_b represent transformed constant winding inductances of the motor, n_p is the number of permanent magnet rotor pole pairs, and K_B is a positive constant representing the electromotive force coefficient. The electrical dynamics of the BLDC motor are given by [2] [3]

$$L_a L_{as} \dot{I}_a + R_a I_a + K_{\tau 1} I_b \dot{q} + K_{\tau 2} L_{as} \dot{q} = L_{as} V_a \tag{9.3}$$

and

$$L_b L_{bs} \dot{I}_b + R_b I_b - K_{\tau 1} I_a \dot{q} = L_{bs} V_b \tag{9.4}$$

where the auxiliary positive constants are given by

$$L_{as} = \frac{L_b - L_a}{L_b}, \quad L_{bs} = \frac{L_b - L_a}{L_a}, \quad R_a = R L_{as}, \quad R_b = R L_{bs}, \tag{9.5}$$

R is the constant winding resistance in each phase, and $V_a(t)$, $V_b(t)$ are the transformed winding voltages.

Remark 9.1

The dynamic model given in (9.1) through (9.5) is the same as that given in Chapter 4. However, the mechanical subsystem given in (9.1) is not divided by the constant $K_{\tau 2}$ as was done in Chapter 4. Additionally, the electrical subsystem dynamics given in (9.3) and (9.4) have been multiplied by the constants L_{as} and L_{bs}, respectively, to facilitate the output feedback controller development. As was done in Chapter 4, the control consideration for the zero axis is neglected since it is stable and independent of the other axes dynamics.

Remark 9.2

Since the above two-phase transformed dynamics are position-dependent, any control approach which is based on this dynamic representation implicitly requires load (rotor) position measurements. That is, rotor position measurements are explicitly utilized when the actual three-phase voltage input is applied to the terminals of the motor.

9.3 Control Objective

Given the dynamics of (9.1) through (9.5) and exact-model knowledge, the objective in this chapter is to design voltage control inputs which ensure load position tracking via output feedback. That is, we only measure load position while observing load velocity and stator currents. With this objective in mind, we define the position tracking error to be

$$e = q_d - q \tag{9.6}$$

where $q_d(t)$ represents the desired load position. For control purpose, we will assume the desired position trajectory and its first, second, and third time derivatives exist and remain bounded. Since we will observe load velocity, we define the velocity observation error as follows

$$\dot{\tilde{q}} = \dot{q} - \dot{\hat{q}} \tag{9.7}$$

where $\dot{\hat{q}}(t)$ denotes the observed load velocity. Since we will also observe the stator winding currents, we define the current observation errors as follows

$$\tilde{I}_a = I_a - \hat{I}_a \quad \text{and} \quad \tilde{I}_b = I_b - \hat{I}_b \tag{9.8}$$

with $\hat{I}_a(t)$ and $\hat{I}_b(t)$ denoting the observed stator currents. An important requirement in the overall control strategy will be to ensure that the observation errors $\dot{\tilde{q}}(t)$, $\tilde{I}_a(t)$, and $\tilde{I}_b(t)$ are all driven to zero.

Since measurements of load velocity and electrical currents are not available, the observed load velocity and the observed electrical currents are utilized in the formulation of the voltage control input. In some ways, the design of the output feedback controller is similar to the design of the full state feedback (FSFB), exact-model knowledge controller presented in Chapter 4. For example, the FSFB controller utilized the following filtered tracking error definition

$$r = \dot{e} + \alpha e = \dot{q}_d - \dot{q} + \alpha e \tag{9.9}$$

where α is a positive control constant. Note that the above definition of the filtered tracking error given by (9.9) depends on measurements of load velocity (*i.e.*, $\dot{q}(t)$) and load position. Since the OFB controller can only use load position measurements, we define the observed filtered tracking error as

$$\hat{r} = \dot{q}_d - \dot{\hat{q}} + \alpha e. \tag{9.10}$$

Note that the above definition of the observed filtered tracking error given by (9.10) only depends on the measurement of load position.

A second similarity between the FSFB controller and the OFB controller is the notion of current tracking. That is, in Chapter 4, the current tracking errors were defined as

$$\eta_a = I_{da} - I_a \quad \text{and} \quad \eta_b = I_{db} - I_b \tag{9.11}$$

where $I_{da}(t)$ and $I_{db}(t)$ were used to denote the desired current trajectories. Note that the above definitions of the current tracking error given in (9.11) depend on measurements of the electrical currents (*i.e.*, $I_a(t)$ and $I_b(t)$). Since the OFB controller can only use load position measurements, we define the observed current tracking errors as

$$\hat{\eta}_a = I_{da} - \hat{I}_a \quad \text{and} \quad \hat{\eta}_b = I_{db} - \hat{I}_b \tag{9.12}$$

where $I_{da}(t)$ and $I_{db}(t)$ are again used to denote the desired current trajectories which must be designed to move the mechanical load along a desired position trajectory. Note that the above definitions of the observed current tracking errors given by (9.12) do not depend on measurements of electrical current.

Remark 9.3

It should be noted that the selections of the desired current trajectories $I_{da}(t)$ and $I_{db}(t)$ designed to drive $\hat{r}(t)$ to zero will be different from those used to drive $r(t)$ to zero in Chapter 4. That is, in Chapter 4, the desired current trajectory $I_{db}(t)$ was selected to be zero and $I_{da}(t)$ was used to drive $r(t)$ to zero. However, in this chapter, $I_{da}(t)$ will be selected to be a positive constant while $I_{db}(t)$ is designed to force $\hat{r}(t)$ to zero. This deviation from previous work is generated by the requirements of the stability analysis. That is, due to the structure of the BLDC electrical dynamics and hence the electrical observer dynamics, we must judiciously design the desired current trajectories to directly cancel out *selected* nonlinearities during the observer stability analysis.

Before we begin the design of the voltage control inputs, the overall voltage input control strategy can be summarized, in a heuristic manner, using the above auxiliary definitions. That is, the role of the voltage control inputs will be to force $\hat{I}_a(t)$ to track $I_{da}(t)$ and $\hat{I}_b(t)$ to track $I_{db}(t)$ in the electrical subsystems while the combined action of the observer and controller will cause $I_a(t)$ to track $\hat{I}_a(t)$ and $I_b(t)$ to track $\hat{I}_b(t)$; hence, the overall effect is that $I_a(t)$ will be forced to track $I_{da}(t)$ and $I_b(t)$ will be forced to track $I_{db}(t)$. The result of the actual currents tracking the desired current trajectories is that the desired torque will in effect be applied to the mechanical subsystem, which then will cause $q(t)$ to track $q_d(t)$ and $\dot{\hat{q}}(t)$ to track $\dot{q}_d(t) + \alpha e(t)$ in the mechanical subsystem. The combined action of the velocity observer and the desired current trajectories will then force $\dot{q}(t)$ to track $\dot{\hat{q}}(t)$. In short, the combined function of the composite observer-controller is to force $I_a(t)$ to track $I_{da}(t)$, $I_b(t)$ to track $I_{db}(t)$ and as a result force $q(t)$ to track $q_d(t)$ and force $\dot{q}(t)$ to track $\dot{q}_d(t)+\alpha e(t)$. Thus the observers and the voltage control inputs work together to cause tracking of the internal system states to the designed trajectories and thereby achieve tracking in the output state $q(t)$ (*i.e.*, the load position trajectory).

9.4 Observer Formulation

Based on the structure of the dynamics given by (9.1) through (9.5), the stator current observers are defined as follows

$$L_a L_{as} \dot{\hat{I}}_a = -R_a \hat{I}_a - K_{\tau 1} \hat{I}_b \dot{\hat{q}} - K_{\tau 2} L_{as} \dot{\hat{q}} + L_{as} V_a \tag{9.13}$$

and

$$L_b L_{bs} \dot{\hat{I}}_b = -R_b \hat{I}_b + K_{\tau 1} \hat{I}_a \dot{\hat{q}} + L_{bs} V_b. \tag{9.14}$$

Based on the structure of (9.1) and the subsequent stability analysis, the second-order load velocity observer is defined as

$$\dot{\hat{q}} = y + k_1 M^{-1} \tilde{q} \tag{9.15}$$

where

$$\tilde{q} = q - \hat{q}$$

in which $\hat{q}(t)$ is the observed position signal. The auxiliary variable $y(t)$ of (9.15) is updated on-line according to the following expression

$$\dot{y} = M^{-1} \left(\left(K_{\tau 1} \hat{I}_b + K_{\tau 2} \right) \hat{I}_a - B \dot{\hat{q}} - f(q) \right), \tag{9.16}$$

in which k_1 is a positive control gain that is explicitly defined as

$$k_1 = k_s + k_n + k_n \left(K_{\tau 2}^2 (1 - L_{as})^2 + 4K_{\tau 1}^2 I_{da}^2 + 5K_{\tau 1}^2 \right) \tag{9.17}$$

where k_s, k_n are additional positive control gains, and $I_{da}(t)$ was defined in (9.12). The form of k_1 in (9.17) is motivated by the ensuing stability analysis where the term $k_1 M^{-1} \tilde{q}$ will appear as $k_1 M^{-1} \dot{\tilde{q}}$ to "damp out" terms associated with the unmeasurable $\dot{\tilde{q}}(t)$.

Remark 9.4

After differentiating (9.15) with respect to time, substituting (9.16) for $\dot{y}(t)$, and multiplying the resulting expression by M, we obtain the following second-order equation for the velocity observer dynamics

$$M \ddot{\hat{q}} = \left(K_{\tau 1} \hat{I}_b + K_{\tau 2} \right) \hat{I}_a - B \dot{\hat{q}} - f(q) + k_1 \dot{\tilde{q}} . \tag{9.18}$$

Given the above load velocity and stator current observers, we now form the observation error systems to facilitate the subsequent stability analysis. First, we subtract (9.18) from (9.1) to obtain the load velocity observation error system in the form

$$\begin{aligned} M \ddot{\tilde{q}} &= -B \dot{\tilde{q}} - k_1 \dot{\tilde{q}} + \tfrac{1}{2} K_{\tau 1} \left(I_b + \hat{I}_b \right) \tilde{I}_a \\ &\quad + \tfrac{1}{2} K_{\tau 1} \left(I_a + \hat{I}_a \right) \tilde{I}_b + K_{\tau 2} \tilde{I}_a . \end{aligned} \tag{9.19}$$

Subtracting the electrical observers given by (9.13) and (9.14) from the electrical dynamics of (9.3) and (9.4) yields the stator current observation error systems in the following advantageous forms

$$\begin{aligned} L_a L_{as} \dot{\tilde{I}}_a &= -R_a \tilde{I}_a - K_{\tau 2} L_{as} \dot{\tilde{q}} - \\ &\quad \tfrac{1}{2} K_{\tau 1} \left(I_b + \hat{I}_b \right) \dot{\tilde{q}} - \tfrac{1}{2} K_{\tau 1} \left(\dot{q} + \dot{\hat{q}} \right) \tilde{I}_b \end{aligned} \tag{9.20}$$

and

$$L_b L_{bs} \dot{\tilde{I}}_b = -R_b \tilde{I}_b + \frac{1}{2} K_{\tau 1} \left(I_a + \hat{I}_a \right) \dot{\tilde{q}} + \frac{1}{2} K_{\tau 1} \left(\dot{q} + \dot{\hat{q}} \right) \tilde{I}_a . \tag{9.21}$$

To analyze the performance of the above observation error systems of (9.19) through (9.21), we define the following non-negative function

$$V_o = \frac{1}{2} M \dot{\tilde{q}}^2 + \frac{1}{2} L_a L_{as} \tilde{I}_a^2 + \frac{1}{2} L_b L_{bs} \tilde{I}_b^2 . \tag{9.22}$$

Taking the time derivative of (9.22) and substituting (9.19) through (9.21), we obtain

$$\begin{aligned} \dot{V}_o &= -B\dot{\tilde{q}}^2 - R_a\tilde{I}_a^2 - R_b\tilde{I}_b^2 - k_1\dot{\tilde{q}}^2 \\ &\quad + K_{\tau 1}\left(I_a + \hat{I}_a\right)\dot{\tilde{q}}\,\tilde{I}_b + K_{\tau 2}\left(1 - L_{as}\right)\tilde{I}_a\,\dot{\tilde{q}}\,. \end{aligned} \tag{9.23}$$

Given the definitions of $\tilde{I}_a(t)$ and $\hat{\eta}_a(t)$ in (9.8) and (9.12), the term $I_a + \hat{I}_a$ in (9.23) can be rewritten as $\tilde{I}_a + 2I_{da} - 2\hat{\eta}_a$. By using this fact and substituting directly for k_1 defined in (9.17), $\dot{V}_o(t)$ can be upper bounded as follows

$$\begin{aligned} \dot{V}_o \leq\; & -R_a\tilde{I}_a^2 - R_b\tilde{I}_b^2 + \left[K_{\tau 1}\tilde{I}_b\tilde{I}_a\,\dot{\tilde{q}} - k_n K_{\tau 1}^2\,\dot{\tilde{q}}^2\right] \\ & + \left[-2K_{\tau 1}\hat{\eta}_a\tilde{I}_b\,\dot{\tilde{q}} - 4k_n K_{\tau 1}^2\,\dot{\tilde{q}}^2\right] - \left(k_s + k_n\right)\dot{\tilde{q}}^2 \\ & + \left[K_{\tau 2}\left(1 - L_{as}\right)\dot{\tilde{q}}\,\tilde{I}_a - k_n K_{\tau 2}^2\left(1 - L_{as}\right)^2\dot{\tilde{q}}^2\right] \\ & + \left[2K_{\tau 1}I_{da}\tilde{I}_b\,\dot{\tilde{q}} - 4k_n K_{\tau 1}^2 I_{da}^2\,\dot{\tilde{q}}^2\right]. \end{aligned} \tag{9.24}$$

Each bracketed term in (9.24) can be considered to be a nonlinear damping pair; hence, we can apply Lemma 1.8 in Chapter 1 to these terms to obtain the new upper bound for $\dot{V}_o(t)$ in the following form

$$\begin{aligned} \dot{V}_o &\leq -\left(R_a - \frac{1}{k_n}\right)\tilde{I}_a^2 - \left(R_b - \frac{1}{k_n}\right)\tilde{I}_b^2 \\ &\quad - \left(k_s + k_n\right)\dot{\tilde{q}}^2 + \frac{1}{k_n}\tilde{I}_b^2\tilde{I}_a^2 + \left[\frac{1}{k_n}\tilde{I}_b^2\hat{\eta}_a^2\right]. \end{aligned} \tag{9.25}$$

Remark 9.5

From (9.25), we can see that if the bracketed term is neglected, and if the control gain k_n is selected sufficiently large then $\dot{V}_o(t)$ will be negative definite for a given set of initial conditions, and hence the observation error terms $\dot{\tilde{q}}\,(t)$, $\tilde{I}_a(t)$, and $\tilde{I}_b(t)$ will converge to zero exponentially fast. That is, the observed velocity will track the actual velocity and the observed currents will track the actual currents. As we will show later during the combined observer-controller stability analysis, the last term in (9.25) can be easily dealt with by injecting an appropriate negative definite term into the observed current tracking error dynamics for $\hat{\eta}_a(t)$.

Remark 9.6

We have arranged equations (9.19), (9.20) and (9.21) to show explicit dependence on the observation errors (*i.e.*, $\dot{\tilde{q}}(t)$, $\tilde{I}_a(t)$, $\tilde{I}_b(t)$). This provides additional insight into the observer operation. Roughly speaking, the form of (9.19) motivates us to classify (9.15) and (9.16) as a closed-loop observer because the gain k_1 can be adjusted to make $\dot{\tilde{q}}(t)$ go to zero faster (assuming $\tilde{I}_a(t)$ and $\tilde{I}_b(t)$ are identically zero). In the same sense, the form of (9.20) and (9.21) motivates us to classify (9.13) and (9.14) as open-loop observers as there is no gain that can be adjusted to speed up the convergence of the observation error (again assuming $\tilde{I}_b = \dot{\tilde{q}} = 0$ in (9.20) and $\tilde{I}_a = \dot{\tilde{q}} = 0$ in (9.21)).

9.5 Voltage Control Inputs Design

Having designed the requisite observers in the previous section, we now turn our attention to the design of voltage level inputs which will drive the position tracking error to zero.

9.5.1 Position Tracking Error Dynamics

In order to obtain the position tracking error dynamics, we take the time derivative of (9.6) to yield

$$\dot{e} = \dot{q}_d - \dot{q}. \tag{9.26}$$

Since there is no control input in (9.26), we add and subtract the observed filtered tracking error term $\hat{r}(t)$, defined in (9.10), to the right-hand side of (9.26) to yield

$$\dot{e} = \dot{q}_d - \hat{r} + \hat{r} - \dot{q}. \tag{9.27}$$

After substituting the right-hand side of (9.10) for the first occurrence of $\hat{r}(t)$ in (9.27), we obtain the closed-loop load position error dynamics in the form

$$\dot{e} = -\alpha e + \hat{r} - \dot{\tilde{q}} \tag{9.28}$$

where (9.7) has been utilized. From (9.28), we can see that if $\hat{r}(t)$ and $\dot{\tilde{q}}(t)$ were both equal to zero, then the position tracking error would converge to zero exponentially fast. From our previous analysis of the observation error systems, we have a good feeling about our ability to drive $\dot{\tilde{q}}(t)$ to zero. Hence, motivated by the above discussion and the form of (9.28), our controller must ensure that $\hat{r}(t)$ is driven to zero.

To accomplish this new control objective, we need to construct the open-loop dynamics for $\hat{r}(t)$. To this end, we multiply both sides of (9.10) by M

and then take the time derivative to yield

$$M\,\dot{\hat{r}} = M\left(\ddot{q}_d + \alpha\left(\dot{q}_d - \dot{q}\right)\right) - M\,\ddot{\hat{q}}\,. \tag{9.29}$$

Substituting the second-order load velocity observer dynamics of (9.18) into (9.29) and rearranging the resulting expression in an advantageous manner, we obtain

$$M\,\dot{\hat{r}} = \Omega_1 + \Omega_2\,\dot{\tilde{q}} - \left(K_{\tau 1}\hat{I}_b + K_{\tau 2}\right)\hat{I}_a \tag{9.30}$$

where (9.7) has been utilized, and $\Omega_1(q, \dot{\hat{q}}, t)$, Ω_2 contain measurable terms and are explicitly given by

$$\Omega_1 = M\left(\ddot{q}_d + \alpha\dot{q}_d\right) + \left(B - M\alpha\right)\dot{\hat{q}} + f(q) \tag{9.31}$$

and

$$\Omega_2 = -\left(M\alpha + k_1\right) \tag{9.32}$$

with k_1 defined in (9.17). Since there is no control input on the right-hand side of (9.30), we add and subtract the desired current trajectories (*i.e.*, $I_{db}(t)$ and $I_{da}(t)$ defined in (9.11)) to yield the following open-loop dynamic equation for $\hat{r}(t)$ as

$$\begin{aligned} M\,\dot{\hat{r}} \;=\;& \Omega_1 + \Omega_2\,\dot{\tilde{q}} - \left(K_{\tau 1}I_{db} + K_{\tau 2}\right)I_{da} \\ & + \left(\frac{1}{2}K_{\tau 1}\left(I_{db} + \hat{I}_b\right) + K_{\tau 2}\right)\hat{\eta}_a \\ & + \frac{1}{2}K_{\tau 1}\left(I_{da} + \hat{I}_a\right)\hat{\eta}_b \end{aligned} \tag{9.33}$$

where $\hat{\eta}_a(t)$ and $\hat{\eta}_b(t)$ were explicitly defined in (9.11).

9.5.2 Voltage Input Controller

Given (9.33), we now design the desired current trajectory $I_{db}(t)$ to force the observed filtered tracking error $\hat{r}(t)$ to zero as follows

$$I_{db} = \frac{1}{K_{\tau 1}I_{da}}\left(\Omega_1 - K_{\tau 2}I_{da} + k_s\hat{r} + k_n\hat{r} + k_n\Omega_2^2\hat{r}\right). \tag{9.34}$$

Substituting $I_{db}(t)$ of (9.34) for the first occurrence of $I_{db}(t)$ in (9.33) yields the closed-loop dynamics for $\hat{r}(t)$ as follows

$$\begin{aligned} M\,\dot{\hat{r}} =\;& -k_s\hat{r} + \left[\Omega_2\,\dot{\tilde{q}} - k_n\Omega_2^2\hat{r}\right] - k_n\hat{r} \\ & + \left[\left(\frac{1}{2}K_{\tau 1}\left(I_{db} + \hat{I}_b\right) + K_{\tau 2}\right)\hat{\eta}_a\right] \\ & + \left[\frac{1}{2}K_{\tau 1}\left(I_{da} + \hat{I}_a\right)\hat{\eta}_b\right]. \end{aligned} \tag{9.35}$$

To ensure that the observed current tracking errors $\hat{\eta}_a(t)$ and $\hat{\eta}_b(t)$ go to zero, we now construct the corresponding open-loop error dynamics. To this end, we multiply both sides of the second equality in (9.12) by $L_b L_{bs}$ and take the time derivative to obtain the following expression

$$L_b L_{bs} \dot{\hat{\eta}}_b = L_b L_{bs} \dot{I}_{db} - L_b L_{bs} \dot{\hat{I}}_b . \tag{9.36}$$

After taking the time derivative of (9.34) to obtain $\dot{I}_{db}(t)$ and substituting the resulting expression into (9.36), we have

$$L_b L_{bs} \dot{\hat{\eta}}_b = \frac{L_b L_{bs}}{K_{\tau 1} I_{da}} \left(\dot{\Omega}_1 + k_s \dot{\hat{r}} + k_n \dot{\hat{r}} + k_n \Omega_2^2 \dot{\hat{r}} \right) - L_b L_{bs} \dot{\hat{I}}_b . \tag{9.37}$$

Continuing with the construction of the open-loop dynamics, we substitute the time derivative of (9.31) for $\dot{\Omega}_1$ into (9.37) to obtain

$$\begin{aligned} L_b L_{bs} \dot{\hat{\eta}}_b = & \frac{L_b L_{bs}}{K_{\tau 1} I_{da}} \left(M \dddot{q}_d + M \alpha \ddot{q}_d + (B - \alpha M) \ddot{\hat{q}} \right. \\ & \left. + \frac{\partial f(q)}{\partial q} \dot{q} \right) + \frac{L_b L_{bs}}{K_{\tau 1} I_{da}} \left(k_s + k_n + k_n \Omega_2^2 \right) \dot{\hat{r}} \\ & - L_b L_{bs} \dot{\hat{I}}_b . \end{aligned} \tag{9.38}$$

Finally, to complete the description, we substitute the velocity observer dynamics of (9.18) for $\ddot{\hat{q}}(t)$, the observed filtered tracking error dynamics of (9.30) for $\dot{\hat{r}}(t)$, and the stator current observer dynamics of (9.14) for $\dot{\hat{I}}_b(t)$ all into (9.38). We then group the measurable and unmeasurable terms in the resulting expression to obtain

$$L_b L_{bs} \dot{\hat{\eta}}_b = \Omega_3 + \Omega_4 \dot{q} + \Omega_5 \dot{\tilde{q}} - L_{bs} V_b \tag{9.39}$$

where $\Omega_3(t)$, $\Omega_4(t)$, and $\Omega_5(t)$ are auxiliary, measurable functions defined by

$$\begin{aligned} \Omega_3 = & \frac{L_b L_{bs}}{K_{\tau 1} I_{da}} \left(M \dddot{q}_d + \alpha M \ddot{q}_d \right) + R_b \hat{I}_b - K_{\tau 1} \hat{I}_a \dot{\hat{q}} \\ & + \frac{L_b L_{bs} (B - \alpha M)}{M K_{\tau 1} I_{da}} \left(\left(K_{\tau 1} \hat{I}_b + K_{\tau 2} \right) \hat{I}_a - B \dot{\hat{q}} - f(q) \right) \\ & + \frac{L_b L_{bs}}{M K_{\tau 1} I_{da}} \left(k_s + k_n + k_n \Omega_2^2 \right) \\ & \cdot \left(\Omega_1 - \left(K_{\tau 1} \hat{I}_b + K_{\tau 2} \right) \hat{I}_a \right), \end{aligned} \tag{9.40}$$

$$\Omega_4 = \frac{L_b L_{bs}}{K_{\tau 1} I_{da}} \frac{\partial f(q)}{\partial q}, \tag{9.41}$$

and

$$\Omega_5 = \frac{L_b L_{bs} k_1 (B - \alpha M)}{M K_{\tau 1} I_{da}} + \frac{L_b L_{bs}}{M K_{\tau 1} I_{da}} \left(k_s + k_n + k_n \Omega_2^2\right) \Omega_2. \tag{9.42}$$

Now, based on the structure of (9.39) and the subsequent stability analysis, we propose the following voltage input to drive the current tracking error (*i.e.*, $\hat{\eta}_b(t)$) to zero

$$\begin{aligned} V_b = & \frac{1}{L_{bs}} \Big(k_s \hat{\eta}_b + \Omega_3 + \Omega_4 \dot{q} + k_n (\Omega_4 + \Omega_5)^2 \hat{\eta}_b \\ & + \frac{1}{2} K_{\tau 1} \left(I_{da} + \hat{I}_a \right) \hat{r} \Big). \end{aligned} \tag{9.43}$$

Substituting (9.43) into (9.39) yields the closed-loop dynamics for $\hat{\eta}_b$ in the form

$$\begin{aligned} L_b L_{bs} \dot{\hat{\eta}}_b = & -k_s \hat{\eta}_b + \left[(\Omega_4 + \Omega_5) \dot{\tilde{q}} - k_n (\Omega_4 + \Omega_5)^2 \hat{\eta}_b \right] \\ & - \frac{1}{2} K_{\tau 1} \left(I_{da} + \hat{I}_a \right) \hat{r}. \end{aligned} \tag{9.44}$$

Continuing with the backstepping procedure, we now construct the open-loop dynamics for $\hat{\eta}_a(t)$. To this end, we multiply both sides of the first equality in (9.12) by $L_a L_{as}$ and take the time derivative; further, we then substitute the right-hand side of (9.13) to obtain the following expression

$$L_a L_{as} \dot{\hat{\eta}}_a = R_a \hat{I}_a + K_{\tau 1} \hat{I}_b \dot{q} + K_{\tau 2} L_{as} \dot{q} - L_{as} V_a \tag{9.45}$$

where we have used the fact that I_{da} is a constant. Now, based on the structure of (9.45) and the subsequent stability analysis, we propose the following voltage input to drive the current tracking error (*i.e.*, $\hat{\eta}_a(t)$) to zero

$$\begin{aligned} V_a = & \frac{1}{L_{as}} \left(k_s \hat{\eta}_a + R_a \hat{I}_a + K_{\tau 1} \hat{I}_b \dot{q} + K_{\tau 2} L_{as} \dot{q} \right) \\ & + \frac{1}{L_{as}} \left(\frac{1}{2} K_{\tau 1} \left(I_{db} + \hat{I}_b \right) + K_{\tau 2} \right) \hat{r}. \end{aligned} \tag{9.46}$$

Substituting (9.46) into (9.45) yields the closed-loop dynamics for $\hat{\eta}_a$ in the form

$$L_a L_{as} \dot{\hat{\eta}}_a = -k_s \hat{\eta}_a - \left(\frac{1}{2} K_{\tau 1} \left(I_{db} + \hat{I}_b \right) + K_{\tau 2} \right) \hat{r}. \tag{9.47}$$

We note that the last term in (9.47) is used to directly cancel out the second bracketed term in (9.35) during the subsequent position tracking analysis. It should be noted that the dynamics given by (9.28), (9.35), (9.44), and (9.47) represent the electromechanical closed-loop system for which the stability analysis is performed while the controllers given by (9.43), and (9.46) represent the controllers which are implemented at the voltage terminals of motor.

9.5.3 Stability of the Position Tracking Error System

In order to analyze the performance of the closed-loop position tracking error systems, we define the following non-negative function

$$V_p = \frac{1}{2}e^2 + \frac{1}{2}M\hat{r}^2 + \frac{1}{2}L_a L_{as}\hat{\eta}_a^2 + \frac{1}{2}L_b L_{bs}\hat{\eta}_b^2. \tag{9.48}$$

After taking the time derivative of (9.48) and substituting the closed-loop position error systems given by (9.28), (9.35), (9.47), and (9.44) into the resulting expression, we have

$$\begin{aligned}\dot{V}_p = \; & e\left(-\alpha e + \hat{r} - \dot{\tilde{q}}\right) \\ & + \hat{r}\left(-k_s\hat{r} + \left[\Omega_2\,\dot{\tilde{q}} - k_n\Omega_2^2\hat{r}\right] - k_n\hat{r}\right) \\ & + \hat{r}\left(\left(\frac{1}{2}K_{\tau 1}\left(I_{db} + \hat{I}_b\right) + K_{\tau 2}\right)\hat{\eta}_a + \right. \\ & \left.\frac{1}{2}K_{\tau 1}\left(I_{da} + \hat{I}_a\right)\hat{\eta}_b\right) + \hat{\eta}_b\left(\left[-k_n(\Omega_4 + \Omega_5)^2\hat{\eta}_b\right]\right. \\ & \left.(\Omega_4 + \Omega_5)\,\dot{\tilde{q}} - k_s\hat{\eta}_b + -\frac{1}{2}K_{\tau 1}\left(I_{da} + \hat{I}_a\right)\hat{r}\right) \\ & + \hat{\eta}_a\left(-k_s\hat{\eta}_a - \left(\frac{1}{2}K_{\tau 1}\left(I_{db} + \hat{I}_b\right) + K_{\tau 2}\right)\hat{r}\right).\end{aligned} \tag{9.49}$$

After cancelling the common terms in (9.49), we have

$$\begin{aligned}\dot{V}_p = \; & -\alpha e^2 - k_s\hat{r}^2 - k_s\hat{\eta}_a^2 - k_s\hat{\eta}_b^2 \\ & -e\,\dot{\tilde{q}} + \left[e\hat{r} - k_n\hat{r}^2\right] + \left[\Omega_2\hat{r}\,\dot{\tilde{q}} - k_n\Omega_2^2\hat{r}^2\right] \\ & + \left[(\Omega_4 + \Omega_5)\hat{\eta}_b\,\dot{\tilde{q}} - k_n(\Omega_4 + \Omega_5)^2\hat{\eta}_b^2\right].\end{aligned} \tag{9.50}$$

After applying Lemma 1.8 of Chapter 1 to the bracketed terms in (9.50), we can obtain an upper bound for $\dot{V}_p(t)$ as follows

$$\begin{aligned} \dot{V}_p \leq & -\left(\alpha - \frac{1}{k_n}\right) e^2 - k_s \hat{r}^2 - k_s \hat{\eta}_a^2 - k_s \hat{\eta}_b^2 \\ & + \left[\left|\dot{\tilde{q}}\right| |e|\right] + \left[\frac{2}{k_n} \dot{\tilde{q}}^2\right]. \end{aligned} \tag{9.51}$$

Remark 9.7

From the form of (9.51), we can see that if $\dot{\tilde{q}}\ (t)$ was equal to zero for all time, and the controller gain α was selected to be sufficiently large then we could guarantee that $e(t)$, $\hat{r}(t)$, $\hat{\eta}_a(t)$, and $\hat{\eta}_b(t)$ would all go to zero. As we will see later during the analysis of the composite observer-controller error systems, the last two bracketed terms in (9.51) can be cancelled or combined with other terms in (9.25) to obtain the desired stability result.

9.6 Stability of the Composite Error System

In this section, we will analyze the performance of the proposed controller used in conjunction with the proposed observers. The performance of the closed-loop error systems is illustrated by the following theorem.

Theorem 9.1

Given the nonlinear observers and the voltage inputs given in the previous sections, we can obtain a semi-global exponential load position tracking result of the form

$$\|x(t)\| \leq \kappa_1 \|x(0)\| \underline{e}^{-\kappa_2 t} \tag{9.52}$$

where κ_1 and κ_2 are positive scalar constants and

$$x = \begin{bmatrix} \dot{\tilde{q}} & \tilde{I}_a & \tilde{I}_b & e & \hat{r} & \hat{\eta}_a & \hat{\eta}_b \end{bmatrix}^T \in \Re^7. \tag{9.53}$$

To ensure the above stability result, the controller gains must be adjusted to satisfy the following sufficient condition

$$k_n > \max\left\{\frac{\lambda_2 \|x(0)\|^2}{\lambda_1 \lambda_3}, \frac{1}{R_a}, \frac{2}{k_s}, \frac{1}{R_b}, \frac{2}{\alpha}\right\}. \tag{9.54}$$

Proof.

To analyze the performance of the composite observer-controller error systems, we combine the non-negative functions $V_o(t)$ in (9.22) and $V_p(t)$ in (9.48) to formulate the following non-negative function

$$V = V_o + V_p. \tag{9.55}$$

After taking the time derivative of (9.55) and substituting the upper bound for $\dot{V}_o(t)$ in (9.25) and the upper bound for $\dot{V}_p(t)$ in (9.51) into the resulting expression, we have

$$\begin{aligned}\dot{V} \leq\ & -\left(R_a - \frac{1}{k_n}\right)\tilde{I}_a^2 - \left(R_b - \frac{1}{k_n}\right)\tilde{I}_b^2 - \left(\alpha - \frac{1}{k_n}\right)e^2 \\ & -k_s\hat{r}^2 - k_s\hat{\eta}_a^2 - k_s\hat{\eta}_b^2 - \left(k_s - \frac{2}{k_n}\right)\dot{\tilde{q}}^2 \\ & +\frac{1}{k_n}\tilde{I}_b^2\tilde{I}_a^2 + \frac{1}{k_n}\tilde{I}_b^2\hat{\eta}_a^2 + \left[\left|\dot{\tilde{q}}\right||e| - k_n\dot{\tilde{q}}^2\right].\end{aligned} \tag{9.56}$$

Applying Lemma 1.8 in Chapter 1 to the last term in (9.56) yields a new upper bound expression for $\dot{V}(t)$ as

$$\begin{aligned}\dot{V} \leq\ & -\left(R_a - \frac{1}{k_n}\right)\tilde{I}_a^2 - \left(R_b - \frac{1}{k_n}\right)\tilde{I}_b^2 - \left(k_s - \frac{2}{k_n}\right)\dot{\tilde{q}}^2 \\ & -\left(\alpha - \frac{2}{k_n}\right)e^2 - k_s\hat{r}^2 - k_s\hat{\eta}_a^2 - k_s\hat{\eta}_b^2 + \frac{1}{k_n}\tilde{I}_b^2\tilde{I}_a^2 + \frac{1}{k_n}\tilde{I}_b^2\hat{\eta}_a^2\end{aligned} \tag{9.57}$$

which can be further bounded as

$$\dot{V} \leq -\lambda_3\|x\|^2 + \frac{1}{k_n}\|x\|^4 \tag{9.58}$$

where $x(t)$ is defined in (9.53) and

$$\lambda_3 = \min\left\{k_s, \left(k_s - \frac{2}{k_n}\right), \left(\alpha - \frac{2}{k_n}\right), \left(R_a - \frac{1}{k_n}\right), \left(R_b - \frac{1}{k_n}\right)\right\} \tag{9.59}$$

(Note λ_3 is positive given that (9.54) is satisfied). Given the definition of $V(t)$ in (9.55), we have

$$\frac{1}{2}\lambda_1\|x\|^2 \leq V \leq \frac{1}{2}\lambda_2\|x\|^2 \tag{9.60}$$

where

$$\begin{aligned}&\lambda_1 = \min\{M,\ L_aL_{as},\ L_bL_{bs},\ 1\}, \\ \text{and}\quad &\lambda_2 = \max\{M,\ L_aL_{as},\ L_bL_{bs},\ 1\}.\end{aligned} \tag{9.61}$$

From (9.58) and (9.60), we have the following upper bound for $\dot{V}(t)$

$$\dot{V} \leq -\left(\lambda_3 - \frac{2V}{k_n\lambda_1}\right)\frac{2V}{\lambda_2} \tag{9.62}$$

which can be upper bounded by

$$\dot{V} \le -2\kappa_2 V \qquad \text{for} \quad V(t) < \frac{\lambda_1\lambda_3 k_n}{2} \tag{9.63}$$

where κ_2 is some positive constant. Since $V(t)$ of (9.55) is positive definite, (9.63) implies that $V(t) \le V(0)$; therefore, we have

$$\dot{V} \le -2\kappa_2 V \qquad \text{for} \quad V(0) < \frac{\lambda_1\lambda_3 k_n}{2}. \tag{9.64}$$

After applying (9.60) to (9.64), we can obtain

$$\dot{V} \le -2\kappa_2 V \qquad \text{for} \quad \|x(0)\|^2 < \frac{\lambda_1\lambda_3 k_n}{\lambda_2}. \tag{9.65}$$

Hence, if the control gain k_n is selected such that

$$k_n > \frac{\lambda_2 \|x(0)\|^2}{\lambda_1\lambda_3} \tag{9.66}$$

then

$$\dot{V} \le -2\kappa_2 V. \tag{9.67}$$

After applying Lemma 1.1 in Chapter 1 to (9.67), the result of (9.52) and (9.54) follows. □

Remark 9.8

An important property of the output feedback controller is that all system signals remain bounded. The result of the above analysis, as summarized in (9.52), is that $x(t) \in L_\infty^7$ and therefore the variables $e(t)$, $\hat{r}(t)$, $\tilde{\eta}_a$, $\tilde{\eta}_b$, $\tilde{I}_a$, $\tilde{I}_b$, $\dot{\tilde{q}} \in L_\infty$. A direct result of the definitions of $\tilde{q}\ (t)$, $e(t)$, and $\hat{r}(t)$ given in (9.7), (9.6), and (9.10) is that $q(t) \in L_\infty$, $\dot{q}(t) \in L_\infty$, and $\dot{\hat{q}}\ (t) \in L_\infty$. Equation (9.34) can be written to show the dependence of $I_{db}(t)$ on bounded variables as

$$I_{db} = g_1\left(q_d, \dot{q}_d, \ddot{q}_d, q, \dot{\hat{q}}\right)$$

where $g_1(\cdot) \in L_\infty$ due to the structure of the desired current trajectory and the fact that it is only dependent on bounded quantities. Thus, we can conclude that $I_{db}(t) \in L_\infty$. Given that $I_{db}(t) \in L_\infty$, the definitions of $\tilde{\eta}_a(t)$, $\tilde{\eta}_b(t)$, $\tilde{I}_a(t)$, and $\tilde{I}_b(t)$ in (9.12) and (9.8) can be used to show that $I_a(t) \in L_\infty$, $I_b(t) \in L_\infty$, $\hat{I}_a(t) \in L_\infty$ and $\hat{I}_b(t) \in L_\infty$. Equations (9.46) and (9.43) can now be written to show the dependence of $V_a(t)$ and $V_b(t)$ on bounded variables as

$$V_a = g_2\left(q_d, \dot{q}_d, \ddot{q}_d, \dddot{q}_d, q, \dot{\hat{q}}, \hat{I}_a, \hat{I}_b\right)$$

and

$$V_b = g_3\left(q_d, \dot{q}_d, \ddot{q}_d, \dddot{q}_d, q, \dot{\hat{q}}, \hat{I}_a, \hat{I}_b\right)$$

where $g_2(\cdot) \in L_\infty$ and $g_3(\cdot) \in L_\infty$ due to the structure of the voltage input controllers and the fact that they are only dependent on bounded quantities. Thus we can conclude that $V_a(t) \in L_\infty$ and $V_b(t) \in L_\infty$. With regard to the observer which is actually implemented, we know that $q(t) \in L_\infty$, $\dot{\hat{q}}\ (t) \in L_\infty$, $\hat{I}_a(t) \in L_\infty$ and $\hat{I}_b(t) \in L_\infty$; therefore, (9.16) implies that $\dot{y}(t) \in L_\infty$. Now using the fact that $\dot{\tilde{q}}\ (t)$ is exponentially stable as described in (9.52), it can be concluded from Lemma 1.9 in Chapter 1 that $\tilde{q}(t) \in L_\infty$. Hence, (9.7) can be utilized to show that $\hat{q}(t) \in L_\infty$. Since of $\tilde{q}(t) \in L_\infty$ and $\dot{\hat{q}}\ (t) \in L_\infty$, (9.15) may be used to conclude that $y(t) \in L_\infty$. Finally, the above information, (9.1), (9.3), and (9.4) imply that $\ddot{q}(t) \in L_\infty$, $\dot{I}_a(t) \in L_\infty$ and $\dot{I}_b(t) \in L_\infty$. It has now been shown that all signals in the system remain bounded during closed-loop operation.

Remark 9.9

From (9.54), we can see that the controller gain k_n can be increased to cover any set of initial conditions; therefore, the stability result is referred to as SGES [1]. It is important to note that convergence rate provided by κ_2 can not be increased arbitrarily due to the structure of the observers given by (9.13) and (9.14). In addition, we note that a more comprehensive stability analysis which yields exact values for κ_1 and κ_2 in (9.52) is given in [5].

Remark 9.10

The selection of the desired current trajectories $I_{da}(t)$ and $I_{db}(t)$ in this chapter (*i.e.*, $I_{da}(t)$ is a constant and $I_{db}(t)$ is designed to drive $\hat{r}(t)$ to zero) may not be practical since for most BLDC motors, the torque constant $K_{\tau 2}$ is much larger than $K_{\tau 1}$. From the right-hand side of (9.1) (*i.e.*, the torque generated by the motor electrical dynamics), we can see that the small $K_{\tau 1}$ always multiplies $I_b(t)$, which will be forced to follow $I_{db}(t)$. Therefore, in order to drive the mechanical load using $I_b(t)$, large $I_{db}(t)$ may be required which could result in the voltage input exceeding the rated value.

9.7 Simulation Results

The velocity and current observers presented in Section 9.4 and the voltage control inputs presented in Section 9.5 were simulated for a BLDC motor driving a position-dependent load. The position-dependent load function was selected as

$$f(q) = N\sin(q).$$

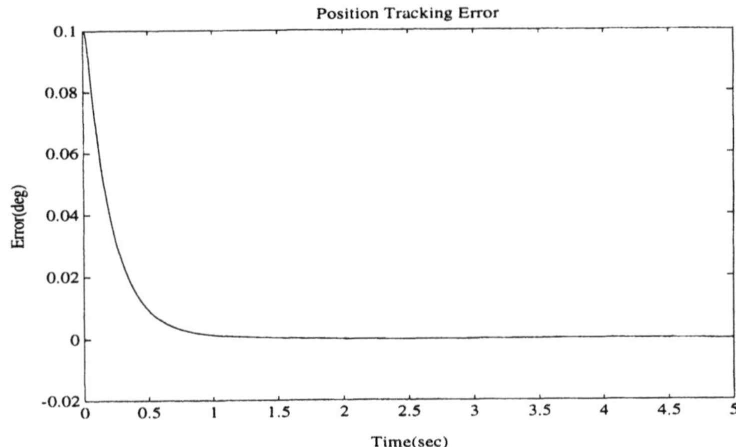

Figure 9.1: Position Tracking Error

The system parameters defined in (9.1) through (9.5) are given by

$$L_a = 1.325 \times 10^{-3} H, \quad L_b = 1.925 \times 10^{-3} H,$$

$$K_B = 0.02502 V \cdot \sec / rad, \quad n_p = 4, \quad M = 0.5 kg \cdot m^2,$$

$$N = 5 Nm, \quad B = 0.01 N \cdot m \cdot \sec / rad, \quad R = 0.9 \Omega.$$

The constant commanded current $I_{da}(t)$ is set to $0.5A$. The desired position trajectory is selected as

$$q_d = \pi \sin^2(\pi t) \; rad.$$

The initial position is set to $-0.1rad$ in order to demonstrate the exponential nature of the tracking result. The initial values for the other states are all set to zero. The control gains are set to

$$k_s = 1 \quad \text{and} \quad k_n = 0.001.$$

The resulting position tracking error is shown in Figure 9.1 and the actual voltage inputs (*i.e.*, transformed back into the actual stator coordinates) are shown in Figure 9.2. From these figures, we can see that the position tracking error goes to zero exponentially fast and the voltage control inputs stay bounded.

9.8 Notes

The elimination of state measurements for BLDC motor control has been addressed by several researchers. For example, Becerra *et al.* presented

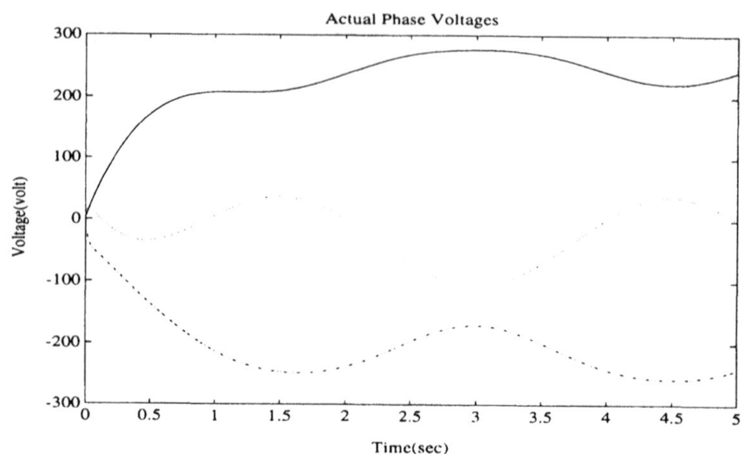

Figure 9.2: Actual Phase Voltages

a four-quadrant sensorless brushless drive in [6]. In [7], Matsui *et al.* developed a controller for a BLDC motor without velocity measurements. Furuhashi *et al.* also presented a sensorless velocity controller for BLDC motors using an adaptive sliding mode velocity observer in [8]. Using an integrator backstepping approach, position tracking controllers have been presented in [9] to eliminate current measurements and in [10] to eliminate velocity measurements, respectively. In [5], a BLDC output feedback controller, which formed the basis of this chapter, was presented.

Bibliography

[1] M. Krstic, I. Kanellakopoulos, and P. Kokotovic, *Nonlinear and Adaptive Control Design*, John Wiley & Sons, 1995.

[2] N. Hemati and M. Leu, "A Complete Model Characterization of Brushless DC Motors", *IEEE Transactions on Industry Applications*, Vol. 28, No. 1, pp. 172-180, Jan/Feb. 1992.

[3] P. C. Krause, *Analysis of Electric Machinery*, McGraw-Hill, Inc., 1986.

[4] M. Spong and M. Vidyasagar, *Robot Dynamics and Control*, New York: John Wiley and Sons, Inc., 1989.

[5] J.J. Carroll, "Semi-global Position Tracking Control of Brushless DC Motors Using Output Feedback", *Proc. of IEEE Conference on Decision and Control*, Vol. 3, pp. 3405-3409, 1993.

[6] R. Becerra, T. Jahns, and M. Ehsani, "Four-Quadrant Sensorless Brushless ECM Drive", *Proc. of 6th Annual Applied Power Electronics Conf. and Expo.*, pp. 202-209, March 10-15, 1991.

[7] N. Matsui and M. Shigyo, "Brushless DC Motor Control without Position and Speed Sensors", *IEEE Transactions on Industry Applications,* Vol. 28, No. 1, pp. 120-127, Jan/Feb., 1992.

[8] T. Furuhashi, S. Sangwongwanich, and S. Okuma, "A Position-and-Velocity Sensorless Control for Brushless DC Motors Using an Adaptive Sliding Mode Observer", *IEEE Transactions on Industrial Electronics,* Vol. 39, No. 2, pp. 89-95, April, 1992.

[9] J. J. Carroll, and D. M. Dawson, "Tracking Control of Permanent Magnet Brushless DC Motors using Partial State Feedback", *Proc. of the 2nd IEEE Conference on Control Applications,* Vol. 1, pp. 147-151, Vancouver, Canada, September, 1993.

[10] J. Hu, D. Dawson, and K. Anderson, "Position Control of a Brushless DC Motor without Velocity Measurements", *IEE Electric Power Applications,* Vol. 142, No. 2, Mar., 1995, pp. 113-122.

Chapter 10

SR and BLDC Motor (PSFB)

10.1 Introduction

In this chapter, we use the framework developed in Chapter 8 to develop a load position tracking controller for the electromechanical model of a switched reluctance (SR) motor driving a position-dependent load under the assumptions that load position and stator current measurements are available and that an exact model for the electromechanical system dynamics can be determined. While the output feedback feedback (OFB) controller for permanent magnet stepper (PMS) motor provides a starting point for the design of a partial state feedback (PSFB) controller for SR motor, the SR motor control design is more complicated due to the fact that the multi-phase SR motor/load electromechanical dynamics exhibit: i) nonlinear terms, which are composed of the square of the electrical current multiplied by trigonometric terms, in the static torque transmission equation, and ii) nonlinear terms, which are products of the rotor velocity, electrical current, and trigonometric terms, in the electrical subsystem dynamics. Indeed, it is these differences between the PMS and the SR motor electromechanical dynamics that have prevented the development of an OFB SR controller; however, the observed-integrator backstepping procedure [1] used in Chapter 8 can be utilized to design an SR PSFB controller. Specifically, by utilizing the structure of the SR electromechanical system, we design a nonlinear observer to estimate the load velocity. A Lyapunov-like argument is then used to illustrate that the open-loop observer ensure that the load velocity observation error converges to zero exponentially fast. Based on the structure of this observer, we then use an observed-integrator backstepping procedure, nonlinear damping tools, and the commutation strategy given in Chapter 5 to design a load position tracking controller. A Lyapunov-like argument is then used to prove that all of the signals in the actual electromechanical systems and the observer remain bounded during closed-loop operation; furthermore, the controller ensures that the load position tracks the desired trajectory exponentially fast. In addition to the SR PSFB controller, we discuss a similar PSFB controller for the BLDC motor. Finally, experimental results are utilized to illustrate the performance of the proposed controller for the SR motor and the BLDC motor.

10.2 System Model

To facilitate the controller development, the system model presented in Chapter 5 for a SR motor driving a position-dependent mechanical load is adopted for use in this chapter. The dynamic model of the mechanical subsystem is taken to be a single-link, direct drive robot of the form

$$M\ddot{q} + B\dot{q} + N\sin(q) = \tau \tag{10.1}$$

where the constant parameter M denotes the mechanical inertia of the motor rotor and the connected load, B is the viscous friction coefficient, N denotes the lumped load term, $q(t)$, $\dot{q}(t)$, and $\ddot{q}(t)$ represent the load position, velocity, and acceleration, respectively, and $\tau(t)$ represents the electromechanical coupling torque.

In this chapter, we will still limit the discussion to control design for a four-phase SR motor. To characterize the electrical subsystem dynamics, N_r will be used to represent the total number of rotor saliencies (*i.e.*, poles) and the subscript j will be used to refer to the individual motor phases, for $j = 1, \cdots, 4$. In the four-phase SR motor, the number of control inputs (*i.e.*, voltages) and states (*i.e.*, load position, load velocity, and winding currents) are 4 and 6, respectively. The expression relating the per phase winding voltage and current for the SR motor with a linear magnetic circuit is given by

$$L_j(q)\dot{\gamma}_j = -R_j(q, \dot{q}, \gamma_j) + V_j, \tag{10.2}$$

where

$$L_j(q) = L_0 - L_1\cos(x_j) > 0, \tag{10.3}$$

$$R_j(q, \dot{q}, \gamma_j) = R_0\gamma_j + N_r L_1 \sin(x_j)\gamma_j\dot{q}, \tag{10.4}$$

and

$$x_j = N_r q - \frac{\pi(j-1)}{2}. \tag{10.5}$$

The constant parameters L_0, L_1, and R_0 represent the coefficients of static winding inductance, dynamic winding inductance, and winding resistance, respectively, $\gamma_j(t)$ represent the per phase winding currents, and $V_j(t)$ represent the per phase winding input voltages. The electromechanical torque coupling between the electrical and mechanical subsystems is given by

$$\tau = \sum_{j=1}^{4} \sin(x_j)\gamma_j^2. \tag{10.6}$$

Note that the parameters M, B, and N introduced in (10.1) are defined to include the effects of the torque coefficient constant. That is, these mechanical parameters have been divided by the constant $\frac{N_r L_1}{2}$.

10.3 Control Objective

Given the dynamics of (10.1) through (10.6) and exact model knowledge, the objective in this chapter is to design an input voltage controller which ensures load position tracking via partial state feedback. That is, we only measure load position and stator winding currents while observing load velocity. With this objective in mind, we define the position tracking error as

$$e = q_d - q \tag{10.7}$$

where $q_d(t)$ represents the desired load trajectory. For control purposes, we will assume that the desired position trajectory and its first, second, and third time derivatives exist and remain bounded. This assumption on the "smoothness" of the desired trajectory ensures that the controller, to be defined later, remains bounded (*i.e.*, requires finite control energy). Since we will observe load velocity, we define the velocity observation error as follows

$$\dot{\tilde{q}} = \dot{q} - \dot{\hat{q}} \tag{10.8}$$

where $\dot{\hat{q}}(t)$ denotes the observed load velocity.

Since measurements of load velocity are not available, the observed load velocity is utilized in the formulation of the voltage control inputs. In some ways, the design of the partial-state feedback controller is similar to the design of the full state feedback (FSFB), exact-model knowledge controller presented in Chapter 5. For example, the FSFB controller utilized the following filtered tracking error definition

$$r = \dot{e} + \alpha e = \dot{q}_d - \dot{q} + \alpha e \tag{10.9}$$

where α is a positive control constant. Note that the above definition of the filtered tracking error given by (10.9) depends on measurements of load velocity (*i.e.*, $\dot{q}(t)$) and load position. Since the PSFB controller can only use load position and stator current measurements, we define the observed filtered tracking error as

$$\hat{r} = \dot{q}_d - \dot{\hat{q}} + \alpha e. \tag{10.10}$$

Note that the above definition of the observed filtered tracking error given by (10.10) only depends on the measurement of load position.

A second similarity between the FSFB controller and the PSFB controller is the notion of current tracking error. That is, since the stator currents are available as in the FSFB controller, we will adopt the same notation for current tracking error as that defined in Chapter 5, *i.e.*,

$$\eta_j = \gamma_{dj} - \gamma_j \tag{10.11}$$

where $\gamma_{dj}(t)$ was used to denote the desired current trajectory.

Before we begin the design of the voltage control inputs, the overall voltage input control strategy can be summarized, in a heuristic manner, using the above auxiliary definitions. That is, the role of the voltage control inputs will be to force the actual phase current $\gamma_j(t)$ to track the desired current trajectory $\gamma_{dj}(t)$ in the electrical subsystems. The result of the actual currents tracking the desired current trajectories is that the desired torque will in effect be applied to the mechanical subsystem, which will cause $q(t)$ to track $q_d(t)$ and $\dot{\hat{q}}(t)$ to track $\dot{q}_d(t) + \alpha e(t)$ in the mechanical subsystem. The combined action of the velocity observer and the desired current trajectories will then force $\dot{q}(t)$ to track $\dot{\hat{q}}(t)$. In short, the combined function of the composite observer-controller is to force $\gamma_j(t)$ to track $\gamma_{dj}(t)$ and as a result force $q(t)$ to track $q_d(t)$ and force $\dot{q}(t)$ to track $\dot{q}_d(t) + \alpha e(t)$. Thus the observers and the voltage control inputs work together to cause tracking of the internal system states to the designed trajectories and thereby achieve tracking in the output state $q(t)$ (*i.e.*, the load position trajectory).

10.4 Observer Formulation

As stated earlier, we will only allow measurements of load position and per phase stator winding current. This measurement constraint motivates us to design a dynamic observer for the load velocity. The observer design will allow us to eliminate the need for velocity measurements in the voltage input controller. It should be noted that a similar velocity observer has been utilized in the design of the OFB controllers for the DC motor in Chapter 7, for the PMS motor in Chapter 8, and for the BLDC motor in Chapter 9.

10.4.1 Observer Definition

Based on the structure of (10.1) and the subsequent stability analysis, the second-order load velocity observer is defined as

$$\dot{\hat{q}} = y + k_1 M^{-1}\tilde{q} \tag{10.12}$$

where $\widetilde{q}(t)$ is the observed position error given by

$$\widetilde{q} = q - \hat{q} \tag{10.13}$$

in which $\hat{q}(t)$ is the observed position signal. The auxiliary variable $y(t)$ of (10.12) is updated on-line according to the following expression

$$\dot{y} = M^{-1}\left(\sum_{j=1}^{4}\sin\left(x_j\right)\gamma_j^2 - B\,\dot{\hat{q}} - N\sin(q)\right). \tag{10.14}$$

The positive control gain k_1 of (10.12) is explicitly defined as

$$k_1 = k_s + k_{n5} \tag{10.15}$$

which is a combination of the positive control gain k_s and a positive, scalar damping gain k_{n5}. The gain k_1 is defined in this particular manner to facilitate the nonlinear observer-controller synthesis and the stability of the overall closed-loop system.

10.4.2 Observer Error System

Given the above velocity observer, we now formulate the observation error system to facilitate the subsequent composite stability analysis. First, we differentiate (10.12) with respect to time, substitute from (10.14) for $\dot{y}(t)$, and multiply both sides of the resulting expression by M to obtain the following second-order expression for the velocity observer

$$M\,\ddot{\hat{q}} = -B\,\dot{\hat{q}} - N\sin(q) + \sum_{j=1}^{4}\sin(x_j)\gamma_j^2 + k_1\,\dot{\widetilde{q}}\,. \tag{10.16}$$

Next we subtract the second-order expression for the velocity observer of (10.16) from the mechanical subsystem dynamics of (10.1) to yield the closed-loop velocity observation error dynamics

$$M\,\ddot{\widetilde{q}} = -B\,\dot{\widetilde{q}} - k_1\,\dot{\widetilde{q}}, \tag{10.17}$$

where (10.6) has been utilized for the substitution of $\tau(t)$ and the derivative of (10.8) has been used to substitute $\ddot{\widetilde{q}} = \ddot{q} - \ddot{\hat{q}}$.

10.4.3 Stability of the Observation Error System

The performance of the nonlinear observer when used in an open-loop fashion (*i.e.*, $\dot{\hat{q}}$ has not yet been incorporated into a input voltage control strategy) will now be examined. Given the observer closed-loop error system dynamics of (10.17), we define the following non-negative function

$$V_o = \frac{1}{2}M\,\dot{\widetilde{q}}^2\,. \tag{10.18}$$

After taking the time derivative of (10.18) and substituting for $M\ \ddot{\tilde{q}}$ from (10.17) into the resulting equation, we have

$$\dot{V}_o = -B\ \dot{\tilde{q}}^2 - k_1\ \dot{\tilde{q}}^2 . \tag{10.19}$$

After substituting (10.15) into (10.19), we can now place an upper bound on $\dot{V}_o(t)$ in (10.19) as

$$\dot{V}_o \leq -\left(k_s + k_{n5}\right) \dot{\tilde{q}}^2 . \tag{10.20}$$

It can be easily shown by Lemma 1.1 in Chapter 1 that the velocity observation error term $\dot{\tilde{q}}\ (t)$ converges to zero exponentially fast. That is, the observed velocity $\dot{\hat{q}}\ (t)$ converges to the actual velocity $\dot{q}(t)$ exponentially fast (see Chapter 7 for the details of proof).

10.4.4 Voltage Control Inputs Design

As stated in Section 10.3, the overall objective is to design a voltage level input which ensures global exponential tracking of the load position using only load position and stator current measurements. Since the phase voltages are the physical inputs to the SR motor, the method of observed-integrator backstepping [1] will be utilized to formulate a control strategy at the voltage level. It should be noted, though, that inherent to the integrator backstepping approach is the formulation of intermediate control signals. These control signals are first generated for the output subsystem (*i.e.*, load position in the mechanical subsystem) and are subsequently propagated back through the connected subsystems (*i.e.*, via the torque coupling) to the input subsystem (*i.e.*, voltage inputs in the electrical subsystem). Hence, the above backstepping procedure requires the development of a commutation strategy, and the corresponding desired torque trajectory.

10.4.5 Commutation Strategy

An inherent feature of the SR motor is that it must be electronically commutated to produce a desired motion. The commutation strategy proposed in Chapter 5 will be adopted in this chapter for the PSFB position tracking controller development. In Chapter 5, we proposed the desired per phase winding current, $\gamma_{dj}(t)$, as follows

$$\gamma_{dj} = \sqrt{\frac{\tau_d \sin(x_j) S\left(\sin(x_j)\tau_d\right)}{S_T} + \gamma_{d0}^2}, \tag{10.21}$$

where $\tau_d(t)$ is a desired torque trajectory, to be specified later in the development, designed to ensure load position tracking. The positive scalar design parameter, γ_{d0} of (10.21), is used to set the desired per phase threshold winding current, while the term $S(z)$ of (10.21) is defined to be the following differentiable function for an arbitrary scalar z,

$$S(z) = \begin{cases} 0 & , \ z \leq 0 \\ 1 - e^{-\epsilon_0 z^2} & , \ z > 0 \end{cases} \tag{10.22}$$

where ϵ_0 is a positive scalar design parameter which determines how closely $S(z)$ approximates the unit-step function. The term S_T of (10.21) is also defined to be a differentiable function as shown

$$S_T = \sum_{j=1}^{4} \sin^2(x_j) S\left(\sin(x_j)\tau_d\right). \tag{10.23}$$

To illustrate the motivation for the form of (10.21), suppose we allow $\gamma_j(t) = \gamma_{dj}(t)$ (*i.e.*, the actual phase current equals the desired phase current) so that we can rewrite (10.6) as

$$\tau = \sum_{j=1}^{4} \sin(x_j)\gamma_{dj}^2. \tag{10.24}$$

After substituting (10.21) into (10.24), we have

$$\tau = \sum_{j=1}^{4} \sin(x_j) \left(\frac{\tau_d \sin(x_j) S\left(\sin(x_j)\tau_d\right)}{S_T} + \gamma_{d0}^2 \right). \tag{10.25}$$

After expanding (10.25) and then utilizing the definition of S_T in (10.23) and the fact that the assumed motor geometry implies (see Chapter 5)

$$\sum_{j=1}^{4} \sin(x_j) = 0, \tag{10.26}$$

we can see from (10.25) that the proposed commutation strategy has been developed such that the actual torque trajectory $\tau(t)$ becomes the desired torque trajectory as shown below

$$\tau = \tau_d. \tag{10.27}$$

Thus, the commutation strategy of (10.21) allows us to directly specify a desired torque trajectory. Of course due to the electrical dynamics, we can not guarantee that $\gamma_j = \gamma_{dj}$; therefore, we are motivated to design a controller at the voltage inputs to force the actual per phase current to track the desired per phase winding current.

Remark 10.1

It has been shown in Chapter 5 and Appendix B, that $\gamma_{dj}(t)$, defined in (10.21) through (10.23), is a $C^1(\tau_d, q)$ function (*i.e.*, first-order differentiable). This ensures that the voltage level inputs, to be defined later, are bounded given bounded arguments.

10.4.6 Position Tracking Error System

In order to formulate the position tracking error dynamics, we first take the time derivative of the position tracking error of (10.7) to yield

$$\dot{e} = \dot{q}_d - \dot{q}. \tag{10.28}$$

Since there is no control input in (10.28), we add and subtract the observed filtered tracking error term, $\hat{r}(t)$, to the right-hand side of (10.28) to yield

$$\dot{e} = \dot{q}_d - \hat{r} + \hat{r} - \dot{q}, \tag{10.29}$$

where the observed filtered tracking error $\hat{r}(t)$ was previously defined in (10.10). Substituting (10.10) into (10.29) for the first occurrence of the observed filtered tracking error $\hat{r}(t)$ yields the closed-loop position tracking error dynamics as shown

$$\dot{e} = -\alpha e + \hat{r} - \dot{\tilde{q}}, \tag{10.30}$$

where the velocity observation error definition of (10.8) has been used. It is obvious from (10.30) that if $\hat{r}(t)$ and $\dot{\tilde{q}}(t)$ were both equal to zero, then the position tracking error would converge to zero exponentially fast. From the previous analysis of the observation error system in Section 10.4, it has been shown that the velocity observation error $\dot{\tilde{q}}(t)$ does in fact converges to zero exponentially fast. Hence, motivated by the above discussion and the form of (10.30), our controller must ensure that $\hat{r}(t)$ is driven to zero. To accomplish this new control objective, we need to construct the open-loop dynamics for $\hat{r}(t)$. Therefore, we differentiate (10.10) with respect to time and then multiply both sides by M to yield

$$M\dot{\hat{r}} = M\ddot{q}_d + M\alpha\dot{e} - M\ddot{\hat{q}}. \tag{10.31}$$

After substituting the second-order load velocity observer dynamics of (10.16) into (10.31) and then utilizing the velocity tracking error $\dot{e}(t)$ of (10.28) and the velocity observation error $\dot{\tilde{q}}(t)$ of (10.8) to rearrange the resulting expression in an advantageous manner, we obtain

$$M\dot{\hat{r}} = \Omega_1 + \Omega_2\dot{\tilde{q}} - \sum_{j=1}^{4}\sin(x_j)\gamma_j^2, \tag{10.32}$$

where the measurable auxiliary terms $\Omega_1(q, \dot{\hat{q}}, t)$ and Ω_2 are defined by

$$\Omega_1 = M\left(\ddot{q}_d + \alpha\dot{q}_d\right) + \left(B - M\alpha\right)\dot{\hat{q}} + N\sin(q) \tag{10.33}$$

and

$$\Omega_2 = -\left(M\alpha + k_1\right) \tag{10.34}$$

where k_1 was defined in (10.15). Since there is no control input on the right-hand side of (10.32), we add and subtract the desired per phase winding current term $\gamma_{dj}(t)$ to (10.32) to obtain

$$M\,\dot{\hat{r}} = \Omega_1 + \Omega_2\,\dot{\tilde{q}} - \sum_{j=1}^{4} \sin(x_j)\gamma_{dj}^2 + \sum_{j=1}^{4} \sin(x_j)\left(\gamma_{dj} + \gamma_j\right)\eta_j, \tag{10.35}$$

where $\eta_j(t)$ was defined previously in (10.11). Substituting the desired current trajectory $\gamma_{dj}(t)$ of (10.21) through (10.23) for the first occurrence of $\gamma_{dj}(t)$ in (10.35), we obtain the following dynamic expression for the observed filtered tracking error $\hat{r}(t)$ as

$$M\,\dot{\hat{r}} = \Omega_1 + \Omega_2\,\dot{\tilde{q}} - \tau_d + \sum_{j=1}^{4} \sin(x_j)\left(\gamma_{dj} + \gamma_j\right)\eta_j \tag{10.36}$$

where the fact

$$\tau_d = \sum_{j=1}^{4} \sin(x_j)\gamma_{dj}^2 \tag{10.37}$$

has been utilized (see (10.24), (10.27), and Chapter 5). In order to force $\hat{r}$ to zero, we design the desired torque signal as

$$\tau_d = \Omega_1 + \left(k_s + k_{n1}\Omega_2^2 + k_{n2}\right)\hat{r}, \tag{10.38}$$

where k_s was previously defined in (10.15) and k_{n1}, k_{n2} are positive scalar damping gains. Substituting the desired torque trajectory of (10.38) into the open-loop observed filtered tracking error dynamics of (10.36), we can formulate the closed-loop tracking error dynamics for $\hat{r}(t)$ as

$$M\,\dot{\hat{r}} = -k_s\hat{r} + \left[\Omega_2\,\dot{\tilde{q}} - k_{n1}\Omega_2^2\hat{r}\right] - k_{n2}\hat{r} + \sum_{j=1}^{4} \sin(x_j)\left(\gamma_{dj} + \gamma_j\right)\eta_j. \tag{10.39}$$

10.4.7 Current Tracking Error System

Now that the desired torque signal tracking problem has been reformulated as a current tracking problem, the next step in the observed integrator backstepping design process is to develop a voltage level input that will force $\eta_j(t)$ to zero. To ensure that $\eta_j(t)$ converges to zero, we must first construct the open-loop dynamics for $\eta_j(t)$. Taking the time derivative of (10.11) and multiplying the resulting expression by $L_j(q)$ yields

$$L_j(q)\dot{\eta}_j = L_j(q)\dot{\gamma}_{dj} - L_j(q)\dot{\gamma}_j. \tag{10.40}$$

It has been shown in Chapter 5 that the time derivative of the desired current trajectory can be described as

$$\dot{\gamma}_{dj} = \Sigma_j N_r \dot{q} + \Pi_j \dot{\tau}_d \tag{10.41}$$

where $\Sigma_j(\tau_d, q)$ and $\Pi_j(\tau_d, q)$ are explicitly defined in Appendix B. Given the desired torque trajectory $\tau_d(t)$ of (10.38), the time derivative of $\tau_d(t)$ can be expressed as

$$\begin{aligned}\dot{\tau}_d = \; & M\left(\dddot{q}_d + \alpha \ddot{q}_d\right) + (B - M\alpha)\ddot{\hat{q}} + N\cos(q)\dot{q} \\ & + \left(k_s + k_{n1}\Omega_2^2 + k_{n2}\right)\dot{\hat{r}}\,. \end{aligned} \tag{10.42}$$

After substituting the second-order velocity observer dynamics of (10.16) for $\ddot{\hat{q}}(t)$ and (10.32) for $\dot{\hat{r}}(t)$ into (10.42), we have

$$\begin{aligned}\dot{\tau}_d = \; & M\left(\dddot{q}_d + \alpha \ddot{q}_d\right) + N\cos(q)\dot{q} + (B - M\alpha)M^{-1} \\ & \cdot \left(-B\dot{\hat{q}} - N\sin(q) + \sum_{j=1}^{4}\sin(x_j)\gamma_j^2 + k_1\dot{\tilde{q}}\right) \\ & + \left(k_s + k_{n1}\Omega_2^2 + k_{n2}\right)M^{-1} \\ & \cdot \left(\Omega_1 + \Omega_2\dot{\tilde{q}} - \sum_{j=1}^{4}\sin(x_j)\gamma_j^2\right). \end{aligned} \tag{10.43}$$

Continuing the process of building the open-loop dynamics for $\eta_j(t)$, we substitute (10.43) into (10.41) and the resulting expression into (10.40) to obtain

$$\begin{aligned}L_j(q)\dot{\eta}_j = \; & L_j(q)\Sigma_j N_r\dot{q} + L_j(q)\Pi_j M\left(\dddot{q}_d + \alpha\ddot{q}_d\right) \\ & + L_j(q)\Pi_j N\cos(q)\dot{q} + \left(L_j(q)\Pi_j(B - M\alpha)M^{-1}\right) \\ & \cdot \left(-B\dot{\hat{q}} - N\sin(q) + \sum_{j=1}^{4}\sin(x_j)\gamma_j^2 + k_1\dot{\tilde{q}}\right) \\ & + \left(L_j(q)\Pi_j\left(k_s + k_{n1}\Omega_2^2 + k_{n2}\right)M^{-1}\right) \\ & \cdot \left(\Omega_1 + \Omega_2\dot{\tilde{q}} - \sum_{j=1}^{4}\sin(x_j)\gamma_j^2\right) - L_j(q)\dot{\gamma}_j. \end{aligned} \tag{10.44}$$

Finally, we substitute the phase current dynamics of (10.2) and (10.4) for $L_j(q)\dot{\gamma}_j$ into (10.44) to complete the open-loop dynamics for the current tracking error as

$$\begin{aligned} L_j(q)\dot{\eta}_j = & \; L_j(q)\Sigma_j N_r \dot{q} + L_j(q)\Pi_j M \left(\dddot{q}_d + \alpha \ddot{q}_d\right) \\ & + L_j(q)\Pi_j N \cos(q)\dot{q} + \left(L_j(q)\Pi_j \left(B - M\alpha\right) M^{-1}\right) \\ & \cdot \left(-B\,\dot{\hat{q}} - N\sin(q) + \sum_{j=1}^{4}\sin(x_j)\gamma_j^2 + k_1\,\dot{\tilde{q}}\right) \\ & + \left(L_j(q)\Pi_j \left(k_s + k_{n1}\Omega_2^2 + k_{n2}\right) M^{-1}\right) \\ & \cdot \left(\Omega_1 + \Omega_2\,\dot{\tilde{q}} - \sum_{j=1}^{4}\sin(x_j)\gamma_j^2\right) \\ & + R_0\gamma_j + N_r L_1 \sin(x_j)\gamma_j \dot{q} - V_j. \end{aligned} \tag{10.45}$$

From (10.45), we can see that the input to the electromechanical system, the phase voltage $V_j(t)$, has appeared in the open-loop dynamics of the current tracking error and can therefore be designed to force $\eta_j(t)$ to zero. To simplify the notation, we rearrange (10.45) into measurable and unmeasurable quantities as shown

$$L_j(q)\dot{\eta}_j = \Omega_{3j} + \Omega_{4j}\dot{q} + \Omega_{5j}\,\dot{\tilde{q}} - \frac{1}{2}\dot{L}_j(q)\eta_j - V_j, \tag{10.46}$$

where $\frac{1}{2}\dot{L}_j(q)\eta_j$ was added and subtracted to the right-hand side of (10.46) in order to facilitate the stability analysis, and the auxiliary measurable nonlinear functions $\Omega_{3j}(t)$, $\Omega_{4j}(t)$, and $\Omega_{5j}(t)$ are defined below

$$\begin{aligned} \Omega_{3j} = & \; L_j(q)\Pi_j M \left(\dddot{q}_d + \alpha \ddot{q}_d\right) + R_0\gamma_j \\ & + \left(L_j(q)\Pi_j \left(B - M\alpha\right) M^{-1}\right) \\ & \cdot \left(-B\,\dot{\hat{q}} - N\sin(q) + \sum_{j=1}^{4}\sin(x_j)\gamma_j^2\right) + \left(L_j(q)\Pi_j\right) \\ & \cdot \left(k_s + k_{n1}\Omega_2^2 + k_{n2}\right) M^{-1} \left(\Omega_1 - \sum_{j=1}^{4}\sin(x_j)\gamma_j^2\right), \end{aligned} \tag{10.47}$$

$$\begin{aligned} \Omega_{4j} = & \; \frac{1}{2}\frac{\partial L_j(q)}{\partial q}\eta_j + L_j(q)\Sigma_j N_r + L_j(q)\Pi_j N\cos(q) \\ & + N_r L_1 \sin(x_j)\gamma_j, \end{aligned} \tag{10.48}$$

and

$$\begin{aligned} \Omega_{5j} = \; & L_j(q)\Pi_j\left(B - M\alpha\right)M^{-1}k_1 \\ & + L_j(q)\Pi_j\left(k_s + k_{n1}\Omega_2^2 + k_{n2}\right)M^{-1}\Omega_2. \end{aligned} \tag{10.49}$$

Based on the structure of the open-loop current tracking error dynamics of (10.46) and the subsequent composite observer-controller stability analysis, we propose a voltage control input of the form

$$\begin{aligned} V_j = \; & \Omega_{3j} + \Omega_{4j}\,\dot{\hat{q}} + \\ & \left(k_s + k_{n3}\left(\Omega_{4j} + \Omega_{5j}\right)^2 + k_{n4}\left[\sin(x_j)\left(\gamma_{dj} + \gamma_j\right)\right]^2\right)\eta_j, \end{aligned} \tag{10.50}$$

where k_{n3} and k_{n4} represent positive, scalar nonlinear damping gains. After substituting the voltage controller $V_j(t)$ of (10.50) into (10.46), we can express the closed-loop current tracking error dynamics for $\eta_j(t)$ as

$$\begin{aligned} L_j(q)\dot{\eta}_j = \; & -k_s\eta_j + \left[\left(\Omega_{4j} + \Omega_{5j}\right)\dot{\tilde{q}} - k_{n3}\left(\Omega_{4j} + \Omega_{5j}\right)^2\eta_j\right] \\ & - \left[k_{n4}\sin^2(x_j)\left(\gamma_{dj} + \gamma_j\right)^2\eta_j\right] - \frac{1}{2}\dot{L}_j(q)\eta_j. \end{aligned} \tag{10.51}$$

The dynamics given by (10.30), (10.39), and (10.51), represent the closed-loop position tracking error systems for which the stability analysis is performed while (10.38), (10.21), and (10.50) represent the control which is implemented at the voltage terminals of the SR motor. Note that the desired torque trajectory $\tau_d(t)$ of (10.38), and the desired current trajectory (10.21) are embedded (in the guise of the variable $\eta_j(t)$) inside of the voltage control input $V_j(t)$.

10.4.8 Position Tracking Error Systems Analysis

The stability of the above closed-loop position tracking error system is now examined. Specifically, given the closed-loop error system dynamics of (10.30), (10.39), and (10.51), we define the following non-negative function

$$V_p = \frac{1}{2}e^2 + \frac{1}{2}M\hat{r}^2 + \frac{1}{2}\sum_{j=1}^{4}L_j(q)\eta_j^2. \tag{10.52}$$

After taking the time derivative of (10.52) and substituting (10.30), (10.39), and (10.51), we obtain

$$\begin{aligned}
\dot{V}_p = \; & e\left(-\alpha e + \hat{r} - \dot{\tilde{q}}\right) + \frac{1}{2}\sum_{j=1}^{4} \dot{L}_j(q)\eta_j^2 \\
& + \hat{r}\left(-k_s\hat{r} + \left[\Omega_2\,\dot{\tilde{q}} - k_{n1}\Omega_2^2\hat{r}\right]\right. \\
& \left. - k_{n2}\hat{r} + \sum_{j=1}^{4}\sin(x_j)\left(\gamma_{dj} + \gamma_j\right)\eta_j\right) + \sum_{j=1}^{4}\eta_j \qquad (10.53) \\
& \cdot\left(-k_s\eta_j + \left[\left(\Omega_{4j} + \Omega_{5j}\right)\dot{\tilde{q}} - k_{n3}\left(\Omega_{4j} + \Omega_{5j}\right)^2\eta_j\right]\right. \\
& \left. - \left[k_{n4}\sin^2(x_j)\left(\gamma_{dj} + \gamma_j\right)^2\eta_j\right] - \frac{1}{2}\dot{L}_j(q)\eta_j\right).
\end{aligned}$$

After cancelling the common terms in (10.53), we have

$$\begin{aligned}
\dot{V}_p = \; & -\alpha e^2 - k_s\hat{r}^2 - k_s\sum_{j=1}^{4}\eta_j^2 - e\,\dot{\tilde{q}} \\
& + \left[e\hat{r} - k_{n2}\hat{r}^2\right] + \left[\Omega_2\hat{r}\,\dot{\tilde{q}} - k_{n1}\Omega_2^2\hat{r}^2\right] \\
& + \sum_{j=1}^{4}\left[\left(\Omega_{4j} + \Omega_{5j}\right)\eta_j\,\dot{\tilde{q}} - k_{n3}\left(\Omega_{4j} + \Omega_{5j}\right)^2\eta_j^2\right] \qquad (10.54) \\
& + \sum_{j=1}^{4}\left[\sin(x_j)\left(\gamma_{dj} + \gamma_j\right)\hat{r}\eta_j\right. \\
& \left. - k_{n4}\sin^2(x_j)\left(\gamma_{dj} + \gamma_j\right)^2\eta_j^2\right]
\end{aligned}$$

where the bracketed terms are nonlinear damping pairs. After applying Lemma 1.8 in Chapter 1 to the bracketed terms, we can place a upper bound on $\dot{V}_p(t)$ of (10.54) as follows

$$\begin{aligned}
\dot{V}_p \leq \; & -\left(\alpha - \tfrac{1}{k_{n2}}\right)e^2 - \left(k_s - \tfrac{4}{k_{n4}}\right)\hat{r}^2 - \\
& k_s\textstyle\sum_{j=1}^{4}\eta_j^2 + |e|\left|\dot{\tilde{q}}\right| + \left(\tfrac{1}{k_{n1}} + \tfrac{4}{k_{n3}}\right)\dot{\tilde{q}}^2 .
\end{aligned} \qquad (10.55)$$

10.4.9 Stability of the Composite Error System

In this section, we will analyze the performance of the composite observer-controller error system. The performance of the closed-loop error systems is illustrated by the following theorem.

Theorem 10.1

Given the nonlinear velocity observer of (10.12), (10.14) and the voltage input control of (10.50), the load position tracking error is exponentially stable in the sense that

$$\|x(t)\| \le \sqrt{\frac{\lambda_2}{\lambda_1}}\,\|x(0)\|\,\underline{e}^{-\lambda t}, \quad \forall t \in [0,\infty] \tag{10.56}$$

where

$$x(t) = \left[\dot{\tilde{q}},\, e,\, \hat{r},\, \eta_1,\, \eta_2,\, \eta_3,\, \eta_4\right]^T \quad \in \Re^7, \tag{10.57}$$

$$\lambda_1 = \min_j \left\{M, L_j(q), 1\right\}, \quad \lambda_2 = \max_j \left\{M, L_j(q), 1\right\}, \tag{10.58}$$

$$\lambda = \frac{\lambda_3}{\lambda_2}, \tag{10.59}$$

and

$$\begin{aligned}\lambda_3 = \ & \min\left\{\left(k_s - \tfrac{1}{k_{n1}} - \tfrac{4}{k_{n3}}\right), \left(\alpha - \tfrac{1}{k_{n2}} - \tfrac{1}{k_{n5}}\right),\right. \\ & \left.\left(k_s - \tfrac{4}{k_{n4}}\right), k_s\right\}.\end{aligned} \tag{10.60}$$

To ensure the above stability result, the controller gains must be adjusted to satisfy the following sufficient condition

$$k_{ni} > \max\left\{\frac{5}{k_s}, \frac{2}{\alpha}\right\} \quad \text{for} \quad i = 1\cdots 5. \tag{10.61}$$

Proof:

In order to analyze the stability of the entire dynamic system, we add the Lyapunov-like functions for the observer given in (10.18) and the controller given in (10.52) to yield the following non-negative function

$$V = V_o + V_p = \frac{1}{2}x^T diag\left\{M, 1, M, L_1(q), L_2(q), L_3(q), L_4(q)\right\} x \tag{10.62}$$

where $x(t)$ was defined in (10.57), and

$$diag\left\{M, 1, M, L_1(q), L_2(q), L_3(q), L_4(q)\right\} \in \Re^{7\times 7} \tag{10.63}$$

denotes a diagonal matrix. From the matrix form of $V(t)$ given on the right-hand side of (10.62), Lemma 1.7 in Chapter 1 can be used to form the following upper and lower bounds for $V(t)$

$$\frac{1}{2}\lambda_1 \|x\|^2 \leq V(t) \leq \frac{1}{2}\lambda_2 \|x\|^2 \tag{10.64}$$

where λ_1 and λ_2 were defined in (10.58) (Note due the form of (10.3) the inductance can be upper and lower bounded by positive constants).

Given (10.20) and (10.55), the time derivative of (10.62) can be upper bounded as

$$\begin{aligned}\dot{V} \leq & -\left(k_s - \frac{1}{k_{n1}} - \frac{4}{k_{n3}}\right)\dot{\tilde{q}}^2 - \left(\alpha - \frac{1}{k_{n2}}\right)e^2 \\ & -\left(k_s - \frac{4}{k_{n4}}\right)\hat{r}^2 - k_s\sum_{j=1}^{4}\eta_j^2 + \left[|e|\left|\dot{\tilde{q}}\right| - k_{n5}\,\dot{\tilde{q}}^2\right].\end{aligned} \tag{10.65}$$

Again, the bracketed term in (10.65) is a nonlinear damping pair. After applying Lemma 1.8 in Chapter 1 to the bracketed term in (10.65), we have

$$\begin{aligned}\dot{V} \leq & -\left(k_s - \frac{1}{k_{n1}} - \frac{4}{k_{n3}}\right)\dot{\tilde{q}}^2 - \left(\alpha - \frac{1}{k_{n2}} - \frac{1}{k_{n5}}\right)e^2 \\ & -\left(k_s - \frac{4}{k_{n4}}\right)\hat{r}^2 - k_s\sum_{j=1}^{4}\eta_j^2\end{aligned} \tag{10.66}$$

which can be written in the following concise form

$$\dot{V} \leq -\lambda_3 \|x\|^2 \tag{10.67}$$

where $x(t)$ and λ_3 were defined previously in (10.57) and (10.60), respectively. From (10.64), it is easy to see that $\|x\|^2 \geq 2V(t)/\lambda_2$; hence, $\dot{V}(t)$ in (10.67) can be further upper bounded as

$$\dot{V}(t) \leq -2\lambda V(t) \tag{10.68}$$

where λ was defined in (10.59). Applying Lemma 1.1 in Chapter 1 to (10.68) yields

$$V(t) \leq V(0)e^{-2\lambda t}. \tag{10.69}$$

From (10.64), we have

$$\frac{1}{2}\lambda_1 \|x\|^2 \leq V(t) \quad \text{and} \quad V(0) \leq \frac{1}{2}\lambda_2 \|x(0)\|^2. \tag{10.70}$$

Substituting (10.70) appropriately into the left-hand and right-hand sides of (10.69) allows us to form the following inequality

$$\frac{1}{2}\lambda_1 \left\|x\right\|^2 \leq \frac{1}{2}\lambda_2 \left\|x(0)\right\|^2 \underline{e}^{-2\lambda t}$$

which can be used to solve for $\|x(t)\|$ to obtain the result given in (10.56).

Remark 10.2

An important property of the proposed PSFB controller is that all system quantities remain bounded. The direct result of (10.56) and (10.57) is that $e(t) \in L_\infty$, $\dot{\tilde{q}}\ (t) \in L_\infty$, $\hat{r}(t) \in L_\infty$, $\eta_1(t) \in L_\infty$, $\eta_2(t) \in L_\infty$, $\eta_3(t) \in L_\infty$, and $\eta_4(t) \in L_\infty$. Given $q_d(t)$, and $\dot{q}_d(t)$ are bounded, we know that $q(t) \in L_\infty$, $\dot{\hat{q}}\ (t) \in L_\infty$, and $\dot{q}(t) \in L_\infty$ from the definitions of $e(t)$, $\hat{r}(t)$, and $\dot{\tilde{q}}\ (t)$ given in (10.7), (10.10), and (10.8). It can now be shown from (10.21) and (10.38) that the desired current trajectory $\gamma_{dj}(t)$ can be written in the following form

$$\gamma_{dj} = g_1\left(q, \dot{\hat{q}}, q_d, \dot{q}_d, \ddot{q}_d\right)$$

where $g_1(\cdot) \in L_\infty$ due to the structure of the desired current trajectory and the fact that it is only dependent on bounded quantities. Thus, we can conclude that $\gamma_{dj}(t) \in L_\infty$. Therefore, the definition of phase current tracking error, $\eta_j(t)$, in (10.11) can be used to show that the actual current $\gamma_1(t)$, $\gamma_2(t)$, $\gamma_3(t)$, and $\gamma_4(t)$ are all bounded. In addition, equation (10.50) can now be written to show the dependence of $V_j(t)$ on bounded variables as

$$V_j = g_2\left(q, \dot{\hat{q}}, q_d, \dot{q}_d, \ddot{q}_d, \dddot{q}_d, \gamma_1, \gamma_2, \gamma_3, \gamma_4\right)$$

where $g_2(\cdot) \in L_\infty$ due to the structure of the voltage input controller and the fact that they are only dependent on bounded quantities. Thus, we can conclude that $V_1(t) \in L_\infty$, $V_2(t) \in L_\infty$, $V_3(t) \in L_\infty$, and $V_4(t) \in L_\infty$. With regard to the observer which is actually implemented, we know that $q(t) \in L_\infty$, $\dot{\hat{q}}\ (t) \in L_\infty$, $\gamma_1(t) \in L_\infty$, $\gamma_2(t) \in L_\infty$, $\gamma_3(t) \in L_\infty$, and $\gamma_4(t) \in L_\infty$; therefore, $\dot{y}(t) \in L_\infty$ from the (10.14). Using the fact that $\dot{\tilde{q}}\ (t)$ is exponentially stable as dictated by (10.56), it can be concluded from Lemma 1.9 in Chapter 1 that $\tilde{q}(t) \in L_\infty$ and hence we can use (10.13) to show $\hat{q}(t)$ is bounded. Furthermore, the auxiliary variable $y(t)$ can be shown to be bounded from (10.12) since $\dot{\hat{q}}\ (t)$ and $\tilde{q}(t)$ are bounded. Therefore, it has now been shown that all signals in the system, the observer, and the controller remain bounded during closed-loop operation.

10.5 PSFB Controller for BLDC Motor

The partial-state feedback controller proposed for the SR motor can be extended to provide tracking control for a BLDC motor turning a position-dependent load. That is, by utilizing a similar approach used for the SR motor PSFB controller development, we can develop a position tracking controller for the BLDC motor using only load position and stator current measurements.

10.5.1 BLDC Motor Model

An electromechanical system model similar to that presented in Chapter 4 for a BLDC motor driving a position-dependent load is adopted here for the PSFB controller design. The model is given by

$$M\ddot{q} + B\dot{q} + N\sin(q) = (K_{\tau 1}I_b + K_{\tau 2})\, I_a, \tag{10.71}$$

$$L_a\dot{I}_a + RI_a + n_pL_bI_b\dot{q} + K_{\tau 2}\dot{q} = V_a, \tag{10.72}$$

and

$$L_b\dot{I}_b + RI_b - n_pL_aI_a\dot{q} = V_b \tag{10.73}$$

where $K_{\tau 1}$ and $K_{\tau 2}$ are two positive torque transmission constants. The other variables in the above BLDC electromechanical system model were previously defined in Chapter 4.

Remark 10.3

The only difference between the above BLDC motor model and the model given in Chapter 4 is that the mechanical subsystem of (10.71) is not divided by the constant $K_{\tau 2}$ as was done in Chapter 4.

10.5.2 Velocity Observer for the BLDC Motor

The velocity observer for the BLDC motor is almost the same as that for the SR motor presented in (10.12) and (10.14). For the BLDC motor, the velocity observer is given by

$$\dot{\hat{q}} = y + M^{-1}k_o\tilde{q} \tag{10.74}$$

where k_o is a positive control gain, and y is an auxiliary variable which is updated on-line according to the following equation

$$\dot{y} = M^{-1}\left(-B\,\dot{\hat{q}} - N\sin(q) + (K_{\tau 1}I_b + K_{\tau 2})\, I_a\right). \tag{10.75}$$

Remark 10.4

Comparing the BLDC motor velocity observer of (10.74) and (10.75) to the SR motor velocity observer in (10.12) and (10.14), we can see the only difference is the torque coupling term. For the SR motor, the developed motor torque is $\sum_{j=1}^{4} \sin(x_j)\gamma_j^2$ while for the BLDC motor, the developed motor torque is $(K_{\tau 1} I_b + K_{\tau 2}) I_a$.

10.5.3 Voltage Control Inputs for the BLDC Motor

Based on the above observed velocity signal and a design [2] approach similar to that used for the SR motor, we propose the following PSFB position tracking controller for the BLDC motor. The position tracking error $e(t)$ and the observed filtered tracking error $\hat{r}(t)$ are defined in (10.7) and (10.10), respectively. The desired current trajectories are given as

$$I_{da} = K_{\tau 2}^{-1}\left(\Omega_1 + k_1\hat{r} + k_{n1}\Omega_2^2\hat{r}\right) \quad \text{and} \quad I_{db} = 0, \tag{10.76}$$

where k_1 is a positive control gain and k_{n1} is a nonlinear damping gain. The auxiliary variables $\Omega_1(q, \dot{\hat{q}}, t)$ and Ω_2, defined in terms of system parameters and measurable states, are explicitly given by

$$\Omega_1 = M\ddot{q}_d + \alpha M\dot{q}_d + B\,\dot{\hat{q}} + N\sin(q) - M\alpha\,\dot{\hat{q}} \tag{10.77}$$

and

$$\Omega_2 = -k_o - M\alpha. \tag{10.78}$$

In a similar manner to the commutation strategy proposed in Chapter 4, the desired current trajectories ensure that the load position tracks the desired load position trajectory. To ensure that the actual currents track the desired current trajectories, we can design the voltage control inputs [2] as follows

$$V_a = \Omega_3 + \Omega_4\,\dot{\hat{q}} + k_{n2}\left(\Omega_4 + \Omega_5\right)^2\eta_a + k_2\eta_a + \left(\frac{1}{2}K_{\tau 1}I_b + K_{\tau 2}\right)\hat{r} \tag{10.79}$$

and

$$V_b = RI_b - n_pL_aI_a\,\dot{\hat{q}} + \left(n_pL_aI_a\right)^2 k_{n3}\eta_b + k_3\eta_b + \frac{1}{2}K_{\tau 1}\left(I_{da} + I_a\right)\hat{r} \tag{10.80}$$

where the current tracking error variables $\eta_a(t)$ and $\eta_b(t)$ are defined by

$$\eta_a = I_{da} - I_a \qquad \eta_b = -I_b, \tag{10.81}$$

k_2, k_3 are positive control gains, k_{n2}, k_{n3} are nonlinear damping gains, and $\Omega_3(t)$, $\Omega_4(t)$, $\Omega_5(t)$ are known scalar functions which are defined in terms of system parameters and measurable state variables as given by the following expressions

$$\begin{aligned}\Omega_3 = \; & RI_a + L_a M K_{\tau 2}^{-1} \dddot{q}_d + L_a M \alpha K_{\tau 2}^{-1} \ddot{q}_d \\ & + L_a M^{-1} K_{\tau 2}^{-1} \left(k_1 + k_{n1}\Omega_2^2\right)\left(\Omega_1 - K_{\tau 1} I_a I_b - K_{\tau 2} I_a\right) \\ & + L_a M^{-1} K_{\tau 2}^{-1} \left(B - \alpha M\right) \\ & \left(-B \dot{\hat{q}} - N \sin(q) + K_{\tau 1} I_a I_b + K_{\tau 2} I_a\right),\end{aligned} \tag{10.82}$$

$$\Omega_4 = n_p I_b L_b + K_{\tau 2} + L_a K_{\tau 2}^{-1} N \cos(q), \tag{10.83}$$

and

$$\Omega_5 = L_a M^{-1} K_{\tau 2}^{-1} \left(k_1 + k_{n1}\Omega_2^2\right)\Omega_2 + L_a M^{-1} K_{\tau 2}^{-1} \left(B - \alpha M\right) k_o. \tag{10.84}$$

Remark 10.5

The same stability result can be obtained for the BLDC motor observer-controller error systems as that of the SR motor observer-controller error systems [2]. That is, using the above velocity observer and controller, we can achieve exponential load position tracking performance.

10.6 Experimental Results

The proposed PSFB controllers for the SR motor and the BLDC motor were successfully implemented on a motor test bed described in the earlier chapters (*e.g.* Chapter 5).

10.6.1 SR Motor Experimental Result

The hardware used to implement the PSFB controller for the SR motor is exactly the same as that in Chapter 5. The desired trajectory for the load position is selected as

$$q_d(t) = \frac{\pi}{2} \sin(2t)\left(1 - e^{-0.3t^3}\right) \; rad, \tag{10.85}$$

which has the desirable property $q_d(0) = \dot{q}_d(0) = \ddot{q}_d(0) = \dddot{q}_d(0) = 0$. The PSFB controller was found to yield good load position tracking performance with the following controller gain values

$$\alpha = 70, \quad k_{n1} = k_{n2} = 5 \times 10^{-4}, \quad k_{n3} = 2, \quad k_{n4} = 0.001,$$

$$k_{n5} = 15, \quad \gamma_{d0} = 0.4, \quad \epsilon_0 = 0.1.$$

The initial conditions on the electromechanical system dynamics, the observer dynamics, and the intermediate variable $y(t)$ of (10.14) were all set to zero. The resulting position tracking error is given in Figure 10.1. This figure indicates that the maximum position tracking error is within ± 0.3 degrees. Note that due to measurement noise, quantization error in applying the control, and resolution of the encoder the tracking error does not approach zero as predicted by the theory but rather is driven to a small value. The corresponding per phase winding voltages are shown in Figures 10.2 through Figure 10.5.

10.6.2 BLDC Motor Experimental Result

The hardware used to implement the PSFB controller for the BLDC motor is exactly the same as that in Chapter 4. The desired position trajectory is selected as in (10.85). The initial conditions for the system states and observer states were all set to zero. The control gains were selected as

$$\alpha = 160,\ k_1 = 7,\ k_{n1} = 4 \times 10^{-3},\ k_2 = 0.1,\ k_{n2} = 10^{-6},$$

$$k_{n3} = 10,\ \ k_3 = 2.1,\ \ k_o = 65.$$

The resulting position tracking error is shown in Figure 10.6. The actual phase voltages are shown in Figure 10.7 through Figure 10.9. From the position tracking error plot, we can see that the maximum position tracking error is within ± 0.15 degrees. Again, due to measurement noise, quantization error in applying the control, and resolution of the encoder, the tracking error does not approach zero as predicted by the theory but rather is driven to a small value.

Remark 10.6

The selection of the control gains in the experiments does not satisfy the conditions stated in Theorem 10.1. However, these controller gain requirements are only sufficient, conservative conditions generated by the Lyapunov stability argument.

10.7 Notes

In [3], Ehsani *et al* presented a method to eliminate the discrete position sensor and current sensors in a SR motor drive. State observers for variable-reluctance motors were presented by Lumsdaine *et al* in [4]. In [5], Lagerquist presented a closed-loop torque vector control based speed control for a synchronous reluctance motor that estimates rotor speed from the flux-linkage vector velocity. The PSFB controllers presented in [6] and [2] provide the basic contents of this chapter.

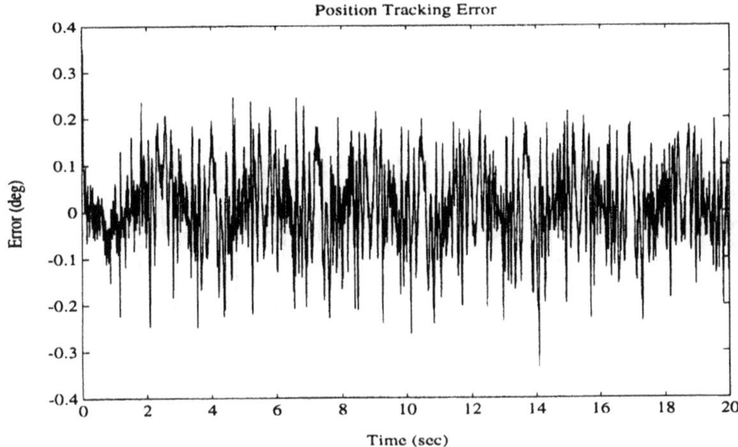

Figure 10.1: Position Tracking Error (SR Motor)

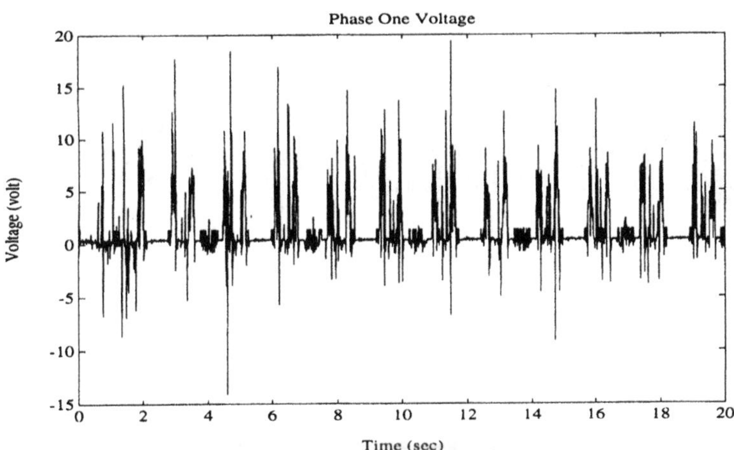

Figure 10.2: Phase One Voltage (SR Motor)

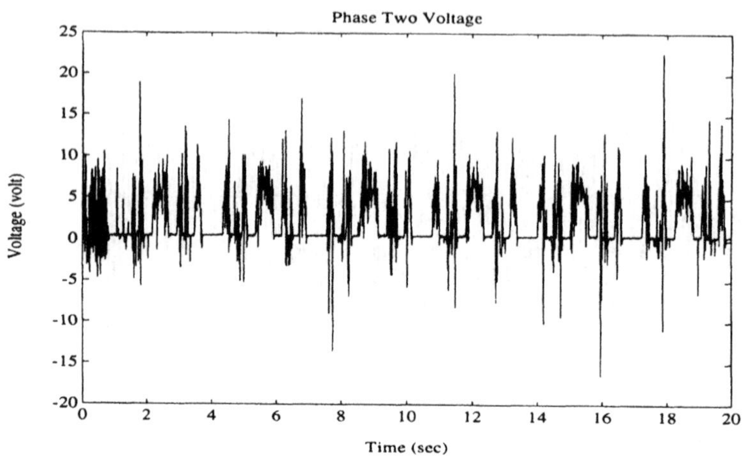

Figure 10.3: Phase Two Voltage (SR Motor)

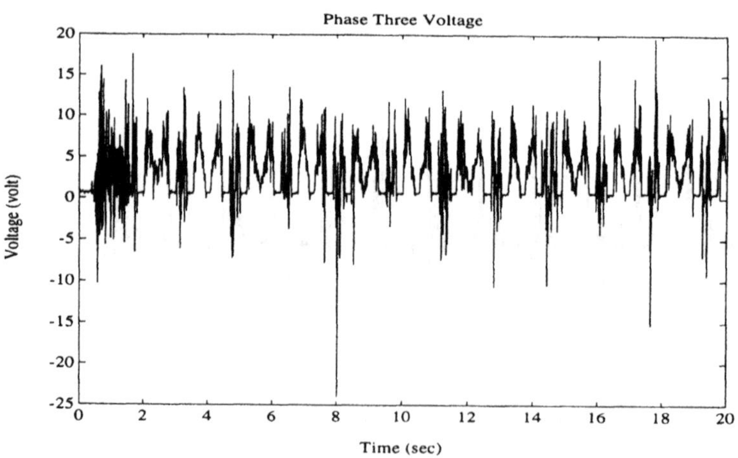

Figure 10.4: Phase Three Voltage (SR Motor)

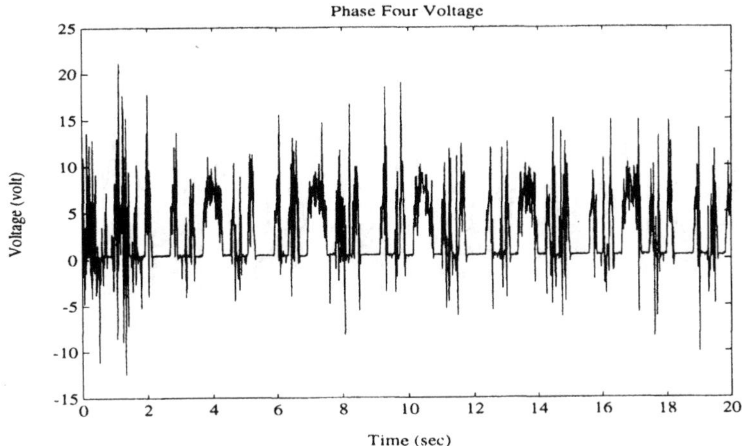

Figure 10.5: Phase Four Voltage (SR Motor)

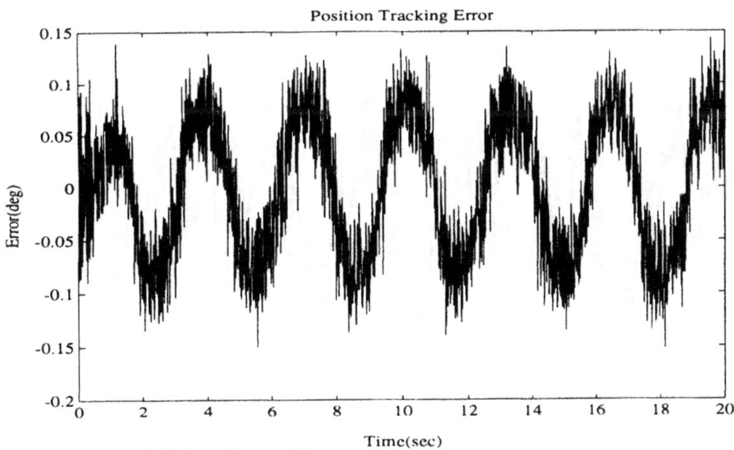

Figure 10.6: Position Tracking Error (BLDC Motor)

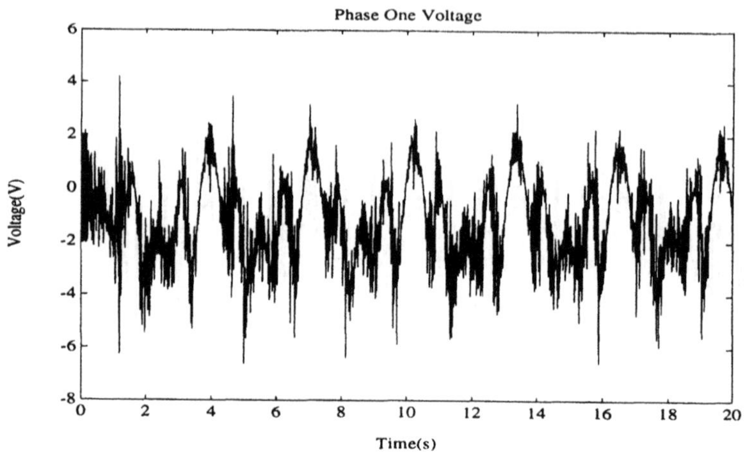

Figure 10.7: Phase One Voltage (BLDC Motor)

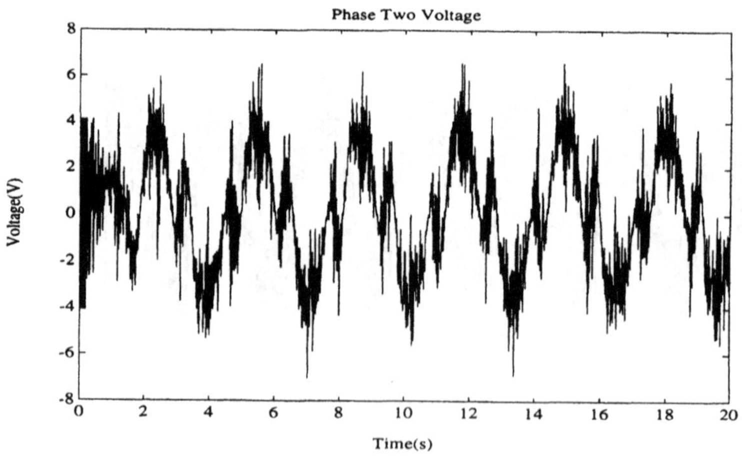

Figure 10.8: Phase Two Voltage (BLDC Motor)

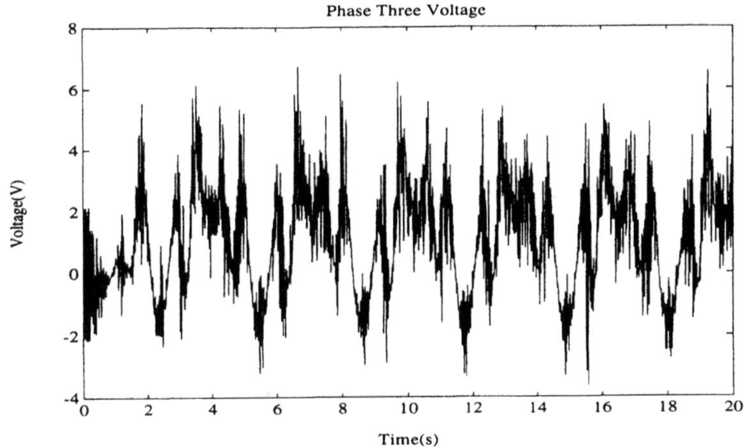

Figure 10.9: Phase Three Voltage (BLDC Motor)

Bibliography

[1] M. Krstic, I. Kanellakopoulos, and P. Kokotovic, *Nonlinear and Adaptive Control Design,* John Wiley & Sons, 1995.

[2] J. Hu, D. Dawson, and K. Anderson, "Position Control of a Brushless DC Motor without Velocity Measurements", *IEE Electric Power Applications*, Vol. 142, No. 2, Mar., 1995, pp. 113-122.

[3] M. Ehsani, I. Husain, and A. Kulkarni, "Elimination of Discrete Position Sensor and Current Sensor in Switched Reluctance Motor Drives", *IEEE Transactions on Industry Applications*, Vol. 28, No. 1, pp. 128-135, Jan./Feb. 1992.

[4] A. Lumsdaine and J. Lang, "State Observers for Variable-Reluctance Motors", *IEEE Transactions on Industrial Electronics*, Vol. 37, No. 2, pp. 133-142, April 1990.

[5] R. Lagerquist, I. Boldea, and T. Miller, "Sensorless Control of the Synchronous Reluctance Motor", *Proc. IEEE Industrial Applications Society Annual Meeting*, pp. 427-436, 1993.

[6] M. D. Leviner, D. M. Dawson, and J. Hu, "Position Tracking Control of a Switched Reluctance Motor Using Partial State Feedback", *Proc. of the American Control Conference*, Vol. 3, pp. 3095-3099, Baltimore, MD, June, 1994.

Chapter 11

Induction Motor (PSFB-I)

11.1 Introduction

In this chapter, we use the framework developed in Chapter 7 to develop a load position tracking controller for the electromechanical model of an induction motor driving a position-dependent load under the assumptions that load position, load velocity, and stator current measurements are available and that an exact model for the electromechanical system dynamics can be determined. While the output feedback controller (OFB) for brushed dc (BDC) motor provides a starting point for the design of an induction PSFB motor controller, the induction motor control design is more complicated due to the fact that the multi-phase induction motor/load electromechanical dynamics exhibit: i) bilinear terms, which are products of the stator electrical current and the rotor flux, in the static torque transmission equation, and ii) bilinear terms, which are products of the rotor velocity and the stator electrical current (or the rotor flux), in the electrical subsystem dynamics. In addition, in contrast with all of the previous electric machine models, the induction motor possesses an additional set of rotor flux dynamic equations which must be accounted for during the control design procedure. Indeed, it is these differences between the BDC and the induction electromechanical dynamics that have prevented the design of an OFB induction motor controller; however, the observed-integrator backstepping procedure [1] used in Chapter 7 can be utilized to design an partial state feedback (PSFB) controller for induction motor. Specifically, by utilizing the structure of the induction electromechanical system, we design nonlinear observers to estimate the rotor flux. A Lyapunov-like argument is then used to motivate the construction of the observers; however, in contrast with the BDC and PMS OFB control approaches, we can not explicitly show that the observation error goes to zero exponentially fast. Specifically, a general statement on the open-loop rotor flux observation error is prevented by the fact that the rotor flux observation error dynamics contain additional nonlinear control terms which facilitate the design of a singularity-free position tracking controller. Notwithstanding this difference from previous chapters, we then use the structure of these observers, the observed-integrator backstepping procedure, nonlinear damping tools, and the commutation strategy given in Chapter 6 to design a load position tracking controller. In addition, we design the desired flux trajectory and the initial conditions for the rotor flux observer to guarantee that control singularities are avoided. A Lyapunov-like argument is then used to prove that all of the signals in the actual electromechanical systems and the observer remain bounded during closed-loop operation; furthermore, the controller ensures that the load position tracks the desired trajectory exponentially fast into some small ball (*i.e.*, semi-global uniform ultimately bounded (SGUUB) load position tracking). The result is semi-global in the sense that a control parameter must be made sufficiently large relative to the electromechanical initial conditions in order to guarantee that control singularities are avoided. Finally, experimental results are utilized to illustrate the performance of the proposed controller.

11.2 System Model

To facilitate the control development, the system model presented in Chapter 6 for an induction motor driving a position-dependent mechanical load is adopted for use in this chapter. The mechanical subsystem is taken to be a one-link, direct-driven robot of the form

$$M\ddot{q} + B\dot{q} + N\sin(q) = \tau \tag{11.1}$$

where M is the mechanical inertia of the system (including the rotor inertia), B is the coefficient of viscous friction at the joint, N is a positive constant related to the mass of the load and the coefficient of gravity, and $q(t)$, $\dot{q}(t)$, $\ddot{q}(t)$ are scalar functions of time representing the load position, velocity, and acceleration, respectively.

Under the assumptions of equal mutual inductances and a linear magnetic circuit, the electrical subsystem dynamics of an induction motor in the stator-fixed $a-b$ reference frame are given by (see Appendix C)

$$L_I\dot{I}_a + R_I I_a - \alpha_1\psi_a - \alpha_2\psi_b\dot{q} = V_a, \tag{11.2}$$

$$L_I\dot{I}_b + R_I I_b - \alpha_1\psi_b + \alpha_2\psi_a\dot{q} = V_b, \tag{11.3}$$

$$L_r\dot{\psi}_a + R_r\psi_a + \alpha_3\psi_b\dot{q} = K_I I_a, \tag{11.4}$$

and

$$L_r\dot{\psi}_b + R_r\psi_b - \alpha_3\psi_a\dot{q} = K_I I_b \tag{11.5}$$

where $\psi_a(t)$, $\psi_b(t)$, $I_a(t)$, $I_b(t)$, $V_a(t)$, and $V_b(t)$ represent the transformed rotor fluxes, transformed stator currents, and transformed stator voltages of the induction motor, respectively. The positive constants L_I, K_I, α_1, α_2, α_3, and R_I are explicitly related to the motor parameters as shown

$$\alpha_1 = M_e R_r / L_r^2, \quad \alpha_2 = n_p M_e / L_r, \quad \alpha_3 = n_p L_r, \quad K_I = R_r M_e, \tag{11.6}$$
$$L_I = L_s - M_e^2 / L_r, \quad \text{and} \quad R_I = \left(M_e^2 R_r + L_r^2 R_s\right) / L_r^2$$

where L_r, L_s, and M_e are the transformed rotor inductance, stator inductance, and mutual inductance of the motor, respectively, R_r and R_s represent the rotor resistance and stator resistance, respectively, and n_p is the motor pole pair number. The electromechanical coupling torque in (11.1) is given by

$$\tau = \alpha_2 \left(\psi_a I_b - \psi_b I_a\right). \tag{11.7}$$

Remark 11.1

The detailed development for the above stator-fixed reference frame transformed model is presented in Appendix C. The only difference between the above model and the model presented in Chapter 6 is that the mechanical subsystem is not divided by α_2 as was done in Chapter 6.

11.3 Control Objective

Our objective in this chapter is to design a load position tracking controller for the electromechanical model given by (11.1) through (11.7). Only stator current, load position, and load velocity are assumed to be measurable; furthermore, the system parameters are assumed to be known exactly. With this objective in mind, we define the load position tracking error to be

$$e = q_d - q \tag{11.8}$$

where $q_d(t)$ represents the desired load position trajectory. We will assume that $q_d(t)$ and its first, second, and third time derivatives are all bounded functions of time. In addition, to simplify the stability proof, we define the filtered tracking error [2] as

$$r = \dot{e} + \alpha e \tag{11.9}$$

where α is a positive scalar constant control gain. It should be noted that if the control gain α is equal to zero, then the filtered tracking error is equivalent to the velocity tracking error; hence, it is easy to see how any controller which ensures good filtered tracking error performance can be used for controlling the load velocity (*i.e.*, a standard motor control problem). In addition, it is easy to show that with α selected to be positive and if the filtered tracking error is SGUUB then the load position tracking error is SGUUB (see Lemma 1.11 of Chapter 1).

11.4 Nonlinear Flux Observer

Since rotor flux is assumed to be unmeasurable, we first design observers to estimate the actual rotor flux. Based on the rotor flux dynamics of (11.4) and (11.5), and the subsequent analysis, we define the nonlinear, rotor flux observers as [3]

$$L_r \dot{\hat{\psi}}_a = -R_r\hat{\psi}_a - \alpha_3\hat{\psi}_b\dot{q} + K_I I_a - \alpha_2 I_b r \tag{11.10}$$

and

$$L_r \dot{\hat{\psi}}_b = -R_r\hat{\psi}_b + \alpha_3\hat{\psi}_a\dot{q} + K_I I_b + \alpha_2 I_a r. \tag{11.11}$$

After subtracting (11.10) and (11.11) from (11.4) and (11.5), respectively, we obtain the following rotor flux observation error systems

$$L_r \dot{\tilde{\psi}}_a = -R_r\tilde{\psi}_a - \alpha_3\tilde{\psi}_b\dot{q} + \alpha_2 I_b r \tag{11.12}$$

and

$$L_r \dot{\tilde{\psi}}_b = -R_r\tilde{\psi}_b + \alpha_3\tilde{\psi}_a\dot{q} - \alpha_2 I_a r \tag{11.13}$$

where the rotor flux observation error terms, $\tilde{\psi}_a(t)$ and $\tilde{\psi}_b(t)$, are defined as

$$\tilde{\psi}_a = \psi_a - \hat{\psi}_a \quad \text{and} \quad \tilde{\psi}_b = \psi_b - \hat{\psi}_b. \tag{11.14}$$

To analyze the performance of the above rotor flux observers, we define the following non-negative function

$$V_o = \frac{1}{2}L_r\tilde{\psi}_a^2 + \frac{1}{2}L_r\tilde{\psi}_b^2. \tag{11.15}$$

The time derivative of $V_o(t)$ is given by

$$\dot{V}_o = L_r\tilde{\psi}_a \dot{\tilde{\psi}}_a + L_r\tilde{\psi}_b \dot{\tilde{\psi}}_b \,. \tag{11.16}$$

After substituting the rotor flux observation error systems of (11.12) and (11.13) into (11.16), we have

$$\dot{V}_o = -R_r\left(\tilde{\psi}_a^2 + \tilde{\psi}_b^2\right) + \left[\alpha_2 r\left(\tilde{\psi}_a I_b - \tilde{\psi}_b I_a\right)\right]. \tag{11.17}$$

Remark 11.2

While the appearance of most of the terms in the rotor flux observer dynamics are due to the structure of the corresponding electrical dynamics, we note that there are two additional terms (*i.e.*, the last two terms on the right-hand side of (11.10) and (11.11)). By utilizing (11.17), we can see that if the bracketed term is neglected then $\dot{V}_o(t)$ will be negative definite, and hence, it is easy to show that the flux observation errors will converge to zero. Note that we can classify the proposed rotor flux observer as an open-loop observer since the rate of convergence can not be altered by a controller gain. It turns out that the bracketed term in (11.17) is designed, via the observer construction, to cancel out a similar term which arises during the development of the load position tracking controller. That is, two like terms with opposite signs are cancelled during the analysis of the composite observer-controller, closed-loop error system.

11.5 Voltage Input Design

Based on the observed rotor flux information, we now turn our attention to the design of a voltage input controller to ensure load position tracking.

11.5.1 Closed-Loop Filtered Tracking Error System

Given the definition of the filtered tracking error $r(t)$ of (11.9), we can write the mechanical subsystem dynamics of (11.1) as

$$M\dot{r} = M\left(\ddot{q}_d + \alpha\dot{e}\right) + B\dot{q} + N\sin(q) - \alpha_2\left(\psi_a I_b - \psi_b I_a\right) \tag{11.18}$$

where (11.7) and (11.8) have been utilized. From the right-hand side of (11.7), we can see that the torque produced by the electrical subsystem is a function of rotor flux which is an unmeasurable quantity; hence, to facilitate the observed backstepping procedure, we rewrite (11.18) as

$$\begin{aligned} M\dot{r} = \; & M\left(\ddot{q}_d + \alpha\dot{e}\right) + B\dot{q} + N\sin(q) \\ & -\alpha_2\left(\tilde{\psi}_a I_b - \tilde{\psi}_b I_a\right) - \alpha_2\left(\hat{\psi}_a I_b - \hat{\psi}_b I_a\right) \end{aligned} \tag{11.19}$$

where (11.14) has been utilized. Since the above filtered tracking error dynamics lack a control input, we add and subtract a desired torque signal $\tau_d(t)$ to the right-hand side of (11.19) to obtain

$$\begin{aligned} M\dot{r} = \; & M\left(\ddot{q}_d + \alpha\dot{e}\right) + B\dot{q} + N\sin(q) \\ & -\alpha_2\tau_d + \alpha_2\hat{\eta}_\tau - \alpha_2\left(\tilde{\psi}_a I_b - \tilde{\psi}_b I_a\right) \end{aligned} \tag{11.20}$$

where $\hat{\eta}_\tau(t)$ is used to represent the estimated torque tracking error and is explicitly defined by

$$\hat{\eta}_\tau = \tau_d - \left(\hat{\psi}_a I_b - \hat{\psi}_b I_a\right). \tag{11.21}$$

Motivated by the structure of (11.20), we design the desired torque signal $\tau_d(t)$ as follows

$$\tau_d = \alpha_2^{-1}\left(M\left(\ddot{q}_d + \alpha\dot{e}\right) + B\dot{q} + N\sin(q) + k_1 r\right) \tag{11.22}$$

where k_1 is a positive control gain. Substituting $\tau_d(t)$ of (11.22) into (11.20) yields the closed-loop dynamics for the filtered tracking error system as follows

$$M\dot{r} = -k_1 r + \alpha_2\hat{\eta}_\tau - \left[\alpha_2\left(\tilde{\psi}_a I_b - \tilde{\psi}_b I_a\right)\right]. \tag{11.23}$$

Remark 11.3

To explain the origins of some of the terms in (11.23), we note that the bracketed term in (11.23) will be directly cancelled by the bracketed term in (11.17) during the composite observer-controller stability analysis. Therefore, assuming that the bracketed term is zero and $\hat{\eta}_\tau(t)$ is small, then we can see that load tracking would be achieved. Based on the above observations and the form of (11.23), we are motivated to continue the backstepping procedure and develop a controller which ensures that $\hat{\eta}_\tau(t)$ is driven to a small value.

To ensure that $\hat{\eta}_\tau(t)$ remains small, we first construct the open-loop dynamics for $\hat{\eta}_\tau(t)$. To this end, we multiply both sides of (11.21) by $L_I L_r$ and take the time derivative of the resulting expression to obtain

$$\begin{aligned} L_I L_r \dot{\hat{\eta}}_\tau = \; & L_I L_r \dot{\tau}_d - L_I L_r\left(\dot{\hat{\psi}}_a I_b + \hat{\psi}_a \dot{I}_b\right) \\ & + L_I L_r\left(\dot{\hat{\psi}}_b I_a + \hat{\psi}_b \dot{I}_a\right). \end{aligned} \tag{11.24}$$

After substituting the time derivative of (11.22) for $\dot{\tau}_d(t)$, the flux observer estimation dynamics from (11.10) and (11.11) for $\dot{\hat{\psi}}_a(t)$ and $\dot{\hat{\psi}}_b(t)$, and the current dynamics from (11.2) and (11.3) for $\dot{I}_a(t)$ and $\dot{I}_b(t)$ into (11.24), we have

$$L_I L_r \dot{\hat{\eta}}_\tau = w_\tau + \Omega_1\psi_a + \Omega_2\psi_b - \left(-L_r\hat{\psi}_b V_a + L_r\hat{\psi}_a V_b\right) \tag{11.25}$$

where the auxiliary measurable terms $w_\tau(t)$, $\Omega_1(t)$, and $\Omega_2(t)$ are given by

$$\begin{aligned} w_\tau \;=\; & L_r L_I \alpha_2^{-1}\left[M\left(\dddot{q}_d + \alpha\ddot{q}_d\right) + k_1\left(\ddot{q}_d + \alpha\dot{e}\right)\right] \\ & - L_r L_I \alpha_2^{-1} M^{-1}\left(B\dot{q} + N\sin(q)\right)\left(B - \alpha M - k_1\right) \\ & - L_I I_b\left(-R_r\hat{\psi}_a - \alpha_3\hat{\psi}_b\dot{q} + K_I I_a - \alpha_2 I_b r\right) \\ & + L_I I_a\left(-R_r\hat{\psi}_b + \alpha_3\hat{\psi}_a\dot{q} + K_I I_b + \alpha_2 I_a r\right) \\ & + L_r L_I \alpha_2^{-1} N\cos(q)\dot{q} + L_r R_I\hat{\psi}_a I_b - L_r R_I\hat{\psi}_b I_a, \end{aligned} \tag{11.26}$$

$$\Omega_1 = L_I L_r M^{-1} (B - \alpha M - k_1) I_b + \alpha_1 L_r \hat{\psi}_b + L_r \alpha_2 \hat{\psi}_a \dot{q}, \tag{11.27}$$

and

$$\Omega_2 = -L_I L_r M^{-1} (B - \alpha M - k_1) I_a - \alpha_1 L_r \hat{\psi}_a + L_r \alpha_2 \hat{\psi}_b \dot{q}. \tag{11.28}$$

During the construction of the open-loop dynamics given in (11.25), the mechanical subsystem dynamics of (11.1) and (11.7) have been substituted for all occurrences of $\ddot{q}(t)$. Based on the structure of (11.25), we now design the following voltage control input relationship to ensure that the torque tracking error $\hat{\eta}_\tau(t)$ remains small

$$\begin{aligned} -L_r \hat{\psi}_b V_a + L_r \hat{\psi}_a V_b &= w_\tau + \Omega_1 \hat{\psi}_a + \Omega_2 \hat{\psi}_b + k_2 \hat{\eta}_\tau \\ &\quad + k_{n1} \Omega_1^2 \hat{\eta}_\tau + k_{n1} \Omega_2^2 \hat{\eta}_\tau + \alpha_2 r \end{aligned} \tag{11.29}$$

where k_2 is a positive control gain, and k_{n1} is a positive nonlinear damping gain. Substituting the right-hand side of (11.29) into (11.25) yields the closed-loop, torque tracking error dynamics in the form

$$\begin{aligned} L_I L_r \dot{\hat{\eta}}_\tau &= -k_2 \hat{\eta}_\tau - \alpha_2 r + \left[\Omega_1 \tilde{\psi}_a - k_{n1} \Omega_1^2 \hat{\eta}_\tau \right] \\ &\quad + \left[\Omega_2 \tilde{\psi}_b - k_{n1} \Omega_2^2 \hat{\eta}_\tau \right]. \end{aligned} \tag{11.30}$$

Remark 11.4

To explain the origins of some of the terms in (11.30), we note that the term $-\alpha_2 r$ will be used to cancel out the $\alpha_2 \hat{\eta}_\tau$ term in (11.23) during the filtered tracking error stability analysis. The bracketed terms in (11.30) are nonlinear damping pairs, and hence, their upper bounds can be combined with other terms during the composite observer-controller analysis to obtain the desired stability result.

11.5.2 Analysis of the Closed-Loop Filtered Tracking Error System

To analyze the performance of the closed-loop position tracking error systems given by (11.23) and (11.30), we define the following non-negative function

$$V_p = \frac{1}{2} M r^2 + \frac{1}{2} L_I L_r \hat{\eta}_\tau^2. \tag{11.31}$$

After taking the time derivative of (11.31) and substituting (11.23) and (11.30) into the resulting expression, we have

$$\begin{aligned} \dot{V}_p = &\ -k_1 r^2 - \alpha_2 r \left(\tilde{\psi}_a I_b - \tilde{\psi}_b I_a \right) \\ &\ -k_2 \hat{\eta}_\tau^2 + \Omega_1 \hat{\eta}_\tau \tilde{\psi}_a + \Omega_2 \hat{\eta}_\tau \tilde{\psi}_b - k_{n1} \Omega_1^2 \hat{\eta}_\tau^2 - k_{n1} \Omega_2^2 \hat{\eta}_\tau^2 \end{aligned} \tag{11.32}$$

which can be upper bounded as

$$\begin{aligned}\dot{V}_p \leq & \left[|\Omega_2|\,|\hat{\eta}_\tau|\left|\tilde{\psi}_b\right| - k_{n1}\Omega_2^2\hat{\eta}_\tau^2\right] - \alpha_2 r\left(\tilde{\psi}_a I_b - \tilde{\psi}_b I_a\right) \\ & + \left[|\Omega_1|\,|\hat{\eta}_\tau|\left|\tilde{\psi}_a\right| - k_{n1}\Omega_1^2\hat{\eta}_\tau^2\right] - k_1 r^2 - k_2\hat{\eta}_\tau^2.\end{aligned} \tag{11.33}$$

By applying Lemma 1.8 in Chapter 1 to the bracketed nonlinear damping pairs in (11.33), we can place a new upper bound on $\dot{V}_p(t)$ as follows

$$\begin{aligned}\dot{V}_p \leq & -k_1 r^2 - k_2\hat{\eta}_\tau^2 \\ & -\alpha_2 r\left(\tilde{\psi}_a I_b - \tilde{\psi}_b I_a\right) + \frac{1}{k_{n1}}\tilde{\psi}_a^2 + \frac{1}{k_{n1}}\tilde{\psi}_b^2.\end{aligned} \tag{11.34}$$

As we will see during the composite observer-controller stability analysis, the terms on the second line of (11.34) will be combined or cancelled with other terms to obtain the desired stability result.

11.6 Flux Controller Development

In addition to the load position tracking error objective, we must ensure all of the signals in the observer and the controller remain bounded during closed-loop operation. As demonstrated in Chapter 6, one method for accomplishing this additional objective is to force the pseudo-magnitude (*i.e.*, $\hat{\psi}_a^2 + \hat{\psi}_b^2$) of the estimated rotor flux to track a bounded positive function. To facilitate this additional tracking requirement, we define the estimated flux tracking error $\hat{\eta}_\psi(t)$ as follows

$$\hat{\eta}_\psi = \psi_d - \frac{1}{2}\left(\hat{\psi}_a^2 + \hat{\psi}_b^2\right) \tag{11.35}$$

where $\psi_d(t)$ is a positive scalar function of time which is selected to be second-order differentiable with respect to time.

Remark 11.5

To motivate how the form of (11.35) can be used to assist one in illustrating the boundedness of the system signals, we note that if a voltage control input relationship could be designed to ensure that $\hat{\eta}_\psi(t)$ remains bounded then it would be easy to show from (11.35) that the observed flux, $\hat{\psi}_a(t)$ and $\hat{\psi}_b(t)$, would remain bounded. Hence, if we can ultimately use the composite observer-controller stability analysis to show that $\tilde{\psi}_a(t)$ and $\tilde{\psi}_b(t)$ remain bounded, then the rotor flux variables $\psi_a(t)$ and $\psi_b(t)$ can be shown to remain bounded via the use of (11.14).

11.6.1 Closed-Loop Flux Tracking Error System

To obtain the open-loop dynamics for the estimated flux tracking error $\hat{\eta}_\psi(t)$, we multiply both side of (11.35) by L_r and take the time derivative of the resulting expression to yield

$$L_r \dot{\hat{\eta}}_\psi = L_r \dot{\psi}_d - L_r \left(\hat{\psi}_a \dot{\hat{\psi}}_a + \hat{\psi}_b \dot{\hat{\psi}}_b \right). \tag{11.36}$$

After substituting the flux observer dynamics of (11.10) and (11.11) into (11.36), we have

$$\begin{aligned} L_r \dot{\hat{\eta}}_\psi = \; & L_r \dot{\psi}_d + R_r \left(\hat{\psi}_a^2 + \hat{\psi}_b^2 \right) - \\ & K_I \left(\hat{\psi}_a I_a + \hat{\psi}_b I_b \right) + \alpha_2 r \left(\hat{\psi}_a I_b - \hat{\psi}_b I_a \right). \end{aligned} \tag{11.37}$$

Upon utilizing (11.35) and (11.21), we can rewrite (11.37) as

$$\begin{aligned} L_r \dot{\hat{\eta}}_\psi = \; & L_r \dot{\psi}_d + 2R_r \left(\psi_d - \hat{\eta}_\psi \right) + \\ & \alpha_2 r \left(\tau_d - \hat{\eta}_\tau \right) - K_I \left(\hat{\psi}_a I_a + \hat{\psi}_b I_b \right). \end{aligned} \tag{11.38}$$

In order to ensure that $\hat{\eta}_\psi(t)$ remains small, we add and subtract a fictitious controller $u_\psi(t)$ on the right-hand side of (11.38) to yield

$$L_r \dot{\hat{\eta}}_\psi = L_r \dot{\psi}_d + 2R_r \left(\psi_d - \hat{\eta}_\psi \right) + \alpha_2 r \left(\tau_d - \hat{\eta}_\tau \right) - K_I u_\psi + K_I \hat{\eta}_I \tag{11.39}$$

where $\hat{\eta}_I(t)$ is an auxiliary tracking error variable given by

$$\hat{\eta}_I = u_\psi - \left(\hat{\psi}_a I_a + \hat{\psi}_b I_b \right). \tag{11.40}$$

Motivated by the structure of (11.39) and the subsequent stability analysis, we design $u_\psi(t)$ as follows

$$u_\psi = K_I^{-1} \left(L_r \dot{\psi}_d + \alpha_2 r \tau_d + 4R_r^2 \hat{\eta}_\psi \frac{\psi_d^2}{\epsilon} + k_3 \hat{\eta}_\psi + k_{n2} \alpha_2^2 r^2 \hat{\eta}_\psi \right) \tag{11.41}$$

where k_3 is a positive control gain, and ϵ, k_{n2} are positive nonlinear damping gains. Substituting (11.41) into (11.39) yields the closed-loop dynamics for the estimated flux tracking error $\hat{\eta}_\psi(t)$ as follows

$$\begin{aligned} L_r \dot{\hat{\eta}}_\psi \; = \; & -2R_r \hat{\eta}_\psi - k_3 \hat{\eta}_\psi + \left[2R_r \psi_d - 4R_r^2 \hat{\eta}_\psi \frac{\psi_d^2}{\epsilon} \right] \\ & + \left[-\alpha_2 r \hat{\eta}_\tau - k_{n2} \alpha_2^2 \hat{\eta}_\psi r^2 \right] + K_I \hat{\eta}_I. \end{aligned} \tag{11.42}$$

Remark 11.6

The bracketed terms in (11.42) are nonlinear damping pairs, and hence, their upper bounds can be combined with other terms during the composite observer-controller analysis to obtain the desired stability result. It should also be noted that the design of $u_\psi(t)$ given in (11.41) represents a significant design change from the approach given in [3]. Indeed, it is the high-gain construction of $u_\psi(t)$, coupled with a specific selection for the desired flux trajectory $\psi_d(t)$ and the initial conditions for the estimated rotor flux, which allows us to show in a subsequent section that control singularities can be avoided for any set of electromechanical initial conditions (*e.g.*, even if $\psi_a(0) = \psi_b(0) = 0$).

From the structure of (11.42) and the above discussion, we are compelled to design a voltage input control relationship to ensure that $\hat{\eta}_I(t)$ remains small. To obtain the open-loop dynamics for $\hat{\eta}_I(t)$, we multiply both sides of (11.40) by $L_r L_I$ and take the time derivative to yield

$$L_r L_I \,\dot{\hat{\eta}}_I = L_r L_I \dot{u}_\psi - L_r L_I \left(\dot{\hat{\psi}}_a \, I_a + \dot{\hat{\psi}}_b \, I_b\right) - L_r L_I \left(\hat{\psi}_a \dot{I}_a + \hat{\psi}_b \dot{I}_b\right). \quad (11.43)$$

Substituting the time derivative of (11.41) which involves substitution from (11.37), the flux observer dynamics of (11.10) and (11.11), and the current dynamics of (11.2) and (11.3) into (11.43) yields

$$L_r L_I \,\dot{\hat{\eta}}_I = w_I + \Omega_4 \psi_a + \Omega_5 \psi_b - \left(L_r \hat{\psi}_a V_a + L_r \hat{\psi}_b V_b\right) \quad (11.44)$$

where the measurable auxiliary terms $w_I(t)$, $\Omega_4(t)$, and $\Omega_5(t)$ are given by

$$\begin{aligned}\Omega_4 = \; & -L_I L_r K_I^{-1} M^{-1} \left(\alpha_2 \tau_d + 2k_{n2} \alpha_2^2 r \hat{\eta}_\psi\right) \alpha_2 I_b \\ & + L_I L_r K_I^{-1} M^{-1} r \alpha_2 I_b \left(B - \alpha M - k_1\right) \\ & - \alpha_1 L_r \hat{\psi}_a + \alpha_2 L_r \hat{\psi}_b \dot{q},\end{aligned} \quad (11.45)$$

$$\begin{aligned}\Omega_5 = \; & L_I L_r K_I^{-1} M^{-1} \left(\alpha_2 \tau_d + 2k_{n2} \alpha_2^2 r \hat{\eta}_\psi\right) \alpha_2 I_a \\ & - L_I L_r K_I^{-1} M^{-1} r \alpha_2 I_a \left(B - \alpha M - k_1\right) \\ & - \alpha_1 L_r \hat{\psi}_b - \alpha_2 L_r \hat{\psi}_a \dot{q}.\end{aligned} \quad (11.46)$$

and

$$\begin{aligned} w_I = \ & L_I L_r K_I^{-1}\Big[L_r\ddot{\psi}_d + Nr\cos(q)\dot{q} \\ & +\left(\alpha_2\tau_d + 2k_{n2}\alpha_2^2 r\hat{\eta}_\psi\right)\Big]\left(\ddot{q}_d + \alpha\dot{e}\right) \\ & + L_I L_r K_I^{-1} M^{-1}\left(\alpha_2\tau_d + 2k_{n2}\alpha_2^2 r\hat{\eta}_\psi\right)\left(B\dot{q} + N\sin(q)\right) \\ & + L_I L_r K_I^{-1}\left[\frac{8\psi_d\dot{\psi}_d}{\epsilon}R_r^2\hat{\eta}_\psi + Mr\left(\dddot{q}_d + \alpha\ddot{q}_d\right)\right] \\ & - L_I L_r K_I^{-1} M^{-1} r\left(B\dot{q} + N\sin(q)\right)\left(B - \alpha M - k_1\right) \\ & + L_I L_r K_I^{-1} k_1 r\left(\ddot{q}_d + \alpha\dot{e}\right) \\ & + L_I L_r K_I^{-1}\dot{\psi}_d\left(4R_r^2\frac{\psi_d^2}{\epsilon} + k_3 + k_{n2}\alpha_2^2 r^2\right) \\ & + L_I K_I^{-1} R_r\left(4R_r^2\frac{\psi_d^2}{\epsilon} + k_3 + k_{n2}\alpha_2^2 r^2\right)\left(\hat{\psi}_a^2 + \hat{\psi}_b^2\right) \\ & - L_I\left(4R_r^2\frac{\psi_d^2}{\epsilon} + k_3 + k_{n2}\alpha_2^2 r^2\right)\left(\hat{\psi}_a I_a + \hat{\psi}_b I_b\right) \\ & + L_I K_I^{-1}\alpha_2 r\left(4R_r^2\frac{\psi_d^2}{\epsilon} + k_3 + k_{n2}\alpha_2^2 r^2\right)\left(\hat{\psi}_a I_b - \hat{\psi}_b I_a\right) \\ & - L_I I_a\left(-R_r\hat{\psi}_a - \alpha_3\hat{\psi}_b\dot{q} + K_I I_a\right) \\ & - L_I I_b\left(-R_r\hat{\psi}_b + \alpha_3\hat{\psi}_a\dot{q} + K_I I_b\right) \\ & + L_r R_I\hat{\psi}_a I_a + L_r\hat{\psi}_b R_I I_b, \end{aligned} \tag{11.47}$$

During the construction of the open-loop dynamics given in (11.44), the mechanical subsystem dynamics of (11.1) and (11.7) have been substituted for all occurrences of $\ddot{q}(t)$. Based on the structure of (11.44), we propose the following voltage input control relationship to ensure that $\hat{\eta}_I(t)$ remains small

$$\begin{aligned} L_r\hat{\psi}_a V_a + L_r\hat{\psi}_b V_b \ = \ & w_I + \Omega_4\hat{\psi}_a + \Omega_5\hat{\psi}_b + k_4\hat{\eta}_I \\ & + K_I\hat{\eta}_\psi + k_{n3}\Omega_4^2\hat{\eta}_I + k_{n3}\Omega_5^2\hat{\eta}_I \end{aligned} \tag{11.48}$$

where k_4 is a positive control gain, and k_{n3} is a positive nonlinear damping gain. Substituting the voltage input controller of (11.48) into (11.44) yields the closed-loop error dynamics for $\hat{\eta}_I(t)$ as follows

$$\begin{aligned} L_r L_I \dot{\hat{\eta}}_I &= -k_4\hat{\eta}_I - K_I\hat{\eta}_\psi + \left[\Omega_4\tilde{\psi}_a - k_{n3}\Omega_4^2\hat{\eta}_I\right] \\ &\quad + \left[\Omega_5\tilde{\psi}_b - k_{n3}\Omega_5^2\hat{\eta}_I\right]. \end{aligned} \tag{11.49}$$

Remark 11.7

With regard to the form of (11.49), we note that the term $-K_I\hat{\eta}_\psi$ will be used to cancel out the $K_I\hat{\eta}_I$ in (11.42) during the flux tracking stability analysis. The bracketed terms in (11.49) are nonlinear damping pairs, and hence, their upper bounds can be combined with other terms during the composite observer-controller analysis to obtain the desired stability result.

11.6.2 Analysis of the Closed-Loop Flux Tracking Error System

To analyze the closed-loop flux tracking error systems, we define the following non-negative function

$$V_\psi = \frac{1}{2}L_r\hat{\eta}_\psi^2 + \frac{1}{2}L_I L_r\hat{\eta}_I^2. \tag{11.50}$$

After substituting the closed-loop dynamics from (11.42) and (11.49) into the time derivative of (11.50), we have

$$\begin{aligned} \dot{V}_\psi = &\left[\Omega_5\hat{\eta}_I\tilde{\psi}_b - k_{n3}\Omega_5^2\hat{\eta}_I^2\right] - k_3\hat{\eta}_\psi^2 \\ &+ \left[2R_r\psi_d\hat{\eta}_\psi - 4R_r^2\hat{\eta}_\psi^2\frac{\psi_d^2}{\epsilon}\right] \\ &+ \left[\Omega_4\hat{\eta}_I\tilde{\psi}_a - k_{n3}\Omega_4^2\hat{\eta}_I^2\right] - k_4\hat{\eta}_I^2 - 2R_r\hat{\eta}_\psi^2. \\ &+ \left[-\alpha_2 r\hat{\eta}_\psi\hat{\eta}_\tau - k_{n2}\alpha_2^2\hat{\eta}_\psi^2 r^2\right] \end{aligned} \tag{11.51}$$

After applying the nonlinear damping argument of Lemma 1.8 in Chapter 1 to the bracketed terms in (11.51), we can form the following upper bound for $\dot{V}_\psi(t)$

$$\begin{aligned} \dot{V}_\psi &\leq -k_3\hat{\eta}_\psi^2 - k_4\hat{\eta}_I^2 - 2R_r\hat{\eta}_\psi^2 \\ &\quad + \epsilon + \frac{1}{k_{n2}}\hat{\eta}_\tau^2 + \frac{1}{k_{n3}}\tilde{\psi}_a^2 + \frac{1}{k_{n3}}\tilde{\psi}_b^2. \end{aligned} \tag{11.52}$$

As we will see during the composite observer-controller stability analysis, the terms on the second line of (11.52) will be combined with other terms to obtain the desired stability result.

11.7 Selection of Desired Flux Trajectory

Now that we have completed our control design procedure, we can utilize (11.29) and (11.48) to solve for the transformed stator voltage inputs $V_a(t)$ and $V_b(t)$ via the following matrix equation

$$\begin{bmatrix} V_a \\ \\ V_b \end{bmatrix} = C^{-1} \begin{bmatrix} \text{right-hand side of (11.48)} \\ \\ \text{right-hand side of (11.29)} \end{bmatrix} \tag{11.53}$$

where

$$C = \begin{bmatrix} L_r \hat{\psi}_a & L_r \hat{\psi}_b \\ \\ -L_r \hat{\psi}_b & L_r \hat{\psi}_a \end{bmatrix}. \tag{11.54}$$

From (11.54), we can see that the matrix C is not globally invertible since the determinate of the matrix is computed as

$$\det(C) = L_r^2 \left(\hat{\psi}_a^2 + \hat{\psi}_b^2 \right) \tag{11.55}$$

which equals to zero if

$$\hat{\psi}_a = \hat{\psi}_b = 0;$$

however, the desired flux trajectory $\psi_d(t)$ can be designed to guarantee the existence of the inverse of C (*i.e.*, the voltage input controller will not blow up for some values of rotor flux). Specifically, we define $\psi_d(t)$ to be a positive scalar function of the form

$$\psi_d(t) = (\delta_0 - \psi_{ds}) \frac{2\underline{e}^{-\lambda t/2}}{1 + \underline{e}^{-\lambda t}} + \psi_{ds} \tag{11.56}$$

where δ_0 is a positive scalar design constant which sets the desired initial value of $\psi_d(t)$, ψ_{ds} is a positive scalar design constant which sets the desired "steady state" value of $\psi_d(t)$, and

$$\lambda = \frac{\min\left(k_1,\ k_2 - \dfrac{1}{k_{n2}},\ k_3 + 2R_r,\ k_4,\ R_r - \dfrac{1}{k_{n1}} - \dfrac{1}{k_{n3}} \right)}{\dfrac{1}{2} \max(M,\ L_r,\ L_r L_I)} \tag{11.57}$$

(As illustrated in the subsequent theorem statement, the controller gains are adjusted to ensure that λ of (11.57) is positive). After we state the main result of the chapter, we will give sufficient, explicit conditions on the size of δ_0 and ψ_{ds} which ensure that the voltage input controller does not exhibit a singularity.

In addition, we select the initial conditions for the flux observers of (11.10) and (11.11) as

$$\hat{\psi}_a(0) = \hat{\psi}_b(0) = \sqrt{\delta_0}. \tag{11.58}$$

In the next section, we will show that the selection of $\psi_d(t)$ given in (11.56) and the choice of rotor flux estimator initial conditions given by (11.58) ensure that the matrix C is invertible for any set of electromechanical initial conditions (*i.e.*, load position, load velocity, rotor flux and stator current). That is, we show that the operating condition of $\hat{\psi}_a = \hat{\psi}_b = 0$ can always be avoided.

11.8 Stability Analysis

In this section, we give a theorem delineating the stability of the closed-loop electromechanical error systems developed for the flux observation error system in Section 11.4, the filtered tracking error system in Section 11.5, and the estimated flux tracking error system in Section 11.6. We then give two design conditions which are used to show that (11.56) and (11.58) guarantee that voltage input controller is free of singularities.

Theorem 11.1

Given the voltage input controllers of (11.29) and (11.48) and the non-linear observers of (11.10) and (11.11), we obtain SGUUB load position tracking in the sense that

$$\|r(t)\| \leq \|x(t)\| \leq \left[A + B\underline{e}^{-\lambda t}\right]^{1/2} \tag{11.59}$$

where λ was defined in (11.57),

$$A = \frac{\epsilon}{\lambda_1 \lambda}, \qquad B = \frac{\lambda_2}{\lambda_1} \|x(0)\|^2 - \frac{\epsilon}{\lambda_1 \lambda}, \tag{11.60}$$

$$\lambda_1 = \frac{1}{2} \min\left(M,\ L_r,\ L_r L_I\right), \quad \lambda_2 = \frac{1}{2} \max(M,\ L_r,\ L_r L_I), \tag{11.61}$$

$$\lambda_3 = \min\left(k_1,\ k_2 - \frac{1}{k_{n2}},\ k_3 + 2R_r,\ k_4,\ R_r - \frac{1}{k_{n1}} - \frac{1}{k_{n3}}\right), \tag{11.62}$$

and $x(0)$ is the initial value of the closed-loop state vector $x(t)$ which is defined as

$$x = \left[r,\ \tilde{\psi}_a,\ \tilde{\psi}_b,\ \hat{\eta}_\tau,\ \hat{\eta}_I,\ \hat{\eta}_\psi\right]^T \in \Re^6. \tag{11.63}$$

In order to ensure that λ_3 defined in (11.62) remains positive, the controller gains are selected such that

$$k_2 > \frac{1}{k_{n2}} \quad \text{and} \quad \frac{1}{k_{n1}} + \frac{1}{k_{n3}} < R_r. \tag{11.64}$$

Proof:

To analyze the performance of the composite observer-controller error system, we now combine the functions $V_o(t)$, $V_p(t)$, and $V_\psi(t)$ given by (11.15), (11.31), and (11.50), respectively to form the following non-negative function

$$V = V_o + V_p + V_\psi. \tag{11.65}$$

Since $V(t)$ in (11.65) is positive definite, we can state that

$$\lambda_1 \|x\|^2 \leq V \leq \lambda_2 \|x\|^2 \tag{11.66}$$

where λ_1, λ_2 were defined in (11.61), and $x(t)$ was defined in (11.63). Taking the time derivative of (11.65) and substituting (11.17), (11.34), and (11.52) yields the upper bound on $\dot{V}(t)$ as follows

$$\begin{aligned}\dot{V} \ \leq \ & -k_1 r^2 - \left(k_2 - \frac{1}{k_{n2}}\right)\hat{\eta}_\tau^2 - (k_3 + 2R_r)\,\hat{\eta}_\psi^2 - k_4\hat{\eta}_I^2 \\ & - \left(R_r - \frac{1}{k_{n1}} - \frac{1}{k_{n3}}\right)\tilde{\psi}_a^2 - \left(R_r - \frac{1}{k_{n1}} - \frac{1}{k_{n3}}\right)\tilde{\psi}_b^2 + \epsilon\end{aligned}$$

which can be written in the following concise form

$$\dot{V} \leq -\lambda_3 \|x\|^2 + \epsilon \tag{11.67}$$

where λ_3 was defined in (11.62), and ϵ was introduced in (11.41). From (11.66) and (11.67), we can use Lemma 1.10 in Chapter 1 to state the result of (11.59). □

Theorem 11.2

Given the desired flux trajectory of (11.56) and the initial flux estimate conditions of (11.58), we can select the control parameters δ_0 and ψ_{ds} defined in (11.56) to ensure that the matrix C defined in (11.54) is invertible for all finite electromechanical initial conditions (*i.e.*, load position, load velocity, rotor flux, and stator current). The sufficient conditions for the desired flux parameters are given by

$$\psi_{ds} > \sqrt{A} \tag{11.68}$$

and

$$\delta_0 > \max\left\{\sqrt{\frac{\epsilon}{\lambda_1\lambda}} + \psi_{ds},\ \left(\frac{c_2\sqrt{\frac{\lambda_2}{\lambda_1}} + \sqrt{c_2^2\frac{\lambda_2}{\lambda_1} + 4\left(c_1\sqrt{\frac{\lambda_2}{\lambda_1}} + \psi_{ds}\right)}}{2}\right)^2\right\} \tag{11.69}$$

where

$$c_1 = |r(0)| + |\tau_d(0)| + |u_\psi(0)| + |\psi_a(0)| + |\psi_b(0)|, \tag{11.70}$$

$$c_2 = 2 + |I_b(0) - I_a(0)| + |I_a(0) + I_b(0)|, \tag{11.71}$$

A, λ_1, λ_2, λ, and ϵ were defined in (11.60), (11.61), (11.57) and (11.41), respectively, $\tau_d(t)$ was defined in (11.22), and $u_\psi(t)$ was defined in (11.41).

Proof:

First note that the matrix C will be invertible if its determinant is always positive. From (11.55), the determinant of the matrix C is given by

$$\det(C) = L_r^2\left(\hat{\psi}_a^2 + \hat{\psi}_b^2\right) = 2L_r^2\left(\psi_d - \hat{\eta}_\psi\right) \tag{11.72}$$

where $\hat{\eta}_\psi$ was defined in (11.35). Since $\psi_d(t)$ is a strictly positive function as defined by (11.56), we can see from (11.72) that $\det(C)$ will always be positive if

$$\psi_d > \sqrt{A} + \sqrt{|B|}\underline{e}^{-\lambda t/2} \geq \sqrt{A + B\underline{e}^{-\lambda t}} \geq \|x\| \geq \left\|\hat{\eta}_\psi\right\| \tag{11.73}$$

where the bound on $\|x(t)\|$ was defined in (11.59), $x(t)$ was defined in (11.63), and A, B were defined in (11.60). From (11.56), we can see that (11.73) can be satisfied if

$$\delta_0 > \sqrt{\left|\frac{\lambda_2}{\lambda_1}\|x(0)\|^2 - \frac{\epsilon}{\lambda_1\lambda}\right|} + \psi_{ds} \tag{11.74}$$

and (11.68) holds.

We now examine two cases to show that (11.69) is a sufficient condition on the size of the control parameter δ_0 to ensure the validity of (11.74). Case 1: suppose $\dfrac{\lambda_2}{\lambda_1}\|x(0)\|^2 < \dfrac{\epsilon}{\lambda_1\lambda}$ in (11.74), then a sufficient condition to ensure the validity of (11.74) is given by

$$\delta_0 > \sqrt{\frac{\epsilon}{\lambda_1\lambda}} + \psi_{ds}; \tag{11.75}$$

and hence, we have the first element of the condition in (11.69). Case 2: suppose $\dfrac{\lambda_2}{\lambda_1}\|x(0)\|^2 \geq \dfrac{\epsilon}{\lambda_1\lambda}$ in (11.74), then a sufficient condition to ensure the validity of (11.74) is given by

$$\delta_0 > \sqrt{\frac{\lambda_2}{\lambda_1}}\|x(0)\| + \psi_{ds}, \tag{11.76}$$

where $\|x(0)\|^2$ can be written in the following form

$$\begin{aligned}\|x(0)\|^2 &= r^2(0) + \left(\psi_a(0) - \sqrt{\delta_0}\right)^2 + \left(\psi_b(0) - \sqrt{\delta_0}\right)^2 \\ &\quad + \left(\tau_d(0) - \sqrt{\delta_0}\left(I_b(0) - I_a(0)\right)\right)^2 \\ &\quad + \left(u_\psi(0) - \sqrt{\delta_0}\left(I_a(0) + I_b(0)\right)\right)^2\end{aligned} \tag{11.77}$$

by use of (11.63), (11.14), (11.35), (11.56), (11.58), (11.21), and (11.40) (Note that we use the fact that $\hat{\eta}_\psi(0) = 0$ which is a result of $\psi_d(0) = \delta_0$ and the choice of flux estimate initial conditions). Given that $\hat{\eta}_\psi(0) = 0$ and $\psi_d(t)$ of (11.56) has been selected to ensure that $\dot{\psi}_d(0) = 0$, it is easy to show that $\tau_d(0)$ and $u_\psi(0)$ are independent of δ_0; hence, all of the quantities $r(0)$, $\psi_a(0)$, $\psi_b(0)$, $I_a(0)$, $I_b(0)$, $\tau_d(0)$, and $u_\psi(0)$ on the right-hand side of (11.77) are independent on δ_0. Since all dependence of δ_0 has been explicitly shown in (11.77), we can use standard inequality properties to bound $\|x(0)\|$ in the form

$$\|x(0)\| < c_1 + c_2\sqrt{\delta_0} \tag{11.78}$$

where c_1, c_2 are defined in (11.70) and (11.71), respectively. After substituting the bound for $\|x(0)\|$ given by (11.78) into (11.76) and solving for δ_0 via the quadratic formula, we obtain the second term in the condition of (11.69). □

Remark 11.8

Since $r(t) \in L_\infty$, $\hat{\eta}_\tau(t) \in L_\infty$, $\hat{\eta}_I(t) \in L_\infty$, $\hat{\eta}_\psi(t) \in L_\infty$, $\tilde{\psi}_a(t) \in L_\infty$, $\tilde{\psi}_b(t) \in L_\infty$ by the result of (11.59) and (11.63) (and hence the quantities $q(t) \in L_\infty$, $\dot{q}(t) \in L_\infty$, $\hat{\psi}_a(t) \in L_\infty$, $\hat{\psi}_b(t) \in L_\infty$, $\left(\hat{\psi}_a I_b - \hat{\psi}_b I_a\right) \in L_\infty$, $\left(\hat{\psi}_a I_a + \hat{\psi}_b I_b\right) \in L_\infty$, $\tau_d(t) \in L_\infty$, and $u_\psi(t) \in L_\infty$), we can multiply $\left(\hat{\psi}_a I_b - \hat{\psi}_b I_a\right)$ by $\hat{\psi}_a$, $\left(\hat{\psi}_a I_a + \hat{\psi}_b I_b\right)$ by $\hat{\psi}_b$ and add the results together to obtain that $I_b\left(\hat{\psi}_a^2 + \hat{\psi}_b^2\right) \in L_\infty$, which shows $I_b(t) \in L_\infty$, (and similarly $I_a(t) \in L_\infty$) for all time (To show $I_a(t) \in L_\infty$ and $I_b(t) \in L_\infty$, we are using the fact that $\hat{\psi}_a^2 + \hat{\psi}_b^2 \neq 0$ for all time which is a direct result of Theorem 11.2). Therefore, the right-hand sides of (11.29) and (11.48) are bounded functions of time. Using the fact that C^{-1} exists, we can see from (11.53) that the voltage control inputs $V_a(t) \in L_\infty$ and $V_b(t) \in L_\infty$. In addition, from the system dynamics of (11.1), (11.2), (11.3), (11.4) and (11.5), we can see that $\ddot{q}(t) \in L_\infty$, $\dot{I}_a(t) \in L_\infty$, $\dot{I}_b(t) \in L_\infty$, $\dot{\psi}_a(t) \in L_\infty$, and $\dot{\psi}_b(t) \in L_\infty$. Therefore, all signals in the system, observer, and the controller remain bounded.

Remark 11.9

Since the filtered tracking error $r(t)$ is bounded as given by (11.59), we can utilize Lemma 1.11 in Chapter 1 to show that the load position tracking error $e(t)$ is bounded by

$$\begin{aligned} \|e(t)\| \leq \; & \underline{e}^{-\alpha t} \|e(0)\| + \frac{\sqrt{A}}{\alpha}\left(1 - \underline{e}^{-\alpha t}\right) \\ & + \frac{2\sqrt{|B|}}{2\alpha - \lambda}\left(\underline{e}^{-\lambda t/2} - \underline{e}^{-\alpha t}\right) \end{aligned} \tag{11.79}$$

where α, λ, A, and B were defined in (11.9), (11.57), and (11.60), respectively. Note that the control parameters can be arbitrarily adjusted to give a desired ultimate bound for the load position tracking error. From (11.79), we can see that the "steady-state" tracking error will stay inside of the bound $\sqrt{\frac{\epsilon}{\lambda_1 \lambda \alpha}}$ which illustrates the parameter ϵ can be used to decrease the size of the tracking error. However, from (11.41), we can see that the control gains must be increased to decrease ϵ. That is, the proposed voltage input controller becomes a high-gain controller for small values of ϵ.

Remark 11.10

The desired flux trajectory defined in (11.56) is a decreasing function with an initial value given by δ_0 and a steady-state value given by ψ_{ds}. It should be noted that the transient tracking error will last longer for larger values of δ_0. However, the conditions on δ_0 and ψ_{ds} given in Theorem 11.2 are generated from the Lyapunov-like analysis, and hence these conditions can be considered to be very conservative. In practice, δ_0 and ψ_{ds} can be tuned to achieve acceptable load position tracking performance while also ensuring that the input voltages and stator winding currents stay within the rated values of a given machine (See the experimental results given in this chapter).

Remark 11.11

In [4], Ortega et al presented a torque tracking controller which did not require measurement of rotor variables. While it is possible to extend the work presented in [4] to position/velocity tracking control applications, this extension is not obvious. Indeed, this extension is discussed in [5], and it is noted that the potential drawback, in extending the controller proposed in [4] [5] from the torque tracking problem to the position/velocity tracking problem, is that to establish global stability, the authors rely on the presence of mechanical damping [5] (Note this condition can be relaxed to establish a local stability result as done in [6]). In the approach given in this chapter, we establish a semi-global tracking result with no restrictions on the mechanical damping.

11.9 Experimental Results

The load position tracking controller proposed in the previous sections has been successfully applied to an induction motor driving a single-link robotic load. The basic hardware setup was described in the experimental section of Chapter 2. The only hardware modifications to the basic setup are two additional linear amplifiers and two additional current sensors since the motor used in the experiment is a three-phase induction motor. The motor is a Baldor model M3541, 2-pole, three-phase induction motor with a 1024 line encoder mounted at the back of the motor. The load velocity signal is obtained by use of a backwards difference algorithm applied to the load position signal with the resulting signal being filtered by a second-order digital filter. The experiment was run at a sampling time of 0.75ms.

The single-link robot is designed as a metal bar link. Therefore, the parameters M and N of (11.1) can be expressed as

$$M = J_m + \frac{mL_0^2}{3}, \quad N = \frac{mGL_0}{2}$$

where J_m is the rotor inertia, m is the link mass, L_0 is the link length, and G is the gravity coefficient. The 3 phase induction motor used for the experiments has 2 poles, a rated speed of 3450 RPM, a rated current of 2.7A, and a rated voltage of 230V. Through standard test procedures, the electromechanical system parameters for the system dynamics of (11.1) through (11.7) were determined to be

$$\begin{array}{lll} J_m = 2.1 \times 10^{-4} kg \cdot m^2, & m = 0.4014 kg, & L_0 = 0.305 m, \\ L_r = 0.306 H, & L_s = 0.243 H, & M_e = 0.225 H, \\ R_s = 3.05 \Omega, & n_p = 1, & B = 0.015 N \cdot m \cdot s/rad, \\ G = 9.81 kg \cdot m/\sec^2, & & R_r = 2.12 \Omega. \end{array}$$

The desired position trajectory is selected as the following smooth start sinusoid

$$q_d(t) = \frac{\pi}{2}\sin(5t)\left(1 - \underline{e}^{-0.1t^3}\right) rad$$

where $q_d(0) = \dot{q}_d(0) = \ddot{q}_d(0) = \dddot{q}_d(0) = 0$. The desired flux trajectory of (11.56) was selected as

$$\psi_d = 0.3\frac{2\underline{e}^{-0.25t}}{1 + \underline{e}^{-0.5t}} + 1.7\, Wb \cdot Wb.$$

As dictated by (11.58), the initial rotor flux estimates conditions were set to

$$\hat{\psi}_a(0) = \hat{\psi}_b(0) = \sqrt{2}\, Wb.$$

The best tracking performance was established using the following control gain values

$$\alpha = 105, \ k_1 = 8, \ k_2 = 7, \ k_3 = 3.5, \ k_4 = 3.8,$$

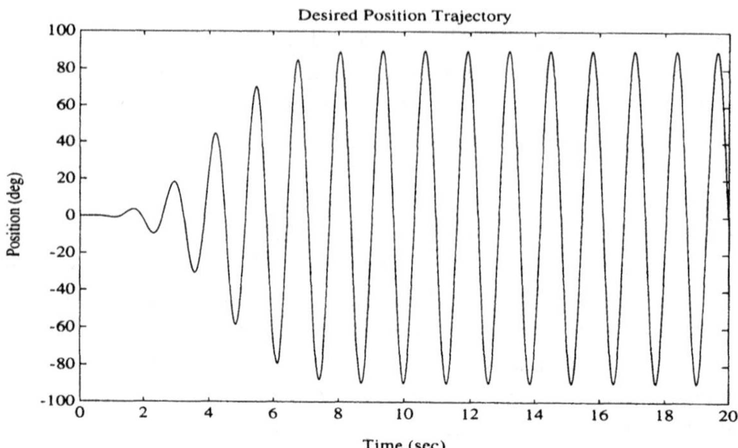

Figure 11.1: Desired Position Trajectory

$$k_{n1} = 0.0005, \ k_{n2} = 0.0005, \ k_{n3} = 0.001, \ \text{ and } \ \epsilon = 10.$$

The desired position trajectory is shown in Figure 11.1. The resulting load position tracking error is shown in Figure 11.2, the actual phase voltages are shown in Figure 11.3 through Figure 11.5. From the load position tracking error plot given in Figure 11.2, we can see that the "steady state" load position tracking error is within ± 0.3 degrees.

Remark 11.12

The selection of the control parameters in the experiment does not satisfy the conditions given in Theorem 11.1 and Theorem 11.2; however, these requirements are only sufficient, conservative conditions generated by the Lyapunov-like stability argument.

11.10 Notes

Since the rotor quantities are not usually measurable, some researchers have addressed the problem of induction motor control without rotor flux (or rotor current) measurements. In [7], Verghese *et al.* proposed observers for flux estimation which showed that corrective feedback can be used to speed up convergence of the flux estimates while also reducing the sensitivity of the estimates to parameter variations. In [8], Stephan *et al.* developed an adaptive controller which can be used for real-time estimation of the parameters and the rotor fluxes. Aided by the use of a nonlinear flux observer, Kanellakapoulos *et al.* [9] [3] presented an exact model based controller

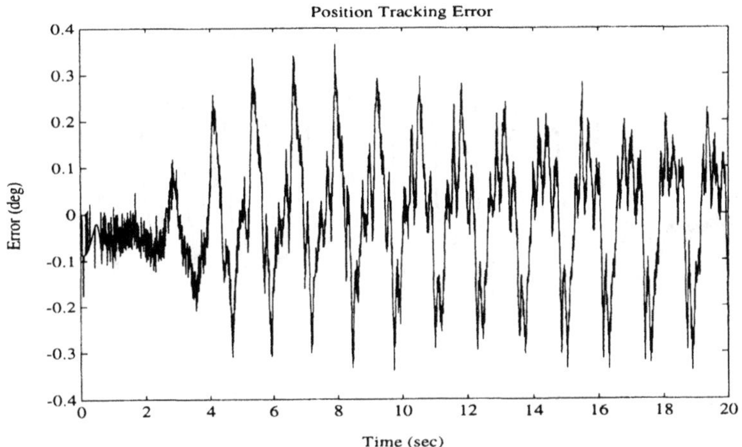

Figure 11.2: Position Tracking Error

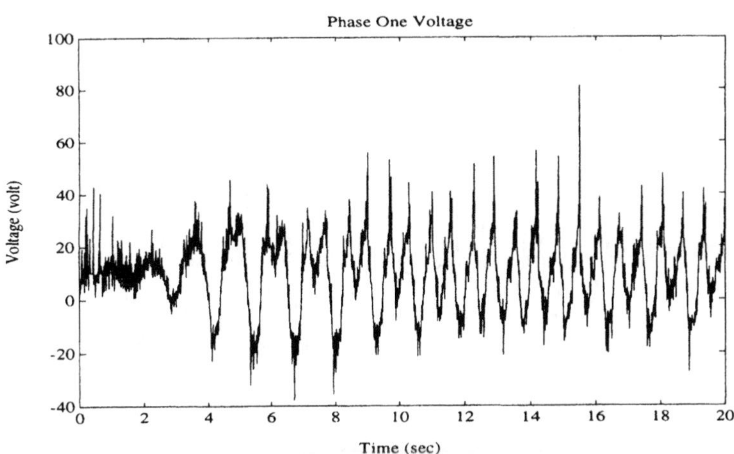

Figure 11.3: Actual Phase One Voltage

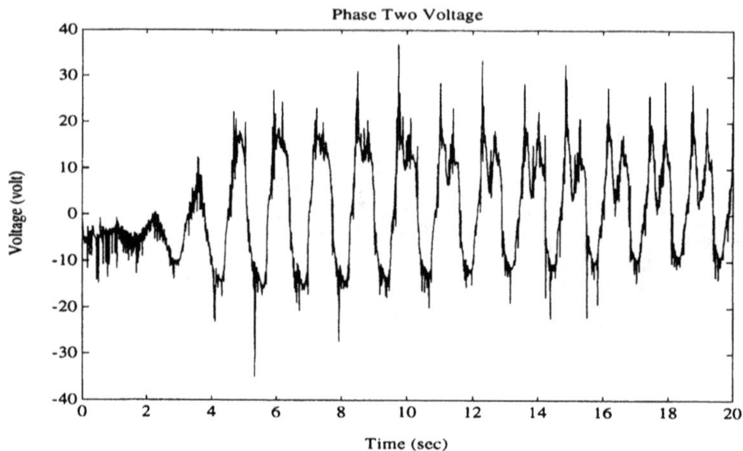

Figure 11.4: Actual Phase Two Voltage

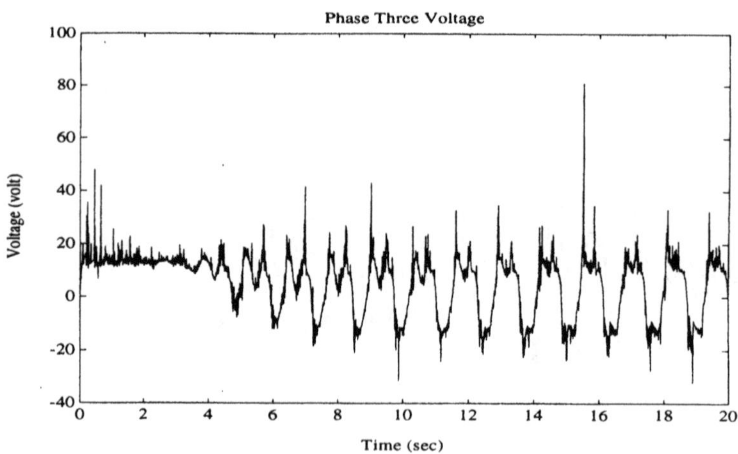

Figure 11.5: Actual Phase Three Voltage

which yielded an exponential stability result for the velocity tracking error. In [9] and [3], the stability result is local in the sense that the initial conditions of the tracking error system must be sufficiently small to ensure that control singularities are avoided. In [10], Ortega *et al.* presented an observer-based adaptive approach which ensures global asymptotic regulation of the generated torque with full or partial state measurements. In [4], Ortega *et al.* extended their torque regulation result in [10] to the torque tracking case without rotor current measurements. Ortega *et al.* also presented an singularity-free, velocity tracking controller in [11] which did not require rotor current measurements. In [15], Dawson *et al.* modified the control structure and stability analysis presented in [11] to illustrate how adaptive backstepping control arguments can be used to design a singularity-free rotor position/velocity tracking controller. In [12], Hu *et al.* presented an exact-model knowledge position tracking controller which estimates rotor velocity and rotor flux while yielding a local exponential stability result for the rotor position tracking error. In [6], Canudas *et al.* proposed utilizing the induction motor for robotic actuation. This intriguing proposal involved the use of an exact model knowledge, nonlinear load position tracking controller which utilized rotor current estimates. The closed-loop system was shown to be locally exponential stability in the sense that the initial conditions must be sufficiently small such that high order nonlinearities (*i.e.*, the Coriolis and centripetal terms) can be dominated by low-order control terms. In [13], Hu *et al.* designed a controller which eliminated rotor flux measurements and achieved semiglobal uniform ultimate bounded link tracking for n-link robot manipulators actuated by induction motors. In [14], Hu *et al.* extended the controller presented in [15] to the multi-degree of freedom case.

In [16], Yang proposed an observed-based nonlinear tracking controller for induction motors which did not require measurements of rotor currents. In [17], Khalil *et al.* designed a continuous approximation of a variable structure speed tracking controller which required stator current, rotor position, rotor velocity, and rotor acceleration measurements and achieved a uniform ultimate bounded stability result. In [18], Nicklasson *et al.* presented a comprehensive procedure for designing passivity based controllers for a class of Blondel-Park transformable electric machines. In [19], Vedagharba *et al.* designed a rotor velocity/rotor flux tracking controller which can be utilized to reduce the power loss in velocity setpoint applications.

Some recent work which attacks the rotor resistance issue can found in [20] where Marino *et al.* proposed an adaptive observer for the rotor flux which compensates for rotor resistance variation. In [21], Marino *et al.* designed a controller which compensated for unknown rotor resistance despite the lack of rotor flux measurements; however, the controller neglected the stator current dynamics. In [22], Hu *et al.* illustrated how the adaptive observer structure in [20] could be modified to design a controller which compensated for rotor resistance uncertainty; however, the controller exhibited a singularity when the magnitude of the estimated rotor flux was equal to zero. In [23], Kwan *et al.* used the rotor flux calculation method proposed in [24] to develop a robust adaptive controller which compensates for uncertainty associated with the mechanical subsystem and the rotor resistance parameters. In [25], Vedagharba *et al.* used the rotor flux calculation method proposed in [24] to design an adaptive singularity-free controller which compensated for uncertainty associated with the mechanical subsystem and the rotor resistance parameters[1]. Recently, Marino *et al.* [26] designed a singularity-free, adaptive controller which compensated for rotor resistance and mechanical subsystem parametric uncertainty while achieving asymptotic rotor position/flux tracking.

Bibliography

[1] M. Krstic, I. Kanellakopoulos, and P. Kokotovic, *Nonlinear and Adaptive Control Design,* John Wiley & Sons, 1995.

[2] J. Slotine and W. Li, *Applied Nonlinear Control*, Englewood Cliff, NJ: Prentice Hall Co., 1991

[3] I. Kanellakopoulos, P. Krein, and F. Disilvestro, "A New Controller Observer Design for Induction Motor Control", *Proc. of the 1992 ASME Winter Meeting*, DSC-Vol. 43, pp. 43-47, 1992.

[4] R. Ortega, C. Canudas and S. Seleme, "Nonlinear Control of Induction Motors: Torque Tracking with Unknown Load Disturbance", *IEEE Transactions on Automatic Control*, Vol. 38, No. 11, pp. 1675-1680, November 1993.

[5] Espinosa-Perez G., and R. Ortega, "State Observers are Unnecessary for Induction Motor Control, *System and Control Letters,* Vol. 23, No. 5, 1994, pp. 315-323.

[6] C. Canudas, R. Ortega, and S.I. Seleme, "Robot Motion Control using Induction Motor Drives", *Proc. of the IEEE International Conference on Robotics and Automation*, Vol. 2, pp. 533-538, 1993.

[7] G. C. Verghese and S.R. Sanders, "Observers for Flux Estimation in Induction Machines", *IEEE Transaction on Industrial Electronics*, Vol. 35, No. 1, pp. 85-94, February 1988.

[8] J. Stephan, M. Bodson, and J. Chiasson, "An Adaptive Method for Induction Motor Control", *Proc. of the 1993 American Control Conference*, Vol. 1, pp. 655-659, June, 1993.

[1] We should note that the controller presented in [25] requires that the electromechanical system be initially at rest.

[9] I. Kanellakopoulos, P. Krein, and F. Disilvestro, "Nonlinear Flux Observer Based Control of Induction Motors", *Proc. of the 1992 American Control Conference*, pp. 1700-1704, 1992.

[10] R. Ortega and G. Espinosa, "Torque Regulation of Induction Motors", *Automatica*, Vol. 29, No.3, pp. 621-633, 1993.

[11] R. Ortega, P. Nicklasson, and G. Espinosa-Perez, "On Speed Control of Induction Motors", *Automatica*, Vol. 32, No. 3, March 1996, pp. 455-460.

[12] J. Hu, D.M. Dawson, and Y. Qian, "Position Tracking of an Induction Motor via Partial State Feedback", *Automatica*, Vol. 31, No. 7, July, 1995, pp. 989-1000.

[13] J. Hu, D.M. Dawson, and Y. Qian, "Position Tracking for Robot Manipulators Driven by Induction Motors without Flux Measurements", *IEEE Transactions on Robotics and Automation*, Vol. 12, No. 3, 1996, pp. 419 - 438.

[14] J. Hu, D. Dawson, P. Vedagarbha, and H. Canbolat, "A Global Adaptive Link Position Tracking Controller for Robot Manipulators Driven by Induction Motors", *International Journal of System Science*, Vol. 28, No. 7, 1997, pp. 625-642.

[15] D. Dawson, J. Hu, and P. Vedagarbha, "An Adaptive Controller for a Class of Induction Motor Systems", *Proc. of the IEEE Conference on Decision and Control*, New Orleans, Dec. 1995, pp. 1567-1572.

[16] J. Yang, W. Yu, and L. Fu, "Nonlinear Observer-Based Tracking Control of Induction Motors with Unknown Load", *IEEE Transactions on Industrial Electronics*, Vol. 42, No. 6, Dec. 1995, pp. 579-585.

[17] H. Khalil and E. Strangas, "Robust Speed Control of Induction Motors Using Position and Current Measurements", *IEEE Transactions on Automatic Control*, Vol. 41, No. 8, Aug. 1996, pp. 1216-1220.

[18] P. Nicklasson, R. Ortega, and G. Espinosa-Perez, "Passivity-Based Control of a Class of Blondel-Park Transformable Electric Machines" *IEEE Transactions on Automatic Control*, Vol. 42, No. 5, May 1997, pp. 629-647.

[19] P. Vedagarbha, D. M. Dawson, and T. Burg, "Rotor Velocity/Flux Tracking Control of Induction Motors with Improved Efficiency", *Mechatronics - An International Journal*, Vol. 7, No.2, 1997, pp. 105-127.

[20] R. Marino, S. Peresada, and P. Valigi, "Exponentially Convergent Rotor Resistance Estimation for Induction Motors", *IEEE Transactions on Industrial Electronics,* Vol. 42, No. 5, Oct. 1995, pp. 508-515.

[21] R. Marino, S. Peresada, and P. Tomei, "Output Feedback Control of Current-Fed Induction Motors with Unknown Rotor Resistance," *IEEE Trans. on Control Systems Technology,* Vol. 4, No. 4, July 1996, pp. 336-346.

[22] J. Hu and D. M. Dawson, "Adaptive Control of Induction Motor Systems Despite Rotor Resistance Uncertainty", *Automatica,* Vol. 32, No. 8, 1996, pp. 1127-1143.

[23] C. Kwan, F. L. Lewis, and K. S. Yeung, "Adaptive Control of Induction Motors Without Flux Measurements", *Automatica,* Vol. 32, No. 6, pp. 903-908, 1996.

[24] R. Soto and K. Yeung, "Sliding-Mode Control of an Induction Motor Without Flux Measurements", *IEEE Transactions on Industry Applications,* Vol. 31, No. 4, Jul. 1995, pp. 744-750.

[25] P. Vedagarbha, D. M. Dawson and T. Burg, "Adaptive Control for a Class of Induction Motors via an On-Line Flux Calculation Method", *Proc. of the IEEE Conference on Control Applications,* Sep. 1996, Dearborn, MI, pp. 620-625.

[26] R. Marino, S. Peresada, and P. Tomei, "Global Adaptive Output Feedback Control of Induction Motors with Uncertain Rotor Resistance," *Proceedings of the 35th Conference on Decision and Control,* Kobe, Japan, Dec. 1996, pp. 4701-4706.

Chapter 12

Adaptive PSFB Control

12.1 Introduction

This chapter presents the design and implementation of a family of adaptive, partial state feedback controllers that can be applied to several of the motors discussed in the preceding chapters. This class of controllers will introduce additional nonlinear control design techniques, most significant of which are the use of a first-order, high-pass filter to eliminate the need for velocity measurements and the use of a signal based on the desired trajectory as a feedforward term in the control. The first-order high-pass filter is a model independent means of producing a velocity related signal from position information, in contrast to the model-based velocity observers in the previous chapters which were of higher order and were designed to mimic the mechanical dynamics. The controller will also make use of previously discussed techniques such as nonlinear damping and integrator backstepping to achieve global asymptotic position tracking.

The motivation for sensor reduction has been discussed in previous chapters. However, it is worth reiterating that it is typically the load velocity which is most difficult to measure accurately. The two common methods for obtaining velocity information (*e.g.*, direct tachometer readings or *ad-hoc* differentiation of the position measurement) may fail to provide accurate, noise-free velocity measurements. One of the most common methods for "differentiating" the position signal is the discrete backwards difference algorithm; however, since many adaptive control techniques are predicated on the use of a continuous time Lyapunov-like argument, it is not obvious how to integrate such an algorithm into a more rigorously-design control approach. In addition, angular position can be accurately determined by optical encoder type transducers while current information can be accurately acquired using Hall-effect type transducers; hence, we consider the elimination of load velocity measurements to be a top priority for the position tracking control design of electrical machines.

The concept of exact model knowledge, OFB control was demonstrated in previous chapters for certain motors; however, there does not appear to be a method for designing adaptive, OFB controllers for any of the multi-phase machines. Thus, we are compelled to investigate the possibility of developing an adaptive, partial state feedback controller for the multi-phase electric machines which eliminates the need for load velocity measurements. Specifically, we will present a control approach that can be tailored to the dynamics of a specific motor in order to yield high performance position tracking yet is sufficiently general such that it can be modified for different motor types (*e.g.*, PMS, SR, and BLDC motors). The control design begins with the description of electromechanical models for each motor. The form of the electromechanical models and the need to operate without velocity measurements motivates a control algorithm that is a fusion of the following well-established nonlinear control design tools and motor control techniques: i) the desired compensation adaptation law (DCAL) [1], ii) dynamic filtering of position measurements to produce a velocity related signal, iii) design of the electronic commutator, iv) the adaptive integrator backstepping technique [2], and v) the nonlinear damping tool [2]. These not necessarily compatible tools are combined through a Lyapunov-like analysis to design a controller for the PMS motor. The same design process is then followed to develop the SR and BLDC controllers, here only the modifications to the PMS controller are shown. Finally, the ability of these controllers to achieve desired position trajectory following is demonstrated experimentally.

12.2 System Model

As shown in the previous chapters, the dynamics for most of the electric motors considered in this book can be separated into three distinct parts: i) a second-order dynamic mechanical subsystem, which for the purposes of this discussion includes a position dependent load and the motor rotor, ii) a first-order dynamic electrical subsystem which includes all of the motor's relevant electrical effects, and iii) a static relationship which represents the conversion of electrical energy into mechanical energy. In this section we show that the mechanical subsystem can be generalized to apply to all the motor types considered thus far. The electrical dynamics are then given for the PMS, SR, and BLDC motors in preparation for the controller design.

12.2.1 Mechanical Subsystem Model

To simplify the subsequent control development, we assume that the mechanical subsystem model is given by

$$\acute{M}\ddot{q} + \acute{B}\dot{q} + \acute{f}(q) = K_\tau \tau \tag{12.1}$$

where $q(t)$, $\dot{q}(t)$, and $\ddot{q}(t)$ represent the load position, velocity, and acceleration, respectively. In the mechanical subsystem of (12.1), $\acute{M}$ is the constant lumped inertia of the rotor-load system, $\acute{B}$ is the dynamic friction constant, K_τ is the torque coupling constant, $\tau(t)$ is the torque produced by the electrical subsystem dynamics, and the load function $\acute{f}(q)$ includes the forces acting on the load as a result of the connected load and the motor construction.

To simplify the development of the subsequent model-based adaptive controller, we assume that the load function $\acute{f}(q)$ can be written for the PMS motor as

$$\acute{f}(q) = \acute{N}\sin(q) + \acute{K}_D \sin(4N_r q) \tag{12.2}$$

where $\acute{K}_D$ is the detente torque constant, N_r is the integer number of rotor teeth, and $\acute{N}$ is a lumped gravity constant determined from the geometry and mass distribution of the specific load.

Remark 12.1

Many types of load function models can be incorporated into the proposed control design provided they are linear parameterizable and first-order differentiable; however, the controller and/or the analysis may have to be reformulated. The pendulum type portion of the load model (*i.e.*, the term $\acute{N}\sin(q)$) given by (12.2) was selected to ease the construction of a common experimental load for the PMS, the BLDC, and the SR machines. Since we are interested in using the PMS motor for relatively low speed operation (*e.g.*, robotics), we have also included the detente torque term (*i.e.*, the term $\acute{K}_D \sin(4N_r q)$) in the model. This additional sinusoidal term in the mechanical subsystem is caused by the permanent magnet on the rotor aligning itself along the directions of minimum reluctance.

To facilitate the control development, we substitute (12.2) into (12.1), divide the resulting expression by K_τ, and parameterize the resulting mechanical subsystem as follows

$$M\ddot{q} + B\dot{q} + f(q) = Y\theta = \tau \tag{12.3}$$

where $Y(q, \dot{q}, \ddot{q}) \in \Re^{1\times 4}$ is the mechanical regression vector [3] defined as

$$Y = \begin{bmatrix} \ddot{q} & \dot{q} & \sin(q) & \sin(4N_r q) \end{bmatrix}, \tag{12.4}$$

and $\theta \in \Re^4$ is a vector of constant scaled mechanical subsystem parameters given by

$$\theta = \begin{bmatrix} M & B & N & K_D \end{bmatrix}^T \tag{12.5}$$

where

$$M = \frac{\acute{M}}{K_\tau}, \quad B = \frac{\acute{B}}{K_\tau}, \quad N = \frac{\acute{N}}{K_\tau}, \quad K_D = \frac{\acute{K}_D}{K_\tau},$$

and $\acute{M}$, $\acute{B}$, $\acute{N}$, $\acute{K}_D$ were previously defined in (12.1), and (12.2), respectively.

12.2.2 Electrical Subsystem Models

The dynamics of the electric motors that can be linked to the generalized mechanical subsystem in (12.1) are given in this section. It should be noted that all of the models given below assume that the magnetic circuit is linear and the stator windings are sinusoidally distributed.

PMS Motor Electrical Model

For a 2-phase PMS motor, the two input control voltages will produce two phases of currents referred to individually as $I_j(t)$. The torque produced by the electrical currents is given by the expression [4]

$$\tau = -\sum_{j=1}^{2} \sin(x_j) I_j \tag{12.6}$$

where

$$x_j = N_r q - \frac{\pi(j-1)}{2}. \tag{12.7}$$

The jth electrical subsystem characterizes the dynamic relationship between the jth input phase voltage, $v_j(t)$, and the jth phase current as delineated by [4]

$$L_j \dot{I}_j + R_j I_j - K_m \sin(x_j)\dot{q} = v_j \quad \text{for} \quad j = 1, 2. \tag{12.8}$$

The constant electrical subsystem parameters in (12.6) through (12.8) include the per-phase self-inductance L_j, the per-phase resistance R_j, and the back-emf coefficient K_m. The PMS motor torque coupling coefficient for the representation given in (12.1) is defined in terms of the electrical subsystem parameters as [4]

$$K_\tau = K_m. \tag{12.9}$$

It should be emphasized that the model of (12.6) through (12.8) utilizes the common PMS motor assumption which neglects the variation of stator inductance with position.

SR Motor Electrical Model

For a 4-phase SR motor, the four input control voltages will produce four phases of currents referred to individually as $\gamma_j(t)$. The torque produced by the electrical currents is given by the expression [5]

$$\tau = \sum_{j=1}^{4} \sin(x_j)\gamma_j^2 \tag{12.10}$$

where

$$x_j = N_r q - \frac{\pi\,(j-1)}{2} \tag{12.11}$$

and the integer N_r denotes the number of rotor saliencies (poles). The electrical dynamics of the jth phase in the SR motor are given by [5]

$$L_j(q)\dot{\gamma}_j + R\gamma_j + N_r L_1 \sin(x_j)\gamma_j\dot{q} = V_j \quad \text{for} \quad j = 1, \ldots, 4 \tag{12.12}$$

where the positive, position dependent inductance is given by

$$L_j(q) = L_0 - L_1 \cos(x_j) \quad \text{for} \quad j = 1, \ldots, 4, \tag{12.13}$$

R represents the phase resistance constant, $V_j(t)$ represents the input voltage, L_1 is the dynamic winding inductance constant, and L_0 is the static winding inductance constant. The SR motor torque coupling coefficient for the representation given in (12.1) is defined in terms of the electrical subsystem parameters as [5]

$$K_\tau = \frac{N_r L_1}{2}. \tag{12.14}$$

It should be emphasized that the model of (12.10) through (12.12) is based on the common SR motor assumption that the mutual inductances between the phases can be neglected.

BLDC Motor Electrical Model

For a 3-phase brushless DC motor, the controller will be designed for the rotor fixed transformed model [6] (*i.e.*, the motor dynamics have been transformed into a more convenient 2-phase equivalent circuit representation in order to simplify the control design procedure). For the 2-phase transformed model of the BLDC, the two input control voltages will produce two phases of currents referred to individually as $I_a(t)$ and $I_b(t)$ [1]. The torque produced by the electrical currents is given by the expression [7]

$$\tau = (K_{\tau n} I_b + 1) I_a \tag{12.15}$$

where $K_{\tau n}$ is a torque conversion constant.

The two electrical subsystems for the BLDC motor which produce I_a and I_b from the input voltages $V_a(t)$ and $V_b(t)$ are given as follows [7]

$$L_a\dot{I}_a + RI_a + n_p L_b I_b \dot{q} + K_{\tau 2}\dot{q} = V_a \tag{12.16}$$

and

$$L_b\dot{I}_b + RI_b - n_p L_a I_a \dot{q} = V_b \tag{12.17}$$

where the constant parameter n_p represents the integer number of rotor pole pairs, and the electrical parameters L_a, L_b, R, and $K_{\tau 2}$ represent the constant transformed phase inductances, the phase resistance constants,

[1] The subscripts a and b assign the electrical quantities to the quadrature or direct axes [6], respectively.

and a second torque transmission constant, respectively. Based on the form of (12.15), the BLDC torque coupling coefficient for the representation given in (12.1) and (12.15) is defined in terms of the electric subsystem parameters as [7] $K_\tau = K_{\tau 2}$ and $K_{\tau n} = \dfrac{K_{\tau 1}}{K_{\tau 2}}$ where $K_{\tau 1}$ is a another torque transmission constant.

Remark 12.2

As mentioned earlier, the proposed adaptive partial state feedback algorithm requires structural knowledge of the electromechanical dynamics; however, the proposed controller will compensate for parametric uncertainty[2]. Since the system parameters are all assumed to be constant it reasonable to assume that upper and lower bounds can be specified for each unknown parameter. These bounds will be denoted by an overbar or underbar for a given parameter. For example, the inertia parameter defined in (12.3) is bounded by

$$\underline{M} \leq M \leq \overline{M} \tag{12.18}$$

where $\underline{M}$ and $\overline{M}$ are known positive constants.

12.3 Control Objective

Given partial state measurement (*i.e.*, q and I_j for the PMS motor), the control objective is to develop load position tracking controllers for the electromechanical dynamics of a given motor. To begin the development, we define the load position tracking error $e(t)$ as

$$e = q_d - q \tag{12.19}$$

where $q_d(t)$ denotes the desired load position trajectory. The controller development will demand that the third time derivative of the desired position trajectory be bounded.

An important step in eliminating load velocity measurements is to use the form of the regression vector given by (12.4) and the tracking control objective to form a desired regression vector [1] which depends only on the desired trajectory and its time derivatives[3]. The desired mechanical regression vector, denoted by $Y_d(q_d, \dot{q}_d, \ddot{q}_d) \in \Re^{1\times 4}$, is formed by replacing $q(t)$, $\dot{q}(t)$, and $\ddot{q}(t)$ in the mechanical regression vector of (12.4) with the

[2] For proper commutation of the motor, we will assume that the number of rotor teeth or the number of rotor saliencies denoted by N_r, and the integer number of pole pairs denoted by n_p are known exactly.

[3] It should also be noted that in addition to eliminating load velocity measurements, the desired regression matrix can usually be precomputed off-line [1]; hence, on-line computation is reduced.

corresponding desired trajectory signals $q_d(t)$, $\dot{q}_d(t)$, and $\ddot{q}_d(t)$ as shown below

$$Y_d = \begin{bmatrix} \ddot{q}_d & \dot{q}_d & \sin(q_d) & \sin(4N_r q_d) \end{bmatrix}. \tag{12.20}$$

As part of the integrator backstepping approach, an intermediate desired torque signal $\tau_d(t)$ is formulated as if this signal is the actual control input to mechanical subsystem dynamics. The desired current trajectories are then designed to produce this torque signal via the relationship between the phase currents and the torque produced by the electrical subsystems. Hence, a secondary control objective is to ensure that the actual torque tracks the desired torque signal. To quantify this auxiliary control objective, we define the torque tracking error $\eta_\tau(t)$ as follows

$$\eta_\tau = \tau_d - \tau. \tag{12.21}$$

Remark 12.3

The form of the mechanical subsystem dynamics given in (12.1) will vary only slightly for each of the electric machines considered in this chapter. Since the SR or the BLDC motor electromechanical models usually do not include a detente torque term such as that given in (12.2), the parameterization of (12.3) reduces to

$$Y = \begin{bmatrix} \ddot{q} & \dot{q} & \sin(q) & 0 \end{bmatrix} \in \Re^{1\times 4} \tag{12.22}$$

and

$$\theta = \begin{bmatrix} M & B & N & 0 \end{bmatrix}^T \in \Re^4 \tag{12.23}$$

while the desired mechanical regression vector $Y_d(t)$ of (12.20) is modified to reflect the change in $Y(q, \dot{q}, \ddot{q})$ as

$$Y_d = \begin{bmatrix} \ddot{q}_d & \dot{q}_d & \sin(q_d) & 0 \end{bmatrix} \in \Re^{1\times 4}. \tag{12.24}$$

12.4 PMS Motor Controller

In this section, the specific problem of controlling the PMS motor is addressed. First, a dynamic filter is introduced to produce a velocity related error signal. Second, the desired torque signal is specified, and the closed-loop mechanical error dynamics are derived and analyzed. Third, a commutation strategy is designed that specifies desired current signals based on the desired torque signal. Fourth, a voltage input control strategy is formulated to produce the desired current signal, and the closed-loop electrical dynamics are derived and analyzed. Finally, a theorem regarding the load position tracking performance is presented.

12.4.1 Pseudo-Velocity Filter

A pseudo-velocity filter [8] is proposed to generate a velocity related signal for use in the closed-loop controller. The filter can be heuristically thought of as a high-pass filter which acts upon the position tracking error signal, $e(t)$, to produce a pseudo-velocity tracking error signal, $e_f(t)$, that will supplant the need for velocity measurements in the controller[4]. The filter output is generated from the position tracking error signal according to

$$\dot{p} = -(\alpha_2 + \alpha_2 k)p + (-k\alpha_1 + \alpha_2 + \alpha_2 k + \alpha_2 k^2)e, \quad p(0) = ke(0), \tag{12.25}$$

$$e_f = -ke + p \tag{12.26}$$

where the auxiliary variable $p(t)$ is used to separate the filter into two implementable parts. The positive controller gains α_1 and α_2 are used to weight $e(t)$ and $e_f(t)$, respectively, and the control gain k will be designed during the stability analysis to "damp-out" state disturbance terms[5].

To better illustrate how α_1 and α_2 are used to weight $e(t)$ and $e_f(t)$ in the control, we now proceed to develop a set of dynamic equations that will eventually be used in the proof of the theorem regarding the position tracking performance. First, a dynamic equation describing the pseudo-velocity signal is developed from the two filter equations in (12.25) and (12.26) by differentiating (12.26) to yield

$$\dot{e}_f = -k\dot{e} + \dot{p}. \tag{12.27}$$

After substituting (12.25) for $\dot{p}(t)$ into (12.27) and then using (12.26) to eliminate $p(t)$, we can express the pseudo-velocity error dynamics in the following form

$$\dot{e}_f = -k\eta - \alpha_2 e_f + \alpha_2 e \tag{12.28}$$

where the filtered tracking error[6] term, $\eta(t)$, is defined as follows

$$\eta = \dot{e} + \alpha_2 e_f + \alpha_1 e. \tag{12.29}$$

From the form of (12.29), we can interpret $\eta(t)$ as being the weighted sum of the velocity tracking error, the pseudo-velocity tracking error, and the position tracking error. That is, the filter of (12.25) and (12.26) has been designed to foster the weighting of the signals $e(t)$ and $e_f(t)$ in the filtered tracking error signal $\eta(t)$. It should also be noted that the expression given

[4] Several recent robotic papers ([9], [10], [11], [12], [13], [14], [15]) have proposed the use of first-order and second-order filters to eliminate velocity measurements in the design of robotic manipulator controller.

[5] These state disturbance terms are spawned by the use of the desired regression vector of (12.20) in the feedforward controller instead of a regression vector which depends on q and $\dot{q}$.

[6] In earlier chapters, a simpler type of filtered tracking error signal was utilized to design full-state feedback adaptive controllers.

in (12.29) can be used to describe the position tracking error dynamics for use in the subsequent stability analysis as

$$\dot{e} = \eta - \alpha_2 e_f - \alpha_1 e. \tag{12.30}$$

12.4.2 Desired Torque Signal

The construction of the filtered tracking error signal $\eta(t)$ defined in (12.29) allows us to simplify the subsequent Lyapunov-like stability analysis. Specifically, the closed-loop tracking error system for the mechanical subsystem and the pseudo-velocity filter can be written as three first-order differential equations (*i.e.*, (12.28) for $e_f(t)$, (12.30) for $e(t)$, and the time derivative of (12.29) for the dynamics of $\eta(t)$). To obtain the dynamics for $\eta(t)$, we differentiate (12.29) with respect to time and multiply the resulting expression by M to yield

$$M\dot{\eta} = M\ddot{q}_d + M\alpha_2\dot{e}_f + M\alpha_1\dot{e} - M\ddot{q} \tag{12.31}$$

where (12.19) has been utilized to substitute $M\ddot{e} = M\ddot{q}_d - M\ddot{q}$. After substituting for $M\ddot{q}$ from (12.3) into (12.31), the dynamics for $\eta(t)$ can be written as

$$M\dot{\eta} = M\ddot{q}_d + B\dot{q}_d + f(q) + M\alpha_2\dot{e}_f + (M\alpha_1 - B)\,\dot{e} - \tau \tag{12.32}$$

where (12.19) has been used to substitute $B\dot{q} = B\dot{q}_d - B\dot{e}$. Finally, (12.30) and (12.28) are substituted for $\dot{e}$ and $\dot{e}_f$, respectively into (12.32) and then the term $Y_d\theta$ of (12.5) and (12.20) is added and subtracted to the resulting right-hand side to yield

$$M\dot{\eta} = \chi - kM\alpha_2\eta + Y_d\theta - \tau \tag{12.33}$$

where the auxiliary term $\chi(e, e_f, \eta, t)$ has been introduced as

$$\begin{aligned} \chi = \quad & M\ddot{q}_d + B\dot{q}_d + (M\alpha_1 - B)(-\alpha_1 e - \alpha_2 e_f + \eta) \\ & + M\alpha_2(-\alpha_2 e_f + \alpha_2 e) + f(q) - Y_d\theta. \end{aligned} \tag{12.34}$$

Since the dynamics of $\eta(t)$ given by (12.33) lack a control input, we inject the desired torque control signal which was previously eluded to in (12.21). Specifically, the desired torque signal, $\tau_d(t)$, is added and subtracted to the right-hand side of (12.33) to produce

$$M\dot{\eta} = \chi - kM\alpha_2\eta + Y_d\theta - \tau_d + \eta_\tau \tag{12.35}$$

where the torque tracking error $\eta_\tau(t)$ was previously defined in (12.21). It should be noted that the terms on the right-hand side of (12.33) have been grouped to foster the design of an *adaptive* desired torque signal which does not require load velocity measurements. That is, the filter of (12.25) and (12.26) and the filtered tracking error signal $\eta(t)$ of (12.29) have been constructed to produce the term $-kM\alpha_2\eta$ in (12.35) which "damps-out"[7] the state disturbance term $\chi(\cdot)$ explicitly defined in (12.34). In addition, the construction of the term $Y_d\theta$ used in (12.35) (*i.e.*, $Y_d(\cdot)$ only depends on the desired trajectory) facilitates the design of a desired torque signal which eliminates the need for load velocity measurements.

Based on the structure of (12.35) and the subsequent stability analysis, we propose the following desired torque signal [8]

$$\tau_d = Y_d\hat{\theta} - ke_f + e \tag{12.36}$$

where k is the same controller gain previously introduced in (12.25) and (12.26), and $\hat{\theta}(t)$ is generated on-line according to the update law

$$\hat{\theta} = \int_0^t \Gamma\left(Y_d^T(\sigma)\left(\alpha_1 e(\sigma) + \alpha_2 e_f(\sigma)\right) - \dot{Y}_d^T(\sigma)e(\sigma)\right)d\sigma + \Gamma Y_d^T e \tag{12.37}$$

where $\Gamma \in \Re^{4\times 4}$ is a diagonal matrix of positive adaptation gains. The closed-loop dynamics for $\eta(t)$ can now be written by substituting the desired torque signal of (12.36) into (12.35) to yield

$$M\dot{\eta} = \chi - kM\alpha_2\eta + Y_d\tilde{\theta} + ke_f - e + \eta_\tau \tag{12.38}$$

where the mismatch between the actual mechanical parameter vector and the mechanical parameter estimate is defined as

$$\tilde{\theta} = \theta - \hat{\theta}. \tag{12.39}$$

After differentiating (12.39) and (12.37) with respect to time, we can formulate the parameter estimation error dynamics as follows

$$\dot{\tilde{\theta}} = -\Gamma Y_d^T\eta = -\dot{\hat{\theta}} \tag{12.40}$$

where (12.29) has been utilized.

To illustrate that the above desired torque signal achieves load position tracking in the mechanical subsystem while also preparing for the subsequent composite stability analysis, we define the non-negative function

$$V_M = \frac{1}{2}e^2 + \frac{1}{2}e_f^2 + \frac{1}{2}M\eta^2 + \frac{1}{2}\tilde{\theta}^T\Gamma^{-1}\tilde{\theta} \tag{12.41}$$

[7] As we will see later, the controller gain k can be designed to allow the corresponding term for χ in the stability analysis to be combined with other terms to yield asymptotic load position tracking.

whose first time derivative is given by

$$\dot{V}_M = e\dot{e} + e_f\dot{e}_f + M\eta\dot{\eta} + \tilde{\theta}^T\Gamma^{-1}\dot{\tilde{\theta}}. \tag{12.42}$$

After the error dynamics from (12.28), (12.30), (12.40), and (12.38) are substituted into (12.42), the resulting expression can be simplified to produce

$$\dot{V}_M = -\alpha_1 e^2 - \alpha_2 e_f^2 + \chi\eta - kM\alpha_2\eta^2 + \eta_\tau\eta \tag{12.43}$$

which can be upper bounded as follows

$$\dot{V}_M \leq -\alpha_1 e^2 - \alpha_2 e_f^2 + |\chi|\,|\eta| - kM\alpha_2\,|\eta|^2 + |\eta_\tau|\,|\eta|. \tag{12.44}$$

Given the structure of $\chi(\cdot)$ defined in (12.34), $Y_d\theta$ defined in (12.5) and (12.20), and $f(q)$ obtained from (12.2), we can write $\chi(\cdot)$ as

$$\begin{aligned} \chi = \;& e\left(-\alpha_1(M\alpha_1 - B) + M\alpha_2^2\right) + \eta\left(M\alpha_1 - B\right) \\ & + e_f\left(-\alpha_2(M\alpha_1 - B) - M\alpha_2^2\right) \\ & + [K_D\sin(4N_r q) + N\sin(q) - \\ & K_D\sin(4N_r q_d) - N\sin(q_d)] \end{aligned} \tag{12.45}$$

To facilitate the analysis, we now use the Mean Value Theorem [16] to bound the bracketed term in (12.45) as

$$\begin{aligned} & \overline{K}_D\left|\sin(4N_r q) - \sin(4N_r q_d)\right| + \\ & \overline{N}\left|\sin(q) - \sin(q_d)\right| \leq 4\overline{K}_D N_r\,|e| + \overline{N}\,|e| \end{aligned} \tag{12.46}$$

where the bounding constants were defined in (12.18). We can make use of (12.46) and standard properties of the absolute value operator to upper bound (12.45) as

$$\begin{aligned} |\chi| \leq \;& |e|\left(\overline{M}\left|\alpha_2^2 - \alpha_1^2\right| + \overline{B}\alpha_1\right) + |\eta|\left(\overline{M}\alpha_1 + \overline{B}\right) \\ & + |e_f|\left(\overline{M}\alpha_1\alpha_2 + \overline{B}\alpha_2 + \overline{M}\alpha_2^2\right) + 4\overline{K}_D N_r\,|e| + \overline{N}\,|e| \end{aligned} \tag{12.47}$$

Based on (12.47), $|\chi|$ can be further upper bounded as follows

$$|\chi| \leq \zeta_o\sqrt{e^2 + e_f^2 + \eta^2} \tag{12.48}$$

where the bounding constant ζ_o is given by

$$\begin{aligned} \zeta_o = \;& 3\left\{\overline{M}\left(\left|\alpha_2^2 - \alpha_1^2\right| + \alpha_1\alpha_2 + \alpha_2^2 + \alpha_1\right)\right. \\ & \left. + \overline{B}\left(\alpha_1 + \alpha_2 + 1\right) + \overline{N} + 4N_r\overline{K}_D\right\}. \end{aligned} \tag{12.49}$$

Based on the form of (12.44) and (12.49), we design the controller gain k as follows

$$k = \frac{1}{\underline{M}\alpha_2}\left(\zeta_o^2 k_{N1} + \alpha_3\right) \tag{12.50}$$

where k_{N1} and α_3 are additional positive controller gains. To illustrate the motivation for the form of (12.50), we substitute (12.48) and (12.50) into (12.44) to form the new upper bound for $\dot{V}_M(t)$ as

$$\begin{aligned} \dot{V}_M \leq & -\alpha_1 e^2 - \alpha_2 e_f^2 - \alpha_3 \eta^2 + |\eta_\tau| \, |\eta| \\ & + \left[\zeta_o |\eta| \sqrt{e^2 + e_f^2 + \eta^2} - \zeta_o^2 k_{N1} |\eta|^2\right] \end{aligned} \tag{12.51}$$

where (12.18) has been utilized. After applying the nonlinear damping tool given in Lemma 1.8 of Chapter 1 to the bracketed pair of terms in (12.51), a new upper bound for $\dot{V}_M(t)$ can be formulated as

$$\begin{aligned} \dot{V}_M \leq & -\left(\alpha_1 - \frac{1}{k_{N1}}\right) e^2 - \left(\alpha_2 - \frac{1}{k_{N1}}\right) e_f^2 \\ & -\left(\alpha_3 - \frac{1}{k_{N1}}\right) \eta^2 + |\eta_\tau| \, |\eta| \, . \end{aligned} \tag{12.52}$$

If the voltage control input can be designed to guarantee exact torque tracking (*i.e.*, $\eta_\tau = 0$) in (12.52), then it would be easy to show using the Lemmas in Chapter 1 that $e(t)$, $e_f(t)$, and $\eta(t)$ all go to zero asymptotically fast (*i.e.*, with a proper choice of the gain k_{N1}). Hence, the desired torque signal of (12.36), the pseudo-velocity filter of (12.25) and (12.26), and the update law of (12.37) have been designed to ensure load position tracking despite parametric uncertainty and the lack of load velocity measurements. That is, if the desired torque signal could be directly applied to the mechanical subsystem, it would produce asymptotic load position tracking. The problem in the following sections now becomes one of designing the voltage control inputs such that the electrical subsystems produce the currents that in turn produce the desired torque signal. In other words the desired torque signal will now be used as the tracking objective for the voltage control input design.

Remark 12.4

The constant term ζ_o will not include the effects of the detente torque for the BLDC or the SR motors. That is, the term $f(q)$ of (12.2) simplifies to

$$f(q) = N \sin(q); \tag{12.53}$$

therefore,

$$|f(q) - f(q_d)| \leq \overline{N} \, |e| \, . \tag{12.54}$$

Based on (12.54), it is easy to show that for the SR and BLDC motors, ζ_o of (12.49) can be recalculated as

$$\begin{aligned}\zeta_o = \; & 3\left\{\overline{M}\left(\left|\alpha_2^2 - \alpha_1^2\right| + \alpha_1\alpha_2 + \alpha_2^2 + \alpha_1\right)\right. \\ & \left. +\overline{B}\left(\alpha_1 + \alpha_2 + 1\right) + \overline{N}\right\}.\end{aligned} \tag{12.55}$$

12.4.3 PMS Motor Voltage Input Controller

We now design the voltage control inputs such that the electrical subsystem produces the desired torque. Since the outputs of the electrical subsystems are the phase currents, a commutation strategy will be used to reformulate the desired torque signal into a set of desired current signals. Based on the structure of the torque transmission terms in (12.6), the desired currents, $I_{dj}(t)$, are specified as follows

$$I_{dj} = -\tau_d \sin(x_j) \quad \text{for} \quad j = 1, 2 \tag{12.56}$$

where $\tau_d(t)$ is the desired torque signal defined in (12.36). The voltage inputs will now be designed to force the actual currents to track the desired current signals and thereby produce the desired torque signal which ensures load position tracking in the mechanical subsystem. To quantify this new tracking objective, the phase current tracking error, $\eta_{Ij}(t)$, is defined as

$$\eta_{Ij} = I_{dj} - I_j. \tag{12.57}$$

To motivate the form of the above commutation strategy, we note that (12.56) has been selected to ensure that

$$\tau_d = -\sum_{j=1}^{2} I_{dj} \sin(x_j) \tag{12.58}$$

which can easily be verified by substituting the right-hand side of (12.56) into (12.58) and using simple trigonometric identities. Hence, using (12.6) the torque tracking error of (12.21) can be written in terms of the current tracking errors as follows

$$\eta_\tau = \tau_d - \tau = \tau_d + \sum_{j=1}^{2} I_{dj} \sin(x_j) - \sum_{j=1}^{2} I_{dj} \sin(x_j) + \sum_{j=1}^{2} I_j \sin(x_j) \tag{12.59}$$

which upon substitution of (12.58) yields

$$\eta_\tau = -\sum_{j=1}^{2} \eta_{Ij} \sin(x_j) \tag{12.60}$$

where (12.57) has been utilized. Thus, we see from (12.60) that if the current tracking errors approach zero then so will the torque tracking error.

Since the voltage input control design now centers on ensuring that the current tracking errors are driven to zero, the dynamics for the phase current tracking error are now derived. First, we differentiate (12.57) with respect to time, multiply the resulting expression by L_j, and then substitute from the electrical dynamics of (12.8) for $L_j\dot{I}_j$ to yield

$$L_j\dot{\eta}_{Ij} = L_j\dot{I}_{dj} + R_jI_j - K_m\sin(x_j)\dot{q} - v_j. \tag{12.61}$$

The term $\dot{I}_{dj}(t)$ in (12.61) can be found from the time derivatives of (12.56) and (12.36) as follows

$$\dot{I}_{dj} = -\tau_d N_r\dot{q}\cos(x_j) - \left(\dot{Y}_d\hat{\theta} + Y_d\,\dot{\hat{\theta}} - k\dot{e}_f + \dot{e}\right)\sin(x_j). \tag{12.62}$$

After substituting (12.62) for $\dot{I}_{dj}$, the left-hand side of (12.40) for $\dot{\hat{\theta}}$, (12.30) for $\dot{e}$, and (12.28) for $\dot{e}_f$ into (12.61), the current tracking error dynamics can be simplified and written into the concise form

$$L_j\dot{\eta}_{Ij} = Y_{Ej}\theta_{Ej} + L_j\eta\Omega_j + K_m\eta\sin(x_j) - v_j \tag{12.63}$$

where the vector of electrical parameters $\theta_{Ej} \in \Re^3$ is defined as

$$\theta_{Ej} = \begin{bmatrix} L_j & R_j & K_m \end{bmatrix}^T, \tag{12.64}$$

the electrical regressor matrix $Y_{Ej}(e, e_f, t) \in \Re^{1\times 3}$ is defined as

$$Y_{Ej} = \begin{bmatrix} Y_{Ej1} & Y_{Ej2} & Y_{Ej3} \end{bmatrix} \tag{12.65}$$

with components given by

$$\begin{aligned} Y_{Ej1} = \;& -\sin(x_j)(\dot{Y}_d\hat{\theta} + k\alpha_2 e_f - k\alpha_2 e - \alpha_1 e - \alpha_2 e_f) \\ & -\tau_d\cos(x_j)N_r(\dot{q}_d + \alpha_1 e + \alpha_2 e_f), \\ Y_{Ej2} = \;& I_j, \\ Y_{Ej3} = \;& -\sin(x_j)(\dot{q}_d + \alpha_1 e + \alpha_2 e_f), \end{aligned}$$

and the auxiliary variable $\Omega_j(t) \in \Re^1$ is defined as

$$\Omega_j = -\sin(x_j)(Y_d\Gamma Y_d^T + k^2 + 1) + \tau_d\cos(x_j)N_r. \tag{12.66}$$

Based on the structure of (12.63) and the subsequent stability analysis, we propose the following voltage control inputs

$$v_j = Y_{Ej}\hat{\theta}_{Ej} + \Omega_j^2 k_{N2}\eta_{Ij} + k_{N3}\eta_{Ij} + k_{N4}\eta_{Ij} + k_s\eta_{Ij} \tag{12.67}$$

where k_s, k_{N2}, k_{N3}, and k_{N4} are positive constant control gains, and $\hat{\theta}_{Ej}(t) \in \Re^3$ is the dynamic estimate of the parameter vector θ_{Ej}. The parameter estimate vector $\hat{\theta}_{Ej}(t)$ is changed on-line by the following update law

$$\hat{\theta}_{Ej} = \int_0^t \Gamma_{Ej} Y_{Ej}^T(\sigma)\eta_{Ij}(\sigma)\, d\sigma \tag{12.68}$$

where $\Gamma_{Ej} \in \Re^{3\times 3}$ is a diagonal matrix of positive adaptation gains. The closed-loop dynamics for $\eta_{Ij}(t)$ can now be formed by substituting the voltage control input of (12.67) into (12.63) to yield

$$\begin{aligned} L_j \dot{\eta}_{Ij} = \; & Y_{Ej}\tilde{\theta}_{Ej} + L_j \eta \Omega_j + K_m \eta \sin(x_j) \\ & -\Omega_j^2 k_{N2}\eta_{Ij} - k_{N3}\eta_{Ij} - k_{N4}\eta_{Ij} - k_s \eta_{Ij} \end{aligned} \tag{12.69}$$

where the mismatch between the actual electrical parameter vector and the electrical parameter estimate is defined as

$$\tilde{\theta}_{Ej} = \theta_{Ej} - \hat{\theta}_{Ej}. \tag{12.70}$$

After differentiating (12.70) and (12.68) with respect to time, we can formulate the parameter estimation error dynamics as follows

$$\dot{\tilde{\theta}}_{Ej} = -\Gamma_{Ej} Y_{Ej}^T \eta_{Ij} = -\dot{\hat{\theta}}_{Ej}\,. \tag{12.71}$$

To illustrate that the above voltage control input achieves current tracking in the electrical subsystem while also preparing for the composite stability analysis in the following section, we define the non-negative function

$$V_E = \frac{1}{2}\sum_{j=1}^{2} L_j \eta_{Ij}^2 + \frac{1}{2}\sum_{j=1}^{2} \tilde{\theta}_{Ej}^T \Gamma_{Ej}^{-1} \tilde{\theta}_{Ej} \tag{12.72}$$

whose first time derivative is given by

$$\dot{V}_E = \sum_{j=1}^{2} L_j \eta_{Ij} \dot{\eta}_{Ij} + \sum_{j=1}^{2} \tilde{\theta}_{Ej}^T \Gamma_{ej}^{-1} \dot{\tilde{\theta}}_{Ej}\,. \tag{12.73}$$

The error dynamics from (12.69) and (12.71) are substituted into (12.73) and the resulting expression is simplified to produce

$$\begin{aligned}\dot{V}_E = & -\sum_{j=1}^{2} k_s\eta_{Ij}^2 - \sum_{j=1}^{2} k_{N4}\eta_{Ij}^2 \\ & + \left[\sum_{j=1}^{2}\left(L_j\eta\Omega_j\eta_{Ij} - \Omega_j^2 k_{N2}\eta_{Ij}^2\right)\right] \\ & + \left[\sum_{j=1}^{2}\left(K_m\sin(x_j)\eta_{Ij}\eta - k_{N3}\eta_{Ij}^2\right)\right]\end{aligned} \tag{12.74}$$

After applying the nonlinear damping tool (Chapter 1, Lemma 1.8) to the bracketed terms in (12.74), a new upper bound for $\dot{V}_E(t)$ can be formulated as

$$\dot{V}_E \leq -\sum_{j=1}^{2} k_s\eta_{Ij}^2 - \sum_{j=1}^{2} k_{N4}\eta_{Ij}^2 + \sum_{j=1}^{2}\frac{\bar{L}_j^2}{k_{N2}}\eta^2 + 2\frac{\bar{K}_m^2}{k_{N3}}\eta^2 \tag{12.75}$$

where inequalities similar to (12.18) have been utilized.

If the desired torque signal has been designed to guarantee that the filtered tracking error is zero (*i.e.*, $\eta = 0$) in (12.75), then it would be easy to show using the Lemmas in Chapter 1 that $\eta_{Ij}(t)$ goes to zero asymptotically fast. Hence, the voltage control input of (12.67) and the adaptation law of (12.68) have been designed to ensure current tracking despite parametric uncertainty and the lack of load velocity measurements. In the next section, we will illustrate how the non-negative functions given by (12.41) and (12.72) and the upper bounds on their respective time derivatives can be added together to yield asymptotic load position tracking for the interconnected electromechanical system.

12.4.4 Composite PMS Motor Controller Analysis

Theorem 12.1

The desired torque signal given by (12.25), (12.26), (12.36), and (12.37), and the corresponding voltage control input of (12.68) and (12.67) provides asymptotic load position tracking for the 2-phase PMS motor in the sense that

$$\lim_{t\to\infty} z(t) = 0$$

where

$$z = \begin{bmatrix} e & e_f & \eta & \eta_{I1} & \eta_{I2} \end{bmatrix}^T \in \Re^5 \tag{12.76}$$

provided that the controller gains are constrained by

$$\alpha_1 > \frac{1}{k_{N1}}, \qquad \alpha_2 > \frac{1}{k_{N1}}, \tag{12.77}$$

and

$$\alpha_3 > \frac{1}{k_{N1}} + \frac{2}{k_{N4}} + 2\frac{\bar{K}_m^2}{k_{N3}} + \sum_{i=1}^{2} \frac{\bar{L}_j^2}{k_{N2}}. \tag{12.78}$$

Proof: First, we define the non-negative function

$$V = V_M + V_E \tag{12.79}$$

where $V_M(t)$ and $V_E(t)$ were previously defined in (12.41) and (12.72). As illustrated by (12.52) and (12.75), the upper bound on the time derivative of (12.79) is given by

$$\begin{aligned} \dot{V} \leq & -\left(\alpha_1 - \frac{1}{k_{N1}}\right) e^2 - \left(\alpha_2 - \frac{1}{k_{N1}}\right) e_f^2 \\ & -\left(\alpha_3 - \frac{1}{k_{N1}} - 2\frac{\bar{K}_m^2}{k_{N3}} - \sum_{j=1}^{2} \frac{\bar{L}_j^2}{k_{N2}}\right) \eta^2 \\ & -\sum_{j=1}^{2} k_s \eta_{Ij}^2 + \left[|\eta_\tau| \, |\eta| - \sum_{j=1}^{2} k_{N4} \eta_{Ij}^2 \right]. \end{aligned} \tag{12.80}$$

From (12.60), it is easy to see that $|\eta_\tau|$ can be bounded as follows

$$|\eta_\tau| \leq \sum_{j=1}^{2} |\eta_{Ij}| ; \tag{12.81}$$

hence, upon substituting (12.81) into (12.80), Lemma 1.8 in Chapter 1 can be applied to the bracketed term in (12.80) to yield the final upper bound for $\dot{V}(t)$ as

$$\begin{aligned} \dot{V} \leq & -\left(\alpha_1 - \frac{1}{k_{N1}}\right) e^2 - \left(\alpha_2 - \frac{1}{k_{N1}}\right) e_f^2 - \sum_{j=1}^{2} k_s \eta_{Ij}^2 \\ & -\left(\alpha_3 - \frac{1}{k_{N1}} - \frac{2}{k_{N4}} - 2\frac{\bar{K}_m^2}{k_{N3}} - \sum_{j=1}^{2} \frac{\bar{L}_j^2}{k_{N2}}\right) \eta^2. \end{aligned} \tag{12.82}$$

The facts that $V(t)$ in (12.79) is positive definite and that $\dot{V}(t)$ in (12.82) is negative semi-definite as a result of (12.77) and (12.78) allow us to state that $V(t) \in L_\infty$. The direct implication is that $e(t) \in L_\infty$, $e_f(t) \in L_\infty$, $\eta(t) \in L_\infty$, $\eta_{I1}(t) \in L_\infty$, $\eta_{I2} \in L_\infty$, $\tilde{\theta}(t) \in L_\infty^4$, $\tilde{\theta}_{E1}(t) \in L_\infty^3$, and $\tilde{\theta}_{E2}(t) \in L_\infty^3$. The fact that $e(t) \in L_\infty$ along with the assumed bound on $q_d(t)$ implies that $q(t) \in L_\infty$. Likewise, (12.30) implies that since $e(t) \in L_\infty$, $e_f(t) \in L_\infty$, and $\eta(t) \in L_\infty$ then $\dot{e}(t) \in L_\infty$; hence, the bound on $\dot{q}_d(t)$ implies $\dot{q}(t) \in L_\infty$. Since $\tilde{\theta}(t) \in L_\infty^4$, $\tilde{\theta}_{E1}(t) \in L_\infty^3$, and $\tilde{\theta}_{E2}(t) \in L_\infty^3$, (12.39) and (12.70) can be used with the fact that θ, θ_{E1}, and θ_{E2} are constant vectors to illustrate that $\hat{\theta}(t) \in L_\infty^4$, $\hat{\theta}_{E1}(t) \in L_\infty^3$, and $\hat{\theta}_{E2}(t) \in L_\infty^3$. In (12.28), it can be seen that $e(t) \in L_\infty$, $e_f(t) \in L_\infty$, and $\eta(t) \in L_\infty$ imply that $\dot{e}_f(t) \in L_\infty$. Equation (12.20) can be used to show that $Y_d(\ddot{q}_d, \dot{q}_d, q_d) \in L_\infty$. The desired torque signal in (12.36) is seen to be a bounded function of bounded inputs as illustrated by

$$\tau_d = g_1(\ddot{q}_d, \dot{q}_d, q_d, \hat{\theta}, q, e_f)$$

from which we conclude $\tau_d(t) \in L_\infty$; hence, for the desired current signal in (12.56) it can be seen that $I_{d1}(t) \in L_\infty$ and $I_{d2}(t) \in L_\infty$. Now $I_{d1}(t) \in L_\infty$, $I_{d2}(t) \in L_\infty$, $\eta_{I1}(t) \in L_\infty$, and $\eta_{I2}(t) \in L_\infty$ imply from (12.57) that $I_1(t) \in L_\infty$ and $I_2(t) \in L_\infty$. The desired regressor matrix $Y_d(\ddot{q}_d, \dot{q}_d, q_d)$ in (12.20) can be differentiated and the assumed bound on $\dddot{q}_d(t)$ can be utilized to show that $\dot{Y}_d(\dddot{q}_d, \ddot{q}_d, \dot{q}_d, q_d) \in L_\infty^4$. The input control voltages of (12.67) can be written to show this bounded function of bounded inputs as

$$v_j = (\dddot{q}_d, \ddot{q}_d, \dot{q}_d, q_d, q, \hat{\theta}, e_f, \hat{\theta}_{Ej}, I_{dj}, I_j)$$

from which we conclude $v_1(t) \in L_\infty$ and $v_2(t) \in L_\infty$. Finally, from (12.26), $p(t) \in L_\infty$ and from (12.25) $\dot{p}(t) \in L_\infty$. Now from (12.38), we conclude that $\dot{\eta}(t) \in L_\infty$. From the derivative of (12.30), it can be seen that since $\dot{\eta}(t) \in L_\infty$, $\dot{e}(t) \in L_\infty$, and $\dot{e}_f(t) \in L_\infty$ then $\ddot{e}(t) \in L_\infty$, which along with the assumed bound on $\ddot{q}_d(t)$ implies $\ddot{q}(t) \in L_\infty$. In (12.61), it can be seen that $\dot{\eta}_{I1}(t) \in L_\infty$ and $\dot{\eta}_{I2}(t) \in L_\infty$; therefore, $\dot{z}(t) = \begin{bmatrix} \dot{e} & \dot{e}_f & \dot{\eta} & \dot{\eta}_{I1} & \dot{\eta}_{I2} \end{bmatrix}^T \in L_\infty^5$. In addition, it easy to show from (12.79) and (12.82) that $z(t) \in L_2^5$ (*i.e.*, as done for all of the previous full state feedback adaptive controllers); hence, Theorem 12.1 is a direct result of Corollary 1.1 in Chapter 1. □

Remark 12.5

Since the torque tracking error, $\eta_\tau(t)$, is statically related to the current tracking error, it was not explicitly included in the above analysis. However, from (12.81), we see that if the current tracking errors approach zero (which was demonstrated above) then so will the torque tracking error.

Remark 12.6

It should be noted that the mechanical subsystem given in (12.3) can be transformed into the output-feedback canonical form; hence, the method outlined in [17] can be used to formulate an adaptive desired torque signal which only depends on load position measurements. Roughly speaking, the advantage of the method described in [17] lies in the fact that additional high-gain control terms (similar to those used in (12.36) and (12.50)) are not required to obtain asymptotic load position tracking; however, the disadvantages are increased computation and the difficulty involved in modifying the approach to deal with other load functions. That is, the adaptive control partial state-feedback controller used in this chapter has been extended to robot manipulator systems actuated by electric machines [8] while to our knowledge the approach given in [17] has not.

Remark 12.7

Theorem 12.1 can be extended to provide a statement on velocity tracking. From (12.30), the following bound can be written for the velocity tracking error

$$|\dot{e}| = |\dot{q}_d - \dot{q}| = |\eta - \alpha_2 e_f - \alpha_1 e| \leq |\eta| + \alpha_2 |e_f| + \alpha_1 |e| .$$

The implication of the above inequality is that, based on the asymptotic tracking of each term on the right-hand side of the inequality, asymptotic velocity tracking is also achieved. Therefore, the proposed controller can be used as a velocity tracking controller by reformulating the desired velocity tracking objective as a position tracking objective.

12.5 SR Motor Controller

An important property of the previously described control technique is the ease with which it can be customized to alternate motor types. The steps presented in this section to tailor the above approach to the SR motor are: i) modify the desired torque signal and the mechanical parameter update law by removing the detente torque term from the mechanical subsystem dynamics, ii) design a new commutation strategy that reflects the SR motor torque production equation, and iii) design the electrical parameter update law and voltage input controller to produce the desired phase currents based on the SR electrical subsystem dynamics.

12.5.1 SR Motor Desired Torque Signal

The desired torque signal in (12.36), the mechanical parameter update law in (12.37), and the pseudo-velocity filter in (12.25) and (12.26) can be used directly if it is noted that the SR mechanical subsystem is parameterized according to (12.3) by θ and $Y(q, \dot{q}, \ddot{q})$ defined in (12.23) and (12.22), respectively.

12.5.2 SR Motor Voltage Input Controller

Since the torque produced by the SR motor is governed by (12.10), a new commutation strategy must be designed to accommodate the new relationship between the desired torque and the desired electrical currents. The commutation strategy presented in [18] and [19] is adopted for this purpose. Specifically, the desired phase currents are generated according to

$$\gamma_{dj} = \sqrt{\frac{\tau_d \sin(x_j) S\left(\sin(x_j)\tau_d\right)}{S_T} + \gamma_{co}^2} \qquad \text{for} \quad j = 1, \ldots, 4 \tag{12.83}$$

where τ_d was defined in (12.36), and γ_{dj} is the desired current for the jth electrical subsystem. The remaining terms in (12.83) are the means by which the task of generating the desired torque is distributed among the phases. The terms $S(\cdot)$ and S_T are first-order differentiable functions given by

$$S(\sigma) = \begin{cases} 0 & \text{for} \quad \sigma \leq 0 \\ 1 - \underline{e}^{-\epsilon_0 \sigma^2}) & \text{for} \quad \sigma > 0 \end{cases} \tag{12.84}$$

and

$$S_T = \sum_{j=1}^{4} \sin^2(x_j)\, S\left(\sin(x_j)\,\tau_d\right) \tag{12.85}$$

in which σ is an arbitrary scalar, and ϵ_0, γ_{co} are positive design constants. The corresponding phase current tracking error is defined as

$$\eta_{Ij} = \gamma_{dj} - \gamma_j. \tag{12.86}$$

The form of the commutation strategy in (12.83) through (12.85) is motivated by the need to ensure that

$$\tau_d = \sum_{j=1}^{4} \sin(x_j)\, \gamma_{dj}^2 \tag{12.87}$$

which can be verified by substituting (12.83) through (12.85) into (12.87) and applying the appropriate trigonometric identities. Hence using (12.10), the torque tracking error of (12.21) can be rewritten for the SR motor as

$$\eta_\tau = \tau_d - \tau = \tau_d - \sum_{j=1}^{4} \sin(x_j)\gamma_{dj}^2 + \sum_{j=1}^{4} \sin(x_j)\gamma_{dj}^2 - \sum_{j=1}^{4} \sin(x_j)\gamma_j^2 \tag{12.88}$$

which upon substitution of (12.87) can be simplified to

$$\eta_\tau = \sum_{j=1}^{4} \sin(x_j)\left(\gamma_{dj} + \gamma_j\right)\eta_{Ij} \tag{12.89}$$

where (12.86) has been utilized.

Remark 12.8

As a result of (12.89), the $\eta_\tau(t)$ term in (12.38) will require a term in the voltage input controller of the form $\sin^2(x_j)\left(\gamma_{dj}+\gamma_j\right)^2\eta_{Ij}$ to damp-out the interconnection term during the stability analysis. This control term is different than the $k_{N4}\eta_{Ij}$ term used in the PMS voltage input of (12.67) due to the differences in the torque production mechanisms.

Remark 12.9

In [19] and [18], the commutation strategy of (12.83) was fused into a full-state feedback controller. Roughly speaking, not all commutation strategies can be fused into the backstepping procedure. The primary design requirements that the commutation strategy of (12.83) satisfy are [19]: i) $\gamma_{dj}(t)$ is first-order differentiable, ii) $\gamma_{dj}(t)$ remains bounded given that $q(t)$ and $\tau_d(t)$ are bounded, and iii) $\dot{\gamma}_{dj}(t)$ remains bounded given that $q(t)$, $\dot{q}(t)$, $\tau_d(t)$, and $\dot{\tau}_d(t)$ are bounded.

With the desired per-phase current specified as above, the final design step is to formulate the voltage control inputs such that the electrical phase currents track the desired current trajectories. Specifically, the current tracking error of (12.86) is differentiated with respect to time and the result is multiplied by $L_j(q)$ to yield

$$L_j(q)\dot{\eta}_{Ij}=L_j(q)\dot{\gamma}_{dj}-L_j(q)\dot{\gamma}_j. \tag{12.90}$$

To facilitate the construction of (12.90), the time derivative of $\gamma_{dj}(t)$ can be written using the partial derivative notation

$$\dot{\gamma}_{dj}=\frac{\partial\gamma_{dj}}{\partial x_j}\dot{x}_j+\frac{\partial\gamma_{dj}}{\partial\tau_d}\dot{\tau}_d\triangleq N_r\Sigma_j\dot{q}+\Pi_j\dot{\tau}_d \tag{12.91}$$

where $\Sigma_j(t)$ and $\Pi_j(t)$ (see Appendix B) are known functions of $q(t)$ and $\tau_d(t)$, and the $\dot{\tau}_d(t)$ term can be found by differentiating (12.36) with respect to time as shown

$$\dot{\tau}_d=\dot{Y}_d\hat{\theta}+\left(Y_d\Gamma Y_d^T+k^2+1\right)\eta+\left(\alpha_2 k-\alpha_2\right)e_f-\left(k\alpha_2+\alpha_1\right)e \tag{12.92}$$

where (12.28), (12.30), and (12.40) have been utilized. Equations (12.12), (12.90), (12.91) and (12.92) are now combined to yield the open-loop current tracking error dynamics as

$$L_j(q)\dot{\eta}_{Ij}=Y_{Ej}\theta_E-\left(\Omega_{1j}L_0+\Omega_{2j}L_1\right)\eta-\frac{1}{2}\dot{L}_j(q)\eta_{Ij}-V_j \tag{12.93}$$

where the auxiliary terms $\Omega_{1j}(t)\in\Re^1$ and $\Omega_{2j}(t)\in\Re^1$ are given by

$$\Omega_{1j}=\Pi_j Y_d\Gamma Y_d^T+\Pi_j(k^2+1)+N_r\Sigma_j \tag{12.94}$$

and

$$\begin{aligned}\Omega_{2j} = &\ \cos(x_j)\Pi_j Y_d \Gamma Y_d^T + \cos(x_j)\Pi_j(k^2+1) \\ &- \cos(x_j)N_r\Sigma_j + N_r\sin(x_j)\gamma_j + \frac{1}{2}\sin(x_j)\eta_{Ij}N_r. \end{aligned} \tag{12.95}$$

The electrical subsystem has been parameterized using $Y_{Ej} \in \Re^{1\times 3}$ given by

$$Y_{Ej} = \begin{bmatrix} Y_{Ej1} & Y_{Ej2} & Y_{Ej3} \end{bmatrix} \in \Re^{1\times 3} \tag{12.96}$$

with constituent elements

$$\begin{aligned} Y_{Ej1} = &\ N_r\Sigma_j\dot{q}_d + \Pi_j(\dot{Y}_d\hat{\theta} + k\alpha_2 e_f - k\alpha_2 e - \alpha_1 e - \alpha_2 e_f) \\ &+ N_r\Sigma_j(\alpha_1 e + \alpha_2 e_f), \\ Y_{Ej2} = &\ -\cos(x_j)N_r\Sigma_j\dot{q}_d + N_r\sin(x_j)\gamma_j\dot{q}_d \\ &- \cos(x_j)\Pi_j(\dot{Y}_d\hat{\theta} + k\alpha_2 e_f - k\alpha_2 e - \alpha_1 e - \alpha_2 e_f) \\ &- (\cos(x_j)N_r\Sigma_j + \sin(x_j)N_r\gamma_j)(\alpha_1 e + \alpha_2 e_f) \\ &+ \frac{1}{2}\sin(x_j)\eta_{Ij}N_r(\dot{q}_d + \alpha_1 e + \alpha_2 e_f), \\ Y_{Ej3} = &\ \gamma_j, \end{aligned}$$

and with the constant electrical parameter vector $\theta_E \in \Re^3$ defined as

$$\theta_E = \begin{bmatrix} L_0 & L_1 & R \end{bmatrix}^T. \tag{12.97}$$

Remark 12.10

When the Lyapunov-like function for the link tracking analysis is formed, it will contain a term of the form $\frac{1}{2}\sum_{j=1}^{4} L_j(q)\eta_{Ij}^2$, which when differentiated according to the product rule will yield a term of the form $\frac{1}{2}\sum_{j=1}^{4} \dot{L}_j(q)\eta_{Ij}^2$. Hence, the term $\frac{1}{2}\dot{L}_j(q)\eta_{Ij}$ is added and subtracted to the right hand-side of (12.93) to compensate for this term in the stability analysis. Additionally, the dynamics for $\dot{e}_f(t)$ from (12.28) and $\dot{e}(t)$ found from (12.30) were used along with $\dot{q} = \dot{q}_d - \eta + \alpha_1 e + \alpha_2 e_f$ to separate terms in (12.93) into measurable and unmeasurable components (*i.e.*, the measurable components are included in the electrical regressor matrix $Y_{Ej}(t)$).

Given the form of (12.93), the voltage control input can now be designed as

$$\begin{aligned} V_j = \; & k_s\eta_{Ij} + Y_{Ej}\hat{\theta}_E + k_{N2}\Omega_{1j}^2\eta_{Ij} + k_{N3}\Omega_{2j}^2\eta_{Ij} \\ & + k_{N4}\sin^2(x_j)(\gamma_{dj} + \gamma_j)^2\eta_{Ij} \end{aligned} \tag{12.98}$$

where $\hat{\theta}_E(t)$ is the dynamic estimate of the electrical parameters θ_E generated on-line according to

$$\hat{\theta}_E = \int_0^t \Gamma_E \sum_{j=1}^{4} \left(Y_{Ej}^T(\sigma)\eta_{Ij}(\sigma)\right) d\sigma \tag{12.99}$$

and k_s, k_{Ni} are positive, scalar control gains, and $\Gamma_E \in \Re^{3\times 3}$ is a diagonal, positive definite gain matrix. Through the use of the non-negative function

$$V = V_M + \frac{1}{2}\sum_{j=1}^{4} L_j(q)\eta_{Ij}^2 + \frac{1}{2}\tilde{\theta}_E^T\Gamma_E^{-1}\tilde{\theta}_E \tag{12.100}$$

where $V_M(t)$ was defined in (12.41), the voltage input controller of (12.98) and (12.99) can be shown to yield the same link tracking result as the PMS motor controller.

12.6 BLDC Motor Controller

In this section, the steps taken to develop the PMS motor controller are reformulated for the BLDC motor dynamics.

12.6.1 BLDC Motor Desired Torque Signal

The desired torque signal in (12.36), the mechanical parameter update law in (12.37), and the pseudo-velocity filter in (12.25) and (12.26) can be used directly if it is noted that the BLDC mechanical subsystem is parameterized according to (12.3) by θ and $Y(q,\dot{q},\ddot{q})$ defined in (12.23) and (12.22), respectively.

12.6.2 BLDC Motor Input Voltage Controller

Since the torque produced by the BLDC motor is governed by (12.15), we can use a simple commutation strategy which dictates that one of the electrical phases actuates the load while the other phase current is driven to *zero*. With such a commutation strategy in mind, the desired current trajectories, denoted by $I_{da}(t)$ and $I_{db}(t)$, are specified as

$$I_{da} = \tau_d \qquad \text{and} \qquad I_{db} = 0 \tag{12.101}$$

where $\tau_d(t)$ was defined in (12.36). The corresponding current tracking error terms are defined as

$$\eta_{Ia} = I_{da} - I_a \qquad \text{and} \qquad \eta_{Ib} = I_{db} - I_b. \tag{12.102}$$

To motivate the selection of the desired currents given by (12.101), we use (12.15) to write the torque tracking error of (12.21) for the BLDC motor as

$$\eta_\tau = \tau_d - \tau = \tau_d - I_{da} + I_{da} - K_{\tau n} I_b I_a - I_a \tag{12.103}$$

which upon substitution of (12.101) can be simplified to

$$\eta_\tau = \left(\frac{1}{2}K_{\tau n} I_b + 1\right)\eta_{Ia} + \frac{1}{2}K_{\tau n}\left(I_{da} + I_a\right)\eta_{Ib} \tag{12.104}$$

where (12.102) has been used.

Remark 12.11

As a result of expression obtained in (12.104), the η_τ term in (12.38) will require terms in the voltage input controller of the form $\left(\frac{1}{2}K_{\tau n} I_b + 1\right)^2 \eta_{Ia}$ and $\frac{1}{4}K_{\tau n}^2\left(I_{da} + I_a\right)^2 \eta_{Ib}$ to damp-out interconnection terms during the stability analysis. This control term is different than the $k_{N4}\eta_{Ij}$ term used in the PMS voltage input of (12.67) due to the differences in the torque production mechanisms.

With the desired per-phase current specified as above, the final design step is to formulate the voltage control inputs such that the electrical phase currents track the desired current trajectories. Hence, the tracking error definitions in (12.101) are differentiated in order to obtain the open-loop current dynamics. After differentiating $\eta_{Ia}(t)$ of (12.102) with respect to time and substituting for $L_a\dot{I}_a$, $\dot{e}_f$, $\dot{e}$, and $\dot{\hat{\theta}}$ from (12.16), (12.28), (12.30), and (12.40), respectively, the phase a open-loop dynamics are found to be

$$L_a\dot{\eta}_{Ia} = Y_{Ea}\theta_E + L_a\eta\left(Y_d\Gamma Y_d^T + k^2 + 1\right) - n_p L_b \eta I_b - K_{\tau 2}\eta - V_a \tag{12.105}$$

where $Y_{Ea} \in \Re^{1\times 4}$ is the regression matrix for the phase a electrical subsystem given by

$$Y_{Ea} = \begin{bmatrix} \dot{Y}_d\hat{\theta} + k\alpha_2 e_f - k\alpha_2 e - \alpha_1 e - \alpha_2 e_f \\ I_a \\ n_p I_b\left(\dot{q}_d + \alpha_1 e + \alpha_2 e_f\right) \\ \dot{q}_d + \alpha_1 e + \alpha_2 e_f \end{bmatrix}^T, \tag{12.106}$$

and the constant electrical parameter vector $\theta_E \in \Re^4$ is defined by

$$\theta_E = \begin{bmatrix} L_a & R & L_b & K_{\tau 2} \end{bmatrix}^T. \tag{12.107}$$

The form of (12.105) suggests that the input voltage control formulated to achieve the desired current tracking in phase a is given by

$$\begin{aligned} V_a = \; & Y_{Ea}\hat{\theta}_E + k_{N2}\left(Y_d\Gamma_M Y_d^T + k^2 + 1\right)^2 \eta_{Ia} + k_{N2}\left(n_p^2 I_b^2 + 1\right)\eta_{Ia} \\ & + k_{N2}\left(\frac{1}{4}\left(I_{db} + I_b\right)^2 + 1\right)\eta_{Ia} + k_a\eta_{Ia} \end{aligned} \tag{12.108}$$

where k_{N2}, k_a are positive constant control gains, and $\hat{\theta}_E(t)$ is a dynamic estimate of θ_E. Similarly, the open-loop dynamics of phase b are determined by differentiating $\eta_{Ib}(t)$ of (12.102) with respect to time and substituting from (12.17) and (12.30) for $L_b\dot{I}_b$ and $\dot{e}$ to yield

$$L_b\dot{\eta}_{Ib} = Y_{Eb}\theta_E + n_p L_a \eta I_{\dot{a}} - V_b \tag{12.109}$$

where the regression matrix for the phase b electrical subsystem $Y_{Eb}(t) \in \Re^{1\times 4}$ is defined by

$$Y_{Eb} = \begin{bmatrix} -n_p I_a \dot{q}_d - n_p I_a\left(\alpha_1 e + \alpha_2 e_f\right) & I_b & 0 & 0 \end{bmatrix}, \tag{12.110}$$

and θ_E was defined in (12.107). The form of (12.109) suggests that the input voltage control to generate the phase b desired current is given by

$$V_b = Y_{Eb}\hat{\theta}_E + k_{N3}\frac{1}{4}\left(I_{da} + I_a\right)^2 \eta_{Ib} + k_{N3} n_p^2 I_a^2 \eta_{Ib} + k_b \eta_{Ib} \tag{12.111}$$

where k_b and k_{N3} are positive gains. Motivated by the form of the closed-loop error system and the previous PMS motor controller design procedure, the estimated electrical parameters used in (12.108) and (12.111) are given by

$$\hat{\theta}_E = \Gamma_E \int_0^t \eta_{Ia}(\sigma) Y_{Ea}^T(\sigma) d\sigma + \Gamma_E \int_0^t \eta_{Ib}(\sigma) Y_{Eb}^T(\sigma) d\sigma \tag{12.112}$$

where $\Gamma_E \in \Re^{5\times 5}$ is a diagonal, positive definite gain matrix. Through the use of the non-negative function

$$V = V_M + \frac{1}{2}L_a\eta_a^2 + \frac{1}{2}L_b\eta_b^2 + \frac{1}{2}\tilde{\theta}_E^T\Gamma_E^{-1}\tilde{\theta}_E \qquad (12.113)$$

where $V_M(t)$ was defined in (12.41), the voltage input controllers of (12.108) and (12.111) can be shown to yield the same link tracking result as the PMS motor controller.

Remark 12.12

The specification of $I_{db}(t)$ given by (12.101) will simplify the control problem by effectively reducing the number of *control inputs* to the mechanical subsystem from two to one. Alternatively, it is possible that $I_{db}(t)$ can be specified to help rotate the mechanical system along the desired position trajectory as in [20]. Specifically, if $I_{db}(t)$ is selected as a positive function of time, an adaptive controller can be developed as in [21]; however, a special adaptation law must be used in order to avoid singularities in the control. As a practical matter, setting $I_{db}(t)$ to zero makes sense in terms of the torque that can actually be attributed to this current. For example, the Baldor BSM3R3-33 BLDC motor exhibits the following nominal values for the torque transmission constant: $K_{\tau n} = 2 \times 10^{-3}/A$. Since $K_{\tau n}$ is 500 times smaller than *one,* the phase b current in (12.15) must be very large in order to have even a small effect on the total amount of torque produced by the motor.

The controllers presented in this chapter dealt with the design of motor controllers where there are multiple electrical inputs and a scalar mechanical subsystem equation having a single output. This control design methodology is also suitable for the case of an n-link, serially connected, direct drive, revolute (RLED) robot where the mechanical subsystem is a vector equation with n outputs [8]. This extension to the RLED problem not only requires consideration of the multiple outputs but also requires the controller to compensate for the time varying inertia matrix and higher-order effects represented by the centripetal/Coriolis terms. In this more general problem the asymptotic tracking result is shown to be limited to an adjustable large set of initial conditions (*i.e.*, semiglobal [2] asymptotic position tracking).

12.7 Experimental Results

In order to substantiate the theoretical developments, the proposed controllers were implemented using the general purpose mechatronics workstation described in previous chapters. This section is devoted to describing the results of the position tracking experiments.

A desired trajectory for the load position was specified by

$$q_d = \frac{\pi}{2}\sin(2t)(1 - e^{-0.3t^3})\ rad$$

which has the desirable property $q_d(0) = \dot{q}_d(0) = \ddot{q}_d(0) = \dddot{q}_d\ (0) = 0$. The adaptive update laws are calculated using a standard trapezoidal integration routine while the initial conditions for all the parameter estimates were set to 50% of their measured values.

PMS Motor The Aerotech 3105MB3 2-phase permanent magnet stepper motor was described in Chapter 3. The control gains that provided the best position tracking are

$$\alpha_1 = 15,\ \alpha_2 = 1.5,\ k = 25,\ k_{N2} = 10^{-5},\ \ k_{N3} = k_{N4} = 0.1,$$

$$k_s = 1.0,\quad \Gamma = diag\{\ 0.0005 \quad 0.005 \quad 0.005 \quad 0.0005\ \},$$

and

$$\Gamma_{E1} = \Gamma_{E2} = diag\{\ 5\times10^{-4} \quad 5\times10^{-4} \quad 5\times10^{-4} \quad 5\times10^{-4}\ \}.$$

The position tracking error corresponding to this set of gains is shown in Figure 12.1. The accompanying motor voltages and currents are shown in Figure 12.2 through Figure 12.5.

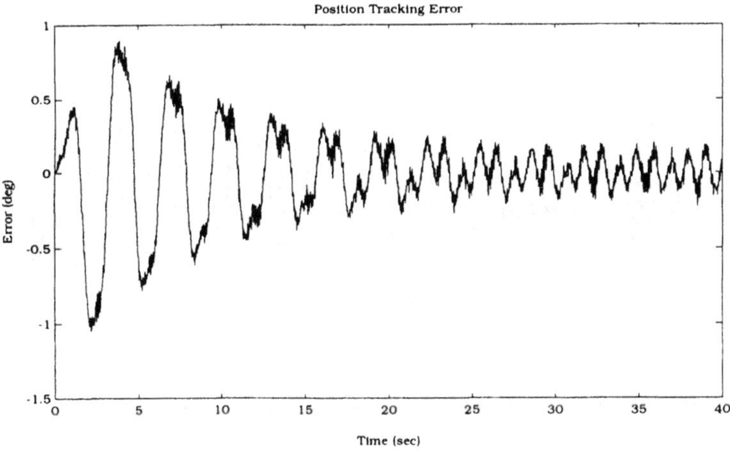

Figure 12.1: PMS Motor - Postion Tracking Error

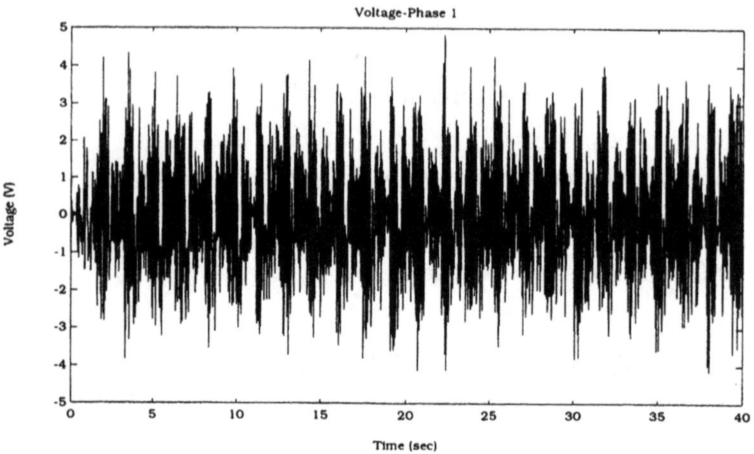

Figure 12.2: PMS Motor - Phase 1 Voltage

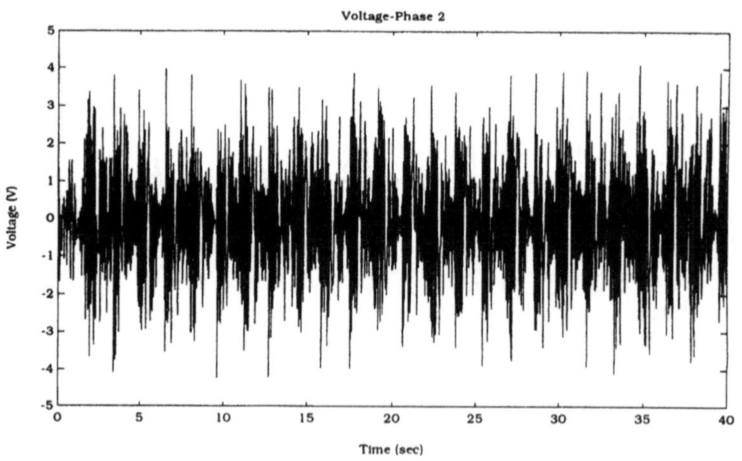

Figure 12.3: PMS Motor - Phase 2 Voltage

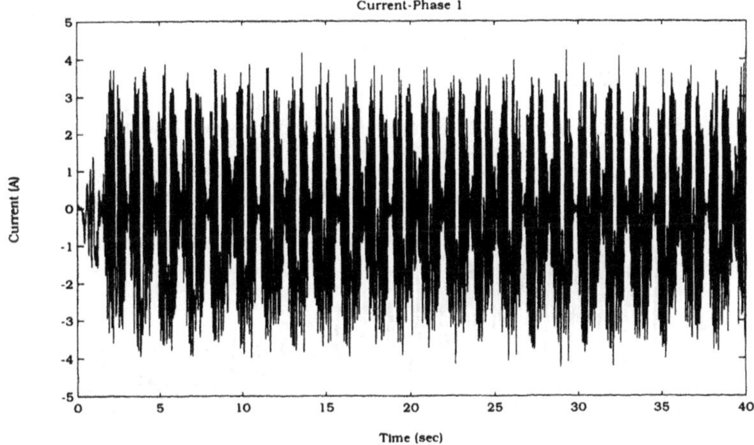

Figure 12.4: PMS Motor - Phase 1 Current

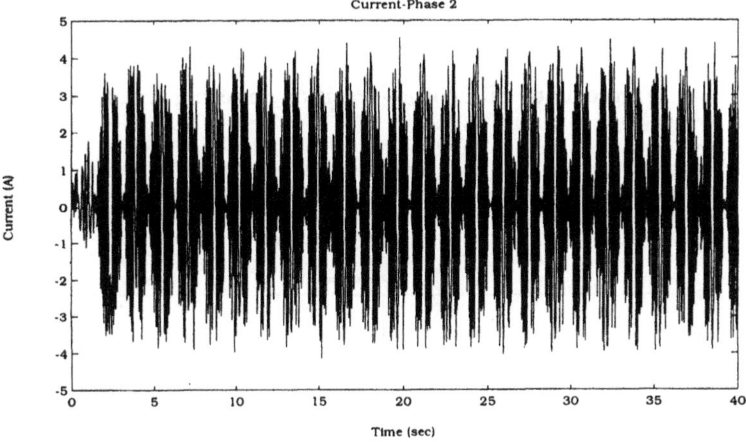

Figure 12.5: PMS Motor - Phase 2 Current

SR Motor The Magna-Physics, Model #90 4-phase SR motor was described in Chapter 5. The control gains that provided the best position tracking are

$$\begin{array}{lllll} \alpha_1 = 15, & \alpha_2 = 1, & k = 55, & k_s = 0.9, & \\ k_{N2} = 10^{-4}, & k_{N3} = 10^{-4}, & k_{N4} = 0.1 & \gamma_{c0} = 0.2, & \epsilon_0 = 0.1 \end{array}$$

$$\Gamma = diag\left\{ \begin{array}{cccc} 1.5 & 50 & 0.5 & 0 \end{array} \right\}, \text{ and } \Gamma_E = diag\left\{ \begin{array}{ccc} 0.05 & 0.05 & 0.1 \end{array} \right\}.$$

The result of the tracking experiment is shown in the tracking error plot in Figure 12.6. Sample phase voltage and current plots are shown in Figure 12.7 and Figure 12.8.

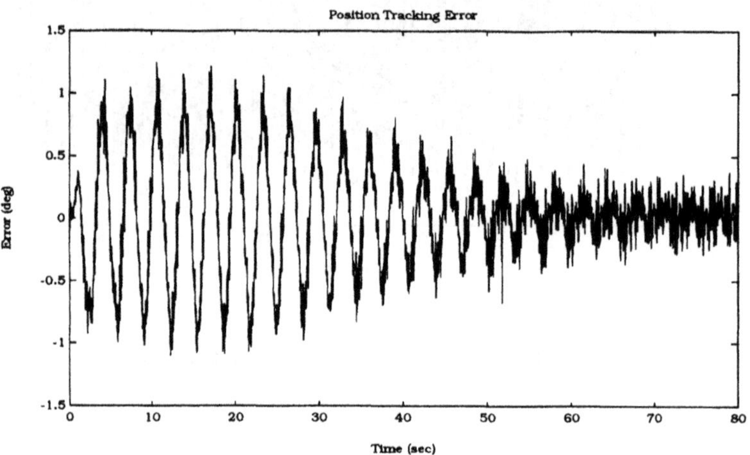

Figure 12.6: SR Motor - Position Tracking Error

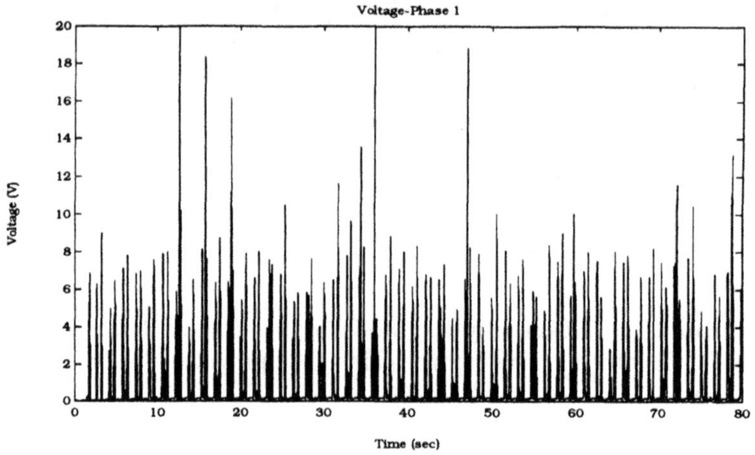

Figure 12.7: SR Motor - Phase 1 Voltage

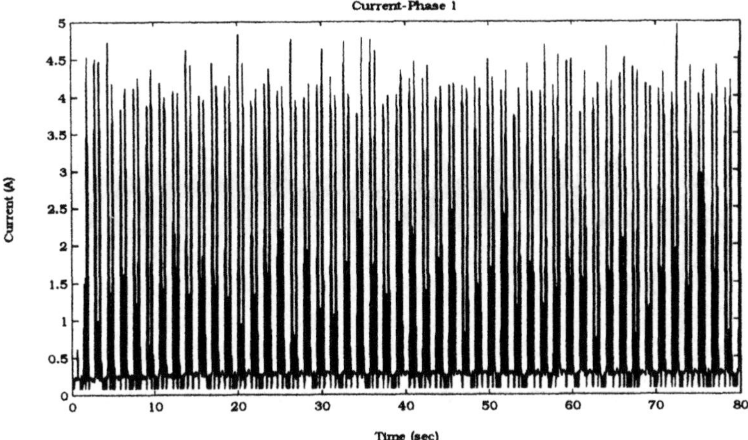

Figure 12.8: SR Motor - Phase 1 Current

BLDC Motor The Baldor BSM 3R 3-phase BLDC motor was described in Chapter 4. The control gains used to achieve the best performance are

$$\begin{array}{llll} \alpha_1 = 10, & \alpha_2 = 0.8, & k = 22, & k_a = 0.4, \quad k_b = 0.7 \\ k_{N2} = 1 \cdot 10^{-4}, & k_{N3} = 1 \cdot 10^{-4}, & & \end{array}$$

$$\Gamma = diag\left\{ \begin{array}{cccc} 0.5 & 1.0 & 80.0 & 0 \end{array} \right\},$$

$$\text{and } \Gamma_E = diag\left\{ \begin{array}{cccc} 0.1 & 0.2 & 0.2 & 0.5 \end{array} \right\}.$$

The position tracking error is shown in Figure 12.9 and plots typical of the applied (*i.e.*, these electrical quantities have not been transformed to the rotor-fixed reference frame) voltages and resulting currents are shown in Figure 12.10 and Figure 12.11, respectively.

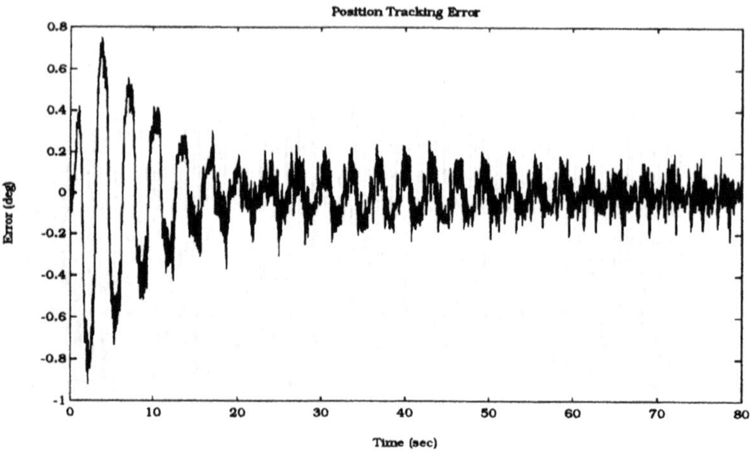

Figure 12.9: BLDC Motor - Position Tracking Error

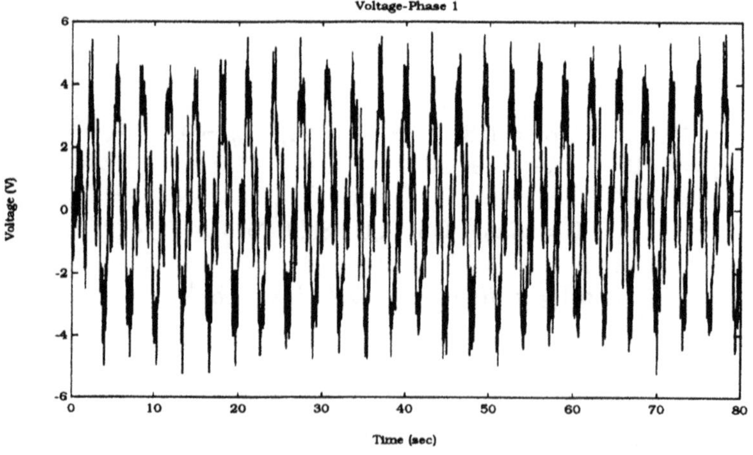

Figure 12.10: BLDC Motor - Phase 1 Voltage

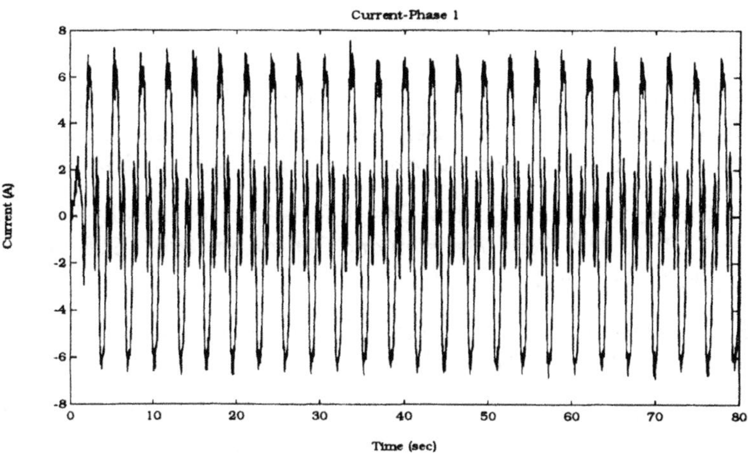

Figure 12.11: BLDC Motor - Phase 1 Current

12.8 Notes

Over the past five years, several classes of adaptive, output-feedback controllers have been developed for general nonlinear systems which can be transformed into the output-feedback canonical form (for example see Kanellakopoulos *et al.* [17]). These controllers can provide for output tracking using only output feedback (*i.e.*, position measurements for the motor control problem). Unfortunately, it seems that the nonlinear torque transmission relationship prohibits the application of this technique in its present form to all but the brushed DC motor. That is, it is not obvious how to design adaptive output feedback controllers for electric machines which can not be transformed into the output-feedback canonical form. The controller described in this chapter represents the specialization of the n-link direct-drive robot controller proposed in [8] to the problem of electric motor control.

Bibliography

[1] N. Sadegh and R. Horowitz, "Stability and Robustness Analysis of a Class of Adaptive Controllers for Robotic Manipulators", *Int. J. Robotics Research*, Vol. 9, No. 3, pp. 74-92, June 1990.

[2] M. Krstic, I. Kanellakopoulos, and P. Kokotovic, *Nonlinear and Adaptive Control Design*, John Wiley & Sons, 1995.

[3] S. Sastry and M. Bodson, *Adaptive Control: Stability, Convergence, and Robustness*, Englewood Cliffs, NJ: Prentice Hall Co., 1989.

[4] M. Bodson and J. Chiasson, "Application of Nonlinear Control Methods to the Positioning of a Permanent Magnet Stepper Motor", *Proc. 28th Conf. Decision and Control*, Tampa, FL, pp. 531-532, Dec. 1989.

[5] D. Taylor, "Adaptive Control Design for a Class of Doubly-Salient Motors", *Proc. 30th Conf. Decision and Control*, pp. 2903-2908, Brighton, England, Dec. 1991.

[6] P. C. Krause and O. Wasynczuk, *Electromechanical Motion Devices*, New York: McGraw-Hill, 1989.

[7] N. Hemanti and M. Leu, "A Complete Model Characterization of Brushless DC Motors", *IEEE Trans. Industry Applications*, Vol. 28, No. 1, pp. 172-180, Jan./Feb. 1992.

[8] T. Burg, D. Dawson, J. Hu, and M. de Queiroz, "An Adaptive Partial State Feedback Controller for RLED Robot Manipulators", *IEEE Transactions on Automatic Control*, Vol 41, No. 7, pp. 1024-1031, July, 1996.

[9] H. Berghuis and H. Nijmeijer, "A Passivity Approach to Controller-Observer Design for Robots", *IEEE Transactions on Robotics and Automation*, Vol. 9, No. 6, pp. 740-754, Dec. 1993.

[10] T. Burg, D. Dawson, and P. Vedagarbha, "Redesign of the DCAL Robot Controller Without Velocity Measurements", *Robotica*, Vol. 15, 1997, pp. 337-346.

[11] K. Kaneko and R. Horowitz, "Repetitive and Adaptive Control of Robot Manipulators with Velocity Estimation", *IEEE Transactions on Robotics and Automation*, Vol. 13, No. 2, pp. 204-217, Apr. 1997.

[12] L. Hsu and F. Lizarralde, "Variable Structure Adaptive Tracking Control of Robot Manipulators without Joint Velocity Measurement", *Proc. IFAC World Congress*, Sydney, Australia, Vol. 1, pp. 145-148, July 1993.

[13] R. Kelly, "A Simple Setpoint Controller by Using Only Position Measurements", *Preprint 1993 IFAC World Congress*, pp. 289-293, Sydney, Australia, July 1993.

[14] Z. Qu, D. Dawson, J. Dorsey, and J. Duffie, "Robust Estimation and Control of Robotic Manipulators", *Robotica*, Vol. 13, pp. 223-231, 1995.

[15] J. Yuan and Y. Stepanenko, "Robust Control of Robotic Manipulators without Velocity Measurements", *Int. J. Robust and Nonlinear Control*, Vol. 1, pp. 203-213, 1991.

[16] R. Ellis and D. Gulick, *Calculus with Analytic Geometry*, New York, NY: Harcourt Brace Jovanovich, Inc., 1982.

[17] I. Kanellakopoulos, P. Kokotovic, and A. Morse, "Adaptive Output-Feedback Control of a Class of Nonlinear Systems", *Proc. 30th Conf. Decision and Control*, pp. 1082-1087, Brighton, England, Dec. 1991.

[18] J. Carroll, D. Dawson, Z. Qu, and M. Leviner, "Robust Tracking Control of Rigid-Link Electrically Driven Robots Actuated by Switched Reluctance Motors", *Proc. ASME Winter Meeting*, Anaheim, CA, DSC-Vol. 42, pp. 97-103, Nov. 1992.

[19] J. Carroll, *Nonlinear Control of Electric Machines*, Ph.D. Dissertation, Clemson University, Clemson, SC, 1993.

[20] J. Carroll and D. Dawson "Adaptive Tracking Control of a Switched Reluctance Motor Turning and Inertial Load", *Proc. American Control Conf.*, San Francisco, CA, pp. 670-674, June 1993.

[21] J. Hu, M. Leviner, D. Dawson, and P. Vedaghabar, "An Adaptive Tracking Controller for Multi-Phase Permanent Magnet Machines", *International Journal of Control*, Vol. 64, No. 6, 1996, pp. 997-1022.

Chapter 13

Sensorless Control of the SEDC

13.1 Introduction

The wound stator DC motor features a stator coil used to establish a magnetic field within the motor. A second wound coil, the rotor coil, contains a current which interacts with the magnetic field produced by the stator coil to produce torque on the rotor coil resulting in mechanical rotation. The magnitude of this torque and speed of rotation will depend on the amount of stator current, and thereby the strength of the magnetic field, along with the magnitude of the rotor current [1]. Different speed and torque characteristics can be produced by changing the stator and rotor currents relative to each other. This modification of the motor characteristics can be done at the time of manufacture by changing the connection of the two coils relative to the electrical supply. Three common connections are commercially available: series connected, shunt connected, and separately excited. It is the separately excited case which is considered in this chapter and will be referred to herein as the separately excited DC (SEDC) motor. The significant feature of the SEDC motor configuration is the ability to produce high starting torque at low operating speeds. These motors are usually chosen for traction applications where high torque at lower operating speeds is typically required [2]. The speed-torque curves of these motors are steep and imply that a small change in load can lead to a significant variation in rotor velocity. Accordingly, speed control applications require a separate low-level control design.

Based on the above description of the SEDC motor, we will consider this motor to have two control inputs: one voltage for the stator coil and one voltage for the rotor coil. Typically, velocity control is achieved by a combination of rotor voltage control, varying the voltage applied to the rotor, and field weakening, whereby the rotor current is held constant and the stator current is varied [1]. The design of a controller to produce the required control voltages is hampered by the fact that the electromechanical dynamics of the SEDC motor is described by a set of nonlinear differential equations ([3], [4]), rendering standard linear techniques somewhat inappropriate. That is, in general, nonlinear techniques must be utilized for the control of such systems.

Since the instrumentation used to measure the motion of the mechanical system is in general more costly than the devices used to measure electrical states such as current, it would be desirable to eliminate the need for the more costly mechanical system sensors. Many recent research papers ([5], [6]) use the terminology sensorless control to describe a control approach which completely eliminates the use of mechanical system sensors. While there have been many claims of sensorless control in the literature, much of this work has not been formulated on a firm mathematical basis. That is, there are no proofs of stability for the resulting closed-loop error system under the proposed sensorless controller. In this chapter we present a sensorless (*i.e.*, the approach only requires current measurements) control method for velocity tracking control of the SEDC motor. Note that this chapter is different from any of the previous chapters in that we specifically address the problem of velocity tracking control.

13.2 System Model

The mathematical model describing the rotor velocity dynamics of the SEDC motor turning an inertial load can be found in [2] or [3] (See Figure 13.1). The mechanical subsystem dynamics for an inertial load actuated by a SEDC motor, including nonlinear saturation effects in the stator magnetic circuit, is given by

$$M\dot{\omega} + B\omega + T_L = g(I_s)I_r \tag{13.1}$$

where $\omega(t)$ represents rotor velocity, and $\dot{\omega}(t)$ denotes rotor acceleration. The mechanical parameter $M = M_0 + M_L$ is the scaled sum of the rotor and load inertias and the parameter $B = B_0 + B_L$ contains the scaled mechanical damping coefficients describing both the motor and the load. The term $g(I_s)$ describes the flux produced by the stator current, $I_s(t)$, and the term T_L denotes the scaled, known constant load (Note that the parameters M, B, and T_L described in (13.1) are defined to include the effects of the torque coefficient constant which characterizes the electromechanical conversion of armature current to torque). The electrical currents $I_r(t)$ and $I_s(t)$, the rotor and stator coil currents, respectively, are generated by the two electrical subsystems according to

$$L_r\dot{I}_r = -R_rI_r - K_Bg(I_s)\omega + v_r \tag{13.2}$$

and

$$\frac{d}{dt}\left[g(I_s)\right] = v_s - R_sI_s \tag{13.3}$$

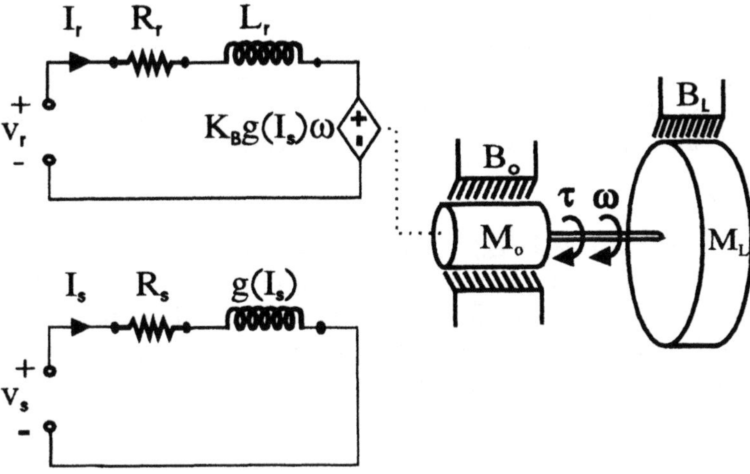

Figure 13.1: Schematic Diagram of an SEDC Motor/Load System

where $v_r(t)$ and $v_s(t)$ are the input rotor and stator coil voltages, respectively. The coefficients L_r and R_r denote the constant inductance and resistance in the rotor electrical subsystem, the coefficient R_s represents the constant resistance in the stator electrical subsystem, and K_B is the constant back-emf parameter.

Remark 13.1

The function $g(I_s)$ is used to describe the relationship between stator current and stator flux. Typically this relationship will be linear over a range in $I_s(t)$ and is often approximated using the inductance of the stator coil as

$$g(I_s) = L_s I_s$$

where L_s is the constant inductance of the stator coil. The above linear relationship does not account for saturation of the stator magnetic circuit outside of the linear region. Saturation refers to the diminishing effect of increased stator current on the production of stator flux, and therefore the diminishing effect of increased stator current on torque production. In order to more accurately model the motor dynamics, a general nonlinear model can be used for this current/flux relationship. In the following controller development, certain reasonable assumptions are made about the form of $g(\cdot)$. These assumptions are summarized as:

A1) $g(I_s)$ is a differentiable, strictly increasing function with respect to I_s

A2) $g(0) = 0$.

From the above assumptions, we know that if $g(I_s)$ can be shown to be bounded under a proposed controller, we know that $I_s(t)$ will remain bounded. Note that the magnetic saturation is typically not a "hard" saturation for normal operating currents and the assumptions A1) and A2) do not necessarily restrict the accuracy of the saturation model.

Remark 13.2

Note that the mechanical subsystem dynamics given in (13.1) are essentially the same as in previous chapters with the exception that the load is not position dependant. To simplify the controller development, we let $\omega(t) = \dot{q}(t)$ in the second-order mechanical dynamics of the previous chapters; hence, the mechanical dynamics are reduced to the first-order model given by (13.1).

13.3 Control Objective

Given exact model knowledge and measurement of electric current only (*i.e.*, $I_r(t)$ and $I_s(t)$), the control objective is to develop a rotor velocity tracking controller for the electromechanical dynamics of (13.1) through (13.3). To facilitate the tracking control formulation, we define the rotor velocity tracking error $e(t)$ as

$$e = \omega_d - \omega \tag{13.4}$$

where $\omega_d(t)$ represents the desired rotor velocity trajectory, and $\omega(t)$ was defined in (13.1). We will assume that $\omega_d(t)$ and its first and second time derivatives are all bounded functions of time.

Since the control problem is constrained by the fact that the only measurable states are the electrical currents, the remaining state variable (*i.e.*, the rotor velocity $\omega(t)$) will be observed. As the term observation implies there will be some variation between the physical and observed quantities. This variation, called observation error, is defined for the velocity observer as

$$\tilde{\omega} = \omega - \hat{\omega} \tag{13.5}$$

where $\hat{\omega}(t)$ denotes the observed rotor velocity. A necessary requirement in the overall control strategy will be to ensure that the observation error $\tilde{\omega}(t)$ is driven to zero.

Remark 13.3

Since we have reduced the mechanical subsystem to a first-order velocity dynamic equation and the control objective is velocity tracking, it will not be necessary to introduce an auxiliary filtered tracking error to simplify the control problem. That is, we will work directly with the velocity tracking error $e(t)$ in the controller development.

13.4 Observer Formulation

Based on the structure of (13.1) through (13.3), the subsequent stability analysis, and the controller/error system development, the following first-order, nonlinear velocity observer is proposed

$$\hat{\omega} = y - \frac{\left(k_s^2 k_{N1} + k_s\right) L_r}{K_B M} \left(\frac{I_r}{g(I_s)}\right) \tag{13.6}$$

where k_s and k_{N1} are constant, positive controller gains, $\hat{\omega}$ is the observed rotor velocity, and the auxiliary variable y is updated via the dynamic equation

$$\begin{aligned} \dot{y} = \; & \frac{\left(k_s^2 k_{N1} + k_s\right)}{K_B M g(I_s)} \left(v_r - R_r I_r\right) \\ & - \frac{\left(k_s^2 k_{N1} + k_s\right) L_r I_r}{K_B M g^2(I_s)} \left(v_s - R_s I_s\right) \\ & - \frac{B}{M}\hat{\omega} - \frac{T_L}{M} + \frac{1}{M} g(I_s) I_r - \frac{\left(k_s^2 k_{N1} + k_s\right)}{M} \hat{\omega}. \end{aligned} \tag{13.7}$$

Remark 13.4

It is important to note that the velocity observer places the constraint on the stator current $g(I_s) \neq 0$ (*i.e.*, by A2) $I_s \neq 0$). In general, this condition on $g(I_s)$ cannot be guaranteed due to the possible excursions of $I_s(t)$ through zero resulting from the electrical dynamics of the stator coil in (13.3) and the proposed voltage controller in the next section. Implementation of this observer requires a filter to generate a modified signal, $g_{filt}(I_s)$, which is utilized in the observer terms requiring division by $g(I_s)$. A possible candidate for such a filter is

$$g_{filt}(I_s) = \begin{cases} g(I_s) & \text{for } |g(I_s)| \geq \alpha \\ sgn\left(g(I_s)\right)\alpha & \text{for } |g(I_s)| < \alpha \end{cases} \tag{13.8}$$

where α is a small positive design parameter and $sgn(\cdot)$ is the signum function. Note that the restriction on $g(I_s)$ is not a result of the structure of the observer given by (13.6) and (13.7); rather, this restriction is a direct result of the form of the electrical dynamic equation given by (13.2). Specifically, if $g(I_s) = 0$ in the back-emf term of (13.2), then rotor velocity can not be observed from the electrical dynamics.

Remark 13.5

After taking the derivative of (13.6) with respect to time, we have

$$\dot{\hat{\omega}} = \dot{y} - \frac{\left(k_s^2 k_{N1} + k_s\right) L_r}{K_B M} \left(\frac{g(I_s)\dot{I}_r - \frac{d}{dt}\left[g(I_s)\right] I_r}{g^2(I_s)} \right). \tag{13.9}$$

After substituting for $\dot{y}(t)$ from (13.7) into (13.9) then multiplying both sides of the resulting expression by M, we have

$$\begin{aligned} M\dot{\hat{\omega}} = {} & \frac{\left(k_s^2 k_{N1} + k_s\right)}{K_B g(I_s)} \left(v_r - R_r I_r\right) \\ & - \frac{\left(k_s^2 k_{N1} + k_s\right) L_r I_r}{K_B g^2(I_s)} \left(v_s - R_s I_s\right) \\ & - B\hat{\omega} - T_L + g(I_s) I_r - \left(k_s^2 k_{N1} + k_s\right)\hat{\omega} \\ & - \frac{\left(k_s^2 k_{N1} + k_s\right) L_r}{K_B} \left(\frac{g(I_s)\dot{I}_r - \frac{d}{dt}\left[g(I_s)\right] I_r}{g^2(I_s)} \right). \end{aligned} \tag{13.10}$$

The last term can be expanded and substitutions can be made for $\dot{I}_r$ and $\frac{d}{dt}\left[g(I_s)\right]$ from (13.2) and (13.3), respectively, to produce the following first-order expression for the rotor velocity observer

$$M\dot{\hat{\omega}} = -B\hat{\omega} - T_L + g(I_s) I_r + k_s^2 k_{N1}\tilde{\omega} + k_s\tilde{\omega}. \tag{13.11}$$

During the subsequent analysis, we will use the rotor velocity observer in the form given by (13.11) to develop the rotor velocity observation error system. However, we should note that (13.11) can only be utilized for analysis purposes since the rotor velocity observer in the above first-order form contains the unmeasurable quantity $\tilde{\omega}(t)$. Hence, the velocity observer expressions given by (13.6) and (13.7) are utilized for actual implementation.

Remark 13.6

While the appearance of most of the terms in the above observer dynamics are due to the structure of the corresponding electromechanical dynamics, we note that there is an additional term which is used for the compensation of other terms during the stability analysis. In particular, the term $\frac{k_s^2 k_{N1} L_r}{K_B M}\left(\frac{I_r}{g(I_s)}\right)$ in (13.6) and the three terms multiplied by

$k_s^2 k_{N1}$ in (13.7) are combined in (13.11) to form the term $k_s^2 k_{N1}\tilde{\omega}$ which is a nonlinear damping term utilized to damp-out a term associated with the unmeasurable quantity $\tilde{\omega}(t)$ in the composite error system stability analysis. This nonlinear damping action, which is directly related to Lemma 1.8 in Chapter 1, will be made clear during the subsequent stability analysis of the observation error dynamics and the velocity tracking error dynamics.

13.4.1 Observation Error Dynamics

In this section, we formulate the observation error dynamics for the rotor velocity observer. A non-negative function is then used to illustrate that the observation error goes to zero exponentially fast when used in open-loop operation (*i.e.*, the observed velocity, $\hat{\omega}(t)$, has not yet been utilized in a voltage control input algorithm). In the next section, we will design a voltage input controller which utilizes this observed velocity in a stable, closed-loop fashion.

The velocity observation error dynamics are derived from the definition of the velocity observation error. Specifically, the velocity observation error of (13.5) is differentiated with respect to time and multiplied by M to yield

$$M\dot{\tilde{\omega}} = M\dot{\omega} - M\dot{\hat{\omega}}. \tag{13.12}$$

Substitution of the mechanical dynamics given in (13.1) for $M\dot{\omega}$ and the first-order velocity observer given in (13.11) for $M\dot{\hat{\omega}}$ into the right-hand side of (13.12) yields

$$M\dot{\tilde{\omega}} = -(B + k_s^2 k_{N1} + k_s)\tilde{\omega} \tag{13.13}$$

where (13.5) has been utilized. Roughly speaking, the form of (13.13) motivates one to classify the observer given by (13.6) and (13.7) as a closed-loop observer because a controller gain (*e.g.*, k_s) can be increased to cause $\tilde{\omega}(t)$ to go to zero faster.

13.4.2 Stability of the Observation Error System

To simplify the subsequent stability analysis of the composite error system and to provide insight into the construction of the observer, we now illustrate that the observer of (13.6) and (13.7) provides for exponentially stable observation of rotor velocity. The theorem given below delineates the performance of the observer in open-loop operation.

Theorem 13.1

The observation error system of (13.13) is exponentially stable in the sense that

$$|\tilde{\omega}(t)| \leq |\tilde{\omega}(0)|\, e^{-\gamma_r t} \quad \forall t \in [0, \infty) \tag{13.14}$$

where

$$\gamma_c = \frac{k_s^2 k_{N1} + k_s}{M} \tag{13.15}$$

provided

$$g(I_s) \neq 0.$$

Proof. First, we define the following non-negative function

$$V_o(t) = \frac{1}{2} M \tilde{\omega}^2 \tag{13.16}$$

which is differentiated with respect to time to yield

$$V_o(t) = \tilde{\omega} M \dot{\tilde{\omega}} . \tag{13.17}$$

After substituting the right-hand side of (13.13) for $M \dot{\tilde{\omega}}$ into (13.17), we have

$$\dot{V}_o(t) = -(B + k_s^2 k_{N1} + k_s)\tilde{\omega}^2. \tag{13.18}$$

From (13.18), the following upper on $\dot{V}_o(t)$ can be formulated as follows

$$\dot{V}_o(t) \leq -(k_s^2 k_{N1} + k_s)\tilde{\omega}^2. \tag{13.19}$$

From (13.16), it easy to see that $\tilde{\omega}^2 = 2V_o(t)/M$; hence, the upper bound on $\dot{V}_o(t)$ in (13.19) can be written as

$$\dot{V}_o(t) \leq -2\gamma_c V_o(t) \tag{13.20}$$

where γ_c was defined in (13.15). Applying Lemma 1.1 in Chapter 1 to (13.20) yields

$$V_o(t) \leq V_o(0)\underline{e}^{-2\gamma_c t}. \tag{13.21}$$

From (13.16), we have

$$\frac{1}{2} M \tilde{\omega}^2(t) = V_o(t) \quad \text{and} \quad V_o(0) = \frac{1}{2} M \tilde{\omega}^2(0). \tag{13.22}$$

Substituting (13.22) appropriately into the left-hand and right-hand sides of (13.21) allows us to form the following inequality

$$\frac{1}{2} M \tilde{\omega}^2(t) \leq \frac{1}{2} M \tilde{\omega}^2(0)\underline{e}^{-2\gamma_c t}. \tag{13.23}$$

Using (13.23), we can solve for $|\tilde{\omega}(t)|$ to obtain the result given in (13.14). □

Remark 13.7

Using the condition $g(I_s) \neq 0$, the result of Theorem 13.1, the structure of the observer, and the electromechanical model, it is straightforward to illustrate that all signals remain bounded during open-loop operation provided that $\omega(t) \in L_\infty$, $\dot{\omega}(t) \in L_\infty$, $v_r(t) \in L_\infty$, $I_r(t) \in L_\infty$, $v_s(t) \in L_\infty$ and $I_s(t) \in L_\infty$. Specifically, from (13.14), we know that $\tilde{\omega} \in L_\infty$; hence, we can use (13.5) to show that $\hat{\omega} \in L_\infty$. Since all the terms on the right-hand side of (13.13) are bounded, we now know that $\dot{\tilde{\omega}}(t) \in L_\infty$; hence, (13.12) can be used to show that $\dot{\hat{\omega}}(t) \in L_\infty$. Based on the above information, (13.6) and (13.7) can be utilized to show that $y(t) \in L_\infty$ and $\dot{y}(t) \in L_\infty$.

Remark 13.8

As illustrated by (13.6) and (13.19), there is a nonlinear damping gain k_{N1} (See Lemma 1.8 in Chapter 1) in the velocity observer. It can easily be established that the result of Theorem 13.1 is still valid even if k_{N1} is set to zero. However, as we will see later, the controller gain k_{N1} is used to damp-out a velocity observation error term during the composite controller-observer analysis presented in a subsequent section.

13.5 Voltage Control Inputs

In this section, we design the voltage control input which drives the velocity tracking error defined in (13.4) to zero. As mentioned in the introduction, the control design is constrained by the fact that only measurements of currents (*i.e.*, $I_r(t)$ and $I_s(t)$) are assumed to be available (*i.e.*, sensorless control). Since measurement of rotor velocity is not available, the observed rotor velocity is utilized in the formulation of the voltage control inputs. In some ways the design of the sensorless controller is similar to the design of the full state feedback controller (FSFB), exact model knowledge controller presented in Chapter 2. For example, the notion of current tracking error is utilized in the following fashion

$$\eta_{Ir} = I_{dr} - I_r \tag{13.24}$$

and

$$\eta_{Is} = g(I_{ds}) - g(I_s) \tag{13.25}$$

where $I_{dr}(t)$ and $I_{ds}(t)$ are used to denote the desired rotor current and stator current trajectories.

Before we begin the design of the voltage control input, the overall voltage input control strategy can be summarized, in a heuristic manner, using the above auxiliary definitions. That is, the role of the voltage control input will be to force $I_r(t)$ to track $I_{dr}(t)$ and $I_s(t)$ to track $I_{ds}(t)$ in the electrical subsystems. The result of the actual current tracking the desired current trajectory is that a desired torque will in effect be applied to the mechanical subsystem, which then will cause $\hat{\omega}(t)$ to track $\omega_d(t)$ in the mechanical subsystem. The combined action of the velocity observer and the desired current trajectory will then force $\omega(t)$ to track $\omega_d(t)$. In short, the combined function of the composite observer-controller is to force $I_r(t)$ to track $I_{dr}(t)$ and $I_s(t)$ to track $I_{ds}(t)$ and as a result force $\omega(t)$ to track $\omega_d(t)$. Thus the observer and the voltage control input work together to cause tracking of the internal system states to the designed trajectories and thereby achieve tracking in the output state $\omega(t)$ (*i.e.*, the rotor velocity trajectory).

13.5.1 Velocity Tracking Error Dynamics

In this section, we develop the velocity tracking error dynamics which are utilized in the subsequent composite stability analysis to show that the velocity tracking error goes to zero exponentially fast. During the development of the error dynamics, we formulate the desired current trajectory and the voltage control input. First, the velocity tracking error of (13.4) can be multiplied throughout by M and the result differentiated with respect to time to yield

$$M\dot{e} = M\dot{\omega}_d - M\dot{\omega}. \tag{13.26}$$

Equation (13.1) can now be used to substitute for $M\dot{\omega}$ to produce

$$M\dot{e} = B\omega + T_L - g(I_s)I_r + M\dot{\omega}_d. \tag{13.27}$$

Since there is no control input on the right-hand side of (13.27), the desired current trajectories (*i.e.*, $I_{dr}(t)$ and $I_{ds}(t)$) in the form $g(I_{ds})I_{dr}$ are added and subtracted on the right-hand side of (13.27) to yield

$$M\dot{e} = B\omega + T_L + M\dot{\omega}_d - g(I_{ds})I_{dr} + g(I_{ds})I_{dr} - g(I_s)I_r\,. \tag{13.28}$$

To facilitate the stability analysis, the last two terms of (13.28) can be written in the form

$$\begin{aligned} M\dot{e} = {} & B\omega + T_L + M\dot{\omega}_d - g(I_{ds})I_{dr} \\ & + \frac{1}{2}\left(\left(I_r + I_{dr}\right)\eta_{I_s} + \left(g(I_s) + g(I_{ds})\right)\eta_{I_r}\right) \end{aligned} \tag{13.29}$$

where the current tracking error terms $\eta_{I_s}(t)$ and $\eta_{I_r}(t)$ defined in (13.24) and (13.25) have been utilized.

We can see from (13.29) that the next step in the design procedure is to design the desired currents $I_{ds}(t)$ and $I_{dr}(t)$ to drive $e(t)$ to zero (under the assumption that $\eta_{I_r}(t)$, $\eta_{I_r}(t)$ and $\tilde{\omega}(t)$ will be driven to zero by the other components in the observer-controller system). Before actually designing these desired currents, we must first address the problem of how to divide the control responsibility between the two control inputs. Strategies from the previous chapters that could be considered are: i) share the control responsibility between the control inputs (*i.e.*, a commutation strategy as done for the PMS or SR motor controllers) or ii) set one of the control inputs to zero or a constant and make the remaining input responsible for control activity (*e.g.*, as done for the BLDC motor controller). Since the two control inputs appear multiplied together in (13.29), the second strategy appears to be most appropriate; therefore, we will specify the stator current to be constant and use the rotor current to control the rotor velocity.

Based on the form of (13.29) the desired current trajectories $I_{dr}(t)$ and $I_{ds}(t)$ are specified as follows

$$I_{dr} = \frac{1}{g(I_{ds})}\left(M\dot{\omega}_d + B\omega_d + T_L + k_s\left(\omega_d - \hat{\omega}\right)\right) \tag{13.30}$$

and

$$I_{ds} = I_{s0} + I_0\left(1 - e^{-\beta t}\right), \tag{13.31}$$

where the positive design parameters I_{s0}, I_0, and β of (13.31) are selected to ensure that $I_{ds}(t)$ is positive for all time. The desired current trajectory I_{dr} can be substituted from (13.30) into (13.29) to yield

$$M\dot{e} = -Be - k_s(\omega_d - \hat{\omega}) + \frac{1}{2}\left(\left(I_r + I_{dr}\right)\eta_{Is} + \left(g(I_s) + g(I_{ds})\right)\eta_{Ir}\right) \tag{13.32}$$

which can be rearranged to yield the velocity tracking error dynamics

$$M\dot{e} = -Be - k_s e - k_s\tilde{\omega} + \frac{1}{2}\left(\left(I_r + I_{dr}\right)\eta_s + \left(g(I_s) + g(I_{ds})\right)\eta_{Ir}\right) \tag{13.33}$$

where (13.5) and (13.4) were utilized. We can see from (13.33) that if $\tilde{\omega}(t)$, $\eta_{Is}(t)$, and $\eta_{Ir}(t)$ were all equal to zero, then the velocity tracking error would go to zero exponentially fast. From our previous analysis of the observation error systems analysis, we have good feeling about our ability to drive $\tilde{\omega}(t)$ to zero. Hence, motivated by the above discussion and the form of (13.33) we must now ensure that the current tracking errors go to zero.

To ensure that the current tracking errors go to zero, we first construct the open-loop dynamics for $\eta_{Is}(t)$ and $\eta_{Ir}(t)$. To this end, we multiply both sides of (13.24) by L_r and take the time derivative to obtain the following expression

$$L_r\dot{\eta}_{Ir} = L_r\dot{I}_{dr} - L_r\dot{I}_r\,. \tag{13.34}$$

The desired rotor current trajectory of (13.30) can be differentiated with respect to time to yield

$$\dot{I}_{dr} = \frac{1}{g(I_{ds})}\left(M\ddot{\omega}_d + B\dot{\omega}_d + k_s\left(\dot{\omega}_d - \dot{\hat{\omega}}\right)\right)$$
$$-\frac{\frac{\partial}{\partial I_{ds}}\left[g(I_{ds})\right]\dot{I}_{ds}}{g^2(I_{ds})}\left(M\dot{\omega}_d + B\omega_d + T_L + k_s\left(\omega_d - \hat{\omega}\right)\right) \tag{13.35}$$

and the desired stator current in (13.31) can be differentiated with respect to time to yield

$$\dot{I}_{ds} = I_0\beta \underline{e}^{-\beta t}. \tag{13.36}$$

The time derivatives given in (13.35) and (13.36) along with the rotor electrical subsystem dynamics in (13.2) can be substituted into (13.34) to yield

$$L_r\dot{\eta}_{I_r} = \frac{L_r}{g(I_{ds})}\left(M\ddot{\omega}_d + B\dot{\omega}_d + k_s(\dot{\omega}_d - \dot{\hat{\omega}})\right) \tag{13.37}$$
$$+R_rI_r + K_Bg(I_s)\omega$$
$$-v_r - \frac{L_r\frac{\partial}{\partial I_{ds}}\left[g(I_{ds})\right]I_0\beta\underline{e}^{-\beta t}}{g^2(I_{ds})}\left(M\dot{\omega}_d + B\omega_d\right)$$
$$-\frac{L_r\frac{\partial}{\partial I_{ds}}\left[g(I_{ds})\right]I_0\beta\underline{e}^{-\beta t}}{g^2(I_{ds})}\left(T_L + k_s(\omega_d - \hat{\omega})\right).$$

The derivative of the velocity observer, $\dot{\hat{\omega}}\,(t)$, found in (13.11) can be substituted into (13.37) to yield

$$L_r\dot{\eta}_{I_r} = \Omega_1 + \Omega_2\tilde{\omega} + \Omega_3\omega - v_r \tag{13.38}$$

where $\Omega_1(t)$, $\Omega_2(t)$, and $\Omega_3(t)$ are measurable, scalar auxiliary variables defined as

$$\Omega_1 = -\frac{L_rk_s}{Mg(I_{ds})}\left(-B\hat{\omega} - T_L + g(I_s)I_r\right) \tag{13.39}$$
$$+\frac{L_r}{g(I_{ds})}\left(M\ddot{\omega}_d + B\dot{\omega}_d + k_s\dot{\omega}_d\right) + R_rI_r$$
$$-\frac{L_r\frac{\partial}{\partial I_{ds}}\left[g(I_{ds})\right]\beta I_0\underline{e}^{-\beta t}}{g^2(I_{ds})}\left(M\dot{\omega}_d + B\omega_d\right)$$
$$-\frac{L_r\frac{\partial}{\partial I_{ds}}\left[g(I_{ds})\right]\beta I_0\underline{e}^{-\beta t}}{g^2(I_{ds})}\left(T_L + k_s\left(\omega_d - \hat{\omega}\right)\right),$$

$$\Omega_2 = -\frac{L_r \left(k_s^2 k_{N1} + k_s\right) k_s}{M g(I_{ds})}\ , \tag{13.40}$$

and

$$\Omega_3 = K_B g(I_s). \tag{13.41}$$

Based on the structure of (13.38) and the subsequent stability analysis, we propose the following rotor voltage control input to drive the rotor current tracking error to zero

$$\begin{aligned} v_r = \; & \Omega_1 + k_s\eta_{Ir} + \Omega_3\hat{\omega} + k_{N2}(\Omega_2 + \Omega_3)^2\eta_{Ir} \\ & + k_{N3}\left(\frac{1}{2}\left(g(I_s) + g(I_{ds})\right)\right)^2 \eta_{Ir} \end{aligned} \tag{13.42}$$

where k_s, k_{N2}, and k_{N3} are positive control gains. The rotor voltage control of (13.42) is now substituted into (13.38) to produce the closed-loop rotor current error dynamics as follows

$$\begin{aligned} L_r\dot{\eta}_{Ir} = \; & -k_s\eta_{Ir} - k_{N3}\left(\frac{1}{2}\left(g(I_s) + g(I_{ds})\right)\right)^2 \eta_{Ir} \\ & + \left[(\Omega_2 + \Omega_3)\tilde{\omega} - k_{N2}\left(\Omega_2 + \Omega_3\right)^2 \eta_{Ir}\right]. \end{aligned} \tag{13.43}$$

The stator current error dynamics are developed by differentiating the stator current error in (13.25) with respect to time to produce

$$\dot{\eta}_{Is} = \frac{\partial}{\partial I_{ds}}\left[g(I_{ds})\right]\dot{I}_{ds} - \frac{d}{dt}\left[g(I_s)\right]. \tag{13.44}$$

The time derivative of $I_{ds}(t)$ in (13.36) is substituted along with the stator electrical subsystem dynamics from (13.3) into (13.44) to yield

$$\dot{\eta}_{Is} = \frac{\partial}{\partial I_{ds}}\left[g(I_{ds})\right] I_0\beta \underline{e}^{-\beta t} + R_s I_s - v_s\ . \tag{13.45}$$

Based on the structure of (13.45) and the subsequent stability analysis, we propose the following stator voltage control input to drive the stator current tracking error to zero

$$\begin{aligned} v_s = \; & k_s\eta_{Is} + \frac{\partial}{\partial I_{ds}}\left[g(I_{ds})\right]\left(\beta I_0 \underline{e}^{-\beta t}\right) \\ & + R_s I_s + k_{N4}\left(\frac{1}{2}\left(I_r + I_{dr}\right)\right)^2 \eta_{Is} \end{aligned} \tag{13.46}$$

where k_s and k_{N4} are positive control gains. Substitution of the stator voltage control of (13.46) into (13.45) produces the closed-loop stator current error dynamics as shown

$$\dot{\eta}_{Is} = -k_s\eta_{Is} - k_{N4}\left(\frac{1}{2}\left(I_r + I_{dr}\right)\right)^2 \eta_{Is}\ . \tag{13.47}$$

Remark 13.9

To explain the origins of some of the terms in (13.43) and (13.47), we note that the auxiliary control terms $k_{N3}\left(\frac{1}{2}\left(g(I_s)+g(I_{ds})\right)\right)^2\eta_{I_r}$ and $k_{N4}\left(\frac{1}{2}\left(I_r+I_{dr}\right)\right)^2\eta_{I_s}$ will be used to damp-out the corresponding terms in (13.33) during the subsequent velocity tracking analysis. The bracketed term in (13.43) is a nonlinear damping pair, and hence, since the previous observation error stability analysis indicates that $\tilde{\omega}(t)$ can be driven to zero, we now see that the form of (13.43) and (13.47) have been specifically constructed to ensure that $\eta_{I_r}(t)$ and $\eta_{I_s}(t)$ are driven to zero.

Remark 13.10

The dynamics given by (13.33), (13.43), and (13.47), represent the closed-loop velocity tracking error system for which the stability analysis is performed while the controller given by (13.42) and (13.46) represents the controller which is implemented at the voltage terminals of the motor. Note that the desired current trajectories $I_{dr}(t)$ and $I_{ds}(t)$ of (13.30) and (13.31) are embedded (in the guise of the variables $\eta_{I_r}(t)$ and $\eta_{I_s}(t)$) inside of the voltage control inputs $v_r(t)$ and $v_s(t)$.

13.5.2 Stability of the Velocity Tracking Error System

In order to analyze the stability of the closed-loop velocity tracking error system, we define the following non-negative function

$$V_p(t)=\frac{1}{2}Me^2+\frac{1}{2}L_r\eta_{I_r}^2+\frac{1}{2}\eta_{I_s}^2. \tag{13.48}$$

The time derivative of (13.48) is given by

$$\dot{V}_p(t)=M\dot{e}e+L_r\dot{\eta}_{I_r}\eta_{I_r}+\dot{\eta}_{I_s}\eta_{I_s}. \tag{13.49}$$

After substituting (13.33), (13.43), and (13.47) into (13.49) for $M\dot{e}$, $L_r\dot{\eta}_{Ir}$, and $\dot{\eta}_{Is}$, respectively, we have

$$\begin{aligned}\dot{V}_p(t) = & -k_s\eta_{Ir}^2 - k_s\eta_{Is}^2 - k_s e^2 - Be^2 + k_s\tilde{\omega}e \\ & + \left[(\Omega_2+\Omega_3)\,\tilde{\omega}\eta_{Ir} - k_{N2}\left(\Omega_2+\Omega_3\right)^2\eta_{Ir}^2\right] \\ & + \left[\frac{1}{2}\left(g(I_s)+g(I_{ds})\right)\eta_{Ir}e \right. \\ & \left. -k_{N3}\left(\frac{1}{2}\left(g(I_s)+g(I_{ds})\right)\right)^2\eta_{Ir}^2\right] \\ & + \left[\frac{1}{2}\left(I_r+I_{dr}\right)\eta_{Is}e - k_{N4}\left(\frac{1}{2}\left(I_r+I_{dr}\right)\right)^2\eta_{Is}^2\right].\end{aligned} \tag{13.50}$$

From (13.50), $\dot{V}_p(t)$ can be upper bounded as follows

$$\begin{aligned}\dot{V}_p(t) \leq & -k_s\eta_{Ir}^2 - k_s\eta_{Is}^2 - k_s e^2 + k_s\left|\tilde{\omega}\right|\left|e\right| \\ & + \left[\left|\Omega_2+\Omega_3\right|\left|\tilde{\omega}\right|\left|\eta_{Ir}\right| - k_{N2}\left(\Omega_2+\Omega_3\right)^2\eta_{Ir}^2\right] \\ & + \left[\frac{1}{2}\left|g(I_s)+g(I_{ds})\right|\left|\eta_{Ir}\right|\left|e\right| \right. \\ & \left. -k_{N3}\left(\frac{1}{2}\left(g(I_s)+g(I_{ds})\right)\right)^2\eta_{Ir}^2\right] \\ & + \left[\frac{1}{2}\left|I_r+I_{dr}\right|\left|\eta_{Is}\right|\left|e\right| - k_{N4}\left(\frac{1}{2}\left(I_r+I_{dr}\right)\right)^2\eta_{Is}^2\right].\end{aligned} \tag{13.51}$$

We now apply Lemma 1.8 in Chapter 1 to the three bracketed terms on the right-hand side of (13.51) to form the following upper bound for $\dot{V}_p(t)$ as

$$\begin{aligned}\dot{V}_p(t) \leq & -k_s\eta_{Ir}^2 - k_s\eta_{Is}^2 - \left(k_s - \frac{1}{k_{N3}} - \frac{1}{k_{N4}}\right)e^2 \\ & + \left[k_s\left|\tilde{\omega}\right|\left|e\right|\right] + \left[\frac{\tilde{\omega}^2}{k_{N2}}\right].\end{aligned} \tag{13.52}$$

Remark 13.11

From the form of (13.52), we can see that if $\tilde{\omega}(t)$ was equal to zero for all time, and the controller gains were selected to satisfy the inequality $k_s > \frac{1}{k_{N3}} + \frac{1}{k_{N4}}$, then we could utilize Lemma 1.1 in Chapter 1 to guarantee that $e(t)$, $\eta_{Ir}(t)$, and $\eta_{Is}(t)$ would all go to zero. As we will see later during the analysis of the composite observer-controller error systems, the bracketed terms in (13.52) can be combined with other terms in (13.19) to obtain the desired stability result. For example, it is easy to see that the last bracketed term in (13.52) can be combined with a similar term in (13.19). Therefore, if the controller gains are selected appropriately, these two terms can be made to produce a non-positive term when the time derivatives of the two Lyapunov-like functions (*i.e.*, $V_o(t)$ of (13.16) and $V_p(t)$ of (13.48)) are added together.

13.6 Stability of the Composite Error System

In this section, we will analyze the performance of the composite observer-controller error systems. The performance of these closed-loop error systems is illustrated by the following theorem.

Theorem 13.2

Given the nonlinear observers of (13.6) and (13.7) and the voltage input control of (13.42) and (13.46), the rotor velocity tracking error is exponentially stable in the sense that

$$\|x(t)\| \leq \sqrt{\frac{\lambda_2}{\lambda_1}}\|x(0)\| \underline{e}^{-\gamma t} \qquad \forall t \in [0, \infty) \tag{13.53}$$

where

$$x(t) = \begin{bmatrix} \tilde{\omega} & e & \eta_{Ir} & \eta_{Is} \end{bmatrix}^T \in \Re^4, \tag{13.54}$$

$$\lambda_1 = \min\{M,\ L_r,\ 1\}, \qquad \lambda_2 = \max\{M,\ L_r,\ 1\}, \tag{13.55}$$

and

$$\gamma = \frac{\min\left\{k_s,\ \left(k_s - \frac{1}{k_{N2}}\right),\ \left(k_s - \frac{1}{k_{N1}} - \frac{1}{k_{N3}} - \frac{1}{k_{N4}}\right)\right\}}{\max\{M,\ L_r,\ 1\}}. \tag{13.56}$$

To ensure the above stability result, the controller gains must be adjusted to satisfy the following sufficient condition

$$k_{Ni} > \frac{3}{k_s} \quad \text{where } i = 1,\ \cdots,\ 4. \tag{13.57}$$

and the following restriction on the stator current must hold

$$g(I_s) \neq 0. \tag{13.58}$$

Proof. First, we define the following non-negative function

$$V(t) = V_o(t) + V_p(t) = \frac{1}{2}x^T diag\{M, M, L_r, 1\}x \tag{13.59}$$

where $x(t)$ was defined in (13.54), $V_o(t)$ was defined in (13.16), $V_p(t)$ was defined in (13.48), and $diag\{M, M, L_r, 1\} \in \Re^{4\times 4}$ is a diagonal matrix. From the matrix form of $V(t)$ given on the right-hand side of (13.59), Lemma 1.7 in Chapter 1 can be use to form the following upper and lower bounds for $V(t)$

$$\frac{1}{2}\lambda_1 \|x(t)\|^2 \leq V(t) \leq \frac{1}{2}\lambda_2 \|x(t)\|^2 \tag{13.60}$$

where λ_1 and λ_2 are defined in (13.55). From (13.19) and (13.52), an upper bound on the time derivative of $V(t)$ defined in (13.59) is given by

$$\begin{aligned} \dot{V}(t) \leq\ & -k_s\eta_{Ir}^2 - k_s\eta_{Is}^2 - \left(k_s - \frac{1}{k_{N3}} - \frac{1}{k_{N4}}\right)e^2 \\ & -\left(k_s - \frac{1}{k_{N2}}\right)\tilde{\omega}^2 + \left[k_s|\tilde{\omega}|\,|e| - k_s^2 k_{N1}\tilde{\omega}^2\right]. \end{aligned} \tag{13.61}$$

After applying Lemma 1.8 in Chapter 1 to the bracketed term in (13.61), Lemma 1.7 in Chapter 1 can be utilized to upper bound $\dot{V}(t)$ in (13.61) as

$$\begin{aligned} \dot{V}(t) \leq\ & -\min\left\{k_s, \left(k_s - \frac{1}{k_{N2}}\right),\right. \\ & \left.\left(k_s - \frac{1}{k_{N1}} - \frac{1}{k_{N3}} - \frac{1}{k_{N4}}\right)\right\}\|x(t)\|^2 \end{aligned} \tag{13.62}$$

where $x(t)$ was defined in (13.54). From (13.60), it easy to see that $\|x(t)\|^2 \geq 2V(t)/\lambda_2$; hence, $\dot{V}(t)$ in (13.62) can be further upper bounded as

$$\dot{V}(t) \leq -2\gamma V(t) \tag{13.63}$$

where γ was defined in (13.56). Applying Lemma 1.1 in Chapter 1 to (13.63) yields

$$V(t) \leq V(0)\underline{e}^{-2\gamma t}. \tag{13.64}$$

From (13.60), we have that

$$\frac{1}{2}\lambda_1 \|x(t)\|^2 \leq V(t) \quad \text{and} \quad V(0) \leq \frac{1}{2}\lambda_2 \|x(0)\|^2. \tag{13.65}$$

Substituting (13.65) appropriately into the left-hand and right-hand sides of (13.64) allows us to form the following inequality

$$\frac{1}{2}\lambda_1 \left\|x(t)\right\|^2 \leq \frac{1}{2}\lambda_2 \left\|x(0)\right\|^2 \mathrm{e}^{-2\gamma t}. \tag{13.66}$$

Using (13.66), we can solve for $\|x(t)\|$ to obtain the result given in (13.53).□

Remark 13.12

From the result given by (13.53) and (13.54), we use standard inequality arguments to form the following upper bounded on the rotor velocity tracking error

$$|e(t)| \leq \sqrt{\frac{\lambda_2}{\lambda_1}} \left\|x(0)\right\| \mathrm{e}^{-\gamma t} \quad \forall t \in [0, \infty);$$

hence, the velocity tracking error goes to zero exponentially fast. In a similar manner to the above inequality, it easy to show that the velocity observation error (*i.e.*, $\tilde{\omega}(t)$) and the current tracking errors (*i.e.*, $\eta_{I_r}(t)$ and $\eta_{I_s}(t)$) go to zero exponentially fast.

Remark 13.13

A necessary property of any controller is that all system quantities remain bounded. The result of the above analysis, as summarized in (13.53), is that $x(t) \in L_\infty^4$ and therefore the variables $\tilde{\omega}(t) \in L_\infty$, $e(t) \in L_\infty$, $\eta_{I_r}(t) \in L_\infty$, and $\eta_{I_s}(t) \in L_\infty$. A direct result of the definition of $\tilde{\omega}(t)$, and $e(t)$ given in (13.5) and (13.4), respectively, is that $\omega(t) \in L_\infty$ and $\hat{\omega}(t) \in L_\infty$. The definition of I_{ds} in (13.31) implies $I_{ds}(t) \in L_\infty$, the assumed properties of $g(\cdot)$ can then be used to conclude that $g(I_{ds}) \in L_\infty$. The definition of $\eta_{I_s}(t)$ in (13.25) and the facts that $g(I_{ds}) \in L_\infty$ and $\eta_{I_s}(t) \in L_\infty$ imply $g(I_s) \in L_\infty$. The properties of $g(\cdot)$ are again invoked and $g(I_s) \in L_\infty$ utilized to conclude $I_s(t) \in L_\infty$. Equation (13.30) can be written to show the dependence of $I_{dr}(t)$ on bounded variables as

$$I_{dr} = g_1\left(\omega_d, \dot{\omega}_d, I_{ds}, \hat{\omega}\right)$$

where $g_1(\cdot) \in L_\infty$ due to the structure of the desired current trajectory and the fact that it is only dependent on bounded quantities. Thus, we can conclude that $I_{dr}(t) \in L_\infty$. The definition of $\eta_{I_r}(t)$ in (13.24) along with the facts $\eta_{I_r}(t) \in L_\infty$ and $I_{dr}(t) \in L_\infty$ imply $I_r(t) \in L_\infty$. Equation (13.46) can now be written to show the dependence of $v_s(t)$ on bounded variables as

$$v_s = g_2\left(I_s, I_r, I_{dr}, I_{ds}\right)$$

which from the structure of (13.46) allows us to conclude that $v_s(t) \in L_\infty$. Equation (13.42) can now be written to show the dependence of $v_r(t)$ on bounded variables as

$$v_r = g_3\left(\omega_d, \dot{\omega}_d, \ddot{\omega}_d, I_s, I_r, I_{dr}, I_{ds}, \hat{\omega}\right)$$

where $g_3(\cdot) \in L_\infty$ due to the structure of the voltage input controller and the fact that it is only dependent on bounded quantities. Thus we can conclude that $v_r(t) \in L_\infty$. With regard to the observer which is actually implemented, we know from (13.7) that

$$\dot{y} = g_3\left(I_s, I_r, v_s, v_r, \hat{\omega}\right);$$

therefore, (13.7) implies that $\dot{y}(t) \in L_\infty$. Now, (13.11) and the above bounded variables can be utilized to show that $\dot{\hat{\omega}}\ (t) \in L_\infty$. Equation (13.6) may now be used to conclude that $y(t) \in L_\infty$. Finally, the above information, (13.1), (13.2), and (13.3) imply that $\dot{\omega}(t) \in L_\infty$, $\dot{I}_r(t) \in L_\infty$, $\frac{d}{dt}\left[g(I_s)\right] \in L_\infty$, and $\dot{I}_s(t) \in L_\infty$. It has now been shown that all signals in the system remain bounded during closed-loop operation.

Remark 13.14

The restriction on the stator current given by (13.58) is not considered significant since the voltage controller of (13.46), (13.31) and (13.25) forces the current to a positive value. Note that this restriction on the stator current is also easy to understand from the standpoint that if $g(I_s) = 0$ then no torque can be delivered from the electrical subsystem to the mechanical subsystem. As discussed in Remark 13.1, a filter will be used on the signal $I_s(t)$ to prevent generation of an unbounded control.

13.7 Experimental Results

The hardware used to implement the sensorless velocity controller presented in this chapter is exactly the same as that described in Chapter 2; however, only the current sensors are used to provide sensory information to the input voltage control algorithm. The Baldor CD3433, 1/3 hp wound field DC motor connected in a separately excited configuration, was used in the experiment. The electromechanical system parameters, before scaling by the torque transmission coefficient, for the motor model in (13.1) through (13.3) were determined to be

$$M = 0.035kg \cdot m^2,\ B = 0.009N \cdot m \cdot \sec/rad,$$

$$L_r = 0.03H,\ R_s = 60.3\Omega,\ R_r = 4.95\Omega,$$

$$K_\tau = 0.415N \cdot m/WbA,\ K_B = 0.415,$$

and $g(I_s)$ was determined experimentally as explained below.

The first step in the experimental work was to develop a model for the saturation effects in the Baldor CD3433 DC motor connected in a separately excited configuration, *i.e.*, develop an analytic expression for $g(I_s)$ based on

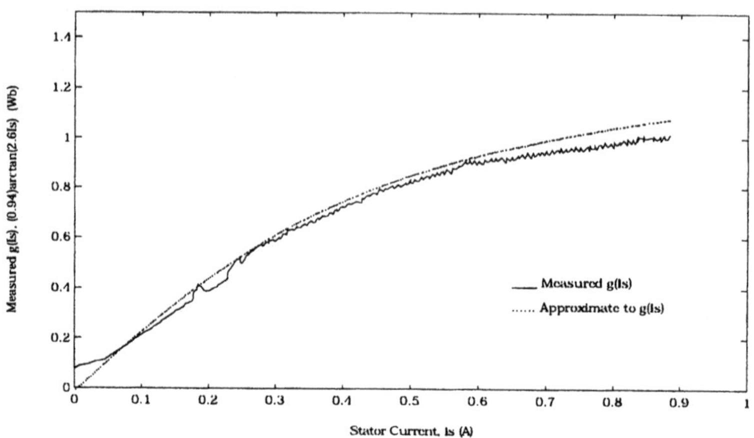

Figure 13.2: Measured and Approximated $g(I_s)$

measured data. The motor was mechanically connected to an inertial load, which was braked using a band-type friction brake set to create a small load torque. A torque sensor, S. Himmelstein and Company Model MCRT 2801T, was placed between the load and the motor in order to measure the left-hand side of equation (13.1). Data points were obtained by applying constant stator and rotor voltages in order to produce a constant rotor velocity. Based on the current readings, the measured torque, and the determined value of K_τ (*i.e.*, the torque transmission coefficient used to scale (13.1)) the value of $g(I_s)$ for a constant $I_s(t)$ was calculated. This process was repeated for a range of I_s to obtain the plot shown in Figure 13.2. The resemblance of the curve to the $arctan(\cdot)$ function, motivated the use of the following function to match the recorded behavior and to meet the required conditions of boundedness and differentiability. The model used to approximate $g(I_s)$ was selected as

$$g(I_s) = 0.941 \arctan(2.6 I_s)$$

and is also plotted in Figure 13.2.

Once the model for $g(I_s)$ was developed the tracking performance was validated. Specifically, the motor was mechanically connected to an inertial load which is an iron wheel attached directly to the motor shaft (for the tracking experiment no external load was added, *i.e.*, $T_L = 0$). A desired trajectory for the rotor velocity was specified as

$$\begin{aligned}\omega_d(t) = \quad & 286.5 \sin\left(\left(0.51 - \underline{e}^{-0.005t^3}\right) t\right) \\ & +573 \left(1 - \underline{e}^{-0.005t^3}\right) \deg / \sec\end{aligned}$$

which has the desirable property $\omega_d(0) = \dot{\omega}_d(0) = \ddot{\omega}_d(0) = 0$. The initial conditions for all of the state variables are set to zero. The control current $I_{ds}(t)$, was specified as

$$I_{ds} = 0.75 + \left(1 - \underline{e}^{-2t}\right) A.$$

After some experimentation, we achieved the best velocity tracking with the control gains set to

$$k_{N1} = 0.4,\ k_{N2} = 5 \cdot 10^{-4},\ k_{N3} = 5 \cdot 10^{-4},$$

$$k_{N4} = 5 \cdot 10^{-3},\ k_s = 2.5$$

and the filter in (13.8) was used on $g\left(I_{ds}\right)$ with $\alpha = 0.1$ to prevent division by zero. The results of the experiment are summarized in Figure 13.3 through Figure 13.5. The plot in Figure 13.3 shows the actual and desired velocities and provides the basis for concluding the control system is functioning as desired (*i.e.*, the velocity tracking error has become small). The actual velocity was obtained using a backward difference algorithm on the position signal from a shaft mounted encoder (it must be emphasized that the position encoder was used as part of the experimental validation but was not used in the controller or observer implementation). Note that due to measurement noise, modeling error, and quantization error in applying the control the tracking error does not approach zero as predicted by the theory but rather is driven to a small value. Figure 13.4 and Figure 13.5 demonstrate that the input voltages stay within the operating range of the motor.

Remark 13.15

The control gains selected for the experiment violate conditions (13.57) for the selection of these gains set forth in Theorem 13.2. However, this condition is a sufficient, conservative condition generated by the Lyapunov-like stability analysis and is used as a guide in selecting the control gains for actual experimentation.

13.8 Notes

The concept of applying nonlinear control to a wound stator DC motor is well established. Commercially available controllers use hardware which implement what can be regarded as de facto nonlinear control due to the use of choppers and limiters. These controllers, however, are based less on a coherent mathematical theory and more on practical experience and experimentation. Use of more traditional nonlinear control techniques can be seen in the sliding mode controllers in [7] and [8] or the adaptive controller in [9]. In another approach to this control problem, feedback linearization

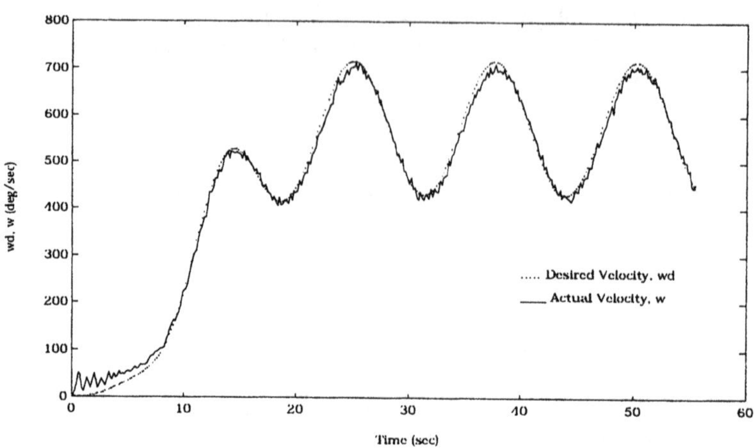

Figure 13.3: Desired and Recorded Velocities

was used to design the nonlinear portion of the electromechanical dynamics to make the remaining system amenable to standard linear techniques (See [10] for example). One of the most advanced systems for controlling the DC motor appears to be a feedback linearizing control used with a velocity observer designed by Chiasson in [2]. Most recently, Chiasson and Bodson [11] applied similar techniques to design a controller for the wound stator DC motor in the shunt configuration. The work presented in this chapter is based on the material found in [12].

Bibliography

[1] P. Sen, Electric Motor Drives and Control - Past, Present, and Future", *IEEE Trans. Industrial Electronics*, Vol. 37, No. 6, pp. 562-575, Dec., 1990.

[2] J. Chiasson, "Nonlinear Differential-Geometric Techniques for Control of a Series DC Motor", *IEEE Trans. on Control Systems Technology*, Vol. 2, No. 1, pp. 35-42, 1994.

[3] P. Krause, *Analysis of Electric Machinery*, McGraw Hill, New York, NY, 1986.

[4] M. Rashid, *Power Electronics - Circuits, Devices and Systems*, 2nd Edition, Prentice Hall, Englewood Cliffs, NJ, 1993.

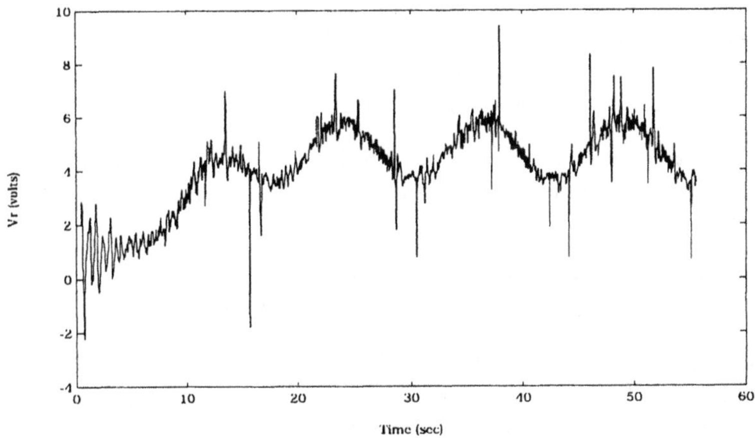

Figure 13.4: Rotor Voltage

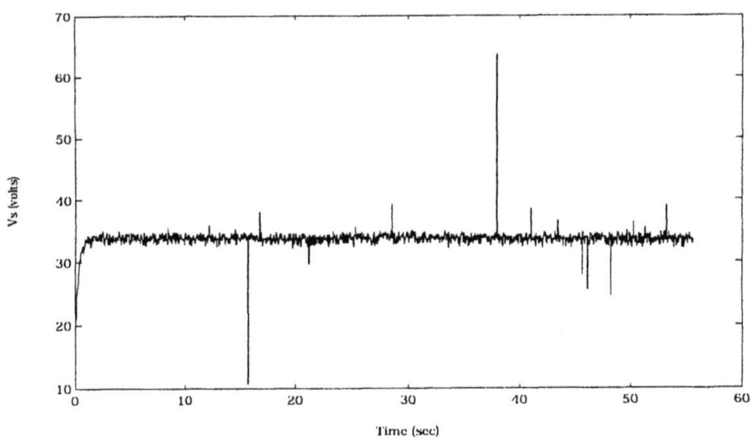

Figure 13.5: Stator Voltage

[5] S. Meshkat and A. Tessarolo, "Part II: Sensorless Brushless DC Motor Control Using DSPs and Kalman Filtering", *PCIM*, pp. 45-50, Aug., 1993.

[6] J. Bass, M. Ehsani, and T. Miller, "Simplified Electronics for Torque Control of Sensorless Switched-Reluctance Motor", *IEEE Trans. Industrial Electronics*, Vol. IE-34, No. 2, pp. 234-239, May, 1987.

[7] F. Harashima, H. Hashimoto, and S. Kondo, "Mosfet converter-fed position servo system with sliding mode control", *IEEE Trans. Industrial Electronics*, Vol. IE-32, No. 4, Aug. 1985.

[8] Y. Dote and R. Hoft, "Microprocessor based sliding mode controller for DC motor drives", Proc. of the *IEEE/IAS Conf.*, pp. 641-645, 1990.

[9] A. Brickwedde, "Microprocessor-Based Adaptive Speed and Position Regulation for Electrical Drives", *IEEE Trans. Industry Applications*, Vol. 1a-21, No. 5, pp. 1154-1161, 1985.

[10] P. Olivier, "Feedback Linearization of DC Motors", *IEEE Trans. Industrial Electronics*, Vol. 38, No. 6, pp. 498-501, Dec., 1991.

[11] J. Chiasson and M. Bodson, "Nonlinear Control of a Shunt DC Motor", *IEEE Trans. on Automatic Control*, Vol. 38, No. 11, pp. 1662-1666, Nov., 1993.

[12] T. Burg, D. Dawson, and P. Vedaghabar, "Sensorless Velocity Control for the Series Connected Wound Stator DC Motor", *Mechatronics - An International Journal*, Vol. 5, No. 4, pp. 349-364, 1995.

Chapter 14

Induction Motor (PSFB - II)

14.1 Introduction

In this chapter, we use the framework developed in Chapter 11 to develop a load position tracking controller for the electromechanical model of an induction motor driving a position-dependent load under the assumptions that only load position and stator current measurements are available and that an exact model for the electromechanical system dynamics can be determined. Specifically, by utilizing the structure of the induction electromechanical system, we design nonlinear observers to estimate the load velocity and the rotor flux. A Lyapunov-like argument is then used to motivate the construction of the observers; however as similarly discussed in Chapter 11, we can not explicitly show that the observation error goes to zero exponentially fast. Specifically, a general statement on the open-loop observation error is prevented by the fact that the observation error dynamics are designed to contain higher-order, nonlinear terms which are a function of the load position tracking error. Notwithstanding this difference from some of the previous material, we then use the structure of these observers, the observed-integrator backstepping procedure, nonlinear damping tools, and the commutation strategy given in Chapter 6 to design a load position tracking controller. A Lyapunov-like argument is then used to prove that all of the signals in the actual electromechanical systems and the observer remain bounded during closed-loop operation; furthermore, the controller ensures that the load position tracks the desired trajectory exponentially fast. Finally, experimental results are utilized to illustrate the performance of the proposed controller.

The main advantage of the proposed controller over the algorithm presented in Chapter 11 is that only measurements of stator current and load position are required. The disadvantage of the proposed approach is that the stability result is local as opposed to the semi-global result of Chapter 11. As illustrated during the error system development, the local stability result is manifested by the appearance of a voltage coefficient matrix pre-multiplying the input voltage vector. This voltage coefficient matrix, which is caused by the use of nonlinear damping terms in the observer dynamics, is singular during some operating conditions. Consequently, the region of nonsingularity for this matrix restricts the initial size of the norm of the composite error vector and hence causes the stability result to be local.

14.2 System Mathematical Model

The transformed system model of an induction motor driving a position-dependent load presented in Chapter 11 will be adopted in this chapter for the partial-state feedback tracking controller design. That is, under the assumption of equal mutual inductance and a linear magnetic circuit, the dynamics of an induction motor driving a robotic load [1] are given by

$$M\ddot{q} + B\dot{q} + N\sin(q) = \tau, \tag{14.1}$$

$$L_I\dot{I}_a - \alpha_1\psi_a - \alpha_2\psi_b\dot{q} + R_I I_a = V_a, \tag{14.2}$$

$$L_I\dot{I}_b - \alpha_1\psi_b + \alpha_2\psi_a\dot{q} + R_I I_b = V_b, \tag{14.3}$$

$$L_\psi\dot{\psi}_a + R_\psi\psi_a + \alpha_3\dot{q}\psi_b = \alpha_1 I_a, \tag{14.4}$$

and

$$L_\psi\dot{\psi}_b + R_\psi\psi_b - \alpha_3\dot{q}\psi_a = \alpha_1 I_b, \tag{14.5}$$

where $\tau(t)$ is a scalar function representing the electromechanical coupling torque given by

$$\tau = \alpha_2\left(\psi_a I_b - \psi_b I_a\right), \tag{14.6}$$

$q(t)$, $\dot{q}(t)$, and $\ddot{q}(t)$ represent the load position, velocity, and acceleration, respectively, $(\psi_a(t)\ \psi_b(t))$, $(I_a(t)\ I_b(t))$, and $(V_a(t)\ V_b(t))$, represent the transformed rotor flux, stator current, and stator voltage, respectively, M is the mechanical inertia of the system (including the rotor inertia), B is the coefficient of viscous friction of the motor, N is a positive constant related to the mass of the load and the coefficient of gravity, L_I, L_ψ, R_ψ, R_I, α_1, α_2, and α_3 are positive constants related to the electromechanical parameters. These parameters are explicitly defined by

$$\alpha_1 = R_r M_e/L_r^2, \quad \alpha_2 = n_p M_e/L_r, \quad \alpha_3 = n_p/L_r,$$

$$\begin{aligned} R_I &= \left(M_e^2 R_r + L_r^2 R_s\right)/L_r^2, \quad L_\psi = 1/L_r, \quad R_\psi = R_r/L_r^2, \\ L_I &= L_s - M_e^2/L_r, \end{aligned}$$

where R_s, R_r, n_p, L_s, L_r, and M_e are the stator resistance, rotor resistance, number of pole pairs, stator inductance, rotor inductance, and mutual inductance parameters, respectively.

Remark 14.1

Note that in order to facilitate the observer/controller development, the rotor flux dynamics has been divided throughout by L_r^2. This algebraic manipulation causes the appearance of α_1 multiplying the right-hand sides of (14.4) and (14.5), and hence facilitates the exact cancellation of certain terms during the stability analysis.

Remark 14.2

The detailed development for the above stator-fixed reference frame transformed model is presented in Appendix C. The mechanical load of (14.1) is a one-link robot [1] which is directly coupled to the induction motor. The main reason for this choice of mechanical load is that several researchers have proposed utilizing the induction motor for robotic applications [2]. For control purposes, only the load position $q(t)$, and the stator currents $I_a(t)$ and $I_b(t)$ are assumed to be measurable.

14.3 Control Objective

Given the dynamics of (14.1) through (14.6) and exact-model knowledge, the objective of this chapter is to design a voltage input which ensures exponential load position tracking while using only partial state feedback. That is, we only measure load position and stator current while observing the load velocity and rotor flux. With this objective in mind, we define the load position tracking error to be

$$e = q_d - q \tag{14.7}$$

where $q_d(t)$ represents the desired load position. For control purposes, we will assume the desired position trajectory and its first, second, and third time derivatives exist and remain bounded. Since we will observe load velocity, we define the load velocity observation error as

$$\dot{\tilde{q}} = \dot{q} - \dot{\hat{q}}\ , \tag{14.8}$$

and hence the corresponding load position observation error is defined as

$$\tilde{q} = q - \hat{q}, \tag{14.9}$$

where $q(t)$, $\dot{q}(t)$ denote the actual load position and velocity while $\hat{q}(t)$, $\dot{\hat{q}}\ (t)$ denote the observed load position and velocity. To facilitate the development of the electrical state observers, we also define the rotor flux and stator current observation error as follows

$$\tilde{I}_a = I_a - \hat{I}_a, \quad \tilde{I}_b = I_b - \hat{I}_b, \quad \tilde{\psi}_a = \psi_a - \hat{\psi}_a, \quad \text{and} \quad \tilde{\psi}_b = \psi_b - \hat{\psi}_b \tag{14.10}$$

with $\hat{I}_a(t)$, $\hat{I}_b(t)$, $\hat{\psi}_a(t)$, $\hat{\psi}_b(t)$ being the observed stator current and rotor flux, respectively.

Remark 14.3

Even though stator current is assumed to be measurable, the analysis is substantially simplified if we observe stator current as well as load velocity and rotor flux. Specifically, the stator current observer can be used in the design of the observed-backstepping controller; therefore, the control design procedure will only require the utilization of the observed dynamics as opposed to the actual dynamics. To expand on this point, we can see from the left-hand side of (14.2) and (14.3) that the actual stator current dynamics contain higher-order, nonlinear terms which are unmeasurable (*i.e.*, $\psi_b\dot{q}$ and $\psi_a\dot{q}$); however, the stator current observer dynamics can be constructed to depend only on measurable quantities (*e.g.*, $\dot{\hat{q}}\,\hat{\psi}_b$ and $\dot{\hat{q}}\,\hat{\psi}_a$). Hence, if we design the load position tracking controller by backstepping on the observed stator current as opposed to the actual stator current, we can avoid the need to deal with some of the unmeasurable, higher-order, nonlinear terms which could appear during the control synthesis procedure.

14.4 Nonlinear Observers

In this section of the chapter, we present the observers for load velocity, rotor flux, and stator current. We then combine the observed dynamics with the actual electromechanical dynamics to form the observation error systems. We then use a Lyapunov-like technique to illustrate the ability of the observers to estimate load velocity, rotor flux, and stator current via measurement of load position and stator current.

14.4.1 Observer Definitions

Based on the structure of the dynamics given by (14.1) through (14.5) and the subsequent stability analysis, the stator current and rotor flux observers are defined as follows

$$L_\psi\,\dot{\hat{\psi}}_a = \Omega_1, \tag{14.11}$$

$$L_\psi\,\dot{\hat{\psi}}_b = \Omega_2, \tag{14.12}$$

$$L_I\,\dot{\hat{I}}_a = \Omega_3 + V_a, \tag{14.13}$$

and

$$L_I\,\dot{\hat{I}}_b = \Omega_4 + V_b \tag{14.14}$$

where the auxiliary terms $\Omega_1(t)$, $\Omega_2(t)$, $\Omega_3(t)$, and $\Omega_4(t)$ are explicitly defined as follows

$$\Omega_1 = -R_\psi\hat{\psi}_a - \alpha_3\,\dot{\hat{q}}\,\hat{\psi}_b + \alpha_1\hat{I}_a + 2\alpha_1\left(I_a - \hat{I}_a\right), \tag{14.15}$$

$$\Omega_2 = -R_\psi \hat{\psi}_b + \alpha_3 \,\hat{\dot{q}}\, \hat{\psi}_a + \alpha_1 \hat{I}_b + 2\alpha_1 \left(I_b - \hat{I}_b \right), \tag{14.16}$$

$$\Omega_3 = \alpha_1 \hat{\psi}_a + \alpha_2 \hat{\psi}_b \,\hat{\dot{q}}\, - R_I \hat{I}_a + k_{n1} \alpha_2^2 \,\hat{\dot{q}}^2 \left(I_a - \hat{I}_a \right), \tag{14.17}$$

$$\Omega_4 = \alpha_1 \hat{\psi}_b - \alpha_2 \hat{\psi}_a \,\hat{\dot{q}}\, - R_I \hat{I}_b + k_{n1} \alpha_2^2 \,\hat{\dot{q}}^2 \left(I_b - \hat{I}_b \right), \tag{14.18}$$

and k_{n1} is a positive nonlinear damping gain.

Remark 14.4

While the appearance of most of the terms in the above observer dynamics are due to the structure of the corresponding actual electrical dynamics, we note that there are several additional terms which are used for the cancellation or compensation of other terms during the stability analysis. For example, the last term in (14.15) and (14.16) is utilized for cancellation of rotor flux, stator current, and load velocity observation error terms which originate from the torque interconnections of (14.6). The last terms in (14.17) and (14.18) are nonlinear damping terms which are utilized to compensate for bilinear terms in the stator current observation error dynamics.

Based on the structure of (14.1), the second-order load velocity observer is defined as

$$\hat{\dot{q}} = y + M^{-1} V_{oq} \left(q - \hat{q} \right) \tag{14.19}$$

where $V_{oq}(t)$ is a nonlinear damping term [3] which is explicitly defined below, $y(t)$ is an auxiliary variable which is updated on-line according to the following expression

$$\begin{aligned} \dot{y} \;=\; & M^{-1} \left[-B \,\hat{\dot{q}}\, - N \sin(q) - \alpha_2 \hat{\psi}_b \hat{I}_a + \alpha_2 \hat{\psi}_a \hat{I}_b \right] \\ & - M^{-1} \dot{V}_{oq} \left(q - \hat{q} \right) - M^{-1} V_{oq} \hat{r}, \end{aligned} \tag{14.20}$$

and $\hat{r}(t)$ is the observed filtered tracking error [4] which is explicitly defined below. The auxiliary term $V_{oq}(t)$ in (14.19) and (14.20) is used to "damp-out" unmeasurable nonlinear terms in the stability analysis and is explicitly defined as

$$V_{oq} = k_{n1} \alpha_3^2 \left(\hat{\psi}_a^2 + \hat{\psi}_b^2 \right) + k_{n1} \alpha_2^2 \left(\hat{I}_a^2 + \hat{I}_b^2 \right) + k_s + k_{n4} \tag{14.21}$$

where k_s is a positive constant control gain, and k_{n4} is a positive nonlinear damping gain. After taking the time derivative of (14.21) and substituting (14.11) through (14.14) into the resulting expression, we can express the term $\dot{V}_{oq}(t)$ in (14.20) as

$$\begin{aligned} \dot{V}_{oq} \;=\; & \frac{2 k_{n1} \alpha_3^2 \left(\hat{\psi}_a \Omega_1 + \hat{\psi}_b \Omega_2 \right)}{L_\psi} \\ & + \frac{2 k_{n1} \alpha_2^2 \left(\hat{I}_a \left(\Omega_3 + V_a \right) + \hat{I}_b (\Omega_4 + V_b) \right)}{L_I}. \end{aligned} \tag{14.22}$$

From (14.22), we can see that $\dot{V}_{oq}(t)$ only depends on measurable signals and the input voltage. The observed filtered tracking error $\hat{r}(t)$ in (14.20) is defined as follows

$$\hat{r} = \dot{q}_d + \alpha e - \dot{\hat{q}} \tag{14.23}$$

where α is a positive constant control gain.

Remark 14.5

After taking the time derivative of (14.19), multiplying the result by M, and substituting (14.20) for $\dot{y}(t)$, we can obtain the following second-order expression for the velocity observer

$$M \ddot{\hat{q}} = -B \dot{\hat{q}} - N\sin(q) - \alpha_2 \hat{\psi}_b \hat{I}_a + \alpha_2 \hat{\psi}_a \hat{I}_b - V_{oq}\hat{r} + V_{oq} \dot{\tilde{q}}. \tag{14.24}$$

For the subsequent analysis, we will use the load velocity observer in the form given by (14.24) to develop the load velocity observation error system. However, we should note that (14.24) can only be utilized for analysis purposes since the load velocity observer in the form of (14.24) contains the unmeasurable quantity $\dot{\tilde{q}}\,(t)$. Hence, the velocity observer expressions given by (14.19) and (14.20) are utilized for actual implementation.

14.4.2 Observation Error Systems

Given the above velocity, current, and flux observers, we now form the observation error systems to facilitate the subsequent stability analysis. First, we subtract (14.24) from (14.1) to obtain the load velocity observation error system in the form

$$\begin{aligned} M \ddot{\tilde{q}} \;&=\; -B \dot{\tilde{q}} - V_{oq} \dot{\tilde{q}} + V_{oq}\hat{r} - \alpha_2 \left(\tilde{\psi}_b \tilde{I}_a + \tilde{\psi}_b \hat{I}_a + \hat{\psi}_b \tilde{I}_a \right) \\ &\quad + \alpha_2 \left(\tilde{\psi}_a \tilde{I}_b + \tilde{\psi}_a \hat{I}_b + \hat{\psi}_a \tilde{I}_b \right) \end{aligned} \tag{14.25}$$

where (14.6) and the definitions in (14.10) have been utilized. Subtracting the electrical observers given by (14.11) through (14.14) from the corresponding electromechanical dynamics of (14.2) through (14.5) yields the stator current and rotor flux electrical observation error systems in the form

$$L_I \dot{\tilde{I}}_a = -R_I \tilde{I}_a + \alpha_1 \tilde{\psi}_a + \alpha_2 \left(\tilde{\psi}_b \dot{\tilde{q}} + \hat{\psi}_b \dot{\tilde{q}} + \tilde{\psi}_b \dot{\hat{q}} \right) - k_{n1} \alpha_2^2 \dot{\tilde{q}}^2 \tilde{I}_a, \tag{14.26}$$

$$L_I \dot{\tilde{I}}_b = -R_I \tilde{I}_b + \alpha_1 \tilde{\psi}_b - \alpha_2 \left(\tilde{\psi}_a \dot{\tilde{q}} + \hat{\psi}_a \dot{\tilde{q}} + \tilde{\psi}_a \dot{\hat{q}} \right) - k_{n1} \alpha_2^2 \dot{\tilde{q}}^2 \tilde{I}_b, \tag{14.27}$$

$$L_\psi \dot{\tilde{\psi}}_a = -R_\psi \tilde{\psi}_a - \alpha_3 \left(\tilde{\psi}_b \dot{\tilde{q}} + \hat{\psi}_b \dot{\tilde{q}} + \tilde{\psi}_b \dot{\hat{q}} \right) - \alpha_1 \tilde{I}_a, \tag{14.28}$$

and

$$L_\psi \dot{\tilde{\psi}}_b = -R_\psi \tilde{\psi}_b + \alpha_3 \left(\tilde{\psi}_a \dot{\tilde{q}} + \hat{\psi}_a \dot{\tilde{q}} + \tilde{\psi}_a \dot{\hat{q}} \right) - \alpha_1 \tilde{I}_b. \tag{14.29}$$

14.4.3 Analysis of the Observation Error Systems

To analyze the performance of the above observation error systems of (14.25) through (14.29), we define the following non-negative function

$$V_o = \frac{1}{2}M\,\dot{\tilde{q}}^2 + \frac{1}{2}L_I\tilde{I}_a^2 + \frac{1}{2}L_I\tilde{I}_b^2 + \frac{1}{2}L_\psi\tilde{\psi}_a^2 + \frac{1}{2}L_\psi\tilde{\psi}_b^2. \tag{14.30}$$

After taking the time derivative of (14.30) and substituting (14.25) through (14.29), we obtain

$$\begin{aligned}
\dot{V}_o = \; & \dot{\tilde{q}}\left(-B\,\dot{\tilde{q}} - V_{oq}\,\dot{\tilde{q}} + V_{oq}\hat{r}\right) \\
& + \dot{\tilde{q}}\left(-\alpha_2\left(\tilde{\psi}_b\tilde{I}_a + \tilde{\psi}_b\hat{I}_a + \hat{\psi}_b\tilde{I}_a\right)\right. \\
& \left. + \alpha_2\left(\tilde{\psi}_a\tilde{I}_b + \tilde{\psi}_a\hat{I}_b + \hat{\psi}_a\tilde{I}_b\right)\right) \\
& + \tilde{I}_a\left(-R_I\tilde{I}_a + \alpha_1\tilde{\psi}_a + \alpha_2\left(\tilde{\psi}_b\,\dot{\tilde{q}} + \hat{\psi}_b\,\dot{\tilde{q}} + \tilde{\psi}_b\,\dot{\hat{q}}\right)\right) \\
& + \tilde{I}_b\left(-R_I\tilde{I}_b + \alpha_1\tilde{\psi}_b - \alpha_2\left(\tilde{\psi}_a\,\dot{\tilde{q}} + \hat{\psi}_a\,\dot{\tilde{q}} + \tilde{\psi}_a\,\dot{\hat{q}}\right)\right) \\
& + \tilde{I}_a\left(-k_{n1}\alpha_2^2\,\dot{\hat{q}}^2\,\tilde{I}_a\right) + \tilde{I}_b\left(-k_{n1}\alpha_2^2\,\dot{\hat{q}}^2\,\tilde{I}_b\right) \\
& + \tilde{\psi}_a\left(-R_\psi\tilde{\psi}_a - \alpha_3\left(\tilde{\psi}_b\,\dot{\tilde{q}} + \hat{\psi}_b\,\dot{\tilde{q}} + \tilde{\psi}_b\,\dot{\hat{q}}\right) - \alpha_1\tilde{I}_a\right) \\
& + \tilde{\psi}_b\left(-R_\psi\tilde{\psi}_b + \alpha_3\left(\tilde{\psi}_a\,\dot{\tilde{q}} + \hat{\psi}_a\,\dot{\tilde{q}} + \tilde{\psi}_a\,\dot{\hat{q}}\right) - \alpha_1\tilde{I}_b\right).
\end{aligned} \tag{14.31}$$

After cancelling the common terms in (14.31), we have

$$\begin{aligned}
\dot{V}_o = \; & \dot{\tilde{q}}\left(-B\,\dot{\tilde{q}} - V_{oq}\,\dot{\tilde{q}} + V_{oq}\hat{r}\right) + \dot{\tilde{q}}\left(-\alpha_2\tilde{\psi}_b\hat{I}_a + \alpha_2\tilde{\psi}_a\hat{I}_b\right) \\
& + \tilde{I}_a\left(-R_I\tilde{I}_a + \alpha_2\tilde{\psi}_b\,\dot{\hat{q}}\right) + \tilde{I}_a\left(-k_{n1}\alpha_2^2\,\dot{\hat{q}}^2\,\tilde{I}_a\right) \\
& + \tilde{I}_b\left(-R_I\tilde{I}_b - \alpha_2\tilde{\psi}_a\,\dot{\hat{q}}\right) + \tilde{I}_b\left(-k_{n1}\alpha_2^2\,\dot{\hat{q}}^2\,\tilde{I}_b\right) \\
& + \tilde{\psi}_a\left(-R_\psi\tilde{\psi}_a - \alpha_3\hat{\psi}_b\,\dot{\tilde{q}}\right) + \tilde{\psi}_b\left(-R_\psi\tilde{\psi}_b + \alpha_3\hat{\psi}_a\,\dot{\tilde{q}}\right).
\end{aligned} \tag{14.32}$$

After substituting the expression for $V_{oq}(t)$ from (14.21) into the $V_{oq}\,\dot{\tilde{q}}$ term in (14.32) and then placing an upper bound on the resulting expression, we have

$$\begin{aligned}\dot{V}_o \leq\ & -B\,\dot{\tilde{q}}^2 - (k_s + k_{n4})\,\dot{\tilde{q}}^2 - R_I \tilde{I}_a^2 - R_I \tilde{I}_b^2 \\ & - R_\psi \tilde{\psi}_a^2 - R_\psi \tilde{\psi}_b^2 + V_{oq}\hat{r}\,\dot{\tilde{q}} \\ & + \dot{V}_1 + \dot{V}_2 + \dot{V}_3 + \dot{V}_4 + \dot{V}_5 + \dot{V}_6\end{aligned} \tag{14.33}$$

where

$$\dot{V}_1 = \alpha_2 |\,\dot{\tilde{q}}\,||\tilde{\psi}_b||\hat{I}_a| - k_{n1}\alpha_2^2\,\dot{\tilde{q}}^2\,\hat{I}_a^2, \tag{14.34}$$

$$\dot{V}_2 = \alpha_2 |\,\dot{\tilde{q}}\,||\tilde{\psi}_a||\hat{I}_b| - k_{n1}\alpha_2^2\,\dot{\tilde{q}}^2\,\hat{I}_b^2, \tag{14.35}$$

$$\dot{V}_3 = \alpha_3 |\,\dot{\tilde{q}}\,||\tilde{\psi}_a||\hat{\psi}_b| - k_{n1}\alpha_3^2\,\dot{\tilde{q}}^2\,\hat{\psi}_b^2, \tag{14.36}$$

$$\dot{V}_4 = \alpha_3 |\,\dot{\tilde{q}}\,||\tilde{\psi}_b||\hat{\psi}_a| - k_{n1}\alpha_3^2\,\dot{\tilde{q}}^2\,\hat{\psi}_a^2, \tag{14.37}$$

$$\dot{V}_5 = \alpha_2 |\tilde{I}_a||\tilde{\psi}_b||\,\dot{q}\,| - k_{n1}\alpha_2^2\,\dot{q}^2\,\tilde{I}_a^2, \tag{14.38}$$

and

$$\dot{V}_6 = \alpha_2 |\tilde{I}_b||\tilde{\psi}_a||\,\dot{q}\,| - k_{n1}\alpha_2^2\,\dot{q}^2\,\tilde{I}_b^2. \tag{14.39}$$

After applying Lemma 1.8 in Chapter 1, each of the nonlinear damping pairs defined in (14.34) through (14.39) can be upper bounded as follows

$$\dot{V}_1 \leq \frac{1}{k_{n1}}\tilde{\psi}_b^2, \quad \dot{V}_2 \leq \frac{1}{k_{n1}}\tilde{\psi}_a^2, \quad \dot{V}_3 \leq \frac{1}{k_{n1}}\tilde{\psi}_a^2,$$

$$\dot{V}_4 \leq \frac{1}{k_{n1}}\tilde{\psi}_b^2, \quad \dot{V}_5 \leq \frac{1}{k_{n1}}\tilde{\psi}_b^2, \quad \text{and } \dot{V}_6 \leq \frac{1}{k_{n1}}\tilde{\psi}_a^2;$$

therefore, $\dot{V}_o(t)$ in (14.33) can be upper bounded as

$$\begin{aligned}\dot{V}_o \leq\ & -(k_s + k_{n4})\,\dot{\tilde{q}}^2 - R_I \tilde{I}_a^2 - R_I \tilde{I}_b^2 - R_\psi \tilde{\psi}_a^2 - R_\psi \tilde{\psi}_b^2 \\ & + \frac{1}{k_{n1}}\tilde{\psi}_b^2 + \frac{1}{k_{n1}}\tilde{\psi}_a^2 + \frac{1}{k_{n1}}\tilde{\psi}_a^2 + \frac{1}{k_{n1}}\tilde{\psi}_b^2 \\ & + \frac{1}{k_{n1}}\tilde{\psi}_b^2 + \frac{1}{k_{n1}}\tilde{\psi}_a^2 + V_{oq}\hat{r}\,\dot{\tilde{q}}\end{aligned} \tag{14.40}$$

which can be then rearranged as follows

$$\begin{aligned}\dot{V}_o \leq\ & -(k_s + k_{n4})\,\dot{\tilde{q}}^2 - R_I \tilde{I}_a^2 - R_I \tilde{I}_b^2 - \left(R_\psi - \frac{3}{k_{n1}}\right)\tilde{\psi}_a^2 \\ & - \left(R_\psi - \frac{3}{k_{n1}}\right)\tilde{\psi}_b^2 + V_{oq}\hat{r}\,\dot{\tilde{q}}\,.\end{aligned} \tag{14.41}$$

Remark 14.6

A quick glance at the load velocity, stator current, and rotor flux observers given in this section of the chapter, provides little motivation for the form of some of the auxiliary nonlinear terms. However, it is the form of (14.41) which has motivated the use of these terms. Indeed, from (14.41), we can see that if the control gain k_{n1} is selected such that $k_{n1} \geq \frac{3}{R_\psi}$ and if the last term $V_{oq}\hat{r}\,\dot{\tilde{q}}$ is neglected then $\dot{V}_o(t)$ will be non-positive, and hence the observation error terms $\dot{\tilde{q}}\,(t)$, $\tilde{I}_a(t)$, $\tilde{I}_b(t)$, $\tilde{\psi}_a(t)$, and $\tilde{\psi}_b(t)$ would converge to zero exponentially fast (This assertion can be established by using Lemma 1.1 in Chapter 1). As discussed in Remark 14.4, the last term in (14.41) (*i.e.*, $V_{oq}\hat{r}\,\dot{\tilde{q}}$) is used to cancel a term which arises during the development of the position tracking controller error systems. That is, two like terms with different signs are cancelled during the analysis of the combined observer-controller, closed-loop error system.

14.5 Position Tracking Controller Development

Now that we have designed the requisite observers in the previous section, we now turn our attention to the design of a voltage level input which will drive the position tracking error to zero.

14.5.1 Position Tracking Error Systems

In order to obtain the position tracking error dynamics, we take the time derivative of (14.7) to yield

$$\dot{e} = \dot{q}_d - \dot{q}. \tag{14.42}$$

Since there is no control input in (14.42), we add and subtract the observed filtered tracking error term $\hat{r}(t)$, defined in (14.23), to the right-hand side of (14.42) to yield

$$\dot{e} = \dot{q}_d - \hat{r} + \hat{r} - \dot{q}. \tag{14.43}$$

After substituting the right-hand side of (14.23) for the first occurrence of $\hat{r}(t)$ only, we obtain the closed-loop load position error dynamics in the form

$$\dot{e} = -\alpha e + \hat{r} - \dot{\tilde{q}} \tag{14.44}$$

where (14.8) has been utilized. From (14.44), we can see that if $\hat{r}(t)$ and $\dot{\tilde{q}}\,(t)$ were both equal to zero, then the position tracking error would converge to zero exponentially fast. From our previous analysis of the observation error systems, we have good feeling about our ability to drive $\dot{\tilde{q}}\,(t)$ to zero. Hence, motivated by the above discussion and the form of (14.44), our controller must ensure that $\hat{r}(t)$ is driven to zero. To accomplish this new control objective, we need to construct the open-loop dynamics for $\hat{r}(t)$. To this end, we multiply both sides of (14.23) by M and then take the time derivative of the resulting expression to yield

$$M\dot{\hat{r}} = M\left(\ddot{q}_d + \alpha\dot{e}\right) - M\ddot{\hat{q}}. \tag{14.45}$$

After substituting the second-order load velocity observer dynamics of (14.24) into (14.45) and rearranging the resulting expression in an advantageous manner, we obtain

$$M\dot{\hat{r}} = w_\tau - V_{oq}\dot{\tilde{q}} - M\alpha\dot{\tilde{q}} - \alpha_2\left(\hat{\psi}_a\hat{I}_b - \hat{\psi}_b\hat{I}_a\right) \tag{14.46}$$

where $w_\tau(t)$ contains all of the measurable terms and is explicitly given by

$$w_\tau = M\ddot{q}_d + M\alpha\dot{q}_d - M\alpha\dot{\hat{q}} + B\dot{\hat{q}} + V_{oq}\hat{r} + N\sin(q) \tag{14.47}$$

where (14.8) was used. Since there is no control input on the right-hand side of (14.46), we add and subtract a desired torque trajectory $\tau_d(t)$ to yield

$$M\dot{\hat{r}} = w_\tau - V_{oq}\dot{\tilde{q}} - M\alpha\dot{\tilde{q}} - \alpha_2\tau_d + \alpha_2\hat{\eta}_\tau \tag{14.48}$$

where $\hat{\eta}_\tau(t)$ is used to represent the observed torque tracking error and is explicitly defined by

$$\hat{\eta}_\tau = \tau_d - \left(\hat{\psi}_a\hat{I}_b - \hat{\psi}_b\hat{I}_a\right). \tag{14.49}$$

The desired torque trajectory $\tau_d(t)$ in (14.48) will now be designed to force the observed filtered tracking error $\hat{r}(t)$ to zero. Specifically, we select $\tau_d(t)$ as follows

$$\tau_d = \frac{1}{\alpha_2}\left[w_\tau + k_s\hat{r} + k_{n5}M^2\alpha^2\hat{r} + k_{n6}\hat{r}\right] \tag{14.50}$$

where k_s is the same positive control gain introduced in (14.21) and k_{n5}, k_{n6} are positive nonlinear damping gains. Substituting $\tau_d(t)$ of (14.50) into (14.48) yields the closed-loop dynamics for $\hat{r}(t)$ as follows

$$M\dot{\hat{r}} = -k_s\hat{r} - V_{oq}\dot{\tilde{q}} - k_{n6}\hat{r} + \left[-M\alpha\dot{\tilde{q}} - k_{n5}M^2\alpha^2\hat{r}\right] + \alpha_2\hat{\eta}_\tau. \tag{14.51}$$

To ensure that $\hat{\eta}_\tau(t)$ goes to zero, we first construct the open-loop dynamics for $\hat{\eta}_\tau(t)$. To this end, we multiply both sides of (14.49) by $L_I L_\psi$ and take the time derivative of the resulting expression to obtain

$$\begin{aligned} L_I L_\psi\dot{\hat{\eta}}_\tau &= L_I L_\psi\dot{\tau}_d - \left(L_\psi\hat{\psi}_a L_I\dot{\hat{I}}_b - L_\psi\hat{\psi}_b L_I\dot{\hat{I}}_a\right) \\ &\quad - \left(L_\psi\dot{\hat{\psi}}_a L_I\hat{I}_b - L_\psi\dot{\hat{\psi}}_b L_I\hat{I}_a\right). \end{aligned} \tag{14.52}$$

After taking the time derivative of (14.50) and substituting the resulting expression into (14.52) for $\dot{r}_d(t)$, we have

$$
\begin{aligned}
L_I L_\psi \dot{\eta}_\tau &= L_I L_\psi \alpha_2^{-1}\left[\dot{w}_\tau + \left(k_s + k_{n5}M^2\alpha^2 + k_{n6}\right)\dot{\hat{r}}\right] \\
&\quad - \left(L_\psi \hat{\psi}_a L_I \dot{\hat{I}}_b - L_\psi \hat{\psi}_b L_I \dot{\hat{I}}_a\right) \\
&\quad - \left(L_\psi \dot{\hat{\psi}}_a L_I \hat{I}_b - L_\psi \dot{\hat{\psi}}_b L_I \hat{I}_a\right).
\end{aligned} \tag{14.53}
$$

Continuing with the construction of the open-loop dynamics, we substitute the time derivative of (14.47) for $\dot{w}_\tau(t)$ into (14.53) to obtain

$$
\begin{aligned}
L_I L_\psi \dot{\eta}_\tau &= L_I L_\psi \alpha_2^{-1}\left[M \dddot{q}_d + M\alpha\ddot{q}_d + (B - M\alpha)\ddot{\hat{q}}\right] \\
&\quad + L_I L_\psi \alpha_2^{-1}\left(\dot{V}_{oq}\hat{r} + N\cos(q)\dot{q}\right) \\
&\quad + L_I L_\psi \alpha_2^{-1}\left(k_s + k_{n5}M^2\alpha^2 + k_{n6} + V_{oq}\right)\dot{\hat{r}} \\
&\quad - \left(L_\psi \hat{\psi}_a L_I \dot{\hat{I}}_b - L_\psi \hat{\psi}_b L_I \dot{\hat{I}}_a\right) \\
&\quad - \left(L_\psi \dot{\hat{\psi}}_a L_I \hat{I}_b - L_\psi \dot{\hat{\psi}}_b L_I \hat{I}_a\right).
\end{aligned} \tag{14.54}
$$

Finally to complete the description, we substitute the velocity observer dynamics of (14.24) for $\ddot{\hat{q}}\,(t)$, the time derivative of $V_{oq}(t)$ from (14.22) for $\dot{V}_{oq}(t)$, the observed filtered tracking error dynamics of (14.46) for $\dot{\hat{r}}\,(t)$, the current and flux observer dynamics of (14.11) through (14.14) into (14.54), and group the measurable and unmeasurable terms in an advantageous manner to obtain the following expression

$$
\begin{aligned}
L_I L_\psi \dot{\eta}_\tau = &\ \Omega_5 + \Omega_6\dot{q} + \Omega_7\ddot{\tilde{q}} - \\
&\left(-L_\psi\hat{\psi}_b - 2k_{n1}\hat{r}\alpha_2\hat{I}_a L_\psi\right)V_a \\
&-\left(L_\psi\hat{\psi}_a - 2k_{n1}\hat{r}\alpha_2\hat{I}_b L_\psi\right)V_b
\end{aligned} \tag{14.55}
$$

where $\Omega_5(t)$, $\Omega_6(t)$, and $\Omega_7(t)$ are measurable functions defined by

$$\begin{aligned}\Omega_5 = \; & \alpha_2^{-1} L_I L_\psi \left(M \dddot{q}_d + M\alpha\ddot{q}_d + \left(BM^{-1} - \alpha\right)\left(-B\,\dot{q} - N\sin(q)\right)\right) \\ & + \alpha_2^{-1} L_I L_\psi \left(BM^{-1} - \alpha\right)\left(-\alpha_2 \hat{\psi}_b \hat{I}_a + \alpha_2 \hat{\psi}_a \hat{I}_b - V_{oq}\hat{r}\right) \\ & + \left(\alpha_2^{-1} L_I L_\psi \hat{r}\right)\left[2k_{n1} L_\psi^{-1} \alpha_3^2 \left(\hat{\psi}_a \Omega_1 + \hat{\psi}_b \Omega_2\right)\right. \\ & \left. + 2k_{n1} L_I^{-1} \alpha_2^2 \left(\hat{I}_a \Omega_3 + \hat{I}_b \Omega_4\right)\right] \\ & + L_I \hat{I}_a \Omega_2 - L_I \hat{I}_b \Omega_1 + L_\psi \hat{\psi}_b \Omega_3 - L_\psi \hat{\psi}_a \Omega_4 \\ & + \alpha_2^{-1} L_I L_\psi M^{-1} \left(V_{oq} + k_s + k_{n5} M^2 \alpha^2 + k_{n6}\right) \\ & \left(w_\tau + \alpha_2 \hat{\psi}_b \hat{I}_a - \alpha_2 \hat{\psi}_a \hat{I}_b\right),\end{aligned}$$

$$\Omega_6 = \alpha_2^{-1} L_I L_\psi N \cos(q),$$

and

$$\begin{aligned}\Omega_7 = \; & -M^{-1} \alpha_2^{-1} L_I L_\psi \left(V_{oq} + M\alpha\right)\left(V_{oq} + k_s + k_{n5} M^2 \alpha^2 + k_{n6}\right) \\ & + \alpha_2^{-1} L_I L_\psi \left(BM^{-1} - \alpha\right) V_{oq}.\end{aligned}$$

Now, based on the structure of (14.55) and the subsequent stability analysis, we propose the following voltage input relationship to drive the observed torque tracking error to zero

$$\begin{aligned} & \left(-L_\psi \hat{\psi}_b - 2k_{n1}\hat{r}\alpha_2 \hat{I}_a L_\psi\right) V_a + \left(L_\psi \hat{\psi}_a - 2k_{n1}\hat{r}\alpha_2 \hat{I}_b L_\psi\right) V_b \\ & = \Omega_5 + \Omega_6\,\dot{q} + k_{n7}\left(\Omega_6 + \Omega_7\right)^2 \hat{\eta}_\tau + k_s \hat{\eta}_\tau + k_{n8}\alpha_2^2 \hat{\eta}_\tau \end{aligned} \tag{14.56}$$

where k_{n7} and k_{n8} are positive nonlinear damping gains. Substituting the right-hand side of (14.56) into (14.55) yields the closed-loop dynamics for $\hat{\eta}_\tau(t)$ in the form

$$L_I L_\psi \,\dot{\hat{\eta}}_\tau = -k_s \hat{\eta}_\tau - k_{n8}\alpha_2^2 \hat{\eta}_\tau + \left[\left(\Omega_6 + \Omega_7\right)\tilde{\dot{q}} - k_{n7}\left(\Omega_6 + \Omega_7\right)^2 \hat{\eta}_\tau\right]. \tag{14.57}$$

Remark 14.7

It is important to note that the voltage input relationship given by the right-hand side of (14.56) constitutes only one of our control objectives, namely that we drive the position tracking error (*i.e.*, $e(t)$), the observed filtered tracking error (*i.e.*, $\hat{r}(t)$), and the observed torque tracking error (*i.e.*, $\hat{\eta}_\tau(t)$) all to zero. The other hidden control objective is that we ensure that all of the signals in the observer and the controller remain bounded. This second control objective, which will be addressed in the next section, will be used to form another voltage input relationship which together with (14.56) can then be used to solve for the input stator voltages $V_a(t)$ and $V_b(t)$. It is also easy to see from the form of (14.56) that singularities in the voltage control input could possibly arise under certain operating conditions; however, we will address this issue after we have completed the control design procedure.

14.5.2 Position Tracking Error System Analysis

In order to analyze the performance of the closed-loop position tracking error systems given by (14.44), (14.51), and (14.57), we define the following non-negative function

$$V_p = \frac{1}{2}e^2 + \frac{1}{2}M\hat{r}^2 + \frac{1}{2}L_I L_\psi \hat{\eta}_\tau^2. \tag{14.58}$$

The time derivative of (14.58) after substitution of the closed-loop position tracking error systems of (14.44), (14.51), and (14.57) is given by

$$\begin{aligned} \dot{V}_p = \quad & -\alpha e^2 + e\hat{r} - \dot{\tilde{q}}\, e - k_s\hat{r}^2 - V_{oq}\,\dot{\tilde{q}}\,\hat{r} - M\alpha\,\dot{\tilde{q}}\,\hat{r} \\ & -k_{n5}M^2\alpha^2\hat{r}^2 - k_{n6}\hat{r}^2 + \alpha_2\hat{\eta}_\tau\hat{r} - k_s\hat{\eta}_\tau^2 + (\Omega_6 + \Omega_7)\,\dot{\tilde{q}}\,\hat{\eta}_\tau \\ & -k_{n7}(\Omega_6 + \Omega_7)^2\hat{\eta}_\tau^2 - k_{n8}\alpha_2^2\hat{\eta}_\tau^2. \end{aligned} \tag{14.59}$$

From (14.59), $\dot{V}_p(t)$ can be upper bounded as follows

$$\begin{aligned} \dot{V}_p \leq \quad & -\alpha e^2 - k_s\hat{r}^2 - k_s\hat{\eta}_\tau^2 + \left[M\alpha\left|\dot{\tilde{q}}\right|\left|\hat{r}\right| - k_{n5}M^2\alpha^2\hat{r}^2\right] \\ & + \left[\left|e\right|\left|\hat{r}\right| - k_{n6}\hat{r}^2\right] + \left[\alpha_2\left|\hat{\eta}_\tau\right|\left|\hat{r}\right| - k_{n8}\alpha_2^2\hat{\eta}_\tau^2\right] \\ & + \left[(\Omega_6 + \Omega_7)\left|\dot{\tilde{q}}\right|\left|\hat{\eta}_\tau\right| - k_{n7}(\Omega_6 + \Omega_7)^2\hat{\eta}_\tau^2\right] \\ & + \left|\dot{\tilde{q}}\right|\left|e\right| - V_{oq}\,\dot{\tilde{q}}\,\hat{r}. \end{aligned} \tag{14.60}$$

We now apply Lemma 1.8 in Chapter 1 to the four bracketed nonlinear damping pairs on the right-hand side of (14.60) to form the following upper bound for $\dot{V}_p(t)$

$$\begin{aligned}\dot{V}_p \leq & -\left(\alpha - \frac{1}{k_{n6}}\right) e^2 - k_s \hat{\eta}_\tau^2 - \left(k_s - \frac{1}{k_{n8}}\right) \hat{r}^2 \\ & + \left[\left|\dot{\tilde{q}}\right| |e|\right] + \left[\left(\frac{1}{k_{n5}} + \frac{1}{k_{n7}}\right) \dot{\tilde{q}}^2\right] - \left[V_{oq}\, \dot{\tilde{q}}\, \hat{r}\right]\end{aligned} \tag{14.61}$$

Remark 14.8

From the form of (14.61), we can see that if $\dot{\tilde{q}}\,(t)$ was equal to zero for all time, and the controller gains were selected to satisfy the inequalities $\alpha > \dfrac{1}{k_{n6}}$ and $k_s > \dfrac{1}{k_{n8}}$, then we could guarantee that $e(t)$, $\hat{\eta}_\tau(t)$, and $\hat{r}(t)$ would all go to zero. As we will see later during the analysis of the composite observer-controller error systems, the last three bracketed terms in (14.61) can be cancelled or combined with other terms in (14.41) to obtain the desired stability result. For example, it is easy to see that the last bracketed term in (14.61) is the same as the last term in (14.41) but with an opposite sign. Therefore, these two terms will be cancelled when the time derivatives of the two Lyapunov-like functions (*i.e.*, $V_o(t)$ of (14.30) and $V_p(t)$ of (14.58)) are added together.

14.6 Flux Controller Development

In addition to position tracking, the other main control objective is to ensure that all of the signals in the observer and the controller remain bounded during closed-loop operation. One method for accomplishing this additional control objective is to force the pseudo-magnitude (*i.e.*, $\hat{\psi}_a^2 + \hat{\psi}_b^2$) of the observed rotor flux to track a desired positive function. That is, we will define the observed flux tracking error, denoted by $\hat{\eta}_\psi(t)$, by the following expression

$$\hat{\eta}_\psi = \psi_d - \frac{1}{2}\left(\hat{\psi}_a^2 + \hat{\psi}_b^2\right) \tag{14.62}$$

where $\psi_d(t)$ is a non-negative scalar function of time which is selected to be second-order differentiable with respect to time.

Remark 14.9

We should point out that a similar type of observed flux tracking error control objective is also formulated during the analysis of more traditional control approaches [5] for the induction motor. The reader is referred to [6] for a discussion on the relationship between the field-oriented control technique and a nonlinear control approach.

14.6.1 Observed Flux Tracking Error Systems

To design a voltage input that will force the observed flux tracking error of (14.62) to zero, we must construct the corresponding open-loop dynamics. To obtain the observed flux tracking error dynamics, we multiply both sides of (14.62) by L_ψ and take the time derivative of the resulting expression to yield

$$L_\psi \dot{\hat{\eta}}_\psi = L_\psi \dot{\psi}_d - \hat{\psi}_a L_\psi \dot{\hat{\psi}}_a - \hat{\psi}_b L_\psi \dot{\hat{\psi}}_b . \tag{14.63}$$

After substituting the flux observer dynamics of (14.11) and (14.12) into (14.63), we have

$$L_\psi \dot{\hat{\eta}}_\psi = L_\psi \dot{\psi}_d - \hat{\psi}_a \Omega_1 - \hat{\psi}_b \Omega_2 . \tag{14.64}$$

After substituting (14.15) and (14.16) for $\Omega_1(t)$ and $\Omega_2(t)$, respectively, into (14.64), we have

$$L_\psi \dot{\hat{\eta}}_\psi = w_\psi - \alpha_1 \left(\hat{\psi}_a \hat{I}_a + \hat{\psi}_b \hat{I}_b \right) - 2\alpha_1 \hat{\psi}_a \tilde{I}_a - 2\alpha_1 \hat{\psi}_b \tilde{I}_b \tag{14.65}$$

where the measurable, auxiliary variable $w_\psi(t)$ is given by

$$w_\psi = L_\psi \dot{\psi}_d + R_\psi \left(\hat{\psi}_a^2 + \hat{\psi}_b^2 \right) . \tag{14.66}$$

Since there is no control input in the above observed flux tracking error dynamics, we add and subtract a fictitious [3] control input, denoted by u_ψ, to the right-hand side of (14.65) to yield

$$L_\psi \dot{\hat{\eta}}_\psi = w_\psi - \alpha_1 u_\psi + \alpha_1 \hat{\eta}_I - 2\alpha_1 \hat{\psi}_a \tilde{I}_a - 2\alpha_1 \hat{\psi}_b \tilde{I}_b \tag{14.67}$$

where $\hat{\eta}_I(t)$ is an auxiliary tracking error variable used in the backstepping procedure and is explicitly given by

$$\hat{\eta}_I = u_\psi - \left(\hat{\psi}_a \hat{I}_a + \hat{\psi}_b \hat{I}_b \right) . \tag{14.68}$$

As we will show later, the dynamics of $\hat{\eta}_I(t)$ can be used to design a voltage input relationship which will drive $\hat{\eta}_I(t)$ to zero and hence drive $\hat{\eta}_\psi(t)$ to zero.

Motivated by the form of the dynamics given in (14.67) and the subsequent stability analysis, we define $u_\psi(t)$ as follows

$$u_\psi = \frac{1}{\alpha_1} \left(w_\psi + k_s \hat{\eta}_\psi + k_{n2} \alpha_1^2 \hat{\psi}_a^2 \hat{\eta}_\psi + k_{n2} \alpha_1^2 \hat{\psi}_b^2 \hat{\eta}_\psi \right) \tag{14.69}$$

where k_{n2} is a positive nonlinear damping gain. After substituting $u_\psi(t)$ of (14.69) into (14.67), we can obtain the closed-loop dynamics for $\hat{\eta}_\psi(t)$ in following form

$$\begin{aligned} L_\psi \dot{\hat{\eta}}_\psi &= -k_s\hat{\eta}_\psi + \alpha_1\hat{\eta}_I + \left[-2\alpha_1\hat{\psi}_a\tilde{I}_a - k_{n2}\alpha_1^2\hat{\psi}_a^2\hat{\eta}_\psi\right] \\ &+ \left[-2\alpha_1\hat{\psi}_b\tilde{I}_b - k_{n2}\alpha_1^2\hat{\psi}_b^2\hat{\eta}_\psi\right]. \end{aligned} \tag{14.70}$$

We note that the last two bracketed terms are nonlinear damping pairs; therefore, since the previous observation error stability analysis indicates that $\tilde{I}_a(t)$ and $\tilde{I}_b(t)$ can both be driven to zero, we are motivated by the form of (14.70) to drive $\hat{\eta}_I(t)$ to zero. That is, we can see that if $\hat{\eta}_I(t)$ is equal to zero then $\hat{\eta}_\psi(t)$ will go to zero.

To obtain the dynamics for $\hat{\eta}_I(t)$, we multiply both sides of (14.68) by $L_\psi L_I$ and take the time derivative of the resulting expression to yield

$$\begin{aligned} L_\psi L_I \dot{\hat{\eta}}_I &= L_\psi L_I\dot{u}_\psi - L_\psi\hat{\psi}_a L_I \dot{\hat{I}}_a - L_\psi\hat{\psi}_b L_I \dot{\hat{I}}_b \\ &- L_\psi \dot{\hat{\psi}}_a L_I\hat{I}_a - L_\psi \dot{\hat{\psi}}_b L_I\hat{I}_b. \end{aligned} \tag{14.71}$$

After substituting the observer dynamics of (14.11) through (14.14) and the time derivative of (14.69) for $\dot{u}_\psi(t)$ into (14.71), we obtain

$$L_\psi L_I \dot{\hat{\eta}}_I = w_I - \left(L_\psi\hat{\psi}_a V_a + L_\psi\hat{\psi}_b V_b\right) \tag{14.72}$$

where the auxiliary, measurable variable $w_I(t)$ is given by

$$\begin{aligned} w_I = \; & 2\Omega_1 L_I R_\psi\hat{\psi}_a\alpha_1^{-1} - \Omega_2 L_I\hat{I}_b + \alpha_1^{-1}L_\psi^2 L_I\ddot{\psi}_d \\ & +2\Omega_1 L_I k_{n2}\alpha_1\hat{\psi}_a\hat{\eta}_\psi - \Omega_1 L_I\hat{I}_a + 2\Omega_2 L_I R_\psi\hat{\psi}_b\alpha_1^{-1} \\ & -\hat{\psi}_b\Omega_2 L_I\alpha_1^{-1}\left(k_s + k_{n2}\alpha_1^2\hat{\psi}_a^2 + k_{n2}\alpha_1^2\hat{\psi}_b^2\right) \\ & +\dot{\psi}_d L_I L_\psi\alpha_1^{-1}\left(k_s + k_{n2}\alpha_1^2\hat{\psi}_a^2 + k_{n2}\alpha_1^2\hat{\psi}_b^2\right) \\ & -L_\psi\hat{\psi}_a\Omega_3 - L_\psi\hat{\psi}_b\Omega_4 + 2\Omega_2 L_I k_{n2}\alpha_1\hat{\psi}_b\hat{\eta}_\psi \\ & -\hat{\psi}_a\Omega_1 L_I\alpha_1^{-1}\left(k_s + k_{n2}\alpha_1^2\hat{\psi}_a^2 + k_{n2}\alpha_1^2\hat{\psi}_b^2\right) \end{aligned} \tag{14.73}$$

Note that (14.64) is used for the substitution of $\dot{\hat{\eta}}_\psi(t)$ to obtain (14.73). Motivated by the form of (14.72), we propose the following voltage input relationship to drive $\hat{\eta}_I(t)$ to zero

$$L_\psi\hat{\psi}_a V_a + L_\psi\hat{\psi}_b V_b = w_I + k_s\hat{\eta}_I + k_{n3}\alpha_1^2\hat{\eta}_I \tag{14.74}$$

where k_s is the same positive control gain introduced in (14.21) and k_{n3} is a positive nonlinear damping gain. The last control term in (14.74) is a nonlinear damping term which is used to damp-out the interconnection term $\alpha_1\hat{\eta}_I$ in (14.70). After substituting (14.74) into (14.72), we obtain the closed-loop dynamics for $\hat{\eta}_I(t)$ in the following form

$$L_\psi L_I \dot{\hat{\eta}}_I = -k_s\hat{\eta}_I - k_{n3}\alpha_1^2\hat{\eta}_I. \tag{14.75}$$

From the form of (14.75), we can indeed see that $\hat{\eta}_I(t)$ will be driven to zero.

14.6.2 Observed Flux Tracking Error Systems Analysis

Given the form of the closed-loop, observed flux tracking error dynamics for $\hat{\eta}_\psi(t)$ in (14.70) and the closed-loop auxiliary tracking error dynamics for $\hat{\eta}_I(t)$ in (14.75), we propose the following non-negative function for analysis of the closed-loop, observed flux tracking error systems

$$V_\psi = \frac{1}{2}L_\psi L_I\hat{\eta}_I^2 + \frac{1}{2}L_\psi\hat{\eta}_\psi^2. \tag{14.76}$$

After taking the time derivative of (14.76) and substituting (14.70) and (14.75), we have

$$\begin{aligned}\dot{V}_\psi = \;& -k_s\hat{\eta}_I^2 - k_{n3}\alpha_1^2\hat{\eta}_I^2 - k_s\hat{\eta}_\psi^2 + \alpha_1\hat{\eta}_\psi\hat{\eta}_I \\ & -2\alpha_1\hat{\psi}_a\tilde{I}_a\hat{\eta}_\psi - k_{n2}\alpha_1^2\hat{\psi}_a^2\hat{\eta}_\psi^2 - \\ & 2\alpha_1\hat{\psi}_b\tilde{I}_b\hat{\eta}_\psi - k_{n2}\alpha_1^2\hat{\psi}_b^2\hat{\eta}_\psi^2\end{aligned} \tag{14.77}$$

which can be upper bounded as follows

$$\begin{aligned}\dot{V}_\psi \;\le\; & -k_s\hat{\eta}_I^2 - k_s\hat{\eta}_\psi^2 + \left[\alpha_1|\hat{\eta}_\psi||\hat{\eta}_I| - k_{n3}\alpha_1^2\hat{\eta}_I^2\right] \\ & + \left[2\alpha_1|\hat{\psi}_a||\tilde{I}_a||\hat{\eta}_\psi| - k_{n2}\alpha_1^2\hat{\psi}_a^2\hat{\eta}_\psi^2\right] \\ & + \left[2\alpha_1|\hat{\psi}_b||\tilde{I}_b||\hat{\eta}_\psi| - k_{n2}\alpha_1^2\hat{\psi}_b^2\hat{\eta}_\psi^2\right].\end{aligned} \tag{14.78}$$

After applying Lemma 1.8 in Chapter 1 to the bracketed nonlinear damping pairs in (14.78), $\dot{V}_\psi(t)$ can be upper bounded as follows

$$\dot{V}_\psi \le -k_s\hat{\eta}_I^2 - \left(k_s - \frac{1}{k_{n3}}\right)\hat{\eta}_\psi^2 + \frac{4\tilde{I}_a^2}{k_{n2}} + \frac{4\tilde{I}_b^2}{k_{n2}}. \tag{14.79}$$

Remark 14.10

Now that we have completed our control design procedure, we can utilize (14.56) and (14.74) to solve for the transformed stator voltage inputs $V_a(t)$ and $V_b(t)$ via the following matrix equation

$$\begin{bmatrix} V_a \\ V_b \end{bmatrix} = C^{-1} \begin{bmatrix} w_I + k_s\hat{\eta}_I + k_{n3}\alpha_1^2\hat{\eta}_I \\ \Omega_5 + \Omega_6\ \dot{\hat{q}} + k_{n7}\left(\Omega_6 + \Omega_7\right)^2 \hat{\eta}_\tau + k_s\hat{\eta}_\tau + k_{n8}\alpha_2^2\hat{\eta}_\tau \end{bmatrix} \tag{14.80}$$

where the 2×2 matrix $C(t)$ is defined by

$$C = \begin{bmatrix} L_\psi\hat{\psi}_a & L_\psi\hat{\psi}_b \\ -L_\psi\hat{\psi}_b - 2k_{n1}\hat{r}\alpha_2\hat{I}_a L_\psi & L_\psi\hat{\psi}_a - 2k_{n1}\hat{r}\alpha_2\hat{I}_b L_\psi \end{bmatrix} \in \Re^{2\times 2}. \tag{14.81}$$

Later, after we complete the analysis of the observer-controller error systems, we will discuss the circumstances which allow the matrix $C(t)$ to be inverted.

14.7 Tracking Performance Analysis

In this section, we will analyze the performance of the composite observer-controller error system. Specifically, the performance of these closed-loop error system is illustrated by the following theorem.

Theorem 14.1

Given the nonlinear observers of (14.11) through (14.20) and the voltage input given in (14.80), we can obtain local exponential load position tracking in the form

$$\|x(t)\| \leq \sqrt{\frac{\lambda_2}{\lambda_1}}\, \|x(0)\|\, e^{-\gamma t} \tag{14.82}$$

where

$$x = \begin{bmatrix} \dot{\tilde{q}} & \tilde{I}_a & \tilde{I}_b & \tilde{\psi}_a & \tilde{\psi}_b & \hat{\eta}_\psi & \hat{\eta}_I & e & \hat{r} & \hat{\eta}_\tau \end{bmatrix}^T \in \Re^{10}, \tag{14.83}$$

$$\begin{aligned} \lambda_1 &= \tfrac{1}{2}\min\left\{M, L_I, L_\psi, L_I L_\psi, 1\right\}, \\ \lambda_2 &= \tfrac{1}{2}\max\left\{M, L_I, L_\psi, L_I L_\psi, 1\right\}, \end{aligned} \tag{14.84}$$

$$\gamma = \frac{\lambda_3}{2\lambda_2}, \tag{14.85}$$

and

$$\begin{aligned}\lambda_3 = \quad & \min\left\{k_s, \alpha - \frac{1}{k_{n4}} - \frac{1}{k_{n6}}, k_s - \frac{1}{k_{n8}}, k_s - \frac{1}{k_{n3}},\right. \\ & \left. k_s - \frac{1}{k_{n5}} - \frac{1}{k_{n7}}, R_\psi - \frac{3}{k_{n1}}, R_I - \frac{4}{k_{n2}}\right\}.\end{aligned} \tag{14.86}$$

To ensure the validity of the result given by (14.82), the controller gains must be adjusted to satisfy the following sufficient condition

$$k_{ni} > \max\left\{\frac{4}{R_I}, \frac{2}{\alpha}, \frac{2}{k_s}, \frac{3}{R_\psi}\right\} \quad \text{for } i = 1, \cdots, 8. \tag{14.87}$$

Proof:

To analyze the performance of the composite observer-controller error system, we now combine the non-negative functions $V_o(t)$, $V_p(t)$, and $V_\psi(t)$, defined in (14.30), (14.58), and (14.76) respectively, to formulate the following non-negative function

$$V = V_o + V_\psi + V_p. \tag{14.88}$$

According Lemma 1.7 in Chapter 1, we can state that

$$\lambda_1 \|x\|^2 \leq V \leq \lambda_2 \|x\|^2 \tag{14.89}$$

where λ_1, λ_2 were defined in (14.84), and $x(t)$ was defined in (14.83). After taking the time derivative of (14.88), we can use (14.41), (14.61), and (14.79) to obtain the following upper bound for $\dot{V}(t)$

$$\begin{aligned}\dot{V} \leq \; & -k_s \dot{\tilde{q}}^2 - R_I \tilde{I}_a^2 - R_I \tilde{I}_b^2 - \left(R_\psi - \frac{3}{k_{n1}}\right)\tilde{\psi}_a^2 \\ & - \left(R_\psi - \frac{3}{k_{n1}}\right)\tilde{\psi}_b^2 + V_{oq}\hat{r}\,\dot{\tilde{q}} - \left(\alpha - \frac{1}{k_{n6}}\right)e^2 \\ & -k_s\hat{\eta}_\tau^2 - \left(k_s - \frac{1}{k_{n8}}\right)\hat{r}^2 + \left(\frac{1}{k_{n5}} + \frac{1}{k_{n7}}\right)\dot{\tilde{q}}^2 \\ & -V_{oq}\hat{r}\cdot\dot{\tilde{q}} - k_s\hat{\eta}_I^2 - \left(k_s - \frac{1}{k_{n3}}\right)\hat{\eta}_\psi^2 \\ & + \frac{4\tilde{I}_a^2}{k_{n2}} + \frac{4\tilde{I}_b^2}{k_{n2}} + \left[\left|\dot{\tilde{q}}\right||e| - k_{n4}\dot{\tilde{q}}^2\right]\end{aligned} \tag{14.90}$$

After cancelling the appropriate common terms, combining the common terms, and applying the nonlinear damping principle to the last bracketed term in (14.90), $\dot{V}(t)$ can be upper bounded as follows

$$\begin{aligned}\dot{V} \leq\; & -\left(\alpha-\frac{1}{k_{n4}}-\frac{1}{k_{n6}}\right)e^2-\left(k_s-\frac{1}{k_{n8}}\right)\hat{r}^2 \\ & -\left(k_s-\frac{1}{k_{n3}}\right)\hat{\eta}_\psi^2-k_s\hat{\eta}_I^2-k_s\hat{\eta}_\tau^2-\left(R_I-\frac{4}{k_{n2}}\right)\tilde{I}_b^2. \\ & -\left(k_s-\frac{1}{k_{n5}}-\frac{1}{k_{n7}}\right)\dot{\tilde{q}}^2-\left(R_\psi-\frac{3}{k_{n1}}\right)\tilde{\psi}_a^2 \\ & -\left(R_\psi-\frac{3}{k_{n1}}\right)\tilde{\psi}_b^2-\left(R_I-\frac{4}{k_{n2}}\right)\tilde{I}_a^2\end{aligned} \tag{14.91}$$

To simplify the notation, we can upper bound $\dot{V}(t)$ of (14.91) in the concise form

$$\dot{V} \leq -\lambda_3 \left\|x\right\|^2 \tag{14.92}$$

where λ_3 and $x(t)$ were defined in (14.86) and (14.83), respectively. Therefore, if we select the controller gains according to (14.87), we can state that λ_3 in (14.86) is positive and hence $\dot{V}(t)$ is non-positive. From (14.89) and (14.92), we can now use Lemma 1.1 in Chapter 1 to state that the result of (14.82).

Remark 14.11

From the structure of (14.49) and (14.68), we can see that $\hat{\eta}_\tau(t)$ and $\hat{\eta}_I(t)$ are continuously differentiable with respect to $\hat{I}_a(t)$ and $\hat{I}_b(t)$. Specifically, from (14.68), we can see that the partial derivatives of $\hat{\eta}_I(t)$ with respect to $\hat{I}_a(t)$ and $\hat{I}_b(t)$ are the quantities $-\hat{\psi}_a(t)$ and $-\hat{\psi}_b(t)$, respectively. In addition, from (14.49), we can see that the partial derivatives of $\hat{\eta}_\tau(t)$ with respect to $\hat{I}_a(t)$ and $\hat{I}_b(t)$ are the quantities $\left(2k_{n1}\alpha_2\hat{r}\hat{I}_a+\hat{\psi}_b\right)$ and $\left(2k_{n1}\alpha_2\hat{r}\hat{I}_b-\hat{\psi}_a\right)$, respectively. Therefore, the Jacobian determinant (pp. 204, [7]) can be expressed as

$$\begin{aligned}\det(D) &= \frac{\partial\left(\hat{\eta}_I,\,\hat{\eta}_\tau\right)}{\partial\left(\hat{I}_a,\,\hat{I}_b\right)}=\left\|\begin{bmatrix}\dfrac{\partial\hat{\eta}_I}{\partial\hat{I}_a} & \dfrac{\partial\hat{\eta}_I}{\partial\hat{I}_b} \\ \dfrac{\partial\hat{\eta}_\tau}{\partial\hat{I}_a} & \dfrac{\partial\hat{\eta}_\tau}{\partial\hat{I}_b}\end{bmatrix}\right\| \\ &= \left\|\begin{bmatrix}-\hat{\psi}_a & -\hat{\psi}_b \\ 2k_{n1}\alpha_2\hat{r}\hat{I}_a+\hat{\psi}_b & 2k_{n1}\alpha_2\hat{r}\hat{I}_b-\hat{\psi}_a\end{bmatrix}\right\| \\ &= \hat{\psi}_a^2+\hat{\psi}_b^2-2k_{n1}\alpha_2\hat{r}\left(\hat{\psi}_a\hat{I}_b-\hat{\psi}_b\hat{I}_a\right).\end{aligned} \tag{14.93}$$

Given (14.62) and (14.49), (14.93) can be rewritten as

$$\det(D) = 2(\psi_d - \hat{\eta}_\psi) - 2k_{n1}\alpha_2\hat{r}\tau_d + 2k_{n1}\alpha_2\hat{r}\hat{\eta}_\tau. \tag{14.94}$$

Since $\psi_d(t)$ is a positive scalar function which can be made arbitrarily large, we can see from (14.94) that in a neighborhood F around $x(t) = 0$ (where $x(t)$ is defined in (14.83)), that $\det(D) > 0$. Therefore, the Inverse Function Theorem (pp. 206, [7]) can be invoked to show that in the neighborhood F, $\hat{I}_a(t)$ and $\hat{I}_b(t)$ can be solved in terms of the variables $\hat{\eta}_\tau(t)$, $\hat{\eta}_I(t)$, $\hat{\psi}_a(t)$, $\hat{\psi}_b(t)$, $\hat{r}(t)$, $\hat{\eta}_\psi(t)$, $q(t)$, $\dot{\hat{q}}(t)$, $\dot{q}_d(t)$, $\ddot{q}_d(t)$, and $\dot{\psi}_d(t)$ (*i.e.*, by using (14.68), (14.49), (14.69), (14.50), (14.66), (14.47), and (14.21)).

Remark 14.12

Since $\dot{V}(t)$ is non-positive as delineated by (14.92), we can state that $V(t)$ in (14.88) is upper bounded by $V(0)$ and lower bounded by zero. Given $V(t)$ is upper bounded and the fact that M, L_I, and L_ψ are all positive definite, we can state from (14.88) that $e(t) \in L_\infty$, $\tilde{\eta}_\psi(t) \in L_\infty$, $\tilde{\eta}_I(t) \in L_\infty$, $\tilde{\eta}_\tau(t) \in L_\infty$, $\tilde{r}(t) \in L_\infty$, $\tilde{\psi}_a(t) \in L_\infty$, $\tilde{\psi}_b(t) \in L_\infty$, $\tilde{I}_a(t) \in L_\infty$, $\tilde{I}_b(t) \in L_\infty$, and $\dot{\tilde{q}}(t) \in L_\infty$ and hence from (14.7), (14.62), (14.23), and (14.8), we can state that the quantities $q(t) \in L_\infty$, $\dot{q}(t) \in L_\infty$, $\hat{\psi}_a(t) \in L_\infty$, $\hat{\psi}_b(t) \in L_\infty$, and $\dot{\hat{q}}(t) \in L_\infty$. Since $\tilde{\psi}_a(t) \in L_\infty$, $\tilde{\psi}_b(t) \in L_\infty$, $\hat{\psi}_a(t) \in L_\infty$, $\hat{\psi}_b(t) \in L_\infty$, we can use (14.10) to state that $\psi_a(t) \in L_\infty$ and $\psi_b(t) \in L_\infty$. From Remark 14.12, we have shown that $\hat{I}_a(t)$ and $\hat{I}_b(t)$ can be solved in terms of the bounded quantities $\hat{\eta}_\tau(t)$, $\hat{\eta}_I(t)$, $\hat{\psi}_a(t)$, $\hat{\psi}_b(t)$, $\hat{r}(t)$, $\hat{\eta}_\psi(t)$, $q(t)$, $\dot{\hat{q}}(t)$, $\dot{q}_d(t)$, $\ddot{q}_d(t)$, and $\dot{\psi}_d(t)$ in the neighborhood F around $x(t) = 0$ and hence $\hat{I}_a(t) \in L_\infty$ and $\hat{I}_b(t) \in L_\infty$. Given that $\tilde{I}_a(t) \in L_\infty$, $\tilde{I}_b(t) \in L_\infty$, $\hat{I}_a(t) \in L_\infty$ and $\hat{I}_b(t) \in L_\infty$, we can use (14.10) to state that the actual currents $I_a(t) \in L_\infty$ and $I_b(t) \in L_\infty$ are bounded. Therefore, from (14.21), (14.69), and (14.50), we can use the above information to state that $V_{oq}(t) \in L_\infty$, $u_\psi(t) \in L_\infty$, and $\tau_d(t) \in L_\infty$ which can then be used to show that the right-hand side of (14.74) and (14.56) are both bounded. Provided the matrix $C(t)$ of (14.81) is invertible (See Remark 14.13 below), we can now use (14.80) to state that the input stator voltages $V_a(t) \in L_\infty$ and $V_b(t) \in L_\infty$. From the error system dynamics, we can now state $\dot{e}(t) \in L_\infty$, $\dot{\tilde{\eta}}_\psi(t) \in L_\infty$, $\dot{\tilde{\eta}}_I(t) \in L_\infty$, $\dot{\tilde{\eta}}_\tau(t) \in L_\infty$, $\dot{\tilde{r}}(t) \in L_\infty$, $\dot{\tilde{\psi}}_a(t) \in L_\infty$, $\dot{\tilde{\psi}}_b(t) \in L_\infty$, $\dot{\tilde{I}}_a(t) \in L_\infty$, $\dot{\tilde{I}}_b(t) \in L_\infty$, and $\ddot{\tilde{q}}(t) \in L_\infty$. Also, since $\dot{\tilde{q}}(t)$ is exponentially stable as dictated by (14.82) and (14.83), we can easily show that $\tilde{q}(t) \in L_\infty$ (See Lemma 1.9 in Chapter 1). Therefore, since $e(t) \in L_\infty$ (and hence $q(t) \in L_\infty$), we can show that $\hat{q}(t) \in L_\infty$. Since $\dot{\hat{q}}(t) \in L_\infty$, $V_{oq}(t) \in L_\infty$, and $\tilde{q}(t) \in L_\infty$, we can see from (14.19) that $y(t) \in L_\infty$. Likewise, we can use (14.20) and the above information to show that $\dot{y}(t) \in L_\infty$. Therefore, we have shown that all signals in the system, the observer, and the controller remain bounded in some neighborhood F around $x(t) = 0$.

Remark 14.13

Our controller requires the inversion of the matrix $C(t)$ defined in (14.81); however, $C(t)$ is not globally invertible since

$$\det(C) = L_\psi^2\left(\hat{\psi}_a^2 + \hat{\psi}_b^2\right) - 2k_{n1}\alpha_2 L_\psi^2 \hat{r}\left(\hat{\psi}_a \hat{I}_b - \hat{\psi}_b \hat{I}_a\right). \tag{14.95}$$

From (14.95), $\det(C)$ can also be expressed as

$$\det(C) = L_\psi^2\left[2(\psi_d - \hat{\eta}_\psi) - 2k_{n1}\alpha_2 \hat{r}\tau_d + 2k_{n1}\alpha_2 \hat{r}\hat{\eta}_\tau\right] \tag{14.96}$$

where (14.49) and (14.62) have been used, respectively. Note, from Remark 14.12, $\tau_d(t)$ has been shown to be bounded in the neighborhood F. Based on (14.82) and the fact that $\psi_d(t) > 0$, we can state that (14.96) implies that $\det(C) > 0$ in some neighborhood F_0 around $x(t) = 0$ where $x(t)$ is defined (14.83). That is, $C(t)$ is invertible for some small set of initial conditions (*i.e.*, $\|x(0)\| \leq \beta$ where β is a positive constant), and therefore the voltage input controller is well-defined for this set of initial conditions. It should be noted that (14.94) and (14.96) are the same except for the constant multiplier L_ψ^2 in (14.96) which means the equalities given by (14.94) and (14.96) impose the same requirements on the desired flux trajectory $\psi_d(t)$. Note this problem with the invertibility of $C(t)$ causes the stability result of Theorem 14.1 to be local.

14.8 Experimental Results

The position tracking controller proposed in this chapter has been successfully applied to an induction motor driving a single-link robotic load. The hardware setup and the electromechanical system parameters are the same as that in Chapter 11.

The desired load position trajectory is selected as the follow smooth start sinusoid

$$q_d(t) = \frac{\pi}{2}\sin(5t)\left(1 - \underline{e}^{-0.01t^3}\right) rad$$

where $q_d(0) = \dot{q}_d(0) = \ddot{q}_d(0) = \dddot{q}_d(0) = 0$. The desired rotor flux trajectory of (14.62) was selected as

$$\psi_d = 0.8\frac{2\underline{e}^{-0.25t}}{1 + \underline{e}^{-0.5t}} + 1\ Wb.$$

The initial rotor flux estimates were set as follows

$$\hat{\psi}_a(0) = \hat{\psi}_b(0) = \sqrt{1.8}\ Wb.$$

The best tracking performance was obtained using the following control gain values

$$k_{n1} = 0.00001, \ k_{n2} = 1, \ k_{n3} = 0.2, \ k_{n4} = 0.5,$$

$$k_{n5} = 0.3, \ k_{n6} = 2, \ k_{n7} = 0.0001, \ k_{n8} = 0.01, \ k_s = 1.8.$$

The resulting load position tracking error is shown in Figure 14.1, the actual phase voltages are shown in Figure 14.2 through Figure 14.4. From the load position tracking error plot given in Figure 14.1, we can see that the "steady state" load position tracking error is approximately ± 0.4 degrees. As illustrated in Remark 14.13, the determinant of the matrix $C(t)$ defined in (14.81) may go to zero during some operation conditions. To prevent this control singularity, a threshold constant of 0.5 for the determinant has been utilized in the experiment. That is, if the determinant of the matrix $C(t)$ is less than 0.5, then it was set to 0.5. The value of the determinant of the matrix $C(t)$ is shown in Figure 14.5.

Remark 14.14

The selection of the control parameters in the experiment does not satisfy the conditions given in Theorem 1; however, these requirements are only sufficient, conservative conditions generated by the Lyapunov-like stability argument.

14.9 Notes

Although a lot of research has been done to eliminate rotor flux measurements for induction motor control (see Chapter 11 for related references), there seems to be little work targeted at eliminating both load velocity and rotor flux measurements at the same time. The paper presented by Hu *et al.* in [8] provided the basic contents of this chapter. In [9], Feemster *et al.* designed a singularity-free, adaptive output feedback controller which ensured global asymptotic rotor velocity tracking and only required rotor velocity measurements. Recently, Feemster *et al.* [10] designed a singularity-free, partial state feedback controller which ensured global exponential rotor position tracking and only required rotor position and stator current measurements. For other recent work in induction motor control, the reader is referred to the references in Chapter 11.

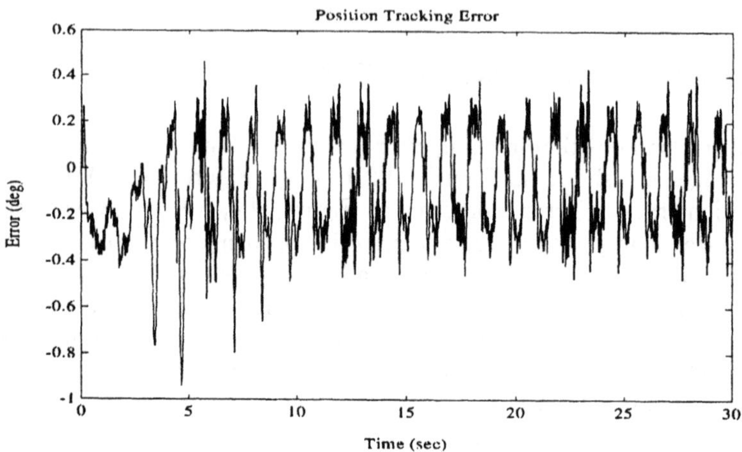

Figure 14.1: Position Tracking Error

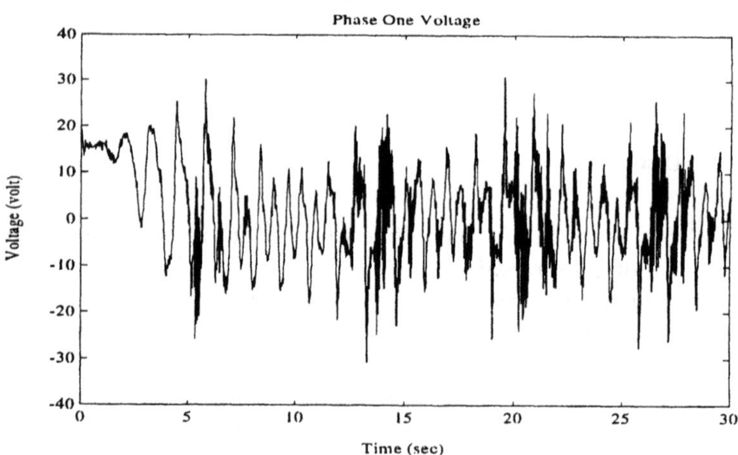

Figure 14.2: Actual Phase One Voltage

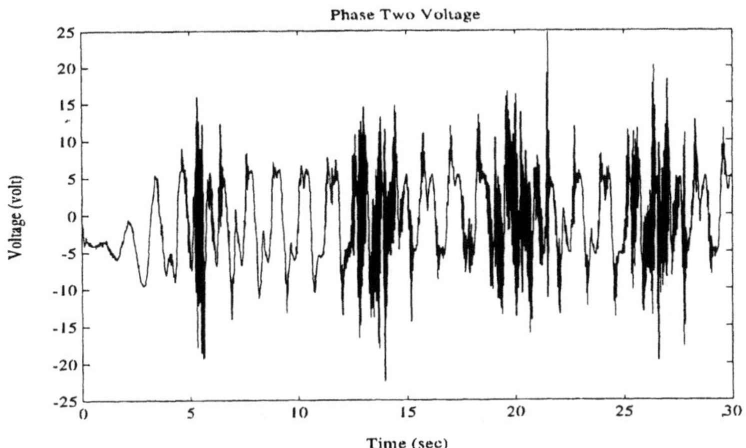

Figure 14.3: Actual Phase Two Voltage

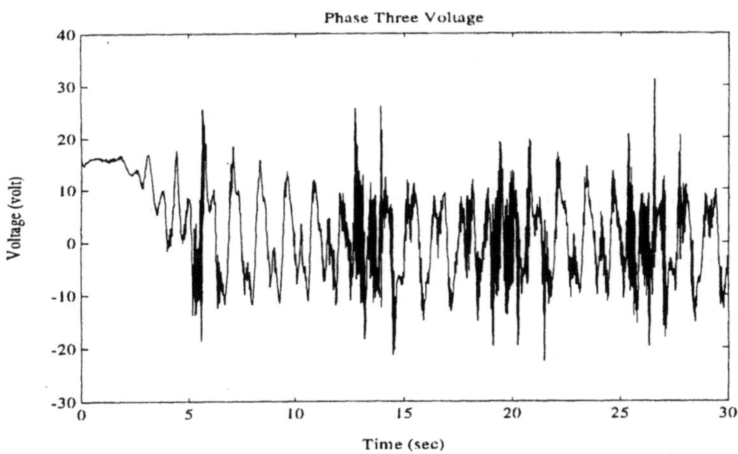

Figure 14.4: Actual Phase Three Voltage

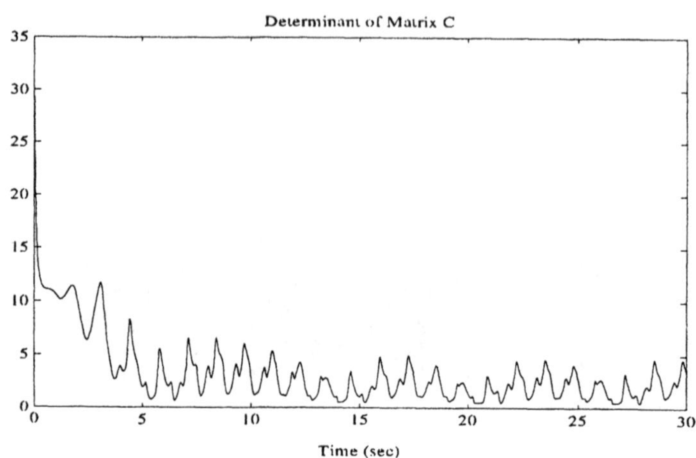

Figure 14.5: Determinant of Matrix C

Bibliography

[1] M. Spong, and M. Vidyasagar, *Robot Dynamics and Control,* New York: John Wiley and Sons, Inc., 1989.

[2] C. Canudas, R. Ortega and S. I. Seleme, "Robot Motion Control using Induction Motor Drives", *Proc. of the IEEE International Conference on Robotics and Automation,* Vol. 2, pp. 533-538, 1993.

[3] M. Krstic, I. Kanellakopoulos, and P. Kokotovic, *Nonlinear and Adaptive Control Design,* John Wiley & Sons, 1995.

[4] J. Slotine, and W. Li, *Applied Nonlinear Control*, Englewood Cliff, NJ: Prentice Hall Co., 1991.

[5] F. Blaschke, "The Principle of Field Orientation Applied to the New Transvector Closed-Loop Control System for Rotation Field Machines", *Siemens Review,* Vol. 39, pp. 217-220, 1972.

[6] R. Marino, S. Peresada, and P. Valigi, "Adaptive Input-Output Linearizing Control of Induction Motors", *IEEE Transactions on Automatic Control,* Vol. 38, No. 2, pp. 208-221, 1993.

[7] J. E. Marsden, *Elementary Classical Analysis,* W. H. Freeman and Company, 1974.

[8] J. Hu, D.M. Dawson, and Y. Qian, "Position Tracking of an Ind[9]uction Motor via Partial State Feedback", *Automatica,* Vol. 31, No. 7, July, 1995, pp. 989-1000.

[9] M. Feemster, P. Vedagarbha, D. Haste, and D. M. Dawson, "Adaptive Output-Feedback Control of Induction Motors", *Proc. of the Conference on Decision and Control*, San Diego, CA December 1997, to appear.

[10] M. Feemster, D. M. Dawson, P. Aquino, and D. Haste, "Position Tracking of the Induction Motor without Rotor Velocity or Rotor Flux Measurements," *American Control Conference*, Philadelphia, PA, June 1998, submitted.

Appendix A

BLDC Rotor-Fixed Transformation

In this appendix, we derive the BLDC motor model given by (4.1) through (4.3). We assume the BLDC motor has reluctance variations (*i.e.*, nonuniform air gap) but no saturation effect (*i.e.*, the flux linkages are linear functions of the phase currents). The electrical dynamics are therefore given by [2]

$$\dot{\Lambda} + R_N I = V \tag{A.1}$$

where $\Lambda \in \Re^3$ represents the flux linkages, $I(t) \in \Re^3$ represents the stator winding currents, $V \in \Re^3$ represents the stator voltage input, and $R_N \in \Re^{3\times 3}$ represents the stator winding resistance defined by

$$\Lambda = [\Lambda_1\ \Lambda_2\ \Lambda_3]^T,\ I = [I_1\ I_2\ I_3]^T,\ V = [V_1\ V_2\ V_3]^T$$

and

$$R_N = RU_{3\times 3} \tag{A.2}$$

where R is a positive constant, and $U_{3\times 3}$ is the 3×3 identity matrix. The subscripts 1, 2, and 3 correspond to the individual stator phase windings. The flux linkages are given by

$$\Lambda = LI + \Gamma \tag{A.3}$$

where $L \in \Re^{3\times 3}$ is a symmetric matrix representing the self and mutual inductances of the stator windings, and $\Gamma \in \Re^3$ represents the permanent magnet flux linkage. The matrix L is defined in [2] as shown

$$L = \begin{bmatrix} L_{c1} & L_{c2} & L_{c3} \end{bmatrix}$$

where

$$L_{c1} = \begin{bmatrix} L_s + L_A - L_B \cos(2\theta) \\ -\frac{1}{2}L_A - L_B \cos\left(2\theta - \frac{2\pi}{3}\right) \\ -\frac{1}{2}L_A - L_B \cos\left(2\theta + \frac{2\pi}{3}\right) \end{bmatrix},$$

$$L_{c2} = \begin{bmatrix} -\frac{1}{2}L_A - L_B \cos\left(2\theta - \frac{2\pi}{3}\right) \\ L_s + L_A - L_B \cos\left(2\theta + \frac{2\pi}{3}\right) \\ -\frac{1}{2}L_A - L_B \cos(2\theta) \end{bmatrix},$$

$$\text{and} \quad L_{c3} = \begin{bmatrix} -\frac{1}{2}L_A - L_B \cos\left(2\theta + \frac{2\pi}{3}\right) \\ -\frac{1}{2}L_A - L_B \cos(2\theta) \\ L_s + L_A - L_B \cos\left(2\theta - \frac{2\pi}{3}\right) \end{bmatrix}.$$

L_s, L_A, L_B are positive constants which represent the nominal self-inductance, mutual inductance, and amplitude variation in inductance due to the nonuniform air gap such that $L_A > L_B$. The term θ is scalar dependent variable defined as

$$\theta = n_p q \tag{A.4}$$

where n_p is the number of permanent magnet pole pairs and q is the rotor position. Γ of (A.3) is given by

$$\Gamma = \left[\begin{array}{ccc} K_E \sin(\theta) & K_E \sin\left(\theta - \frac{2\pi}{3}\right) & K_E \sin\left(\theta + \frac{2\pi}{3}\right) \end{array}\right]^T$$

where K_E is a positive constant representing the electromotive force coefficient (*i.e.*, back EMF).

Substituting (A.3) into (A.1) yields

$$V = R_N I + \frac{d}{dt}\{LI\} + \dot{\Gamma}. \tag{A.5}$$

Equation (A.5) represents a system of differential equations with time varying coefficients. Since we have assumed that the BLDC motor possesses sinusoidally distributed windings as in [1], it is possible to transform the electrical dynamics from the stator's fixed frame of reference to a set of hypothetical windings associated with the rotor. This coordinate change effectively eliminates the time varying inductances from the electrical dynamics of (A.5). To accomplish this transformation, we introduce the following orthogonal Park's [2] transformation $T \in \Re^{3\times 3}$ defined by

$$T = \sqrt{\frac{2}{3}} \begin{bmatrix} \cos(\theta) & \cos\left(\theta - \frac{2\pi}{3}\right) & \cos\left(\theta + \frac{2\pi}{3}\right) \\ \sin(\theta) & \sin\left(\theta - \frac{2\pi}{3}\right) & \sin\left(\theta + \frac{2\pi}{3}\right) \\ \frac{1}{\sqrt{2}} & \frac{1}{\sqrt{2}} & \frac{1}{\sqrt{2}} \end{bmatrix} \in \Re^{3\times 3}. \tag{A.6}$$

Note that this transformation possesses the property that $T^{-1} = T^T$. We define the new coordinates $I_N = TI,\ V_N = TV$ such that

$$I_N = [I_a,\ I_b,\ I_o]^T,\quad V_N = [V_a,\ V_b,\ V_o]^T \tag{A.7}$$

where the subscripts a, b, and o represent hypothetical quadrature, direct, and zero axis windings respectively, which are fixed to the rotor.

Now, we rewrite (A.5) in terms of the hypothetical *abo* coordinates of (A.7) as

$$T^{-1}V_N = R_N T^{-1} I_N + \frac{d}{dt}\left\{LT^{-1}I_N\right\} + \dot{\Gamma}. \tag{A.8}$$

Multiplying (A.8) by T yields

$$V_N = R_N I_N + T\frac{d}{dt}\left\{LT^{-1}\right\} I_N + L_N \dot{I}_N + T\dot{\Gamma} \tag{A.9}$$

where

$$L_N = TLT^{-1} = diag\left\{L_a,\ L_b,\ L_o\right\}, \tag{A.10}$$

with

$$L_a = L_s + \frac{3}{2}(L_A - L_B),\ L_b = L_s + \frac{3}{2}(L_A + L_B),\ L_o = L_s. \tag{A.11}$$

From (A.10), we also have $L = T^T L_N T$ since $T^{-1} = T^T$; therefore, we can rewrite (A.9) as

$$V_N = R_N I_N + T\dot{T}^{-T} L_N I_N + L_N \dot{I}_N + T\dot{\Gamma}. \tag{A.12}$$

Since the term T^T and Γ only depend on the time varying scalar q, we can utilize (A.4) to rewrite (A.12) as

$$V_N = R_N I_N + G_1 I_N \dot{q} + G_2 \dot{q} + L_N \dot{I}_N. \tag{A.13}$$

where $G_1 \in \Re^{3\times 3}$ is a constant matrix, and $G_2 \in \Re^3$ is a constant vector defined as follows

$$G_1 \triangleq T\frac{\partial}{\partial q}\left\{T^T\right\} L_N = n_p \begin{bmatrix} 0 & L_b & 0 \\ -L_a & 0 & 0 \\ 0 & 0 & 0 \end{bmatrix}, \tag{A.14}$$

and

$$G_2 \triangleq T\frac{\partial\Gamma}{\partial q} = \left[\ \sqrt{\tfrac{3}{2}}n_pK_E,\quad 0,\quad 0\ \right]^T. \tag{A.15}$$

It is now possible to substitute (A.2), (A.10), (A.14), and (A.15) into (A.13) to yield (4.1) through (4.3) with the exception of the torque term on the right hand side of (4.1).

Now, we transfer the electromechanical coupling torque to the rotor fixed hypothetical coordinates. The electromechanical coupling torque can be derived from the co-energy function as shown in [2] as

$$\tau = \frac{1}{2}I^T\frac{\partial L}{\partial q}I + \left[\frac{\partial\Gamma}{\partial q}\right]^T I. \tag{A.16}$$

Substituting $L = T^TL_NT$ and $I_N = TI$ into (A.16) yields

$$\tau = \frac{1}{2}I_N^TT\frac{\partial}{\partial q}\left\{T^TL_NT\right\}T^TI_N + \left[\frac{\partial\Gamma}{\partial q}\right]^T T^TI_N \tag{A.17}$$

which can be rewritten as follows

$$\tau = I_N^TT\frac{\partial}{\partial q}\left\{T^T\right\}L_NTT^TI_N + \left[\frac{\partial\Gamma}{\partial q}\right]^T T^TI_N. \tag{A.18}$$

Applying the definitions for G_1 and G_2 from (A.14) and (A.15) to (A.18), we have

$$\tau = I_N^TG_1^TI_N + G_2^TI_N. \tag{A.19}$$

Substituting the right-hand sides of (A.14) and (A.15) into (A.19) yields

$$\tau = \left(K_{\tau 1}I_b + K_{\tau 2}\right)I_a \tag{A.20}$$

which is the right-hand side of (4.1) after division by $K_{\tau 2}$. Note the constants $K_{\tau 1}$ and $K_{\tau 2}$ of (A.20) are explicitly defined as

$$K_{\tau 1} = n_p\left(L_b - L_a\right) \quad \text{and} \quad K_{\tau 2} = \sqrt{\frac{3}{2}}n_pK_E.$$

Bibliography

[1] N. Hemati, "Dynamic Analysis of Brushless Motors Based on Compact Representations of Equations of Motion", *Proc. of IEEE Industrial Application Society Annual Meeting*, Vol. 1, pp. 51-58, September, 1993.

[2] P. C. Krause, *Analysis of Electric Machinery*, McGraw-Hill, Inc., 1986.

Appendix B

Differentiation of γ_{dj}

As noted, γ_{dj} of (5.18) defines a smooth "sharing" of the required control responsibilities between the individual SR motor phases. The existence of $\gamma_{dj}(\tau_d, q)$ is guaranteed [1] for all bounded τ_d and q since

$$\frac{\tau_d \sin(x_j) S\left(\sin(x_j)\tau_d\right)}{S_T} \geq 0,$$

and due the construction of $S(z)$ defined in (5.19). Differentiating γ_{dj} of (5.18) with respect to time yields

$$\dot{\gamma}_{dj} = \Sigma_j \dot{x}_j + \Pi_j \dot{\tau}_d,$$

where

$$\begin{aligned} \Sigma_j = \; & \left[\frac{\partial}{\partial S_T}\gamma_{dj}\right]\left\{\sin^2(x_j)\left[\frac{\partial}{\partial x_j} S\left(\sin(x_j)\tau_d\right)\right]\right. \\ & \left. +2\sin(x_j)\cos(x_j) S\left(\sin(x_j)\tau_d\right)\right\} \\ & + \left[\frac{\partial}{\partial \sin(x_j)}\gamma_{dj}\right]\left[\frac{\partial}{\partial x_j}\sin(x_j)\right] \\ & + \left[\frac{\partial}{\partial S(\cdot)}\gamma_{dj}\right]\left[\frac{\partial}{\partial x_j} S\left(\sin(x_j)\tau_d\right)\right], \end{aligned}$$

and

$$\begin{aligned} \Pi_j = \; & \frac{\partial}{\partial \tau_d}\gamma_{dj} + \left[\frac{\partial}{\partial S(\cdot)}\gamma_{dj}\right]\left[\frac{\partial}{\partial \tau_d} S\left(\sin(x_j)\tau_d\right)\right] \\ & + \left[\frac{\partial}{\partial S_T}\gamma_{dj}\right]\sum_{j=1}^{4}\sin^2(x_j)\left[\frac{\partial}{\partial \tau_d} S\left(\sin(x_j)\tau_d\right)\right], \end{aligned}$$

with the auxiliary terms given by

$$\frac{\partial}{\partial \sin(x_j)}\gamma_{dj} = \frac{\tau_d S\left(\sin(x_j)\tau_d\right)}{2S_T\sqrt{\frac{\tau_d \sin(x_j) S\left(\sin(x_j)\tau_d\right)}{S_T} + \gamma_{c0}^2}},$$

$$\frac{\partial}{\partial x_j}\sin(x_j) = \cos(x_j),$$

$$\frac{\partial}{\partial S(\cdot)}\gamma_{dj} = \frac{\tau_d \sin(x_j)}{2S_T\sqrt{\frac{\tau_d \sin(x_j) S\left(\sin(x_j)\tau_d\right)}{S_T} + \gamma_{c0}^2}},$$

$$\frac{\partial}{\partial S_T}\gamma_{dj} = -\frac{\tau_d \sin(x_j) S\left(\sin(x_j)\tau_d\right)}{2S_T^2\sqrt{\frac{\tau_d \sin(x_j) S\left(\sin(x_j)\tau_d\right)}{S_T} + \gamma_{c0}^2}},$$

$$\frac{\partial}{\partial \tau_d}\gamma_{dj} = \frac{\sin(x_j) S\left(\sin(x_j)\tau_d\right)}{2S_T\sqrt{\frac{\tau_d \sin(x_j) S\left(\sin(x_j)\tau_d\right)}{S_T} + \gamma_{c0}^2}},$$

$$\frac{\partial}{\partial x_j} S\left(\sin(x_j)\tau_d\right) =$$

$$\begin{cases} 0, & \text{for } \sin(x_j)\tau_d \leq 0, \\ 2\epsilon_0 \tau_d^2 \sin(x_j)\cos(x_j)\underline{e}^{-\epsilon_0(\sin(x_j)\tau_d)^2}, & \text{for } \sin(x_j)\tau_d > 0, \end{cases}$$

$$\frac{\partial}{\partial \tau_d} S\left(\sin(x_j)\tau_d\right) =$$

$$\begin{cases} 0, & \text{for } \sin(x_j)\tau_d \leq 0, \\ 2\epsilon_0 \tau_d \sin^2(x_j)\underline{e}^{-\epsilon_0(\sin(x_j)\tau_d)^2}, & \text{for } \sin(x_j)\tau_d > 0, \end{cases}$$

$$\dot{x}_j = N_r \dot{q} \qquad \forall\, j,$$

and $\dot{\tau}_d$ was defined in (5.33). It is important to note that the only possible singularities in the above partial derivative terms occur when $S_T = 0$ which corresponds to $\tau_d = 0$. In [1], the following properties were shown to hold

$$\lim_{\tau_d \to 0} \Sigma_j = 0 \tag{B.1}$$

and

$$\lim_{\tau_d \to 0} \Pi_j = \sin^3(x_j) \left[2\gamma_{co} \sum_{j=1}^{4} \sin^4(x_j) \right]^{-1} . \tag{B.2}$$

Therefore, based on the structure of the above partial derivatives, (B.1), and (B.2), we can state that $\dot{\gamma}_{dj}$ exists and is bounded for bounded $\dot{q}$, τ_d, and $\dot{\tau}_d$. Hence, based on the above equations and (5.36), it is easy to see that $\dot{\gamma}_{dj} \in L_\infty$ if q, $\dot{q}$, $\gamma_j \in L_\infty$.

Bibliography

[1] J. J. Carroll, *Nonlinear Control of Electric Machines*, Ph.D. Dissertation, Clemson University, August 1993.

Appendix C

Stator-Fixed Transformation

The first part of the following model derivation is the standard DQO transformation for a three-phase symmetrical induction motor. The second part is the derivation of the model described in (6.2) through (6.5) (*i.e.*, the stator-fixed reference frame transformed model). In general, the voltage equations may be expressed in machine variables as [1]

$$V_{123s} = R_s I_{123s} + \dot{\psi}_{123s} \tag{C.1}$$

$$V_{123r} = R_r I_{123r} + \dot{\psi}_{123r} \tag{C.2}$$

where the subscript s denotes variables and parameters associated with the stator circuit, the subscript r denotes variables and parameters associated with rotor circuit, R_s is the scalar stator resistance, R_r is the scalar rotor resistance, $V_{123s} \in \Re^3$ represents the stator terminal voltages $V_{123s} = [V_{1s}\ V_{2s}\ V_{3s}]^T$, $V_{123r} \in \Re^3$ represents the rotor voltage $V_{123r} = [V_{1r}\ V_{2r}\ V_{3r}]^T$, $I_{123s} \in \Re^3$ represents the stator currents $I_{123s} = [I_{1s}\ I_{2s}\ I_{3s}]^T$, $\psi_{123s} \in \Re^3$ represents the stator flux linkages $\psi_{123s} = [\psi_{1s}\ \psi_{2s}\ \psi_{3s}]^T$, $I_{123r} \in \Re^3$ represents the rotor currents $I_{123r} = [I_{1r}\ I_{2r}\ I_{3r}]^T$, and $\psi_{123r} \in \Re^3$ represents the rotor flux linkages $\psi_{123r} = [\psi_{1r}\ \psi_{2r}\ \psi_{3r}]^T$.

Based on the assumption of a linear magnetic circuit [1], the flux linkages may be expressed as

$$\begin{bmatrix} \psi_{123s} \\ \psi_{123r} \end{bmatrix} = \begin{bmatrix} L_s & L_{sr} \\ (L_{sr})^T & L_r \end{bmatrix} \begin{bmatrix} I_{123s} \\ I_{123r} \end{bmatrix}, \tag{C.3}$$

where the winding inductances matrices $L_s \in \Re^{3\times3}$, $L_r \in \Re^{3\times3}$, and $L_{sr} \in \Re^{3\times3}$ are given by

$$L_s = \begin{bmatrix} L_{ls} + L_{ms} & -\frac{1}{2}L_{ms} & -\frac{1}{2}L_{ms} \\ -\frac{1}{2}L_{ms} & L_{ls} + L_{ms} & -\frac{1}{2}L_{ms} \\ -\frac{1}{2}L_{ms} & -\frac{1}{2}L_{ms} & L_{ls} + L_{ms} \end{bmatrix} \tag{C.4}$$

$$L_r = \begin{bmatrix} L_{lr} + L_{mr} & -\frac{1}{2}L_{mr} & -\frac{1}{2}L_{mr} \\ -\frac{1}{2}L_{mr} & L_{lr} + L_{mr} & -\frac{1}{2}L_{mr} \\ -\frac{1}{2}L_{mr} & -\frac{1}{2}L_{mr} & L_{lr} + L_{mr} \end{bmatrix} \tag{C.5}$$

and

$$L_{sr} = L_{sl} \begin{bmatrix} \cos(\theta_r) & \cos(\theta_r + \frac{2}{3}\pi) & \cos(\theta_r - \frac{2}{3}\pi) \\ \cos(\theta_r - \frac{2}{3}\pi) & \cos(\theta_r) & \cos(\theta_r + \frac{2}{3}\pi) \\ \cos(\theta_r + \frac{2}{3}\pi) & \cos(\theta_r - \frac{2}{3}\pi) & \cos(\theta_r) \end{bmatrix} \tag{C.6}$$

in which L_{ls} and L_{ms} are the leakage and magnetizing inductances of the stator windings, respectively, L_{lr} and L_{mr} are the leakage and magnetizing inductances of the rotor windings, respectively, L_{sl} is the mutual inductance between stator and rotor windings, and θ_r is the electrical angular displacement of the rotor from the fixed stator phase 1. Under the assumption of a symmetrical induction motor and equal mutual inductance [1], we have

$$L_{ms} = L_{mr} = L_{sl}, \tag{C.7}$$

and

$$\theta_r = n_p q, \quad \text{therefore} \quad \dot{\theta}_r = n_p \dot{q}, \tag{C.8}$$

where n_p is the pole-pair number and q denotes the rotor position.

Now we transform the stator and rotor variables to an arbitrary reference frame. Let subscript q, d, and o represent the hypothetical quadrature, direct, and zero axes, respectively, and assume that the q-axis angular displacement from the fixed stator phase 1 is θ, and therefore the qdo frame is rotating at angular velocity $\dot{\theta}$. Hence in general, we have

$$V_{qdo} = TV_{123}, \quad I_{qdo} = TI_{123}, \quad \psi_{qdo} = T\psi_{123}, \tag{C.9}$$

where $T \in \Re^{3\times 3}$ denotes a transformation matrix which transforms the machine variables to the qdo frame, $V_{qdo} \in \Re^3$ represents the transformed voltage $V_{qdo} = [V_q \; V_d \; V_o]^T$, $I_{qdo} \in \Re^3$ represents the transformed current $I_{qdo} = [I_q \; I_d \; I_o]^T$, and $\psi_{qdo} \in \Re^3$ represents the transformed flux linkage $\psi_{qdo} = [\psi_q \; \psi_d \; \psi_o]^T$.

We now consider the stator transformation. Let T_s denote the stator transformation matrix which transforms the stator variables to the arbitrary reference frame qdo. We select the stator transformation matrix as

$$T_s = \sqrt{\frac{2}{3}} \begin{bmatrix} \cos(\theta) & \cos(\theta - \frac{2}{3}\pi) & \cos(\theta + \frac{2}{3}\pi) \\ \sin(\theta) & \sin(\theta - \frac{2}{3}\pi) & \sin(\theta + \frac{2}{3}\pi) \\ \frac{1}{\sqrt{2}} & \frac{1}{\sqrt{2}} & \frac{1}{\sqrt{2}} \end{bmatrix} \in \Re^{3\times 3} \tag{C.10}$$

where θ is the angular displacement of the reference frame q-axis from the fixed stator phase 1. Note, given the structure of (C.10), we have $T_s^{-1} = T_s^T$. The rotor transformation matrix T_r [1] is selected as

$$T_r = \sqrt{\frac{2}{3}} \begin{bmatrix} \cos(\beta) & \cos(\beta - \frac{2}{3}\pi) & \cos(\beta + \frac{2}{3}\pi) \\ \sin(\beta) & \sin(\beta - \frac{2}{3}\pi) & \sin(\beta + \frac{2}{3}\pi) \\ \frac{1}{\sqrt{2}} & \frac{1}{\sqrt{2}} & \frac{1}{\sqrt{2}} \end{bmatrix} \in \Re^{3\times 3} \qquad \text{(C.11)}$$

where $\beta = \theta - \theta_r$ (note $T_r^{-1} = T_r^T$). From (C.3) and (C.9), we obtain the stator flux variables expressed in the qdo frame as follows

$$\psi_{qdos} = G_{1s} I_{qdos} + G_{2s} I_{qdor}, \qquad \text{(C.12)}$$

where based on (C.4), (C.6), (C.10), and (C.11),

$$G_{1s} = T_s L_s T_s^{-1} = \begin{bmatrix} L_{ls} + M_e & 0 & 0 \\ 0 & L_{ls} + M_e & 0 \\ 0 & 0 & L_{ls} \end{bmatrix} \in \Re^{3\times 3} \qquad \text{(C.13)}$$

and

$$G_{2s} = T_s L_{sr} T_r^{-1} = \begin{bmatrix} M_e & 0 & 0 \\ 0 & M_e & 0 \\ 0 & 0 & 0 \end{bmatrix} \in \Re^{3\times 3} \qquad \text{(C.14)}$$

where $M_e = \frac{3}{2} L_{ms}$. Using (C.9), the stator voltages of (C.1) can be expressed in the qdo frame as

$$V_{qdos} = R_s I_{qdos} + G_{3s} \psi_{qdos} + \dot{\psi}_{qdos} \qquad \text{(C.15)}$$

where

$$G_{3s} = T_s \dot{T}_s^{-1} = \begin{bmatrix} 0 & \dot{\theta} & 0 \\ -\dot{\theta} & 0 & 0 \\ 0 & 0 & 0 \end{bmatrix} \in \Re^{3\times 3}. \qquad \text{(C.16)}$$

Now we consider the transformation of the rotor variables. Using (C.9), the rotor flux expression of (C.3) can be written in the qdo frame as

$$\psi_{qdor} = G_{1r} I_{qdor} + G_{2r} I_{qdos}, \qquad \text{(C.17)}$$

where

$$G_{1r} = T_r L_r T_r^{-1} = \begin{bmatrix} L_{lr} + M_e & 0 & 0 \\ 0 & L_{lr} + M_e & 0 \\ 0 & 0 & L_{lr} \end{bmatrix} \in \Re^{3\times 3} \qquad \text{(C.18)}$$

and

$$G_{2r} = T_r L_{sr}^T T_s^{-1} = \begin{bmatrix} M_e & 0 & 0 \\ 0 & M_e & 0 \\ 0 & 0 & 0 \end{bmatrix} \in \Re^{3\times 3}. \tag{C.19}$$

Similarly, the rotor voltage equation of (C.2) can also be expressed in the *qdo* frame as

$$V_{qdor} = R_r I_{qdor} + G_{3r}\psi_{qdor} + \dot{\psi}_{qdor}, \tag{C.20}$$

where

$$G_{3r} = T_r \dot{T}_r^{-1} = \begin{bmatrix} 0 & \left(\dot{\theta} - \dot{\theta}_r\right) & 0 \\ -\left(\dot{\theta} - \dot{\theta}_r\right) & 0 & 0 \\ 0 & 0 & 0 \end{bmatrix} \in \Re^{3\times 3}. \tag{C.21}$$

Note the voltage equations given by (C.15) and (C.20) are the standard *qdo* model for the induction motor [1].

Based on the above development, we can now derive the induction motor model described in (6.2) through (6.5). Since the state variables are ψ_{qr}, ψ_{dr}, I_{qs}, and I_{ds}, we need to solve for ψ_{qs}, ψ_{ds}, I_{qr}, and I_{dr} of (C.15) and (C.20) in terms of ψ_{qr}, ψ_{dr}, I_{qs}, and I_{ds}. From (C.17), we have

$$I_{qr} = \frac{1}{L_{lr} + M_e}\psi_{qr} - \frac{M_e}{L_{lr} + M_e} I_{qs} \tag{C.22}$$

and

$$I_{dr} = \frac{1}{L_{lr} + M_e}\psi_{dr} - \frac{M_e}{L_{lr} + M_e} I_{ds}. \tag{C.23}$$

Substituting (C.22) and (C.23) into (C.12) yields

$$\psi_{qs} = (L_{lr} + M_e) I_{qs} + \frac{M_e}{L_{lr} + M_e}\psi_{qr} - \frac{M_e^2}{L_{lr} + M_e} I_{qs} \tag{C.24}$$

and

$$\psi_{ds} = (L_{lr} + M_e) I_{ds} + \frac{M_e}{L_{lr} + M_e}\psi_{dr} - \frac{M_e^2}{L_{lr} + M_e} I_{ds}. \tag{C.25}$$

Substituting (C.22) through (C.25) into the voltage equations of (C.15) and (C.20) yields

$$\begin{aligned} V_{qs} &= R_s I_{qs} + \frac{\dot{\theta} M_e}{L_{lr} + M_e}\psi_{dr} + (\dot{\theta} L_{ls} + \frac{\dot{\theta} M_e L_{lr}}{L_{lr} + M_e}) I_{ds} \\ &\quad + \frac{M_e}{L_{lr} + M_e}\dot{\psi}_{qr} + (L_{ls} + \frac{M_e L_{lr}}{L_{lr} + M_e})\dot{I}_{qs}, \end{aligned} \tag{C.26}$$

$$V_{ds} = R_s I_{ds} + \frac{\dot{\theta} M_e}{L_{lr} + M_e}\psi_{qr} - (\dot{\theta} L_{ls} + \frac{\dot{\theta} M_e L_{lr}}{L_{lr} + M_e}) I_{qs} + \frac{M_e}{L_{lr} + M_e}\dot{\psi}_{dr} + (L_{ls} + \frac{M_e L_{lr}}{L_{lr} + M_e})\dot{I}_{ds}, \tag{C.27}$$

$$V_{qr} = \frac{R_r}{L_{lr} + M_e}\psi_{qr} - \frac{M_e R_r}{L_{lr} + M_e} I_{qs} + (\dot{\theta} - \dot{\theta}_r)\psi_{dr} + \dot{\psi}_{qr}, \tag{C.28}$$

$$V_{dr} = \frac{R_r}{L_{lr} + M_e}\psi_{dr} - \frac{M_e R_r}{L_{lr} + M_e} I_{ds} - (\dot{\theta} - \dot{\theta}_r)\psi_{qr} + \dot{\psi}_{dr}, \tag{C.29}$$

and

$$V_{os} = R_s I_{os} + \dot{\psi}_{os}, V_{or} = R_o I_{or} + \dot{\psi}_{or}. \tag{C.30}$$

Since there are no actual voltage inputs in the rotor equations of (C.28) and (C.29), we have $V_{qr} = V_{dr} = 0$. Therefore, we can solve for $\dot{\psi}_{qr}$ and $\dot{\psi}_{dr}$ from (C.28) and (C.29), respectively, and then substitute the resulting expression into (C.26) and (C.27) to yield

$$V_{qs} = R_s I_{qs} + \frac{\dot{\theta} M_e}{L_{lr} + M_e}\psi_{dr} + (\dot{\theta} L_{ls} + \frac{\dot{\theta} M_e L_{lr}}{L_{lr} + M_e}) I_{ds} + (L_{ls} + \frac{M_e L_{lr}}{L_{lr} + M_e})\dot{I}_{qs} - \frac{M_e}{L_{lr} + M_e}(\dot{\theta} - \dot{\theta}_r)\psi_{dr} + \frac{M_e^2 R_r}{(L_{lr} + M_e)^2} I_{qs} - \frac{M_e R_r}{(L_{lr} + M_e)^2}\psi_{qr} \tag{C.31}$$

and

$$V_{ds} = R_s I_{ds} + \frac{\dot{\theta} M_e}{L_{lr} + M_e}\psi_{qr} - (\dot{\theta} L_{ls} + \frac{\dot{\theta} M_e L_{lr}}{L_{lr} + M_e}) I_{qs} + (L_{ls} + \frac{M_e L_{lr}}{L_{lr} + M_e})\dot{I}_{ds} + \frac{M_e}{L_{lr} + M_e}(\dot{\theta} - \dot{\theta}_r)\psi_{qr} + \frac{M_e^2 R_r}{(L_{lr} + M_e)^2} I_{ds} - \frac{M_e R_r}{(L_{lr} + M_e)^2}\psi_{dr}. \tag{C.32}$$

If the reference frame is fixed on the stator (*i.e.*, $\dot{\theta} = 0$), and we define

$$L_r = L_{lr} + M_e, \; L_s = L_{ls} + M_e, \quad L_I = L_s - \frac{M_e^2}{L_r}, \tag{C.33}$$

then we can obtain the induction motor model described in (6.2) through (6.5) by simply interchanging the subscripts q to b and d to a in (C.28), (C.29), (C.31), and (C.32).

The electromechanical coupling torque can be derived using the co-energy principle [1] as shown

$$\tau = n_p (I_{123s})^T \frac{\partial}{\partial \theta_r} [L_{sr}] I_{123r}. \tag{C.34}$$

In the qdo frame, (C.34) becomes

$$\tau = n_p (I_{qdos})^T T_s^{-T} \frac{\partial}{\partial \theta_r} [L_{sr}] T_r^{-1} I_{qdor} \tag{C.35}$$

by using (C.9) and (C.10). After significant algebraic operations, we can show that

$$T_s^{-T} \frac{\partial}{\partial \theta_r} [L_{sr}] T_r^{-1} = \begin{bmatrix} 0 & M_e & 0 \\ -M_e & 0 & 0 \\ 0 & 0 & 0 \end{bmatrix} \in \Re^{3\times3}, \tag{C.36}$$

therefore using (C.36), (C.35) can be written as

$$\tau = n_p M_e \left(I_{qs} I_{dr} - I_{ds} I_{qr} \right). \tag{C.37}$$

Substituting I_{qr} and I_{dr} from (C.22) and (C.23) into (C.37) and setting the subscripts $q = b$ and $d = a$, we obtain the torque equation described in (6.1).

Remark C.1

Note, the zero axis dynamics (C.32) have been neglected in our model description since they are independent of the other axes variables. That is, we can always choose $V_{os} = V_{or} = 0$. Also from (C.37), we can see that the zero axis dynamics do not contribute to torque production.

Bibliography

[1] P. C. Krause, *Analysis of Electric Machinery,* McGraw-Hill, Inc., 1986.

Appendix D

Singularity Problem

In this appendix, we present an adaptive, partial-state feedback, position/velocity tracking controller for the full-order, nonlinear dynamic model representing an induction motor actuating a mechanical subsystem. The proposed controller compensates for parametric uncertainty in the mechanical subsystem while yielding global asymptotic rotor position/velocity tracking. The proposed controller does not require measurement of rotor flux or rotor current; furthermore, the controller does not exhibit any singularities. Experimental results are provided to verify the effectiveness of the approach.

D.1 Introduction

Over the last couple of years, there has been an increasing amount of interest in the use of induction machinery for position/velocity tracking applications. For example, in [1], Marino *et al.* applied adaptive, nonlinear techniques for the design of a full-state feedback, velocity controller for an induction motor driving a load (For earlier work and other related full-state feedback work, see the reviews in [1] and [2]). Recently, several researchers have attacked the position/velocity tracking control problem for the full-order model describing the electromechanical system under the constraint that rotor flux (or rotor current) measurements are not available. For example, in [3] and [4], Kanellakopoulos *et al.* developed a velocity tracking controller which utilized a rotor flux observer in a stable closed-loop fashion; however, as pointed out in [5], the set of control singularities has yet to be investigated. In [6], Hu *et al.* developed a position/velocity tracking controller which only required position and stator current measurements; however, roughly speaking, control singularities rendered a local exponential tracking result. Some recent control strategies for robotic manipulators actuated by induction motors can also be found in [7] and [8]. Recent work which involves the development of nonlinear controllers along with some corresponding experimental validation can be found in [9], [10], and [11]. Most recently, Espinosa-Pereza *et al.* [12] presented a singularity-free, velocity tracking controller which did not require rotor current measurements; however, the performance of the scheme is limited by that the fact the convergence rate of the speed tracking error is determined by the natural mechanical damping of the motor. In [13], Ortega *et al.* used an ingenious filtering procedure to design a singularity-free, position/velocity tracking controller which removed the mechanical damping restriction required in [12].

In this short note, we design a singularity-free, position/velocity tracking for an induction motor driving a mechanical subsystem. Provided the mechanical subsystem can be linear parameterized and only rotor position, rotor velocity, and stator current measurements are available, the controller ensures asymptotic position/velocity tracking despite parametric uncertainty in the mechanical subsystem and the lack of rotor flux measurements. The control strategy relies on a judicious selection of the desired electrical trajectories along with a liberal use of the nonlinear damping tool [14]. A disadvantage of the controller is the need for exact knowledge of the electrical subsystem parameters. To illustrate that this work provides some movement forward in this area, we first note that the high-pass filter used in [13] as a velocity surrogate somewhat limits the class of mechanical subsystem for which the strategy can be used; however, further analysis and/or design modifications may be used to enlarge the class of mechanical subsystems (*e.g.*, a similar high-pass filter was used for the control of robot manipulators in [15]). In addition, we note that the position/velocity tracking result given in [13] is based on a two-part Lyapunov/Gronwall-inequality type argument while the result in this paper relies only on a Lyapunov-type argument; hence, further extensions (*e.g.*, control of uncertainty mechanical systems and/or robot manipulators) fit nicely into the proposed control structure. To illustrate this claim, we design the controller to yield position/velocity tracking for a general, *uncertain* mechanical subsystem.

D.2 Electromechanical Model

To simplify the control development, we will utilize the transformed induction motor model in the stator fixed $a-b$ reference frame [1] [16]. Under the assumptions of equal mutual inductance and a linear magnetic circuit, the electromechanical model of an induction motor actuating a mechanical subsystem will be taken to be of the form

$$M_m\ddot{q} + W_m(q,\dot{q})\theta_m = \alpha_2\left(\psi_a I_b - \psi_b I_a\right), \tag{D.1}$$

$$L_I\dot{I}_a + R_I I_a - \alpha_1\psi_a - \alpha_2\psi_b\dot{q} = V_a, \tag{D.2}$$

$$L_I \dot{I}_b + R_I I_b - \alpha_1 \psi_b + \alpha_2 \psi_a \dot{q} = V_b, \tag{D.3}$$

$$L_r \dot{\psi}_a + R_r \psi_a + \alpha_3 \psi_b \dot{q} = K_I I_a, \tag{D.4}$$

and

$$L_r \dot{\psi}_b + R_r \psi_b - \alpha_3 \psi_a \dot{q} = K_I I_b \tag{D.5}$$

where $q(t)$, $\dot{q}(t)$, $\ddot{q}(t)$ represent the rotor position, velocity, and acceleration, respectively, M_m is the unknown, constant mechanical inertia of the system (including the rotor inertia), $W_m(q, \dot{q}) \in \Re^{1\times p}$ is a known, differentiable regression vector dependent on the rotor position and velocity, $\theta_m \in \Re^p$ is a constant vector containing the unknown mechanical subsystem parameters, $(\psi_a(t)\ \psi_b(t))$, $(I_a(t)\ I_b(t))$, $(V_a(t)\ V_b(t))$ represent the transformed rotor flux, stator current, and stator voltage of the induction motor, respectively, and L_I, K_I, α_1, α_2, α_3, R_I are known, positive constants related to the electric circuit parameters which are explicitly defined as follows

$$\alpha_1 = MR_r/L_r^2, \quad \alpha_2 = n_p M/L_r, \quad \alpha_3 = n_p L_r, \quad K_I = R_r M, \tag{D.6}$$

and

$$L_I = L_s - M^2/L_r, \quad R_I = \left(M^2 R_r + L_r^2 R_s\right)/L_r^2$$

where L_r, L_s, M denote the rotor inductance, stator inductance, and mutual inductance of the motor, respectively, R_r, R_s represent the rotor resistance and stator resistance, respectively, and n_p is the motor pole pair number.

Remark D.1

To facilitate the analysis, we require knowledge of lower and upper bounds for the inertial parameter M_m (*i.e.*, $\underline{M}_m < M_m < \bar{M}_m$ where $\underline{M}_m$ and $\bar{M}_m$ are known positive constants). In addition, we will assume that $W_m^T(\cdot) \in L_\infty^p$ if q, $\dot{q} \in L_\infty$ and $\dot{W}_m^T(\cdot) \in L_\infty^p$ if q, $\dot{q}$, $\ddot{q} \in L_\infty$.

D.3 Problem Formulation

Since our objective is the design of a rotor position/velocity tracking controller for the electromechanical model given by (D.1) through (D.5), we define the rotor position tracking error and the corresponding filtered tracking error [17] as follows

$$e = q_d - q \qquad\qquad r = \dot{e} + \alpha e \tag{D.7}$$

where $q_d(t)$ represents the desired rotor position trajectory supplied by the user, and α is a positive constant control gain (It is assumed that q_d, $\dot{q}_d$, $\ddot{q}_d$, $\dddot{q}_d \in L_\infty$). To facilitate the subsequent control development, we also define the stator current tracking error and the rotor flux tracking error as follows

$$\eta_{Ia} = I_{da} - I_a, \; \eta_{Ib} = I_{db} - I_b,$$
$$\eta_{\psi a} = \psi_{da} - \psi_a, \quad \eta_{\psi b} = \psi_{db} - \psi_b \tag{D.8}$$

where $I_{da}(t)$, $I_{db}(t)$ represent the desired stator current trajectories, and $\psi_{da}(t)$, $\psi_{db}(t)$ represent the desired rotor flux trajectories. The explicit definitions for I_{da}, I_{db}, ψ_{da}, and ψ_{db} will be given during the design procedure given below.

D.3.1 Filtered Tracking Error System

Given the definition of the filtered tracking error given in (D.7) and the electrical tracking error variables given in (D.8), we can write the mechanical subsystem dynamics of (D.1) as

$$\begin{aligned} M_m\dot{r} \;=\;& M_m(\ddot{q}_d + \alpha\dot{e}) + W_m\theta_m - \alpha_2(\psi_{da}I_{db} - \psi_{db}I_{da}) \\ & +\alpha_2\psi_{da}\eta_{Ib} - \alpha_2\psi_{db}\eta_{Ia} + \alpha_2 I_{db}\eta_{\psi a} - \alpha_2 I_{da}\eta_{\psi b} \\ & +\alpha_2\eta_{Ia}\eta_{\psi b} - \alpha_2\eta_{Ib}\eta_{\psi a}. \end{aligned} \tag{D.9}$$

From the form of (D.9) and the subsequent flux tracking control objective, we design the desired current trajectories as follows

$$I_{da} = -\psi_{db}\tau_d + \tfrac{\psi_{da}}{K_I}(R_r + \alpha_2 r\tau_d)$$
$$I_{db} = \psi_{da}\tau_d + \tfrac{\psi_{db}}{K_I}(R_r + \alpha_2 r\tau_d) \tag{D.10}$$

where $\tau_d(t)$ denotes the desired torque trajectory defined as follows

$$\tau_d = \frac{1}{\alpha_2\left(\psi_{da}^2 + \psi_{db}^2\right)}\left(\hat{M}_m(\ddot{q}_d + \alpha\dot{e}) + W_m\hat{\theta}_m + K_s r\right) \tag{D.11}$$

where K_s is a positive control gain, $\hat{M}_m(t) \in \Re^1$ is a dynamic estimate of the inertia term M_m, $\hat{\theta}_m(t) \in \Re^p$ is a dynamic estimate of the parameter vector θ_m (To eliminate overparameterization [18], the update laws for $\hat{M}_m$ and $\hat{\theta}_m$ will be explicitly defined later in the design procedure). After explicitly substituting I_{da}, I_{db}, and τ_d of (D.10) and (D.11), respectively, into only the first line of (D.9) and then simplifying the resulting expression, we obtain the closed-loop dynamics for the filtered tracking error in the following form

$$\begin{aligned} M_m\dot{r} =\;& -K_s r + \tilde{M}_m(\ddot{q}_d + \alpha\dot{e}) + W_m\tilde{\theta}_m + \\ & \alpha_2 I_{db}\eta_{\psi a} - \alpha_2 I_{da}\eta_{\psi b} + \alpha_2\psi_{da}\eta_{Ib} \\ & -\alpha_2\psi_{db}\eta_{Ia} + \alpha_2\eta_{Ia}\eta_{\psi b} - \alpha_2\eta_{Ib}\eta_{\psi a} \end{aligned} \tag{D.12}$$

where $\tilde{M}_m(t) \in \Re^1$ and $\tilde{\theta}_m(t) \in \Re^p$ denote the parameter estimation error defined by

$$\tilde{M}_m = M_m - \hat{M}_m \quad \text{and} \quad \tilde{\theta}_m = \theta_m - \hat{\theta}_m. \tag{D.13}$$

D.3.2 Flux Tracking Error System

After taking the time derivative of the flux tracking error $\eta_{\psi a}$ defined in (D.8), substituting the rotor flux dynamics of (D.4) for $\dot{\psi}_a$, and then multiplying the resulting expression by L_r, we have

$$L_r\dot{\eta}_{\psi a} = L_r\dot{\psi}_{da} + R_r\psi_a + \alpha_3\dot{q}\psi_b - K_I I_a. \tag{D.14}$$

The right-hand side of (D.14) can be written in terms of the current and flux tracking error defined in (D.8) in the following advantageous manner

$$\begin{aligned} L_r\dot{\eta}_{\psi a} &= -R_r\eta_{\psi a} - \alpha_3\dot{q}\eta_{\psi b} + K_I\eta_{Ia} - \alpha_2 I_{db}r \\ &\quad + L_r\dot{\psi}_{da} + R_r\psi_{da} + \alpha_3\dot{q}\psi_{db} - K_I I_{da} + \alpha_2 I_{db}r \end{aligned} \tag{D.15}$$

where the term $\alpha_2 I_{db}r$ has been added and subtracted to compensate for the $\alpha_2 I_{db}\eta_{\psi a}$ term in the first line of (D.12). Based on the second row of the (D.15), we now design the desired flux trajectory ψ_{da} according to the following differential relationship

$$L_r\dot{\psi}_{da} = -R_r\psi_{da} - \alpha_3\dot{q}\psi_{db} + K_I I_{da} - \alpha_2 I_{db}r \qquad \psi_{da}(0) = \sqrt{\frac{\delta_o}{2}}\ ; \tag{D.16}$$

where δ_o is a positive design constant. Based on the structure of (D.16), the closed-loop flux tracking error dynamics for $\eta_{\psi a}$ are represented by the first line of (D.15). In a similar fashion, we can design the desired flux trajectory ψ_{db} according to the following differential relationship

$$L_r\dot{\psi}_{db} = -R_r\psi_{db} + \alpha_3\dot{q}\psi_{da} + K_I I_{db} + \alpha_2 I_{da}r \qquad \psi_{db}(0) = \sqrt{\frac{\delta_o}{2}}; \tag{D.17}$$

hence, the closed-loop flux tracking error dynamics for $\eta_{\psi b}$ can be obtained as follows

$$L_r\dot{\eta}_{\psi b} = -R_r\eta_{\psi b} + \alpha_3\dot{q}\eta_{\psi a} + K_I\eta_{Ib} + \alpha_2 I_{da}r. \tag{D.18}$$

After explicitly substituting the desired current trajectories defined in (D.10) into the desired flux differential relationships of (D.16) and (D.17) and then simplifying the resulting expression, we can obtain the following expressions for the differential equations governing the desired flux trajectory

$$\begin{aligned} L_r\dot{\psi}_{da} &= -\psi_{db}\Omega \\ L_r\dot{\psi}_{db} &= \psi_{da}\Omega \\ \psi_{da}(0) &= \psi_{db}(0) = \sqrt{\tfrac{\delta_o}{2}} \end{aligned} \tag{D.19}$$

where the auxiliary function $\Omega(t)$ is defined as

$$\Omega=\left(\alpha_3\dot{q}+K_I\tau_d+\alpha_2 r\frac{R_r}{K_I}+\alpha_2^2 r^2\frac{\tau_d}{K_I}\right). \tag{D.20}$$

Remark D.2

Based on the structure of (D.19), it is easy to show using standard separation of variables techniques that

$$\psi_{da}^2(t)+\psi_{db}^2(t)=\delta_o \qquad \forall t\in[0,\infty); \tag{D.21}$$

hence, the relationship given by (D.21) illustrates that $\psi_{da}(t),\psi_{db}(t)\in L_\infty$ [19]. In addition, we note that relationship given by (D.21) precludes the possibility of singularities in the desired torque signal given in (D.11); furthermore, this relationship allows us to replace all occurrences of the term $\psi_{da}^2(t)+\psi_{db}^2(t)$ in the control architecture with the constant design parameter δ_o.

To provide motivation for some of the control terms injected during the subsequent voltage control input design, we can perform the following preliminary analysis. Specifically, we define the following non-negative function

$$V_1=\frac{1}{2}M_m r^2+\frac{1}{2}L_r\left(\eta_{\psi a}^2+\eta_{\psi b}^2\right). \tag{D.22}$$

After taking the time derivative of (D.22) along (D.12), (D.18), and the first line of (D.15), we can obtain the following simplified expression

$$\begin{aligned}\dot{V}_1 \;=\;& -K_s r^2-R_r\left(\eta_{\psi a}^2+\eta_{\psi b}^2\right)+\tilde{M}_m\left(\ddot{q}_d+\alpha\dot{e}\right)r+W_m\tilde{\theta}_m r\\ &+\alpha_2 r\psi_{da}\eta_{Ib}-\alpha_2 r\psi_{db}\eta_{Ia}+\left(\alpha_2 r\eta_{Ia}+K_I\eta_{Ib}\right)\eta_{\psi b}\\ &+\left(-\alpha_2 r\eta_{Ib}+K_I\eta_{Ia}\right)\eta_{\psi a}.\end{aligned} \tag{D.23}$$

From the form of (D.23), we can make the following observations: i) the adaptive update laws must be designed to compensate for the last two terms in the first line, ii) the first two terms in the second line are measurable; hence, these terms can be directly cancelled out by injecting the appropriate terms into the current tracking error dynamics, and iii) the last two terms are unmeasurable; however, these terms can be damped-out by injecting the appropriate nonlinear damping [14] control terms into the current tracking error dynamics.

D.3.3 Current Tracking Error System

After taking the time derivative of the current tracking error η_{Ia} defined in (D.8), substituting the current dynamics of (D.2) for $\dot{I}_a$ and the time derivative of I_{da} given by (D.10) for $\dot{I}_{da}$, and then multiplying the resulting expression by L_I, we have

$$\begin{aligned} L_I\dot{\eta}_{Ia} = \; & L_I\left(-\dot{\psi}_{db}\tau_d + \left(\frac{\psi_{da}}{K_I}r\alpha_2 - \psi_{db}\right)\dot{\tau}_d + \right. \\ & \left.\left(\frac{R_r}{K_I} + \frac{r\tau_d\alpha_2}{K_I}\right)\dot{\psi}_{da}\right) + L_I\frac{\psi_{da}}{K_I}\tau_d\alpha_2\dot{r} \\ & + R_I I_a - \alpha_1\psi_a - \alpha_2\psi_b\dot{q} - V_a. \end{aligned} \tag{D.24}$$

By applying the tracking error definitions of (D.8) to (D.24), we can rewrite the open-loop current tracking dynamics for η_{Ia} in the following advantageous form

$$\begin{aligned} L_I\dot{\eta}_{Ia} = \; & L_I\left(-\psi_{da}\frac{\Omega}{L_r}\tau_d + \left(\frac{\psi_{da}}{K_I}r\alpha_2 - \psi_{db}\right)\dot{\tau}_d \right. \\ & \left. -\left(\frac{R_r}{K_I} + \frac{r\tau_d\alpha_2}{K_I}\right)\psi_{db}\frac{\Omega}{L_r}\right) \\ & + L_I\frac{\psi_{da}}{K_I}\tau_d\alpha_2\left(\ddot{q}_d + \alpha\dot{e}\right) - L_I\frac{\psi_{da}}{K_I}\tau_d\alpha_2\ddot{q} + R_I I_a \\ & + \alpha_1\eta_{\psi a} + \alpha_2\eta_{\psi b}\dot{q} - \alpha_1\psi_{da} - \alpha_2\psi_{db}\dot{q} - V_a \end{aligned} \tag{D.25}$$

where (D.7) and (D.19) have been utilized. Given the definition of the desired torque defined in (D.11) and the relationship given by (D.21), we know that τ_d can be expressed as a function of $\hat{M}_m$, $\hat{\theta}_m$, q, $\dot{q}$, δ_o, and t; therefore, $\dot{\tau}_d$ can be expressed as

$$\dot{\tau}_d = \frac{\partial\tau_d}{\partial t} + \frac{\partial\tau_d}{\partial\hat{M}_m}\dot{\hat{M}}_m + \frac{\partial\tau_d}{\partial\hat{\theta}_m}\dot{\hat{\theta}}_m + \frac{\partial\tau_d}{\partial q}\dot{q} + \frac{\partial\tau_d}{\partial\dot{q}}\ddot{q} \tag{D.26}$$

where the above partial derivatives are explicitly given in Section D.7. Since the subsequent adaptive update laws for $\hat{M}_m$ and $\hat{\theta}_m$ will allow for direct substitution for $\dot{\hat{M}}_m$ and $\dot{\hat{\theta}}_m$ into the right-hand side of (D.26), all of the quantities on the right-hand side of (D.26) are measurable except for $\ddot{q}$; however, from (D.1), the rotor acceleration can be expressed as

$$\ddot{q} = -\frac{1}{M_m}W_m\theta_m + \frac{\alpha_2}{M_m}\left(\psi_a I_b - \psi_b I_a\right). \tag{D.27}$$

By utilizing the tracking error definitions given in (D.8), (D.27) can be rewritten as

$$\ddot{q} = -\frac{1}{M_m}W_m\theta_m + \frac{\alpha_2}{M_m}\left(\psi_{da}I_b - \psi_{db}I_a\right) - \frac{\alpha_2}{M_m}\left(\eta_{\psi a}I_b - \eta_{\psi b}I_a\right). \tag{D.28}$$

After substituting (D.28) into (D.26), then substituting the resulting expression for $\dot{\tau}_d$ along with (D.28) for $\ddot{q}$ into (D.25), and then multiplying both sides of the resulting expression by M_m, we can rearrange the open-loop current tracking dynamics for η_{Ia} in the following advantageous manner

$$\begin{aligned} L_I M_m \dot{\eta}_{Ia} = \; & \Omega_1 + M_m \Omega_2 + Y_1 \theta_m + \alpha_1 \eta_{\psi a} M_m \\ & + \alpha_2 \dot{q} \eta_{\psi b} M_m + \Omega_3 \eta_{\psi a} + \Omega_4 \eta_{\psi b} - M_m V_a \end{aligned} \tag{D.29}$$

where $\Omega_1(t)$, $\Omega_2(t)$, $\Omega_3(t)$, $\Omega_4(t) \in \Re^1$ and $Y_1 \in \Re^{1 \times p}$ are auxiliary measurable functions explicitly defined in Section D.7. Based on the open-loop dynamics of (D.29), the structure of (D.23), and the subsequent analysis, we define the voltage input control V_a as

$$\begin{aligned} V_a \; = \; & K_e \eta_{Ia} + \Omega_2 + \hat{M}_m^{-1} \left[\Omega_1 + Y_1 \hat{\theta}_m - \alpha_2 \psi_{db} r \right] \\ & + K_n \left(\alpha_2 r + \Omega_4 \right)^2 \eta_{Ia} + \alpha_1^2 K_n \eta_{Ia} + \alpha_2^2 \dot{q}^2 K_n \eta_{Ia} \\ & + K_n \left(K_I + \Omega_3 \right)^2 \eta_{Ia} \end{aligned} \tag{D.30}$$

where K_e, K_n are two positive control gains. After substituting the control input given by (D.30) into (D.29), we can write the closed-loop dynamics for η_{Ia} as follows

$$\begin{aligned} L_I M_m \dot{\eta}_{Ia} = \; & -K_e M_m \eta_{Ia} + Y_1 \tilde{\theta}_m + \alpha_2 \psi_{db} r \\ & + \left[\alpha_1 \eta_{\psi a} M_m - \alpha_1^2 K_n M_m \eta_{Ia} \right] \\ & + \left[\alpha_2 \dot{q} \eta_{\psi b} M_m - \alpha_2^2 \dot{q}^2 K_n M_m \eta_{Ia} \right] + \\ & \Omega_3 \eta_{\psi a} + \Omega_4 \eta_{\psi b} - K_n M_m \left(K_I + \Omega_3 \right)^2 \eta_{Ia} \\ & - K_n M_m \left(\alpha_2 r + \Omega_4 \right)^2 \eta_{Ia} \\ & - \hat{M}_m^{-1} \left(\Omega_1 + Y_1 \hat{\theta}_m - \alpha_2 \psi_{db} r \right) \tilde{M}_m . \end{aligned} \tag{D.31}$$

Similar to the above procedure, we can express the open-loop current tracking dynamics for η_{Ib} in the following advantageous form

$$\begin{aligned} L_I M_m \dot{\eta}_{Ib} \; = \; & \Omega_5 + M_m \Omega_6 + Y_2 \theta_m + \alpha_1 \eta_{\psi b} M_m - \alpha_2 \dot{q} \eta_{\psi a} M_m \\ & + \Omega_7 \eta_{\psi a} + \Omega_8 \eta_{\psi b} - M_m V_b \end{aligned} \tag{D.32}$$

where $\Omega_5(t)$, $\Omega_6(t)$, $\Omega_7(t)$, $\Omega_8(t) \in \Re^1$ and $Y_2 \in \Re^{1 \times p}$ are auxiliary measurable functions explicitly defined in Section D.7. Based on (D.32), the structure of (D.23), and the subsequent analysis, we define the voltage control input V_b as follows

$$\begin{aligned} V_b &= K_e\eta_{Ib} + \Omega_6 + \hat{M}_m^{-1}\left[\Omega_5 + Y_2\hat{\theta}_m + \alpha_2\psi_{da}r\right] \\ &\quad + K_n\left(-\alpha_2 r + \Omega_7\right)^2\eta_{Ib} + \alpha_1^2 K_n\eta_{Ib} + \alpha_2^2\dot{q}^2 K_n\eta_{Ib} \\ &\quad + K_n\left(K_I + \Omega_8\right)^2\eta_{Ib}. \end{aligned} \tag{D.33}$$

After substituting V_b of (D.33) into the open-loop dynamics of (D.32), we can write the closed-loop current tracking error dynamics for η_{Ib} in the following form

$$\begin{aligned} L_I M_m\dot{\eta}_{Ib} = &-K_e M_m\eta_{Ib} + Y_2\tilde{\theta}_m - \alpha_2\psi_{da}r + \Omega_8\eta_{\psi b} \\ &+\left[\alpha_1\eta_{\psi b}M_m - \alpha_1^2 K_n M_m\eta_{Ib}\right] \\ &+\left[-\alpha_2\dot{q}\eta_{\psi a}M_m - \alpha_2^2\dot{q}^2 K_n M_m\eta_{Ib}\right] \\ &+\Omega_7\eta_{\psi a} - K_n M_m\left(K_I + \Omega_8\right)^2\eta_{Ib} \\ &-K_n M_m\left(-\alpha_2 r + \Omega_7\right)^2\eta_{Ib} \\ &-\hat{M}_m^{-1}\left(\Omega_5 + Y_2\hat{\theta}_m + \alpha_2\psi_{da}r\right)\tilde{M}_m. \end{aligned} \tag{D.34}$$

To provide motivation for the design of the adaptive update laws while also preparing for the statement of the main result, we perform the following preliminary analysis. Specifically, we define the following non-negative function

$$V = V_1 + \frac{1}{2}L_I M_m\left(\eta_{Ia}^2 + \eta_{Ib}^2\right) + \frac{1}{2}\Gamma_1^{-1}\tilde{M}_m^2 + \frac{1}{2}\tilde{\theta}_m^T\Gamma_2^{-1}\tilde{\theta}_m \tag{D.35}$$

where V_1 was defined in (D.22), $\Gamma_1 \in \Re^1$ is a positive constant gain, and $\Gamma_2 \in \Re^{p\times p}$ is a positive definite, diagonal, constant gain matrix. After taking the time-derivative of V of (D.35) and substituting $\dot{V}_1$ from (D.23), the closed-loop error systems from (D.31) and (D.34) for $\dot{\eta}_{Ia}$ and $\dot{\eta}_{Ib}$, respectively, and then cancelling some common terms, we can arrange $\dot{V}$ in the following manner

$$\begin{aligned}
\dot{V} = & -K_s r^2 - R_r\left(\eta_{\psi a}^2 + \eta_{\psi b}^2\right) - K_e M_m \eta_{Ia}^2 - K_e M_m \eta_{Ib}^2 \\
& + \left[\alpha_1 \eta_{\psi a}\eta_{Ia} M_m - \alpha_1^2 K_n M_m \eta_{Ia}^2\right] \\
& + \left[\alpha_2 \dot{q}\eta_{\psi b}\eta_{Ia} M_m - \alpha_2^2 \dot{q}^2 K_n M_m \eta_{Ia}^2\right] \\
& + \left[(\alpha_2 r + \Omega_4)\eta_{\psi b}\eta_{Ia} - K_n M_m (\alpha_2 r + \Omega_4)^2 \eta_{Ia}^2\right] \\
& + \left[(K_I + \Omega_3)\eta_{\psi a}\eta_{Ia} - K_n M_m (K_I + \Omega_3)^2 \eta_{Ia}^2\right] \\
& + \left[\alpha_1 \eta_{\psi b}\eta_{Ib} M_m - \alpha_1^2 K_n M_m \eta_{Ib}^2\right] \\
& + \left[-\alpha_2 \dot{q}\eta_{\psi a}\eta_{Ib} M_m - \alpha_2^2 \dot{q}^2 K_n M_m \eta_{Ib}^2\right] \\
& + \left[(-\alpha_2 r + \Omega_7)\eta_{\psi a}\eta_{Ib} - K_n M_m (-\alpha_2 r + \Omega_7)^2 \eta_{Ib}^2\right] \\
& + \left[(K_I + \Omega_8)\eta_{\psi b}\eta_{Ib} - K_n M_m (K_I + \Omega_8)^2 \eta_{Ib}^2\right] \\
& + \tilde{\theta}_m^T \left(W_m^T r + Y_1^T \eta_{Ia} + Y_2^T \eta_{Ib} - \Gamma_2^{-1}\dot{\hat{\theta}}_m\right) \\
& + \tilde{M}_m \left((\ddot{q}_d + \alpha\dot{e}) r - \Gamma_1^{-1}\dot{\hat{M}}_m\right) \\
& - \tilde{M}_m \hat{M}_m^{-1}\left(\left(\Omega_1 + Y_1\hat{\theta}_m - \alpha_2 \psi_{db} r\right)\eta_{Ia}\right. \\
& \left. + \left(\Omega_5 + Y_2\hat{\theta}_m + \alpha_2 \psi_{da} r\right)\eta_{Ib}\right)
\end{aligned} \tag{D.36}$$

Based on the structure of (D.36), we design the parameter update laws[1] [18] for $\dot{\hat{M}}_m$ and $\dot{\hat{\theta}}_m$ as follows

$$\dot{\hat{\theta}}_m = \Gamma_2\left(W_m^T r + Y_1^T \eta_{Ia} + Y_2^T \eta_{Ib}\right) \tag{D.37}$$

$$\dot{\hat{M}}_m = \begin{cases} \Omega_m & \text{if} \quad \hat{M}_m > \underline{M}_m \\ \Omega_m & \text{if} \quad \hat{M}_m = \underline{M}_m \quad \text{and} \quad \Omega_m \geq 0 \\ 0 & \text{if} \quad \hat{M}_m = \underline{M}_m \quad \text{and} \quad \Omega_m < 0 \end{cases} \tag{D.38}$$

[1] See [19] for further information on the standard projection algorithm used for $\hat{M}_m$.

where $\hat{M}_m(0) = \underline{M}_m$ and the auxiliary term $\Omega_m(t)$ is given by

$$\begin{aligned}\Omega_m &= \Gamma_1\left(\left(\ddot{q}_d + \alpha\dot{e}\right)r - \hat{M}_m^{-1}\left(\Omega_1 + Y_1\hat{\theta}_m - \alpha_2\psi_{db}r\right)\eta_{Ia}\right) \\ &\quad -\Gamma_1\hat{M}_m^{-1}\left(\Omega_5 + Y_2\hat{\theta}_m + \alpha_2\psi_{da}r\right)\eta_{Ib}\end{aligned} \tag{D.39}$$

After substituting (D.38), (D.37), and (D.39) into (D.36) and applying the nonlinear damping argument [14] to the bracketed terms on the second thru the ninth line of (D.36), we can formulate the following upper bound for $\dot{V}$

$$\begin{aligned}\dot{V} \leq\ & -K_s r^2 - K_e M_m\left(\eta_{Ia}^2 + \eta_{Ib}^2\right) - \\ & \left(R_r - \frac{2\bar{M}_m}{K_n} - \frac{2}{K_n\underline{M}_m}\right)\left(\eta_{\psi a}^2 + \eta_{\psi b}^2\right).\end{aligned} \tag{D.40}$$

D.4 Main Result

Based on the above control development, we now state the main result of the paper.

Theorem D.1

If the following sufficient condition

$$K_n > \frac{2\bar{M}_m}{R_r} + \frac{2}{\underline{M}_m R_r} \tag{D.41}$$

is satisfied, then the proposed input voltage control ensures global asymptotic rotor position/velocity tracking of the form

$$\lim_{t\to\infty} e(t),\ \dot{e}(t) = 0. \tag{D.42}$$

Proof:

First, note that V of (D.35) can be bounded as

$$\lambda_1\|z\| \leq V \leq \lambda_2\|z\| \tag{D.43}$$

where

$$\lambda_1 = \frac{1}{2}\min\left\{M_m, L_r, L_I M_m, \Gamma_1^{-1}, \lambda_{\min}\left\{\Gamma_2^{-1}\right\}\right\}, \tag{D.44}$$

$$\lambda_2 = \frac{1}{2}\max\left\{M_m, L_r, L_I M_m, \Gamma_1^{-1}, \lambda_{\max}\left\{\Gamma_2^{-1}\right\}\right\}, \tag{D.45}$$

and

$$z = \begin{bmatrix} r & \eta_{Ia} & \eta_{Ib} & \eta_{\psi a} & \eta_{\psi b} & \tilde{M}_m & \tilde{\theta}_m^T \end{bmatrix}^T \in \Re^{6+p}. \tag{D.46}$$

From (D.40), $\dot{V}$ can be upper bounded in the following form

$$\dot{V} \leq -\lambda_3 \|x\|^2 \tag{D.47}$$

where

$$\lambda_3 = \min\left\{K_s, K_e M_m, R_r - \frac{2}{K_n \underline{M}_m} - \frac{2\bar{M}_m}{K_n}\right\} \tag{D.48}$$

and

$$x = \begin{bmatrix} r & \eta_{Ia} & \eta_{Ib} & \eta_{\psi a} & \eta_{\psi b} \end{bmatrix}^T \in \Re^5. \tag{D.49}$$

If the control gain condition given by (D.41) is satisfied, we know that λ_3 in (D.48) will be positive; therefore, since V is positive definite and $\dot{V}$ is negative semi-definite, we know that $V(t)$ is upper bounded by $V(0)$ hence $z \in L_\infty^{6+p}$ and $x \in L_\infty^5$. In addition, it easy to show from (D.47) that $x \in L_2^5$ [19]. Given $z \in L_\infty^{6+p}$, we know that $r \in L_\infty$ (and hence e, $\dot{e}$, q, $\dot{q}$,$\in L_\infty$ [17]) and that $\tilde{M}_m \in L_\infty$, $\tilde{\theta}_m \in L_\infty^p$ (and hence $\hat{M}_m \in L_\infty$, $\hat{\theta}_m \in L_\infty^p$); hence, (D.19) can used to show that ψ_{da}, $\psi_{db} \in L_\infty$. In addition, since $z \in L_\infty^{6+p}$, we know that $\eta_{\psi a}$, $\eta_{\psi b}$, η_{Ia}, $\eta_{Ib} \in L_\infty$ (and hence ψ_a, ψ_b, I_a, $I_b \in L_\infty$). From the above information and the structure of the control input, it is now easy to show that V_a, $V_b \in L_\infty$; therefore, all of the system signals remain bounded during closed-loop operation. In addition, the closed-loop error systems can also be used to illustrate that $\dot{x} \in L_\infty^5$. Finally, since $x \in L_\infty^5$, $x \in L_2^5$, and $\dot{x} \in L_\infty^5$, Barbalat's Lemma [19] can be used to show that

$$\lim_{t\to\infty} x = 0 \quad \text{which yields} \quad \lim_{t\to\infty} r, \eta_{\psi a}, \eta_{\psi b}, \eta_{Ia}, \eta_{Ib} = 0.$$

Given the definition of the filtered tracking error of (D.7) and the above result, standard linear control arguments [17] can be use to state (D.42). □

Remark D.3

If the mechanical subsystem parameters are known exactly, the proposed control structure can easily be redesigned (*i.e.*, the desired torque trajectory of (D.11) and the voltage inputs of (D.30) and (D.33) would be redesigned such that $\hat{M}_m = M_m$ and $\hat{\theta}_m = \theta_m$ while setting $\dot{\hat{M}}_m = \dot{\hat{\theta}}_m = 0$) to deliver global exponential rotor position/velocity tracking (It should be noted that the rate of convergence will be dependent on the value of rotor resistance). In addition, we note that the stator current transformation for the induction motor can be selected to be independent of rotor position measurements; hence, if $W_m(\cdot)$ of (D.1) is independent of rotor position, we can set α of (D.7) to *zero* in the proposed control structure to yield a control strategy which delivers global asymptotic rotor velocity tracking (*i.e.*, for parametric uncertainty in mechanical subsystem only) or global exponential rotor velocity tracking (*i.e.*, for the case of exact model knowledge) with only rotor velocity and stator current measurements.

D.5 Experimental Results

The hardware setup used to implement the controller consists of the following: 1) a IBM compatible 66MHz 486 PC, 2) a TMS320C30 DSP system board from Spectrum Signal Processing, 3) a DS-2 encoder interface board from Integrated Motions Inc., 4) a Baldor model M3541, 2-pole induction motor (rated speed of 3450 RPM, rated current of 2.7A, and rated voltage of 230V) with a 1024 line encoder, 5) 3 Techron 7571 linear amplifiers, 6) 3 CSLB1AD hall effect current sensors from Microswitch, and 7) assorted interfacing electronics and hardware. The rotor velocity signal was obtained by use of a backwards difference algorithm applied to the rotor position signal with the resulting signal being filtered by a second-order digital filter. The experiment was run at a sampling time of 0.75ms.

Since we are interested in utilizing the induction motor for relatively low-speed, high performance position tracking applications such as robotics, we constructed the mechanical subsystem such that the term $W_m(q,\dot{q})\theta_m$ in (D.1) can be reasonably modeled as

$$W_m(q,\dot{q})\theta_m = B\dot{q} + N\sin(q)$$

where

$$M_m = J_m + \frac{mL_0^2}{3}, \quad N = \frac{mgL_0}{2},$$

B represents the viscous friction coefficient, J_m is the rotor inertia, m is the link mass, L_0 is the link length, and g is the gravity coefficient. By utilizing standard test procedures, the nominal values for the electromechanical system parameters were determined to be

$$\begin{array}{lll} J_m = 1.87\times 10^{-4} kg\cdot m^2, & m = 0.401kg, & L_0 = 0.305m, \\ L_r = 0.306H, & L_s = 0.243H, & M = 0.225H, \\ R_s = 3.05\Omega, & n_p = 1, & B = 0.003N\cdot m\cdot s/rad \\ R_r = 2.12\Omega & & g = 9.81kg\cdot m/\sec^2. \end{array}$$

The desired position trajectory is selected as the following smooth start sinusoid

$$q_d(t) = \frac{\pi}{2}\sin(4t)\left(1 - e^{-0.1t^3}\right) rad$$

where $q_d(0) = \dot{q}_d(0) = \ddot{q}_d(0) = \dddot{q}_d(0) = 0$ while δ_o of (D.21) was set to $4\ Wb\cdot Wb$. The adaptive update laws were calculated using a standard trapezoidal integration routine with the initial parameter estimates set to 80% of their nominal values. The best tracking performance was established using the following control gain values

$$\alpha = 45,\ K_s = 0.1,\ K_e = 7.0,\ K_n = 0.001,$$

$$\Gamma_1 = 0.001,\ \Gamma_2 = diag\{0.0001, 1.5\}.$$

The resulting rotor position tracking error is shown in Figure D.1 while a representative input 3-phase voltage is shown in Figure D.2. From the rotor position tracking error plot given in Figure D.1, we can see that the "steady state" rotor position tracking error is approximately within ±0.35 degrees. Due to digital implementation issues, the lack of velocity measurements, and inexact modeling, the rotor position tracking error does approach zero; rather, the tracking error is driven to a small value.

D.6 Conclusions

In this appendix, we have presented a singularity-free, adaptive, partial-state feedback, position/velocity tracking controller for the full-order, nonlinear dynamic model representing an induction motor actuating a mechanical subsystem. The important aspects of the controller are that global asymptotic rotor position/velocity tracking is achieved with neither rotor flux measurements nor *a priori* knowledge of the mechanical subsystem parameters. Preliminary experimental results seem to indicate that the rugged induction motor may indeed be a viable candidate actuator for robot manipulators. Although it is a laborious endeavor, the controller presented in this paper can be extended for the control of robot manipulators actuated by induction motors by using aspects of the techniques presented in [15] and [20]. This extension will be the subject of upcoming work.

D.7 Auxiliary Terms

$$\frac{\partial \tau_d}{\partial q} = \frac{1}{\alpha_2 \delta_o}\left[\frac{\partial W_m}{\partial q}\hat{\theta}_m - \alpha K_s\right], \qquad \frac{\partial \tau_d}{\partial \hat{\theta}_m} = \frac{W_m}{\alpha_2 \delta_o},$$

$$\frac{\partial \tau_d}{\partial \dot{q}} = \frac{1}{\alpha_2 \delta_o}\left[-\alpha \hat{M}_m + \frac{\partial W_m}{\partial \dot{q}}\hat{\theta}_m - K_s\right], \qquad \frac{\partial \tau_d}{\partial \hat{M}_m} = \frac{\ddot{q}_d + \alpha \dot{e}}{\alpha_2 \delta_o},$$

$$\frac{\partial \tau_d}{\partial t} = \frac{1}{\alpha_2 \delta_o}\left[\alpha K_s \dot{q}_d + \left(\alpha \hat{M}_m + K_s\right)\ddot{q}_d + \hat{M}_m \dddot{q}_d\right],$$

$$\Omega_9 = L_I\left(\frac{\psi_{da}}{K_I} r \alpha_2 - \psi_{db}\right)\frac{\partial \tau_d}{\partial \dot{q}} - L_I \frac{\psi_{da}}{K_I}\tau_d \alpha_2,$$

$$\begin{aligned} \Omega_1 &= \alpha_2 \Omega_9 \left(\psi_{da} I_b - \psi_{db} I_a\right), \ \Omega_3 = -\alpha_2 \Omega_9 I_b, \ \Omega_4 = \alpha_2 \Omega_9 I_a, \\ Y_1 &= -\Omega_9 W_m, \end{aligned}$$

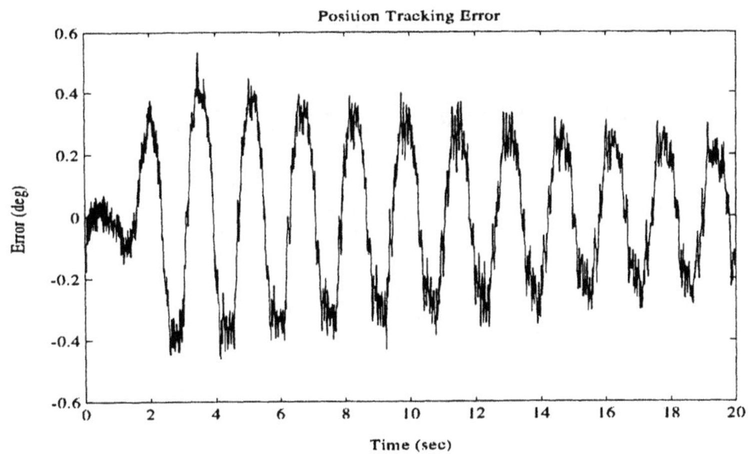

Figure D.1: Position Tracking Error

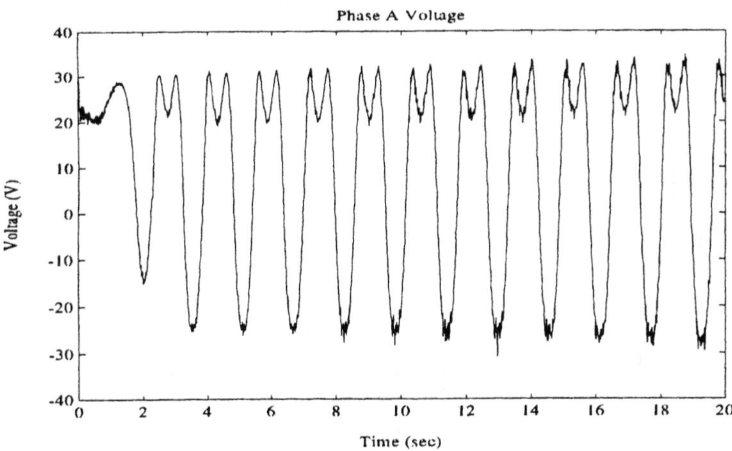

Figure D.2: Phase A Voltage

$$\begin{aligned}
\Omega_2 = {} & L_I\left(-\frac{\Omega}{L_r}\psi_{da}\tau_d - \left(\frac{R_r}{K_I} + \frac{r\tau_d\alpha_2}{K_I}\right)\frac{\Omega}{L_r}\psi_{db}\right. \\
& \left. + \frac{\psi_{da}}{K_I}\tau_d\alpha_2\left(\ddot{q}_d + \alpha\dot{e}\right)\right) + R_I I_a - \alpha_1\psi_{da} - \alpha_2\psi_{db}\dot{q} \\
& + L_I\left(\frac{\psi_{da}}{K_I}r\alpha_2 - \psi_{db}\right)\left(\frac{\partial\tau_d}{\partial t} + \frac{\partial\tau_d}{\partial\hat{M}_m}\dot{\hat{M}}_m\right) \\
& + L_I\left(\frac{\psi_{da}}{K_I}r\alpha_2 - \psi_{db}\right)\left(+\frac{\partial\tau_d}{\partial\hat{\theta}_m}\dot{\hat{\theta}}_m + \frac{\partial\tau_d}{\partial q}\dot{q}\right),
\end{aligned}$$

$$\Omega_{10} = L_I\left(\frac{\psi_{db}}{K_I}r\alpha_2 + \psi_{da}\right)\frac{\partial\tau_d}{\partial\dot{q}} - L_I\frac{\psi_{db}}{K_I}\tau_d\alpha_2,$$

$$\Omega_5 = \alpha_2\Omega_{10}\left(\psi_{da}I_b - \psi_{db}I_a\right), \quad \Omega_7 = -\alpha_2\Omega_{10}I_b,$$

$$\Omega_8 = \alpha_2\Omega_{10}I_a, \quad Y_2 = -\Omega_{10}W_m,$$

$$\begin{aligned}
\Omega_6 = {} & L_I\left(-\frac{\Omega}{L_r}\psi_{db}\tau_d + \left(\frac{R_r}{K_I} + \frac{r\tau_d\alpha_2}{K_I}\right)\frac{\Omega}{L_r}\psi_{da}\right. \\
& \left. + \frac{\psi_{db}}{K_I}\tau_d\alpha_2\left(\ddot{q}_d + \alpha\dot{e}\right)\right) + R_I I_b - \alpha_1\psi_{db} + \alpha_2\psi_{da}\dot{q} \\
& + L_I\left(\frac{\psi_{db}}{K_I}r\alpha_2 + \psi_{da}\right)\left(\frac{\partial\tau_d}{\partial t} + \frac{\partial\tau_d}{\partial\hat{M}_m}\dot{\hat{M}}_m\right) \\
& + L_I\left(\frac{\psi_{db}}{K_I}r\alpha_2 + \psi_{da}\right)\left(+\frac{\partial\tau_d}{\partial\hat{\theta}_m}\dot{\hat{\theta}}_m + \frac{\partial\tau_d}{\partial q}\dot{q}\right).
\end{aligned}$$

Bibliography

[1] R. Marino, S. Peresada, and P. Valigi, "Adaptive Input-Output Linearizing Control of Induction Motors", *IEEE Transactions on Automatic Control*, Vol. 38, No. 2, Feb., 1993, pp. 208-221.

[2] D. Dawson, J. Carroll, and M. Schneider, "Integrator Backstepping Control for a Brush dc Motor Turning a Robotic Load", *IEEE Transactions on Controls Systems Technology*, Vol. 2, No. 3, Sept., 1994, pp. 233-244.

[3] I. Kanellakopoulos, P. Krein, and F. Disilvestro, "A New Controller Observer Design for Induction Motor Control", *ASME Winter Meeting*, Anahiem, CA, DSC-Vol. 43, 1992, pp. 43-47.

[4] I. Kanellakopoulos and P. Krein, "Integral-action Nonlinear Control of Induction Motors, *12th IFAC World Congress*, Sydney, Australia, 1993, pp. 1700-1704.

[5] R. Marino, S. Peresada, and P. Valigi, "Exponentially Convergent Rotor Resistance Estimation for Induction Motors", *IEEE Transactions on Industrial Electronics,* Vol. 42, No. 5, Oct. 1995, pp. 508-515.

[6] J. Hu, D.M. Dawson, and Y. Qian, "Position Tracking of an Induction Motor via Partial State Feedback", *Automatica,* Vol. 31, No. 7, July, 1995, pp. 989-1000.

[7] C. Canudas, R. Ortega, and S. Seleme, "Robot Motion Control using Induction Motor Drives", *IEEE International Conference on Robotics and Automation,* Atlanta, Ga., 1993, pp. 533-538.

[8] J. Hu, D.M. Dawson, and Y. Qian, "Position Tracking for Robot Manipulators Driven by Induction Motors without Flux Measurements", *IEEE Transactions on Robotics and Automation,* Vol. 12, No. 3, 1996, pp. 419 - 438.

[9] T. Raumer, J. Dion, and L. Dugard, "Applied Nonlinear Control of an Induction Motor using Digital Signal Processing", *IEEE Transactions on Control Systems Technology,* Vol. 2, No. 4, pp. 327-335, Dec. 1994.

[10] J. Yang, W. Yu, and L. Fu, "Nonlinear Observer-Based Tracking Control of Induction Motors with Unknown Load", *IEEE Transactions on Industrial Electronics,* Vol. 42, No. 6, Dec. 1995, pp. 579-585.

[11] M. Bodson, J. Chiasson, and R. Novotnak, "High-Performance Induction Motor Control Via Input-Output Linearization", *IEEE Control Systems Magazine,* Vol., 14, No., 4, Aug., 1994, pp. 25-33.

[12] Espinosa-Perez G., and R. Ortega, "State Observers are Unnecessary for Induction Motor Control, *System and Control Letters,* Vol. 23, No. 5, 1994, pp. 315-323.

[13] R. Ortega, P. Nicklasson, and G. Espinosa-Perez, "On Speed Control of Induction Motors", *Automatica,* Vol. 32, No. 3, March 1996, pp. 455-460.

[14] M. Krstic, I. Kanellakopoulos, and P. Kokotovic, *Nonlinear and Adaptive Control Design,* John Wiley & Sons, 1995.

[15] T. Burg, D. Dawson, J. Hu, and M. de Queiroz, "An Adaptive Partial State Feedback Controller for RLED Robot Manipulators", *IEEE Transactions on Automatic Control,* Vol 41, No. 7, pp. 1024-1031, July, 1996.

[16] P. Krause, *Analysis of Electric Machinery*, McGraw Hill, 1986.

[17] J. Slotine and W. Li, *Applied Nonlinear Control*, Englewood Cliff, NJ: Prentice Hall Co., 1991.

[18] M. Krstic, I. Kanellakopoulos, and P. Kokotovic, "Adaptive Nonlinear Control Without Overparameterization", *Systems & Control Letters*, Vol. 19, 1992, pp. 177-185.

[19] S. Sastry and M. Bodson, *Adaptive Control: Stability, Convergence, and Robustness*, Englewoods Cliff, NJ: Prentice Hall Co., 1989.

[20] M.M. Bridges and D.M. Dawson, "Adaptive Control for a Class of Direct Drive Robot Manipulators", *International Journal of Adaptive Control and Signal Processing*, Vol, 10, Jul., 1996, pp. 417-441.

Appendix E

Rotor Resistance Problem

In this paper, we present an adaptive, partial-state feedback, position and/or velocity tracking controller for the full-order, nonlinear dynamic model representing an induction motor actuating a mechanical subsystem. The proposed controller compensates for uncertainty in the form of the rotor resistance parameter and all of the mechanical subsystem parameters while yielding asymptotic rotor position/velocity tracking. The proposed controller does not require measurement of rotor flux or rotor current; however, the controller does exhibit a singularity when the magnitude of the estimated rotor flux is zero. Simulation results are provided to verify the effectiveness of the approach.

E.1 Introduction

Due to the simplistic construction and durability of the induction motor, its use in industrial applications, such as web-handling, overhead cranes, *etc.*, will probably increase over the next decade. Hence, several control researchers have proposed position/velocity tracking controllers for the *full-order* model used to describe the induction motor driving a mechanical load. For example, in [1], Marino *et al.* applied adaptive, nonlinear techniques for the design of a full-state feedback, velocity tracking controller for an induction motor driving a load (For earlier work and other related full-state feedback work, see the reviews in [1], [2], [3]). Recent work which involves the development of nonlinear controllers along with some corresponding experimental validation can be found in [4], [5], and [6]. For other recent induction motor control work involving other aspects such as magnetic optimization for the torque tracking problem, removal of the linear magnetic circuit assumption, and the design of a general torque tracking controller for a class of electric machines, the reader is referred to [7], [8], and [9].

Some of the earlier control techniques suffered from the requirement of full-state feedback. That is, for the induction motor control problem, full-state feedback means that rotor flux (or rotor current) must be measured for actual implementation. Since measurement of rotor flux would tend to counter the argument which promulgates the use of the induction motor due to its simplistic construction, many control researchers have further constrained this control problem to state that rotor flux measurements are not available; moreover, any auxiliary signals which are generated as observers or surrogates for the rotor flux must be utilized in a stable closed-loop fashion. For example, in [10] and [11], Kanellakopoulos *et al.* developed a velocity tracking controller which utilized a rotor flux observer in a stable closed-loop fashion; however, the set of possible control singularities was not discussed. In [12], Hu *et al.* developed a position/velocity tracking controller which only required position and stator current measurements; however, roughly speaking, control singularities rendered a local exponential tracking result. Some recent control strategies for robotic manipulators actuated by induction motors can also be found in [13] and [14]. To address some of the shortcomings in previous work, some researchers have proposed global position/velocity tracking controllers which do not exhibit control singularities while also avoiding rotor flux (or current) measurements. For example, Espinosa-Pereza *et al.* [15] presented a singularity-free, velocity tracking controller which did not require rotor current measurements; however, the convergence rate of the speed tracking error was determined by the natural mechanical damping of the motor. In [16], Ortega *et al.* used an ingenious filtering procedure and a two-part Lyapunov/Gronwall-inequality type argument to develop a controller which retained the desirable characteristics of the algorithm proposed in [15] without the mechanical damping restriction. In [17], Dawson *et al.* proposed a new set of desired electrical trajectories to illustrate how standard Lyapunov-type techniques can be used to design a singularity-free, adaptive rotor position tracking controller for a general, *uncertain* mechanical subsystem. Since all of the previous algorithms assumed exact knowledge of the electrical parameters, the next logical step in the evolution of induction motor control design would be the development of a controller which exhibits some robustness to electrical parameter variation. However, the design of a controller which enjoys this additional, desirable robustness feature is stymied by the fact that rotor flux measurements are not practical. Specifically, as pointed out in [18], it is not obvious how to extend existing nonlinear control techniques such that rotor flux can be observed while simultaneously adapting for parametric uncertainty in the electrical dynamic equations.

Since the design of an adaptive observer for all of the electrical problems seems somewhat intractable at this point in time, Marino *et al.* [18] proposed an adaptive observer for the rotor flux which compensates for rotor resistance variation. More specifically, the observer ensures bounded rotor flux estimates provided the rotor resistance changes slowly with respect to time. In this paper, we modify the adaptive observer structure proposed in [18] to facilitate the design of an adaptive position tracking controller. That is, by systematically injecting additional terms into the rotor flux observer, we illustrate how the rotor flux estimates can be utilized in a stable, closed-loop fashion. Specifically, the proposed controller provides asymptotic rotor position/velocity tracking while compensating for parametric uncertainty in the form of the rotor resistance and a general, *uncertain* mechanical subsystem. While the observer-controller does not require measurement of rotor flux or rotor current, it does exhibit a singularity when the magnitude of estimated rotor flux is zero; however, we note that the singularity is avoided asymptotically since the estimated flux is moved away from the singularity in the limit.

The paper is organized as follows. In Section E.2, the system model and problem statement are mathematically formulated. In Section E.3, the observer is designed and a preliminary stability analysis is given. In Section E.4, we use the observer to form a voltage input control relationship for the position/velocity tracking control objective. In Section E.5, to ensure that all signals remain bounded during closed-loop operation, we design a second voltage input control relationship for a rotor flux tracking control objective. We then perform the composite observer-controller stability analysis. In Section E.6, we state the main result of this chapter while in Section E.7, we present some simulation results. Finally, in Section E.8, we state some conclusions.

E.2 System Model and Problem Statement

E.2.1 Mechanical Subsystem Dynamics

The mechanical subsystem model for the electromechanical system is assumed to be of the form

$$M_m\ddot{q} + W_m(q, \dot{q})\theta_m = \tau \tag{E.1}$$

where $q(t)$, $\dot{q}(t)$, $\ddot{q}(t)$ represent the rotor position, velocity, and acceleration, respectively, M_m is an unknown constant representing the mechanical inertia of the system (including the rotor inertia), $W_m(q, \dot{q})\theta_m$ is used to denote the mechanical load dynamics where $W_m(q, \dot{q}) \in \Re^{1\times p}$ is a known, differentiable regression vector dependent on rotor position and velocity, and $\theta_m \in \Re^p$ is a vector containing the unknown parameters associated with the mechanical load dynamics, and τ represents the torque produced by the electrical subsystem.

E.2.2 Electrical Subsystem Dynamics

The standard induction motor model can be found in [19]. We will utilize the transformed nonlinear model in the stator fixed $a - b$ reference frame [1]. That is, under the assumptions of equal mutual inductance and a linear magnetic circuit, the electrical dynamics of an induction motor can be described by the following differential equations

$$L_o\dot{I}_a = R_r\psi_a - MR_rI_a + \beta_2\psi_b\dot{q} - \beta_1 I_a + \beta_3 V_a, \tag{E.2}$$

$$L_o\dot{I}_b = R_r\psi_b - MR_rI_b - \beta_2\psi_a\dot{q} - \beta_1 I_b + \beta_3 V_b, \tag{E.3}$$

$$L_r\dot{\psi}_a = -R_r\psi_a - \beta_2\psi_b\dot{q} + MR_rI_a, \tag{E.4}$$

and

$$L_r\dot{\psi}_b = -R_r\psi_b + \beta_2\psi_a\dot{q} + MR_rI_b \tag{E.5}$$

where $(\psi_a \ \psi_b)$, $(I_a \ I_b)$, and $(V_a \ V_b)$ represent the transformed rotor flux, transformed stator current, and transformed stator voltage, respectively, R_r represents the unknown constant rotor resistance, R_s represents the known constant stator resistance, L_o, β_1, β_2, and β_3 represent known positive constants related to the motor parameters and are explicitly given by

$$\beta_1 = R_sL_r^2/M, \ \ \beta_2 = n_pL_r, \ \ \beta_3 = L_r^2/M, \ \ L_o = \frac{L_r^2}{M}\left(L_s - M^2/L_r\right) \tag{E.6}$$

where L_r, L_s, and M are the rotor inductance, stator inductance, and mutual inductance of the motor, respectively, and n_p is the known pole pair number. The electromechanical coupling torque τ defined in (E.1) is explicitly given by

$$\tau = \alpha_1\left(\psi_aI_b - \psi_bI_a\right) \tag{E.7}$$

with $\alpha_1 = \dfrac{n_pM}{L_r}$.

Remark E.1

To facilitate the control design, the parameters associated with the mechanical load (*i.e.*, M_m and θ_m), and the rotor resistance R_r are assumed to be unknown constants. All of the other electrical parameters are assumed to be known exactly. In addition, we assume that upper and lower bounds for R_r and M_m are known (*i.e.*, $\underline{R}_r < R_r < \overline{R}_r$, $\underline{M}_m < M_m < \overline{M}_m$ where $\underline{R}_r$, $\overline{R}_r$, $\underline{M}_m$, and $\overline{M}_m$ are known positive constants). Furthermore, we will assume that $W_m^T(q,\dot{q}) \in L_\infty^p$ if q, $\dot{q} \in L_\infty$ [20] and $\dot{W}_m^T(q,\dot{q}) \in L_\infty^p$ if q, $\dot{q}$, $\ddot{q} \in L_\infty$.

E.2.3 Control Objective

Our objective in this paper is to design a rotor position/velocity tracking controller for the electromechanical system given by (E.1) through (E.7). To facilitate the control development, we define the rotor position tracking error as

$$e = q_d - q \tag{E.8}$$

where $q_d(t)$ represents the desired rotor position trajectory. We will assume that q_d and its first, second, and third time derivatives are all bounded functions of time. In addition, we define the filtered tracking error [21] as

$$r = \dot{e} + \alpha e \tag{E.9}$$

where α is a positive scalar constant control gain.

As stated previously, the above position tracking control problem is constrained by the fact that the mechanical subsystem parameters and rotor resistance are assumed to be unknown constants while all the other electrical system parameters are assumed to be known. To facilitate the design of adaptation update laws use to compensate for the parametric uncertainty, we define the parameter estimation error as follows

$$\tilde{M}_m = M_m - \hat{M}_m, \qquad \tilde{\theta}_m = \theta_m - \hat{\theta}_m, \qquad \tilde{R}_r = R_r - \hat{R}_r \tag{E.10}$$

where $\hat{M}_m$, $\hat{\theta}_m \in \Re^p$, and $\hat{R}_r$ denote dynamic estimates of the unknown parameters.

E.3 Observer Design

Since rotor flux is assumed unmeasurable, we will first build rotor flux observers to estimate the rotor flux. In addition, since the rotor resistance is assumed to be an unknown constant, we will design current observers to aid in the construction of a dynamic estimate for the rotor resistance. To facilitate the observer design, we define the rotor flux and stator current observation errors as follows

$$\tilde{\psi}_a = \psi_a - \hat{\psi}_a, \qquad \tilde{\psi}_b = \psi_b - \hat{\psi}_b, \qquad \tilde{I}_a = I_a - \hat{I}_a, \quad \tilde{I}_b = I_b - \hat{I}_b \tag{E.11}$$

where $\hat{\psi}_a$, $\hat{\psi}_b$ represent the estimated rotor flux, and $\hat{I}_a$, $\hat{I}_b$ represent the estimated stator current. In addition to the above observation error quantities, we also define the following auxiliary *weighted* observation quantities [18]

$$\tilde{Z}_a = L_o\tilde{I}_a + L_r\tilde{\psi}_a \qquad\qquad \tilde{Z}_b = L_o\tilde{I}_b + L_r\tilde{\psi}_b. \tag{E.12}$$

To provide for further design freedom during the observer design procedure, we also define the following auxiliary observation error related quantities [18]

$$\eta_a = \tilde{Z}_a - \zeta_a \qquad\qquad \eta_b = \tilde{Z}_b - \zeta_b \tag{E.13}$$

where ζ_a and ζ_b are the measurable states of two first-order auxiliary filters which will be designed to compensate for the unmeasurable quantities, $R_r\psi_a$ and $R_r\psi_b$, in (E.2) and (E.3), respectively. It should be noted that these two terms require special attention in the design procedure given below because each term represents an unknown parameter multiplied by an unmeasurable electrical state.

E.3.1 Observers and Auxiliary Filters

Motivated by the form of the subsequent stability analysis and the structure of (E.2) and (E.3), we propose the following stator current observer

$$L_o \dot{\hat{I}}_a = \hat{R}_r \hat{\psi}_a - \hat{R}_r M I_a + \beta_2 \dot{q} \hat{\psi}_b - \beta_1 I_a + K_o \tilde{I}_a + \beta_3 V_a + u_{o1} + u_{c1} \quad \text{(E.14)}$$

$$L_o \dot{\hat{I}}_b = \hat{R}_r \hat{\psi}_b - \hat{R}_r M I_b - \beta_2 \dot{q} \hat{\psi}_a - \beta_1 I_b + K_o \tilde{I}_b + \beta_3 V_b + u_{o2} + u_{c2} \quad \text{(E.15)}$$

where K_o is a positive observer feedback gain, the u_{ci}'s are auxiliary observer inputs which will be designed later during the construction of the subsequent position tracking control algorithm, and the u_{oi}'s are auxiliary observer inputs designed to facilitate the stability of the stator current observation error systems and are explicitly given by

$$\begin{aligned} u_{o1} &= -\frac{\hat{R}_r}{L_r}\left(L_o \tilde{I}_a - \zeta_a\right) - \frac{\beta_2 \dot{q}}{L_r} L_o \tilde{I}_b, \\ u_{o2} &= -\frac{\hat{R}_r}{L_r}\left(L_o \tilde{I}_b - \zeta_b\right) + \frac{\beta_2 \dot{q}}{L_r} L_o \tilde{I}_a. \end{aligned} \quad \text{(E.16)}$$

Motivated by the form of the subsequent stability analysis and the structure of (E.4) and (E.5), we propose the following rotor flux observers

$$L_r \dot{\hat{\psi}}_a = -\hat{R}_r \hat{\psi}_a - \beta_2 \dot{q} \hat{\psi}_b + \hat{R}_r M I_a + u_{o3} + u_{c3} \quad \text{(E.17)}$$

$$L_r \dot{\hat{\psi}}_b = -\hat{R}_r \hat{\psi}_b + \beta_2 \dot{q} \hat{\psi}_a + \hat{R}_r M I_b + u_{o4} + u_{c4} \quad \text{(E.18)}$$

where the u_{ci}'s are auxiliary observer inputs which will be designed later during the construction of the subsequent position tracking control algorithm, and the u_{oi}'s are auxiliary observer inputs designed to facilitate the stability of the rotor flux observation error system and are explicitly given by

$$\begin{aligned} u_{o3} &= -\frac{\beta_2 \dot{q}}{L_r} \tilde{I}_b - K_o \tilde{I}_a - u_{o1} - u_{c1}, \\ u_{o4} &= \frac{\beta_2 \dot{q}}{L_r} \tilde{I}_a - K_o \tilde{I}_b - u_{o2} - u_{c2}. \end{aligned} \quad \text{(E.19)}$$

Motivated by the form of the subsequent stability analysis and the structure of the rotor flux observers and stator current observers given above, we proposed the following dynamic filter structure for ζ_a and ζ_b

$$\dot{\zeta}_a = \frac{1}{L_r} \tilde{I}_a + \frac{\beta_2 \dot{q}}{L_r} \tilde{I}_b - u_{c3}, \qquad \dot{\zeta}_b = \frac{1}{L_r} \tilde{I}_b - \frac{\beta_2 \dot{q}}{L_r} \tilde{I}_a - u_{c4}. \quad \text{(E.20)}$$

Remark E.2

At this point, we note the specific forms for the auxiliary inputs and the filters given by (E.16), (E.19), and (E.20) are not obvious; however, the structure of the auxiliary terms will become apparent during the subsequent observation error system development and the corresponding stability analysis. The auxiliary inputs denoted by u_{ci}, whose definitions will be explicitly given later, will be designed to cancel out unmeasurable quantities which arise during the position tracking control design. One of the differences between the above observer structure and the one proposed in [18] is that only two first-order auxiliary filters (*i.e.*, (E.20)) are required as opposed to the four first-order filters used in [18].

E.3.2 Observation Error Systems

After taking the time derivative of the stator current observation error $\tilde{I}_a$ defined in (E.11), we have

$$L_o \dot{\tilde{I}}_a = L_o \dot{I}_a - L_o \dot{\hat{I}}_a . \tag{E.21}$$

After substituting the actual current dynamics from (E.2) for $L_o \dot{I}_a$, the current observer dynamics from (E.14) for $L_o \dot{\hat{I}}_a$, and then simplifying the resulting expression, we have

$$L_o \dot{\tilde{I}}_a = R_r \tilde{\psi}_a + \hat{\psi}_a \tilde{R}_r + \beta_2 \dot{q}\, \tilde{\psi}_b - M I_a \tilde{R}_r - K_o \tilde{I}_a - u_{o1} - u_{c1} \tag{E.22}$$

where (E.11) and (E.10) have been utilized. From the form of (E.10), (E.12), and (E.13), it is easy to see that u_{o1} of (E.16) can also be rewritten as

$$u_{o1} = -\frac{R_r}{L_r}\eta_a + \frac{\tilde{R}_r}{L_r}\left(L_o \tilde{I}_a - \varsigma_a\right) - \frac{\beta_2 \dot{q}}{L_r}\tilde{Z}_b + R_r \tilde{\psi}_a + \beta_2 \dot{q}\, \tilde{\psi}_b. \tag{E.23}$$

After substituting (E.23) into (E.22) and then simplifying the resulting expression, we have the final form for the phase a stator current observation error dynamics as follows

$$\begin{aligned} L_o \dot{\tilde{I}}_a = & -K_o \tilde{I}_a + \tfrac{R_r}{L_r}\eta_a + \tfrac{\beta_2 \dot{q}}{L_r}\tilde{Z}_b + \\ & \tilde{R}_r\left(-M I_a + \hat{\psi}_a - \tfrac{L_a}{L_r}\tilde{I}_a + \tfrac{\varsigma_a}{L_r}\right) - u_{c1}. \end{aligned} \tag{E.24}$$

Similar to the above procedure, we can construct the phase b stator current observation error dynamics as follows

$$\begin{aligned} L_o \dot{\tilde{I}}_b = & -K_o \tilde{I}_b + \tfrac{R_r}{L_r}\eta_b - \tfrac{\beta_2 \dot{q}}{L_r}\tilde{Z}_a + \\ & \tilde{R}_r\left(-M I_b + \hat{\psi}_b - \tfrac{L_a}{L_r}\tilde{I}_b + \tfrac{\varsigma_b}{L_r}\right) - u_{c2}. \end{aligned} \tag{E.25}$$

After taking the time derivative of the *weighted* observation error term $\tilde{Z}_a$, defined in (E.12) and (E.11), we have

$$\dot{\tilde{Z}}_a = L_o \dot{I}_a - L_o \dot{\hat{I}}_a + L_r \dot{\psi}_a - L_r \dot{\hat{\psi}}_a . \tag{E.26}$$

After substituting the right-hand sides of (E.2), (E.4), (E.14), and (E.17) for $L_o \dot{I}_a$, $L_r \dot{\psi}_a$, $L_o \dot{\hat{I}}_a$, and $L_r \dot{\hat{\psi}}_a$, respectively, into (E.26), we obtain

$$\dot{\tilde{Z}}_a = -K_o \tilde{I}_a - u_{o1} - u_{c1} - u_{o3} - u_{c3}. \tag{E.27}$$

After substituting (E.19) for u_{o3} into (E.27), the dynamics for $\tilde{Z}_a$ can be simplified into the following form

$$\dot{\tilde{Z}}_a = \frac{\beta_2 \tilde{I}_b \dot{q}}{L_r} - u_{c3}. \tag{E.28}$$

Similar to the above procedure, the *weighted* observation error dynamics for $\tilde{Z}_b$, defined in (E.12) and (E.11), can be obtained as follows

$$\dot{\tilde{Z}}_b = -\frac{\beta_2 \tilde{I}_a \dot{q}}{L_r} - u_{c4}. \tag{E.29}$$

After taking the time derivative of the auxiliary observation error related quantity η_a defined in (E.13), we have

$$\dot{\eta}_a = \dot{\tilde{Z}}_a - \dot{\zeta}_a. \tag{E.30}$$

After substituting for $\dot{\tilde{Z}}_a$ and $\dot{\zeta}_a$ from (E.28) and (E.20), respectively, into (E.30), the dynamics for η_a can be simplified into the following form

$$\dot{\eta}_a = -\frac{1}{L_r} \tilde{I}_a. \tag{E.31}$$

Similar to the above procedure, the dynamics for the auxiliary observation error related quantity η_b, defined in (E.13), can be obtained as follows

$$\dot{\eta}_b = -\frac{1}{L_r} \tilde{I}_b. \tag{E.32}$$

Remark E.3

It is important to emphasize that the form of the observation error systems given by (E.24), (E.25), (E.28), (E.29), (E.31), and (E.32) (and hence the form of the observers and the auxiliary filters given by (E.14), (E.15), (E.17), (E.18), and (E.20)) have been sculpted to mesh with the subsequent, quadratic Lyapunov-like function which is use to analyze the stability of the observation error systems.

E.3.3 Analysis of the Observer Error Systems

To analyze the stability of the proposed observer structure, we define the following non-negative function

$$V_o = \frac{1}{2}L_o\tilde{I}_a^2 + \frac{1}{2}L_o\tilde{I}_b^2 + \frac{1}{2}\tilde{Z}_a^2 + \frac{1}{2}\tilde{Z}_b^2 + \frac{1}{2}R_r\eta_a^2 + \frac{1}{2}R_r\eta_b^2 + \frac{1}{2}\Gamma_r^{-1}\tilde{R}_r^2 \quad \text{(E.33)}$$

where Γ_r is positive scalar adaptation gain. After taking the time derivative of (E.33) and substituting the observer error systems (*i.e.*, (E.24) for $L_o\,\dot{\tilde{I}}_a$, (E.25) for $L_o\,\dot{\tilde{I}}_b$, (E.28) for $\dot{\tilde{Z}}_a$, (E.29) for $\dot{\tilde{Z}}_b$, (E.31) for $\dot{\eta}_a$, and (E.32) for $\dot{\eta}_b$), we have

$$\begin{aligned} \dot{V}_o = & -K_o\tilde{I}_a^2 - K_o\tilde{I}_b^2 - \tilde{I}_a u_{c1} - \tilde{I}_b u_{c2} - \\ & \tilde{Z}_a u_{c3} - \tilde{Z}_b u_{c4} + \tilde{R}_r\left(\Omega_r - \Gamma_r^{-1}\dot{\hat{R}}_r\right) \end{aligned} \quad \text{(E.34)}$$

where the auxiliary measurable term Ω_r is given by

$$\begin{aligned} \Omega_r = & \ \tilde{I}_a\left(-MI_a + \hat{\psi}_a - \tfrac{L_a}{L_r}\tilde{I}_a + \tfrac{\zeta_a}{L_r}\right) \\ & +\tilde{I}_b\left(-MI_b + \hat{\psi}_b - \tfrac{L_u}{L_r}\tilde{I}_b + \tfrac{\zeta_b}{L_r}\right). \end{aligned} \quad \text{(E.35)}$$

Based on the structure of (E.34), we define the following projection-type update law [20] for the rotor resistance

$$\dot{\hat{R}}_r = \begin{cases} \Gamma_r\Omega_r & \text{if } \hat{R}_r > \underline{R}_r \\ \Gamma_r\Omega_r & \text{if } \hat{R}_r = \underline{R}_r \text{ and } \Omega_r \geq 0 \\ 0 & \text{if } \hat{R}_r = \underline{R}_r \text{ and } \Omega_r < 0 \end{cases} \quad \text{(E.36)}$$

where $\hat{R}_r(0) = \underline{R}_r$, and $\underline{R}_r$ was defined in Remark E.1. After substituting the adaptation law of (E.36) into (E.34), we can form the following upper bound for $\dot{V}_o$

$$\dot{V}_o \leq -K_o\tilde{I}_a^2 - K_o\tilde{I}_b^2 - \tilde{I}_a u_{c1} - \tilde{I}_b u_{c2} - \tilde{Z}_a u_{c3} - \tilde{Z}_b u_{c4}. \quad \text{(E.37)}$$

From (E.37), we can see that if the auxiliary observer inputs, denoted by u_{ci}, were set to zero, then $\dot{V}_o$ would be negative semi-definite. Hence, it would be easy to show from (E.33) and (E.37) that $\tilde{I}_a$, $\tilde{I}_b$, $\tilde{Z}_a$, $\tilde{Z}_b$, η_a, η_b, $\tilde{R}_r$, $\hat{R}_r$, $\tilde{\psi}_a$, $\tilde{\psi}_b$, ζ_a, $\zeta_b \in L_\infty$ and that $\tilde{I}_a$ and $\tilde{I}_b$ go to zero asymptotically fast. However, the auxiliary observer inputs, denoted by u_{ci}, will be designed to directly cancel terms during the subsequent composite observer-controller stability analysis.

Remark E.4

A drawback of the above proposed adaptive observer is that the rotor flux observation error does not converge to zero; furthermore, even if the rotor resistance was known exactly, the proposed observer structure with $\hat{R}_r = R_r$ does not provide for convergence of the rotor flux observation error (Note this phenomenon is in contrast to the exponential rotor flux observation error result provided by the rotor-resistant dependent, open-loop observers used in [10]). In addition, we note that in [18], a projection algorithm for the rotor resistance was required during the observer analysis. The projection update law, defined in (E.36), was not required in the above observer analysis; however, the subsequent control design procedure will require that $\hat{R}_r$ always remain positive.

E.4 Position/Velocity Tracking Control Objective

By utilizing the proposed observer structure presented in the previous section, we now design an adaptive position/velocity controller which does not require rotor flux measurements. First, we re-write the dynamics of (E.1) in terms of the filtered tracking error defined in (E.9) as follows

$$M_m \dot{r} = M_m \left(\ddot{q}_d + \alpha \dot{e} \right) + W_m \theta_m - \alpha_1 \left(\psi_a I_b - \psi_b I_a \right) \tag{E.38}$$

where (E.7) has been utilized. Since the above error system lacks a control input and rotor flux is an unmeasurable quantity, we add and subtract the terms $\alpha_1 \tau_d$ and $\alpha_1 \left(\hat{\psi}_a \hat{I}_b - \hat{\psi}_b \hat{I}_a \right)$ to the right-hand side of (E.38) to yield

$$\begin{aligned} M_m \dot{r} &= M_m \left(\ddot{q}_d + \alpha \dot{e} \right) + W_m \theta_m - \alpha_1 \tau_d + \alpha_1 \eta_\tau \\ &\quad + \alpha_1 \left(\hat{\psi}_a \hat{I}_b - \hat{\psi}_b \hat{I}_a \right) - \alpha_1 \left(\psi_a I_b - \psi_b I_a \right) \end{aligned} \tag{E.39}$$

where τ_d is used the represent the desired torque trajectory, and η_τ represents the measurable, *estimated* torque tracking error defined by

$$\eta_\tau = \tau_d - \left(\hat{\psi}_a \hat{I}_b - \hat{\psi}_b \hat{I}_a \right). \tag{E.40}$$

Based on the form of (E.39), we now design the desired torque trajectory as follows

$$\tau_d = \frac{1}{\alpha_1} \left(\hat{M}_m \left(\ddot{q}_d + \alpha \dot{e} \right) + W_m \hat{\theta}_m + K_1 r \right) \tag{E.41}$$

where K_1 is a positive control gain (Note to eliminate overparameterization, the update laws for $\hat{\theta}_m$ and $\hat{M}_m$ will be explicitly defined later in the design procedure). After substituting (E.41) into (E.39), we can simplify the resulting expression as follows

$$\begin{aligned} M_m\dot{r} &= \tilde{M}_m(\ddot{q}_d+\alpha\dot{e})+W_m\tilde{\theta}_m-K_1r+\alpha_1\eta_\tau \\ &\quad +\alpha_1\left(-\hat{\psi}_a\tilde{I}_b-\frac{1}{L_r}\tilde{Z}_aI_b+\frac{L_o}{L_r}\tilde{I}_aI_b\right) \\ &\quad +\alpha_1\left(\hat{\psi}_b\tilde{I}_a+\frac{1}{L_r}\tilde{Z}_bI_a-\frac{L_o}{L_r}\tilde{I}_bI_a\right) \end{aligned} \tag{E.42}$$

where $\tilde{M}_m$ and $\tilde{\theta}_m$ were defined in (E.10) and the definitions for $\tilde{Z}_a$, $\tilde{Z}_b$, $\tilde{I}_a$, and $\tilde{I}_b$ given in (E.12) and (E.11), respectively, have been used to rewrite the second-line in (E.39) in the *advantageous* form given by the second and third line in (E.42). The term advantageous is use here because the form of (E.42), the form of (E.37), and some knowledge of recent control design methodology indicates how the auxiliary observer inputs denoted by u_{ci} in (E.37) might be designed to cancel out measurable and *unmeasurable* terms during the subsequent *composite* observer-controller stability analysis.

If we assume for now that the second and third line of (E.42) can be handled by injecting the appropriate terms into the observation error dynamics via the auxiliary observer inputs, the form of the first-line of (E.42) indicates that the estimated torque tracking error should be driven to zero. The dynamics for η_τ can be obtained by taking the time derivative of (E.40) and multiplying the resulting expression by the constant $L_oL_rM_m$ to yield

$$L_oL_rM_m\dot{\eta}_\tau=L_oL_rM_m\left(\dot{\tau}_d-\dot{\hat{\psi}}_a\,\hat{I}_b-\hat{\psi}_a\,\dot{\hat{I}}_b+\dot{\hat{\psi}}_b\,\hat{I}_a+\hat{\psi}_b\,\dot{\hat{I}}_a\right). \tag{E.43}$$

Given the definition of the desired torque trajectory τ_d defined in (E.41), we know that τ_d is function of the $\hat{\theta}_m$, $\hat{M}_m$, q, $\dot{q}$, and t; therefore, the time derivative of τ_d can be expressed as

$$\dot{\tau}_d=\frac{\partial\tau_d}{\partial\hat{\theta}_m}\dot{\hat{\theta}}_m+\frac{\partial\tau_d}{\partial\hat{M}_m}\dot{\hat{M}}_m+\frac{\partial\tau_d}{\partial q}\dot{q}+\frac{\partial\tau_d}{\partial\dot{q}}\ddot{q}+\frac{\partial\tau_d}{\partial t} \tag{E.44}$$

where the above partial derivative terms are explicitly given in Section E.9. Since the subsequent parameter update laws for $\hat{\theta}_m$ and $\hat{M}_m$ will allow for direct substitution for $\dot{\hat{\theta}}_m$ and $\dot{\hat{M}}_m$, respectively, into the right-hand side of (E.44), the only unmeasurable quantity in (E.44) is the rotor acceleration $\ddot{q}$. However, from (E.1) and (E.7), the rotor acceleration can be written as

$$\begin{aligned}\ddot{q} = & \frac{1}{M_m}(-W_m\theta_m + \\ & \left[\alpha_1\left(\psi_a I_b - \psi_b I_a\right) - \alpha_1\left(\hat{\psi}_a\hat{I}_b - \hat{\psi}_b\hat{I}_a\right)\right]) \\ & + \frac{1}{M_m}\left(\alpha_1\left(\hat{\psi}_a\hat{I}_b - \hat{\psi}_b\hat{I}_a\right)\right)\end{aligned} \quad \text{(E.45)}$$

where term $\alpha_1\left(\hat{\psi}_a\hat{I}_b - \hat{\psi}_b\hat{I}_a\right)$ has been added and subtracted to facilitate the analysis. After writing the bracketed term in (E.45) in terms of the observation quantities defined in (E.12) and (E.11), we can write the rotor acceleration in terms of unknown parameters, measurable quantities, and the observation error as follows

$$\begin{aligned}\ddot{q} = & \frac{1}{M_m}[-W_m\theta_m + \alpha_1 \\ & \left(\hat{\psi}_a\tilde{I}_b - \frac{L_o}{L_r}\tilde{I}_a I_b - \hat{\psi}_b\tilde{I}_a + \frac{L_o}{L_r}\tilde{I}_b I_a\right)] \\ & + \frac{1}{M_m}\left[\alpha_1\left(\hat{\psi}_a\hat{I}_b - \hat{\psi}_b\hat{I}_a\right) + \frac{\alpha_1}{L_r}I_b\tilde{Z}_a - \frac{\alpha_1}{L_r}I_a\tilde{Z}_b\right].\end{aligned} \quad \text{(E.46)}$$

After substituting (E.46) into (E.44) and then substituting the resulting expression for $\dot{\tau}_d$, the right-hand side of (E.17) and (E.18) for $L_r\,\dot{\hat{\psi}}_a$and $L_r\,\dot{\hat{\psi}}_b$, respectively, the right-hand side of (E.14) and (E.15) for $L_o\,\dot{\hat{I}}_a$and $L_o\,\dot{\hat{I}}_b$, respectively, into (E.43), and then rearranging the resulting expression in a manner to highlight the unmeasurable quantities and the unknown parameters, we have

$$\begin{aligned}L_o L_r M_m \dot{\eta}_\tau = & \Omega_1 + M_m\Omega_2 + \Omega_3\tilde{Z}_a + \Omega_4\tilde{Z}_b + Y_1\theta_m \\ & -\alpha_1 r + L_r M_m \beta_3\left(\hat{\psi}_b V_a - \hat{\psi}_a V_b\right)\end{aligned} \quad \text{(E.47)}$$

where the measurable auxiliary variables Ω_1, Ω_2, Ω_3, Ω_4, and $Y_1 \in \Re^{1\times p}$ are explicitly defined below

$$\begin{aligned}\Omega_1 = & \alpha_1 r + L_r L_o \alpha_1 \frac{\partial\tau_d}{\partial\dot{q}}\left(\hat{\psi}_a\hat{I}_b - \hat{\psi}_b\hat{I}_a\right) \\ & + L_r L_o \alpha_1 \frac{\partial\tau_d}{\partial\dot{q}}\left(\hat{\psi}_a\tilde{I}_b - \frac{L_o}{L_r}\tilde{I}_a I_b - \hat{\psi}_b\tilde{I}_a + \frac{L_o}{L_r}\tilde{I}_b I_a\right),\end{aligned} \quad \text{(E.48)}$$

$$\begin{aligned}\Omega_2 = \;& L_o L_r \left(\frac{\partial \tau_d}{\partial t} + \frac{\partial \tau_d}{\partial q}\dot{q} + \frac{\partial \tau_d}{\partial \hat{\theta}_m}\dot{\hat{\theta}}_m + \frac{\partial \tau_d}{\partial \hat{M}_m}\dot{\hat{M}}_m \right) \\ & + L_o \hat{I}_a \left(-\hat{R}_r \hat{\psi}_b + \beta_2 \dot{q} \hat{\psi}_a + \hat{R}_r M I_b + u_{o4} + u_{c4} \right) \\ & - L_o \hat{I}_b \left(-\hat{R}_r \hat{\psi}_a - \beta_2 \dot{q} \hat{\psi}_b + \hat{R}_r M I_a + u_{o3} + u_{c3} \right) \\ & - L_r \hat{\psi}_a \left(\hat{R}_r \hat{\psi}_b - \hat{R}_r M I_b - \beta_2 \dot{q} \hat{\psi}_a \right. \\ & \left. - \beta_1 I_b + K_o \tilde{I}_b + u_{o2} + u_{c2} \right) \\ & + L_o \hat{\psi}_b \left(\hat{R}_r \hat{\psi}_a - \hat{R}_r M I_a + \beta_2 \dot{q} \hat{\psi}_b \right. \\ & \left. - \beta_1 I_a + K_o \tilde{I}_a + u_{o1} + u_{c1} \right),\end{aligned} \tag{E.49}$$

$$\Omega_3 = L_o \alpha_1 \frac{\partial \tau_d}{\partial \dot{q}} I_b, \quad \Omega_4 = -L_o \alpha_1 \frac{\partial \tau_d}{\partial \dot{q}} I_a, \quad Y_1 = -L_o L_r \frac{\partial \tau_d}{\partial \dot{q}} W_m. \tag{E.50}$$

Note the term $-\alpha_1 r$ has been added and subtracted to the right-hand side of (E.47) to cancel the $\alpha_1 \eta_\tau$ in (E.42) during the ensuing stability analysis.

Based on the structure of (E.47) and the subsequent stability analysis, we now design the following voltage input control relationship to force η_τ to zero

$$\begin{aligned}\hat{\psi}_b V_a - \hat{\psi}_a V_b = \;& -K_2 \frac{1}{L_r \beta_3} \eta_\tau - \frac{1}{L_r \beta_3} \Omega_2 - \\ & \frac{1}{L_r \beta_3 \hat{M}_m} \left(Y_1 \hat{\theta}_m + \Omega_1 + u_{c5} \right)\end{aligned} \tag{E.51}$$

where K_2 is positive control gain, and u_{c5} is an auxiliary control input designed to facilitate the subsequent flux tracking control objective (Note it will be explicitly defined during the composite observer-controller stability analysis). To eliminate overparameterization, the update law for $\hat{M}_m$ and $\hat{\theta}_m$ will be explicitly defined later; furthermore, the update law for $\hat{M}_m$ will be designed to ensure $\hat{M}_m > 0$. After substituting (E.51) into (E.47), we can write the closed-loop error system for η_τ into the following form

$$\begin{aligned}L_o L_r M_m \dot{\eta}_\tau \;=\;& -K_2 M_m \eta_\tau + \Omega_3 \tilde{Z}_a + \Omega_4 \tilde{Z}_b + Y_1 \tilde{\theta}_m \\ & - \tilde{M}_m \hat{M}_m^{-1} \left(Y_1 \hat{\theta}_m + \Omega_1 + u_{c5} \right) - \alpha_1 r - u_{c5}\end{aligned} \tag{E.52}$$

where $\tilde{\theta}_m$ and $\tilde{M}_m$ were defined in (E.10).

To analyze the closed-loop tracking error systems for r and η_τ, we define the following non-negative function

$$V_p = V_o + \frac{1}{2}M_m r^2 + \frac{1}{2}L_o L_r M_m \eta_\tau^2 \tag{E.53}$$

where V_o was defined in (E.33). After taking the time-derivative of (E.53) and then substituting the closed-loop error systems from (E.42) and (E.52) for $M_m\dot{r}$ and $L_o L_r M_m \dot{\eta}_\tau$, respectively, we can simplify the resulting expression as follows

$$\begin{aligned} \dot{V}_p = \; & \dot{V}_o - K_1 r^2 - K_2 M_m \eta_\tau^2 - \eta_\tau u_{c5} + \tilde{\theta}_m^T \left(W_m^T r + Y_1^T \eta_\tau\right) \\ & + \tilde{M}_m \left[\left(\ddot{q}_d + \alpha \dot{e}\right) r - \hat{M}_m^{-1}\left(Y_1 \hat{\theta}_m + \Omega_1 + u_{c5}\right)\eta_\tau \right] \\ & + \tilde{Z}_a \left(\Omega_3 \eta_\tau - \frac{\alpha_1 r}{L_r} I_b\right) + \tilde{Z}_b \left(\Omega_4 \eta_\tau + \frac{\alpha_1 r}{L_r} I_a\right) \\ & + \tilde{I}_a \left(\alpha_1 r \hat{\psi}_b + \frac{\alpha_1 r L_o}{L_r} I_b\right) + \tilde{I}_b \left(-\alpha_1 r \hat{\psi}_a - \frac{\alpha_1 r L_o}{L_r} I_a\right). \end{aligned} \tag{E.54}$$

From the form of (E.37) and the form of (E.54), we are motivated to select the auxiliary observer inputs as follows

$$u_{c1} = \alpha_1 r \hat{\psi}_b + \frac{\alpha_1 r L_o}{L_r} I_b, \qquad u_{c2} = -\alpha_1 r \hat{\psi}_a - \frac{\alpha_1 r L_o}{L_r} I_a \tag{E.55}$$

and

$$u_{c3} = \Omega_3 \eta_\tau - \frac{\alpha_1 r}{L_r} I_b + \bar{u}_{c3}, \qquad u_{c4} = \Omega_4 \eta_\tau + \frac{\alpha_1 r}{L_r} I_a + \bar{u}_{c4} \tag{E.56}$$

where the auxiliary observer inputs $\bar{u}_{c3}$ and $\bar{u}_{c4}$ will be defined later during the design procedure. After substituting (E.55) and (E.56) into (E.37) and then substituting the resulting expression into (E.54) for $\dot{V}_o$, we have

$$\begin{aligned} \dot{V}_p = \; & -K_o \tilde{I}_a^2 - K_o \tilde{I}_b^2 - K_1 r^2 - K_2 M_m \eta_\tau^2 \\ & - \tilde{Z}_a \bar{u}_{c3} - \tilde{Z}_b \bar{u}_{c4} - \eta_\tau u_{c5} \\ & + \tilde{M}_m \left[\left(\ddot{q}_d + \alpha \dot{e}\right) r - \hat{M}_m^{-1}\left(Y_1 \hat{\theta}_m + \Omega_1 + u_{c5}\right)\eta_\tau \right] \\ & + \tilde{\theta}_m^T \left(W_m^T r + Y_1^T \eta_\tau\right). \end{aligned} \tag{E.57}$$

From the form of (E.57), we can make the following observations: i) the adaptive update laws must be designed to compensate for the last two lines, and ii) the u_{c5} input in the first line can be used to cancel torque tracking terms during the ensuing flux tracking control objective flux while the auxiliary observer inputs $\bar{u}_{c3}$ and $\bar{u}_{c4}$ can be used to cancel unmeasurable observation error terms.

E.5 Flux Tracking Control Objective

Now that we have designed one voltage control relationship to provide for rotor position tracking, we design a second voltage control relationship to ensure that all of the signals in the observer, controller, and the electromechanical system remain bounded during closed-loop operation. Roughly speaking, one method for accomplishing this task is to force the magnitude of the estimated flux to track a positive function. Specifically, we define the flux tracking error [10] as follows

$$\eta_{\psi} = \psi_d - \frac{1}{2}\left(\hat{\psi}_a^2 + \hat{\psi}_b^2\right) \tag{E.58}$$

where $\psi_d(t)$ is a positive scalar function used to represent the desired *pseudo*-magnitude of the estimated rotor flux. We will assume that ψ_d is constructed such that ψ_d, $\dot{\psi}_d$, and $\ddot{\psi}_d$ are all bounded functions of time.

To obtain the dynamics for η_{ψ}, we take the time derivative of (E.58) and multiply the resulting expression by L_r to yield

$$L_r\dot{\eta}_{\psi} = L_r\dot{\psi}_d - L_r\,\dot{\hat{\psi}}_a\,\hat{\psi}_a - L_r\,\dot{\hat{\psi}}_b\,\hat{\psi}_b. \tag{E.59}$$

From the form of (E.59), we can see that the rotor flux tracking error dynamics requires substitution for $L_r\,\dot{\hat{\psi}}_a$ and $L_r\,\dot{\hat{\psi}}_b$ from the rotor flux observers defined in (E.17) and (E.18), respectively. To facilitate the formation of the dynamics for η_{ψ}, we first construct a more complete description for the rotor flux observers. Specifically, we substitute (E.19) and (E.56) into (E.17) and (E.18) for u_{o3}, u_{o4}, u_{c3}, and u_{c4} which then calls for the substitution of (E.16) and (E.55) for u_{o1}, u_{o2}, u_{c1}, and u_{c2}. After simplifying the resulting expression for the rotor flux observers by isolating all of the explicit occurrences of stator current, we have

$$L_r\,\dot{\hat{\psi}}_a = \Omega_{a1} + \Omega_{a2}\tilde{I}_a + \Omega_{a3}\tilde{I}_b + \Omega_c\hat{I}_b + \hat{R}_r M\hat{I}_a + \Omega_3\eta_\tau + \bar{u}_{c3} \tag{E.60}$$

and

$$L_r\,\dot{\hat{\psi}}_b = \Omega_{b1} + \Omega_{b2}\tilde{I}_a + \Omega_{b3}\tilde{I}_b - \Omega_c\hat{I}_a + \hat{R}_r M\hat{I}_b + \Omega_4\eta_\tau + \bar{u}_{c4} \tag{E.61}$$

where Ω_{ai}, Ω_{bi}, and Ω_c are auxiliary measurable functions which do not depend on stator current and are explicitly defined in Section E.10. The above form for the rotor flux observers allows us to design a voltage level input which forces the rotor flux tracking error to zero while minimizing any possible singularities. Specifically, if the *fictitious* controller [22] which will be subsequently designed to force η_ψ to zero is dependent on stator current, then its time derivative will contain multiple occurrences of the input stator voltage which can ultimately lead to a large control singularity region.

After substituting the right-hand side of (E.60) and (E.61) for $L_r \dot{\hat{\psi}}_a$ and $L_r \dot{\hat{\psi}}_b$, respectively, into (E.59), we have

$$\begin{aligned} L_r\dot{\eta}_\psi = \; & L_r\dot{\psi}_d - \hat{\psi}_a\Omega_{a1} - \hat{\psi}_b\Omega_{b1} - \hat{R}_r M\left(\hat{\psi}_a\hat{I}_a + \hat{\psi}_b\hat{I}_b\right) \\ & -\tilde{I}_a\left(\hat{\psi}_a\Omega_{a2} + \hat{\psi}_b\Omega_{b2}\right) - \tilde{I}_b\left(\hat{\psi}_a\Omega_{a3} + \hat{\psi}_b\Omega_{b3}\right) \\ & -\Omega_c\left(\hat{\psi}_a\hat{I}_b - \hat{\psi}_b\hat{I}_a\right) - \left(\Omega_3\hat{\psi}_a + \Omega_4\hat{\psi}_b\right)\eta_\tau \\ & -\hat{\psi}_a\bar{u}_{c3} - \hat{\psi}_b\bar{u}_{c4}. \end{aligned} \tag{E.62}$$

Since the dynamics of (E.62) lack a control input, we add and subtract a fictitious controller, denoted by u_I, to the right-hand side of (E.62) to obtain

$$\begin{aligned} L_r\dot{\eta}_\psi = \; & \Omega_5 + \tilde{I}_a\left(-\hat{\psi}_a\Omega_{a2} - \hat{\psi}_b\Omega_{b2}\right) \\ & +\tilde{I}_b\left(-\hat{\psi}_a\Omega_{a3} - \hat{\psi}_b\Omega_{b3}\right) - \hat{\psi}_a\bar{u}_{c3} \\ & +\left(\Omega_c - \Omega_3\hat{\psi}_a - \Omega_4\hat{\psi}_b\right)\eta_\tau - \hat{\psi}_b\bar{u}_{c4} \\ & -\hat{R}_r M u_I + \hat{R}_r M \eta_I \end{aligned} \tag{E.63}$$

where (E.40) has been used to write the first term in the third line of (E.62) in terms of η_τ, Ω_5 is an auxiliary measurable function defined as

$$\Omega_5 = L_r\dot{\psi}_d - \hat{\psi}_a\Omega_{a1} - \hat{\psi}_b\Omega_{b1} - \Omega_c\tau_d,$$

and η_I is an auxiliary tracking error variable defined as

$$\eta_I = u_I - \left(\hat{\psi}_a\hat{I}_a + \hat{\psi}_b\hat{I}_b\right). \tag{E.64}$$

Based on the structure of (E.63), we design the fictitious control input u_I as follows

$$\begin{aligned} u_I = \; & \frac{1}{\hat{R}_r M}\left(\Omega_5 + K_n\hat{\psi}_a^2\left(\Omega_{a2}^2 + \Omega_{a3}^2\right)\eta_\psi\right. \\ & \left. +K_n\hat{\psi}_b^2\left(\Omega_{b2}^2 + \Omega_{b3}^2\right)\eta_\psi + K_3\eta_\psi\right) \end{aligned} \tag{E.65}$$

where K_3 and K_n are positive control gains (It is important to note that u_I does not explicitly depend on stator current measurements; furthermore, we have designed the update law for $\hat{R}_r$ to prevent control singularities). After substituting u_I of (E.65) into (E.63), we obtain the following closed-loop error system for η_ψ

$$\begin{aligned} L_r\dot{\eta}_\psi = & -K_3\eta_\psi + \left(\Omega_c - \Omega_3\hat{\psi}_a - \Omega_4\hat{\psi}_b\right)\eta_\tau \\ & + \left[\hat{\psi}_a\left(-\Omega_{a2}\tilde{I}_a - \Omega_{a3}\tilde{I}_b\right) - K_n\hat{\psi}_a^2\left(\Omega_{a2}^2 + \Omega_{a3}^2\right)\eta_\psi\right] \\ & + \left[\hat{\psi}_b\left(\Omega_{b2}\tilde{I}_a - \Omega_{b3}\tilde{I}_b\right) - K_n\hat{\psi}_b^2\left(\Omega_{b2}^2 + \Omega_{b3}^2\right)\eta_\psi\right] \\ & - \hat{\psi}_a\bar{u}_{c3} - \hat{\psi}_b\bar{u}_{c4} + \hat{R}_r M\eta_I \end{aligned} \quad \text{(E.66)}$$

where the bracketed terms are nonlinear damping pairs [22].

Motivated by the form of (E.66), we now design a voltage control input relationship which forces η_I to zero. After taking the time derivative of (E.64), and multiplying the resulting expression by the constant $L_r L_o M_m$, we have

$$\begin{aligned} L_r L_o M_m\dot{\eta}_I = & \; L_o L_r M_m \dot{u}_I - \\ & L_r L_o M_m\left(\dot{\hat{\psi}}_a\,\hat{I}_a + \hat{\psi}_a\,\dot{\hat{I}}_a + \dot{\hat{\psi}}_b\,\hat{I}_b + \hat{\psi}_b\,\dot{\hat{I}}_b\right). \end{aligned} \quad \text{(E.67)}$$

To complete the open-loop description for η_I, we note that u_I of (E.65) is a function of $\hat{R}_r$, $\hat{\psi}_a$, $\hat{\psi}_b$, η_ψ, q, $\dot{q}$, ζ_a, ζ_b, $\hat{\theta}_m$, $\hat{M}_m$, and t; therefore, the time derivative for u_I can be expressed as

$$\begin{aligned} \dot{u}_I = & \frac{\partial u_I}{\partial \hat{R}_r}\dot{\hat{R}}_r + \frac{\partial u_I}{\partial \hat{\psi}_a}\dot{\hat{\psi}}_a + \frac{\partial u_I}{\partial \hat{\psi}_b}\dot{\hat{\psi}}_b + \frac{\partial u_I}{\partial \eta_\psi}\dot{\eta}_\psi \\ & + \frac{\partial u_I}{\partial q}\dot{q} + \frac{\partial u_I}{\partial \dot{q}}\ddot{q} + \frac{\partial u_I}{\partial \zeta_a}\dot{\zeta}_a + \frac{\partial u_I}{\partial \zeta_b}\dot{\zeta}_b \\ & + \frac{\partial u_I}{\partial \dot{q}}\ddot{q} + \frac{\partial u_I}{\partial \hat{\theta}_m}\dot{\hat{\theta}}_m + \frac{\partial u_I}{\partial \hat{M}_m}\dot{\hat{M}}_m + \frac{\partial u_I}{\partial t} \end{aligned} \quad \text{(E.68)}$$

where the above partial derivative terms are all explicitly calculated in Section E.11. After substituting (E.46) for $\ddot{q}$ into (E.68) and then substituting the resulting expression for $\dot{u}_I$, the right-hand side of (E.14) and (E.15) for $L_o\,\dot{\hat{I}}_a$ and $L_o\,\dot{\hat{I}}_b$, respectively, into (E.67), and then rearranging the resulting expression in a manner to highlight the unmeasurable quantities and the unknown parameters, we have

$$L_oL_rM_m\dot{\eta}_I = M_m\Omega_6 + \Omega_7 + \Omega_8\tilde{Z}_a + \Omega_9\tilde{Z}_b + Y_2\theta_m - \hat{R}_rM\eta_\psi - L_r\beta_3M_m\left(\hat{\psi}_aV_a + \hat{\psi}_bV_b\right) \quad \text{(E.69)}$$

where the auxiliary measurable terms Ω_6, Ω_7, Ω_8, Ω_9, and $Y_2 \in \Re^{1\times p}$ are explicitly given below

$$\Omega_6 = L_oL_r\left[\frac{\partial u_I}{\partial \hat{R}_r}\dot{\hat{R}}_r + \frac{\partial u_I}{\partial \hat{\psi}_a}\dot{\hat{\psi}}_a + \frac{\partial u_I}{\partial \hat{\psi}_b}\dot{\hat{\psi}}_b + \frac{\partial u_I}{\partial \eta_\psi}\dot{\eta}_\psi + \frac{\partial u_I}{\partial q}\dot{q}\right] + L_oL_r\left[\frac{\partial u_I}{\partial \zeta_a}\dot{\zeta}_a + \frac{\partial u_I}{\partial \zeta_b}\dot{\zeta}_b + \frac{\partial u_I}{\partial \hat{\theta}_m}\dot{\hat{\theta}}_m + \frac{\partial u_I}{\partial \hat{M}_m}\dot{\hat{M}}_m + \frac{\partial u_I}{\partial t}\right] \quad \text{(E.70)}$$

$$\Omega_7 = L_oL_r\frac{\partial u_I}{\partial \dot{q}}\left[\alpha_1\left(\hat{\psi}_a\hat{I}_b - \hat{\psi}_b\hat{I}_a\right) + \alpha_1\left(\hat{\psi}_a\tilde{I}_b - \frac{L_o}{L_r}\tilde{I}_aI_b\right)\right] - L_oL_r\frac{\partial u_I}{\partial \dot{q}}\alpha_1\left(\hat{\psi}_b\tilde{I}_a - \frac{L_o}{L_r}\tilde{I}_bI_a\right) + \hat{R}_rM\eta_\psi \quad \text{(E.71)}$$

and

$$\Omega_8 = L_o\frac{\partial u_I}{\partial \dot{q}}\alpha_1I_b, \quad \Omega_9 = -L_o\frac{\partial u_I}{\partial \dot{q}}\alpha_1I_a, \quad Y_2 = -L_oL_r\frac{\partial u_I}{\partial \dot{q}}W_m. \quad \text{(E.72)}$$

Note the term $-\hat{R}_rM\eta_\psi$ has been added and subtracted to the right-hand side of (E.69) to cancel the $\hat{R}_rM\eta_I$ in (E.66) during the ensuing stability analysis; furthermore, the right-hand sides of (E.36), (E.17), (E.18), (E.66), and (E.20) can be substituted into (E.70) for $\dot{\hat{R}}_r$, $\dot{\hat{\psi}}_a$, $\dot{\hat{\psi}}_b$, $\dot{\eta}_\psi$, $\dot{\zeta}_a$, $\dot{\zeta}_b$, respectively while the subsequent adaptation laws will allow for substitution for $\dot{\hat{M}}_m$, and $\dot{\hat{\theta}}_m$.

Based on the structure of (E.69) and the subsequent stability analysis, we now design the voltage input control relationship to force η_I to zero as follows

$$\hat{\psi}_aV_a + \hat{\psi}_bV_b = \frac{1}{L_r\beta_3}\left(K_4\eta_I + \Omega_6\right) + \frac{1}{L_r\beta_3\hat{M}_m}\left(Y_2\hat{\theta}_m + \Omega_7 + u_{c6}\right) \quad \text{(E.73)}$$

where K_4 is a positive control gain, and u_{c6} is an auxiliary control input which will be designed to facilitate the subsequent the composite observer-controller stability analysis. To eliminate overparameterization, the update laws for $\hat{M}_m$ and $\hat{\theta}_m$ will be explicitly defined later; furthermore, the update law for $\hat{M}_m$ will be designed to ensure $\hat{M}_m > 0$. After substituting (E.73) into (E.69), we can write the closed-loop error system for η_I in the following form

$$\begin{aligned} L_o L_r M_m \dot{\eta}_I = \quad & -K_4 M_m \eta_I + \Omega_8 \tilde{Z}_a + \Omega_9 \tilde{Z}_b \\ & +Y_2 \tilde{\theta}_m - \hat{R}_r M \eta_\psi - u_{c6} \\ & -\tilde{M}_m \hat{M}_m^{-1} \left(Y_2 \hat{\theta}_m + \Omega_7 + u_{c6} \right) \end{aligned} \tag{E.74}$$

where $\tilde{\theta}_m$ and $\tilde{M}_m$ were defined in (E.10).

E.5.1 Composite Observer-Controller Analysis

To analyze the composite observer-controller tracking error system while also providing for the design of the adaptation laws and the remaining auxiliary observer and control inputs, we define the following non-negative function

$$V_f = V_p + \frac{1}{2} L_r \eta_\psi^2 + \frac{1}{2} L_o L_r M_m \eta_I^2 + \frac{1}{2} \Gamma_1^{-1} \tilde{M}_m^2 + \frac{1}{2} \tilde{\theta}_m^T \Gamma_2^{-1} \tilde{\theta}_m \tag{E.75}$$

where V_p was defined in (E.53), Γ_1 is a positive scalar control gain, $\Gamma_2 \in \Re^{p \times p}$ is a positive definite, diagonal, gain matrix. After taking the time derivative of (E.75), substituting (E.66) and (E.74) into the resulting expression for $L_r \dot{\eta}_\psi$ and $L_o L_r M_m \dot{\eta}_I$, respectively, and then cancelling the common terms, we have

$$\begin{aligned} \dot{V}_f = \quad & \dot{V}_p - K_3 \eta_\psi^2 - K_4 M_m \eta_I^2 + \eta_I \Omega_8 \tilde{Z}_a + \eta_I \Omega_9 \tilde{Z}_b - \eta_\psi \hat{\psi}_a \bar{u}_{c3} \\ & -\eta_\psi \hat{\psi}_b \bar{u}_{c4} + \left[-\hat{\psi}_a \Omega_{a2} \tilde{I}_a \eta_\psi - K_n \hat{\psi}_a^2 \Omega_{a2}^2 \eta_\psi^2 \right] \\ & + \left[-\hat{\psi}_b \Omega_{b2} \tilde{I}_a \eta_\psi - K_n \hat{\psi}_b^2 \Omega_{b2}^2 \eta_\psi^2 \right] - \eta_I u_{c6} \\ & +\tilde{\theta}_m^T \left[Y_2^T \eta_I - \Gamma_2^{-1} \dot{\hat{\theta}}_m \right] + \left(\Omega_c - \Omega_3 \hat{\psi}_a - \Omega_4 \hat{\psi}_b \right) \eta_\psi \eta_\tau \\ & + \left[-\hat{\psi}_a \Omega_{a3} \tilde{I}_b \eta_\psi - K_n \hat{\psi}_a^2 \Omega_{a3}^2 \eta_\psi^2 \right] + \\ & \left[-\hat{\psi}_b \Omega_{b3} \tilde{I}_b \eta_\psi - K_n \hat{\psi}_b^2 \Omega_{b3}^2 \eta_\psi^2 \right] \\ & -\tilde{M}_m \left[\hat{M}_m^{-1} \left(Y_2 \hat{\theta}_m + \Omega_7 + u_{c6} \right) \eta_I + \Gamma_1^{-1} \dot{\hat{M}}_m \right] \end{aligned} \tag{E.76}$$

After applying the nonlinear damping argument [22] to (E.76) and then substituting for $\dot{V}_p$ from (E.57), we can collect common terms to form the following upper bound for $\dot{V}_f$

$$\begin{aligned}\dot{V}_f \leq\ & -\left(K_o - \frac{2}{K_n}\right)\tilde{I}_a^2 - \left(K_o - \frac{2}{K_n}\right)\tilde{I}_b^2 - K_1 r^2 - K_2 M_m \eta_\tau^2 \\ & -K_3\eta_\psi^2 + \tilde{Z}_a\left(\eta_I\Omega_8 - \bar{u}_{c3}\right) - \eta_\psi\hat{\psi}_a\bar{u}_{c3} - \eta_\psi\hat{\psi}_b\bar{u}_{c4} \\ & +\eta_\tau\left(\left(\Omega_c - \Omega_3\hat{\psi}_a - \Omega_4\hat{\psi}_b\right)\eta_\psi - u_{c5}\right) - K_4 M_m \eta_I^2 \\ & +\tilde{\theta}_m^T\left(W_m^T r + Y_1^T\eta_\tau + Y_2^T\eta_I - \Gamma_2^{-1}\dot{\hat{\theta}}_m\right) \\ & -\eta_I u_{c6} + \tilde{Z}_b\left(\eta_I\Omega_9 - \bar{u}_{c4}\right) + \tilde{M}_m\left(\Omega_m - \Gamma_1^{-1}\dot{\hat{M}}_m\right)\end{aligned} \tag{E.77}$$

where the auxiliary function Ω_m is given by

$$\begin{aligned}\Omega_m =\ & \left(\ddot{q}_d + \alpha\dot{e}\right) r - \hat{M}_m^{-1}\left(Y_1\hat{\theta}_m + \Omega_1 + u_{c5}\right)\eta_\tau \\ & -\hat{M}_m^{-1}\left(Y_2\hat{\theta}_m + \Omega_7 + u_{c6}\right)\eta_I.\end{aligned} \tag{E.78}$$

Now based on the third line of (E.77), we design the parameter update laws as follows

$$\dot{\hat{\theta}}_m = \Gamma_2\left(W_m^T r + Y_1^T\eta_\tau + Y_2^T\eta_I\right) \tag{E.79}$$

and

$$\dot{\hat{M}}_m = \begin{cases} \Gamma_1\Omega_m & \text{if } \hat{M}_m > \underline{M}_m \\ \Gamma_1\Omega_m & \text{if } \hat{M}_m = \underline{M}_m \text{ and } \Omega_m \geq 0 \\ 0 & \text{if } \hat{M}_m = \underline{M}_m \text{ and } \Omega_m < 0, \end{cases} \tag{E.80}$$

where $\hat{M}_m(0) = \underline{M}_m$ (Note the above update law for $\hat{M}_m$ ensures $\hat{M}_m > 0$). Now based on the second line of (E.77), we design the auxiliary observer inputs $\bar{u}_{c3}$, $\bar{u}_{c4}$, which were injected at (E.56), and the auxiliary control input u_{c5}, which was injected at (E.51), as follows

$$\bar{u}_{c3} = \eta_I\Omega_8, \qquad \bar{u}_{c4} = \eta_I\Omega_9, \qquad u_{c5} = \left(\Omega_c - \Omega_3\hat{\psi}_a - \Omega_4\hat{\psi}_b\right)\eta_\psi. \tag{E.81}$$

After substituting the auxiliary control inputs given by (E.81) and the update laws given by (E.80) and (E.79) into (E.77), we can collect the common terms to form the new upper bound for $\dot{V}_f$

$$\begin{aligned}\dot{V}_f \leq\ & -\left(K_o - \frac{2}{K_n}\right)\tilde{I}_a^2 - \left(K_o - \frac{2}{K_n}\right)\tilde{I}_b^2 - K_1 r^2 \\ & -K_2 M_m \eta_\tau^2 - K_3 \eta_\psi^2 - K_4 M_m \eta_I^2 \\ & +\eta_I\left(-\eta_\psi \hat{\psi}_a \Omega_8 - \eta_\psi \hat{\psi}_b \Omega_9 - u_{c6}\right).\end{aligned} \tag{E.82}$$

Now based on the second line of (E.82), we design the auxiliary control input u_{c6}, which was injected at (E.73), as follows

$$u_{c6} = \left(-\Omega_8 \hat{\psi}_a - \Omega_9 \hat{\psi}_b\right)\eta_\psi; \tag{E.83}$$

hence, $\dot{V}_f$ of (E.82) can be upper bounded in the following concise form

$$\dot{V}_f \leq -\lambda_3 \|x\|^2 \tag{E.84}$$

where

$$\lambda_3 = \min\left\{K_1, K_2 M_m, K_3, K_4 M_m, K_o - \frac{2}{K_n}\right\} \tag{E.85}$$

and

$$x = \left[\tilde{I}_a,\ \tilde{I}_b,\ r,\ \eta_\tau,\ \eta_\psi,\ \eta_I\right]^T \in \Re^6. \tag{E.86}$$

E.5.2 Voltage Control Input Calculation

Given the voltage control input relationships of (E.51) and (E.73), we can calculate the transformed voltage control inputs V_a and V_b as follows

$$\begin{bmatrix} V_a \\ V_b \end{bmatrix} = C^{-1}\begin{bmatrix} \text{the right hand side of (E.51)} \\ \text{the right hand side of (E.73)} \end{bmatrix} \tag{E.87}$$

where

$$C = \begin{bmatrix} \hat{\psi}_b & -\hat{\psi}_a \\ \hat{\psi}_a & \hat{\psi}_b \end{bmatrix} \in \Re^{2\times 2}. \tag{E.88}$$

Given the definition of C in (E.88), it is easy to see that C is not invertible if $\hat{\psi}_a^2(t) + \hat{\psi}_b^2(t) = 0$; hence, the proposed controller exhibits a control singularity at $\hat{\psi}_a^2(t) + \hat{\psi}_b^2(t) = 0$.

E.6 Main Result

Theorem E.1

If the following sufficient control gain condition

$$K_o > \frac{2}{K_n} \tag{E.89}$$

is satisfied and

$$\hat{\psi}_a^2(t) + \hat{\psi}_b^2(t) \neq 0 \tag{E.90}$$

then the proposed observer-controller ensures asymptotic position/velocity tracking as follows

$$\lim_{t\to\infty} \; e(t),\ \dot{e}(t) = 0. \tag{E.91}$$

Proof:

First, note that V_f of (E.75) can be bounded as

$$\lambda_1 \|z\| \leq V_f \leq \lambda_2 \|z\| \tag{E.92}$$

where

$$\lambda_1 = \frac{1}{2}\min\left\{L_o, 1, R_r, M_m, L_oL_rM_m, L_r, \Gamma_r, \Gamma_1^{-1}, \lambda_{\min}\left\{\Gamma_2^{-1}\right\}\right\} \tag{E.93}$$

$$\lambda_2 = \frac{1}{2}\max\left\{L_o, 1, R_r, M_m, L_oL_rM_m, L_r, \Gamma_r, \Gamma_1^{-1}, \lambda_{\max}\left\{\Gamma_2^{-1}\right\}\right\} \tag{E.94}$$

and

$$z = \left[x^T, \tilde{Z}_a, \tilde{Z}_b, \eta_a, \eta_b, \tilde{R}_r, \tilde{M}_m, \tilde{\theta}_m^T\right]^T \in \Re^{12+p} \tag{E.95}$$

where x was defined in (E.86). If the control gain condition given by (E.89) is satisfied, we know that λ_3 in (E.85) will be positive; therefore, since V_f is positive definite and $\dot{V}_f$ is negative semi-definite, we know that V_f is upper bounded by $V_f(0)$ and lower bounded by zero and hence $z \in L_\infty^{12+p}$ [20] and $x \in L_\infty^6$. Also from (E.84), it is easy to show that $x \in L_2^6$ [20]. Since $r \in L_\infty$, it is easy to show that $e \in L_\infty$, $\dot{e} \in L_\infty$ (hence $q \in L_\infty$, $\dot{q} \in L_\infty$). Given the assumption that θ_m, M_m, and R_r are unknown constants, we know that $\hat{\theta}_m \in L_\infty$, $\hat{M}_m \in L_\infty$, and $\hat{R}_r \in L_\infty$ since $\tilde{\theta}_m \in L_\infty$, $\tilde{M}_m \in L_\infty$, and $\tilde{R}_r \in L_\infty$. Given the definition of η_ψ in (E.58), we know $\hat{\psi}_a \in L_\infty$ and $\hat{\psi}_b \in L_\infty$ since $\eta_\psi \in L_\infty$. Also, since $\tilde{Z}_a \in L_\infty$, $\tilde{Z}_b \in L_\infty$, and $\tilde{I}_a \in L_\infty$, $\tilde{I}_b \in L_\infty$, we know that $\tilde{\psi}_a \in L_\infty$ and $\tilde{\psi}_b \in L_\infty$ (hence $\psi_a \in L_\infty$, $\psi_b \in L_\infty$). In addition, since $\eta_a \in L_\infty$, $\eta_b \in L_\infty$, $\tilde{Z}_a \in L_\infty$, $\tilde{Z}_b \in L_\infty$, we can use (E.13) to show that $\zeta_a \in L_\infty$ and $\zeta_b \in L_\infty$. Utilizing the above information, we are now able to show that τ_d of (E.41) and u_I of (E.65) are both bounded and hence (E.40) and (E.64) can be used to show that $\left(\hat{\psi}_a\hat{I}_b - \hat{\psi}_b\hat{I}_a\right) \in L_\infty$, $\left(\hat{\psi}_a\hat{I}_a + \hat{\psi}_b\hat{I}_b\right) \in L_\infty$. Therefore, we can multiply $\left(\hat{\psi}_a\hat{I}_b - \hat{\psi}_b\hat{I}_a\right)$ by $\hat{\psi}_a$, $\left(\hat{\psi}_a\hat{I}_a + \hat{\psi}_b\hat{I}_b\right)$ by $\hat{\psi}_b$, and add the results together to obtain the bounded quantity $\left(\hat{\psi}_a^2 + \hat{\psi}_b^2\right)\hat{I}_b$ which shows as a result of (E.90) that $\hat{I}_b \in L_\infty$ (and by similar argument, $\hat{I}_a \in L_\infty$). Therefore, $I_a \in L_\infty$ and $I_b \in L_\infty$ since $\tilde{I}_a \in L_\infty$ and $\tilde{I}_b \in L_\infty$. From the right-hand sides of (E.51) and (E.73), we are now able to show that the voltage control inputs are bounded. Furthermore, the closed-loop error systems can now be used to show that $\dot{x} \in L_\infty^6$. Finally, since $x \in L_\infty^6$, $x \in L_2^6$, and $\dot{x} \in L_\infty^6$, Barbalat's Lemma [20] can be used to show that

$$\lim_{t\to\infty} x(t) = 0 \implies \lim_{t\to\infty} \tilde{I}_a,\ \tilde{I}_b,\ r,\ \eta_\tau,\ \eta_\psi,\ \eta_I = 0.$$

Given the definition of the filtered tracking error given in (E.9), we can use standard linear arguments to illustrate (E.91) [21]. □

Remark E.5

The control singularity represented by (E.90) illustrates a drawback of the proposed approach; however, we note that the singularity is avoided asymptotically since $\lim_{t\to\infty} \eta_\psi = 0$ (*i.e.*, the estimated flux is moved away from the singularity in the limit). In addition, we note that since the singularity occurs for the *estimated* flux, it is easy to construct some *ad hoc* means of avoided the singularity during actual implementation. In comparison with the singularity-free control designs proposed in [15], [16], [14], and [17], we note that all of these approaches required exact knowledge of the rotor resistance.

E.7 Simulation Results

To verify the performance of the proposed adaptive observer-controller scheme, we simulated a simple mechanical load driven by an induction motor. Since we are interested in utilizing the induction motor for relatively low-speed, high performance position tracking applications such as robotics, we selected the mechanical load as a single-link robot. Therefore, $W_m(q, \dot{q})\theta_m$ in (E.1) can be reasonably modeled as

$$W_m(q, \dot{q})\theta_m = B\dot{q} + N\sin(q) \tag{E.96}$$

where B represents the viscous friction coefficient. The terms M_m in (E.1) and N in (E.96) are given by

$$M_m = J_m + \frac{mL_0^2}{3}, \qquad N = \frac{mgL_0}{2}$$

where J_m is the rotor inertia, g is the gravity coefficient, m is the link mass, and L_0 is the link length. The simulated induction motor is chosen as a Baldor model M3541, 2 pole induction motor which has a rated speed of 3450 RPM, rated current of 2.7 A, and rated voltage of 230V. The values for the electromechanical system parameters were selected to be

$$\begin{array}{lll} J_m = 1.87 \times 10^{-4} kg \cdot m^2, & m = 0.401 kg, & L_0 = 0.305 m, \\ L_r = 0.306 H, & L_s = 0.243 H, & M = 0.225 H, \\ R_s = 3.05 \Omega, & n_p = 1, & B = 0.003 N \cdot m \cdot s/rad \\ g = 9.81 kg \cdot m/\sec^2 & & R_r = 2.12 \Omega. \end{array}$$

The desired position trajectory is selected as the following smooth start sinusoid

$$q_d(t) = \frac{\pi}{2} \sin(4t) \left(1 - e^{-0.1t^3}\right) rad$$

where $q_d(0) = \dot{q}_d(0) = \ddot{q}_d(0) = \dddot{q}_d(0) = 0$. The desired flux trajectory ψ_d of (E.58) was selected as

$$\psi_d = 0.3 \frac{2e^{-0.25t}}{1 + e^{-0.5t}} + 1.7 \ wb \cdot wb.$$

The initial rotor flux estimates were set to

$$\hat{\psi}_a(0) = \hat{\psi}_b(0) = \sqrt{2} \ wb \tag{E.97}$$

such that $\eta_\psi(0) = 0$. To better mimic actual operation, we selected the rotor resistance as the following slowly time-varying function (*i.e*, we assume $\dot{R}_r \cong 0$)

$$R_r = 2.12 + 4.24(1 - e^{-0.1t}) \ \Omega.$$

The initial value for the estimate of R_r was selected as $\hat{R}_r(0) = \underline{R}_r = 1.06\Omega$ which is 50% of its lowest value. The initial value for the estimate of M_m was also selected to be 50% of its actual value as shown $\hat{M}_m(0) = \underline{M}_m = 0.0065 kg \cdot m^2$. The initial rotor position was set to $-0.3 rad$. The other initial values for the parameter update laws and state variables were all set to zero. The best tracking performance was established using the following control gain values

$$\alpha = 2, \ K_1 = 0.1, \ K_2 = 0.1, \ K_3 = 0.001, \ K_4 = 0.1, \ K_o = 5,$$

$$K_n = 0.01, \ \Gamma_r = 0.01, \ \Gamma_1 = 0.1, \ \Gamma_{21} = \Gamma_{22} = 0.001.$$

The resulting rotor position tracking error is shown in Figure E.1, the voltage control inputs V_a and V_b are shown in Figure E.2 and Figure E.3, respectively. From Figure E.1, we can see that good tracking performance has been achieved under the proposed adaptive observer-based controller.

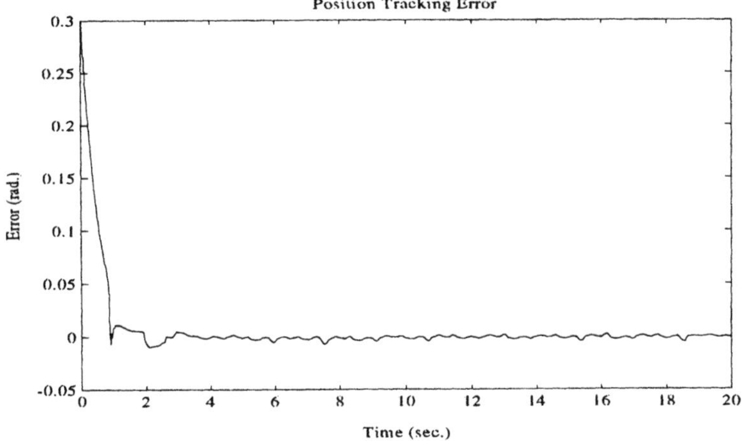

Figure E.1: Position Tracking Error

Remark E.6

The selection of the control parameters in the simulation does not satisfy the conditions given in (E.89); however, this control gain requirement is only sufficient, conservative condition generated by the Lyapunov-like stability argument. It also should be noted that the control singularity of (E.90) never occurred during the simulation given the initial rotor flux estimates of (E.97).

E.8 Conclusions

In this paper, we have presented an adaptive, partial-state feedback, position/velocity tracking controller for the full-order, nonlinear dynamic model representing an induction motor actuating a mechanical subsystem. The proposed controller provides asymptotic rotor position/velocity tracking while compensating for parametric uncertainty in the form of the rotor resistance and a general, *uncertain* mechanical subsystem. While the observer-controller does not require measurement of rotor flux or rotor current, it does exhibit a singularity when the magnitude of the estimated rotor flux is zero; however, the singularity is avoided asymptotically since the estimated flux is moved away from the singularity in the limit. Simulation results are provided to verify the effectiveness of the approach. Future research will involve eliminating the control singularity using similar methods as those outlined in [17] and [16] and performing experimental validation as previously performed in [17] and [12].

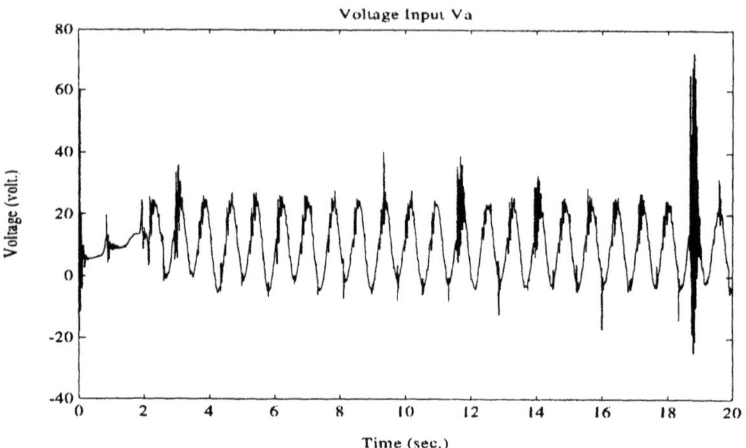

Figure E.2: Phase One Voltage Input

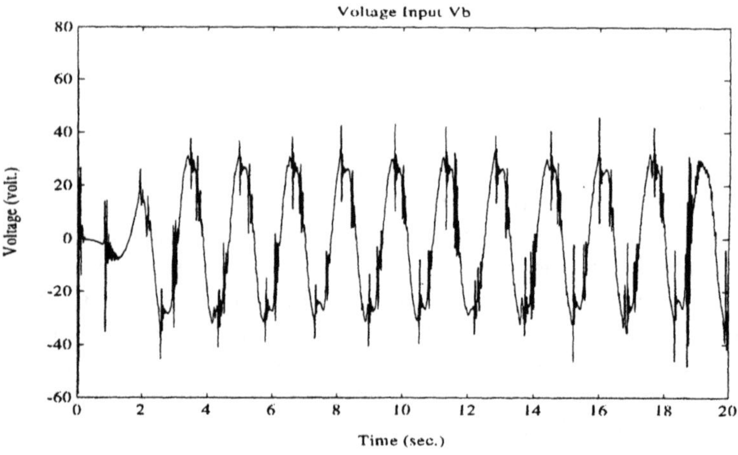

Figure E.3: Phase Two Voltage Input

E.9 Partial Derivatives Terms for $\dot{\tau}_d$

The partial derivative terms defined in (E.44) are explicitly given as follows

$$\frac{\partial \tau_d}{\partial t} = \frac{\alpha K_1}{\alpha_1}\dot{q}_d + \frac{\alpha \hat{M}_m + K_1}{\alpha_1}\ddot{q}_d + \frac{\hat{M}_m}{\alpha_1}\dddot{q}_d, \quad \frac{\partial \tau_d}{\partial q} = \frac{1}{\alpha_1}\left(\frac{\partial W_m}{\partial q}\hat{\theta}_m - \alpha K_1\right)$$

$$\frac{\partial \tau_d}{\partial \dot{q}} = \frac{1}{\alpha_1}\left[-\alpha \hat{M}_m + \frac{\partial W_m}{\partial \dot{q}}\hat{\theta}_m - K_1\right], \quad \frac{\partial \tau_d}{\partial \hat{M}_m} = \frac{\ddot{q}_d + \alpha \dot{e}}{\alpha_1}, \quad \frac{\partial \tau_d}{\partial \hat{\theta}_m} = \frac{W_m}{\alpha_1}.$$

E.10 Definitions of Ω_{ai}, Ω_{bi}, and Ω_c

The measurable terms Ω_{ai}, Ω_{bi}, and Ω_c, defined in the flux observers of (E.60) and (E.61) are given by

$$\Omega_{a1} = -\hat{R}_r\hat{\psi}_a - \beta_2\dot{q}\hat{\psi}_b - \frac{\hat{R}_r}{L_r}\zeta_a - \alpha_1\hat{\psi}_b r$$

$$\Omega_{a2} = \Omega_{b3} = -K_o + \frac{\hat{R}_r L_o}{L_r} + \hat{R}_r M, \quad \Omega_c = -\frac{\alpha_1}{L_r}r\left(1 + L_o\right)$$

$$\Omega_{a3} = \frac{\dot{q}}{L_r}\beta_2\left(L_o - 1\right) - \frac{\alpha_1}{L_r}r\left(1 + L_o\right) = -\Omega_{b2}$$

$$\Omega_{b1} = -\hat{R}_r\hat{\psi}_b + \beta_2\dot{q}\hat{\psi}_a - \frac{\hat{R}_r}{L_r}\zeta_b + \alpha_1\hat{\psi}_a r.$$

E.11 Partial Derivatives Terms for $\dot{u}_I$

The partial derivative terms defined in (E.68) are explicitly given by

$$\begin{aligned}\frac{\partial u_I}{\partial t} = {} & \frac{1 + L_o}{\hat{R}_r M L_r}\alpha_1\ddot{q}_d\left[\tau_d + r\frac{\partial \tau_d}{\partial \dot{q}_d} - 2K_n\hat{\psi}_a^2\eta_\psi\Omega_{a3} + 2K_n\hat{\psi}_b^2\eta_\psi\Omega_{b2}\right] \\ & + \frac{1 + L_o}{\hat{R}_r M L_r}\alpha\dot{q}_d\left[\alpha_1\tau_d + K_1 r \right. \\ & \left. -2\alpha_1 K_n\hat{\psi}_a^2\eta_\psi\Omega_{a3} + 2\alpha_1 K_n\hat{\psi}_b^2\eta_\psi\Omega_{b2}\right] \\ & + \frac{L_r}{\hat{R}_r M}\ddot{\psi}_d + \frac{1 + L_o}{\hat{R}_r M L_r}r\hat{M}_m\dddot{q}_d\end{aligned}$$

$$\frac{\partial u_I}{\partial \zeta_a} = \frac{1}{M L_r}\hat{\psi}_a, \qquad \frac{\partial u_I}{\partial \zeta_b} = \frac{1}{M L_r}\hat{\psi}_b$$

$$\frac{\partial u_I}{\partial q} = \frac{1 + L_o}{\hat{R}_r M L_r}\alpha_1\left[-\alpha\tau_d + r\frac{\partial \tau_d}{\partial q} + 2\alpha K_n\hat{\psi}_a^2\eta_\psi\Omega_{a3} - 2\alpha K_n\hat{\psi}_b^2\eta_\psi\Omega_{b2}\right]$$

$$\begin{aligned}
\frac{\partial u_I}{\partial \dot{q}} = & \frac{1+L_o}{\hat{R}_r M L_r}\alpha_1 \left[-\tau_d + r\frac{\partial \tau_d}{\partial \dot{q}}\right. \\
& \left. +2\alpha K_n \eta_\psi \Omega_{a3}\hat{\psi}_a^2 \left(1+\frac{L_o-1}{\alpha_1(L_o+1)}\beta_2\right)\right] \\
& -\frac{1+L_o}{\hat{R}_r M L_r}\alpha_1 \left[2\alpha K_n \hat{\psi}_b^2 \eta_\psi \Omega_{b2}\left(1+\frac{L_o-1}{\alpha_1(L_o+1)}\beta_2\right)\right]
\end{aligned}$$

$$\begin{aligned}
\frac{\partial u_I}{\partial \hat{\psi}_a} = & \frac{1}{\hat{R}_r M}\left[-\Omega_{a1} + \hat{R}_r\hat{\psi}_a - \hat{\psi}_b(\beta_2\dot{q} + \alpha_1 r)\right. \\
& \left. +2K_n\Omega_{a2}^2\hat{\psi}_a\eta_\psi + 2K_n\Omega_{a3}^2\hat{\psi}_a\eta_\psi\right]
\end{aligned}$$

$$\begin{aligned}
\frac{\partial u_I}{\partial \hat{\psi}_b} = & \frac{1}{\hat{R}_r M}\left[-\Omega_{b1} + \hat{R}_r\hat{\psi}_b + \hat{\psi}_a(\beta_2\dot{q} + \alpha_1 r)\right. \\
& \left. +2K_n\Omega_{b2}^2\hat{\psi}_b\eta_\psi + 2K_n\Omega_{b3}^2\hat{\psi}_b\eta_\psi\right]
\end{aligned}$$

$$\frac{\partial u_I}{\partial \eta_\psi} = \frac{1}{\hat{R}_r M}\left[K_n\Omega_{a2}^2\hat{\psi}_a^2 + K_n\Omega_{a3}^2\hat{\psi}_a^2 + K_n\Omega_{b2}^2\hat{\psi}_b^2 + K_n\Omega_{b3}^2\hat{\psi}_b^2 + K_3\right]$$

$$\frac{\partial u_I}{\partial \hat{M}_m} = \frac{1+L_o}{\hat{R}_r M L_r}\alpha_1 r\frac{\partial \tau_d}{\partial \hat{M}_m}, \qquad \frac{\partial u_I}{\partial \hat{\theta}_m} = \frac{1+L_o}{\hat{R}_r M L_r}\alpha_1 r\frac{\partial \tau_d}{\partial \hat{\theta}_m}$$

$$\begin{aligned}
\frac{\partial u_I}{\partial \hat{R}_r} = & -\frac{u_I}{\hat{R}_r} + \frac{1}{\hat{R}_r M}\left(\hat{\psi}_a^2 + \hat{\psi}_b^2 + \frac{\hat{\psi}_a}{L_r}\zeta_a + \frac{\hat{\psi}_b}{L_r}\zeta_b\right) \\
& +\frac{2}{\hat{R}_r M}K_n\eta_\psi\left(\Omega_{a2}\hat{\psi}_a^2 + \Omega_{b3}\hat{\psi}_b^2\right)\left(\frac{L_o}{L_r} + M\right).
\end{aligned}$$

Bibliography

[1] R. Marino, S. Peresada, and P. Valigi, "Adaptive Input-Output Linearizing Control of Induction Motors", *IEEE Transactions on Automatic Control*, Vol. 38, No. 2, Feb., 1993, pp. 208-221.

[2] D. G. Taylor, "Nonlinear Control of Electric Machines: An Overview", IEEE Control Systems Magazine, Vol. 14, No. 6, December, 1994, pp. 41-51.

[3] D. Dawson, J. Carroll, and M. Schneider, "Integrator Backstepping Control for a Brush dc Motor Turning a Robotic Load", *IEEE Transactions on Controls Systems Technology*, Vol. 2, No. 3, Sept., 1994, pp. 233-244.

[4] T. Raumer, J. Dion, and L. Dugard, "Applied Nonlinear Control of an Induction Motor using Digital Signal Processing", *IEEE Transactions on Control Systems Technology,* Vol. 2, No. 4, pp. 327-335, Dec. 1994.

[5] J. Yang, W. Yu, and L. Fu, "Nonlinear Observer-Based Tracking Control of Induction Motors with Unknown Load", *IEEE Transactions on Industrial Electronics,* Vol. 42, No. 6, Dec. 1995, pp. 579-585.

[6] M. Bodson, J. Chiasson, and R. Novotnak, "High-Performance Induction Motor Control Via Input-Output Linearization", *IEEE Control Systems Magazine,* Vol., 14, No., 4, Aug., 1994, pp. 25-33.

[7] S. Seleme, M. Petersson , and C. Canudas de Wit, "The Torque Tracking of Induction Motors via Magnetic Energy Optimization", *IEEE Conference on Decision and Control,* Lake Buena Vista, FL., Dec., 1994, pp. 1838-1843.

[8] M. Bodson, J. Chiasson, and R. Novotnak, "Nonlinear Servo Control of an Induction Motor Control with Saturation", *IEEE Conference on Decision and Control,* Lake Buena Vista, FL., Dec., 1994, pp. 1832-1837.

[9] P. Nicklasson, R. Ortega, and G. Espinosa-Perez, "Passivity-Based Control of a Class of Blondel-Park Transformable Electric Machines" *IEEE Transactions on Automatic Control,* Vol. 42, No. 5, May 1997, pp. 629-647.

[10] I. Kanellakopoulos, P. Krein, and F. Disilvestro, "A New Controller Observer Design for Induction Motor Control", *ASME Winter Meeting,* Anahiem, CA, DSC-Vol. 43, 1992, pp. 43-47.

[11] I. Kanellakopoulos and P. Krein, "Integral-action Nonlinear Control of Induction Motors, *12th IFAC World Congress,* Sydney, Australia, 1993, pp. 1700-1704.

[12] J. Hu, D.M. Dawson, and Y. Qian, "Position Tracking of an Induction Motor via Partial State Feedback", *Automatica,* Vol. 31, No. 7, July, 1995, pp. 989-1000.

[13] C. Canudas, R. Ortega, and S. Seleme, "Robot Motion Control using Induction Motor Drives", *IEEE International Conference on Robotics and Automation,* Atlanta, Ga., 1993, pp. 533-538.

[14] J. Hu, D.M. Dawson, and Y. Qian, "Position Tracking for Robot Manipulators Driven by Induction Motors without Flux Measurements", *IEEE Transactions on Robotics and Automation,* Vol. 12, No. 3, 1996, pp. 419 - 438.

[15] Espinosa-Perez G., and R. Ortega, "State Observers are Unnecessary for Induction Motor Control, *System and Control Letters,* Vol. 23, No. 5, 1994, pp. 315-323.

[16] R. Ortega, P. Nicklasson, and G. Espinosa-Perez, "On Speed Control of Induction Motors", *Automatica,* Vol. 32, No. 3, March 1996, pp. 455-460.

[17] D. Dawson, J. Hu, and P. Vedagharba, "An Adaptive Control for a Class of Induction Motor Systems", *Proc. of the IEEE Conference on Decision and Control,* New Orleans, LA, Dec., 1995, pp. 1567-1572.

[18] R. Marino, S. Peresada, and P. Valigi, "Exponentially Convergent Rotor Resistance Estimation for Induction Motors", *IEEE Transactions on Industrial Electronics,* Vol. 42, No. 5, Oct. 1995, pp. 508-515.

[19] P. Krause, *Analysis of Electric Machinery,* McGraw Hill, 1986.

[20] S. Sastry and M. Bodson, *Adaptive Control: Stability, Convergence, and Robustness,* Englewoods Cliff, NJ: Prentice Hall Co., 1989.

[21] J. Slotine and W. Li, *Applied Nonlinear Control,* Englewood Cliff, NJ: Prentice Hall Co., 1991.

[22] M. Krstic, I. Kanellakopoulos, and P. Kokotovic, *Nonlinear and Adaptive Control Design,* John Wiley & Sons, 1995.

Index

9 780824 701802

Milton Keynes UK
Ingram Content Group UK Ltd.
UKHW020010071024
449327UK00031B/2722